Urban
Social Geography

An Introduction

Fifth edition

Paul Knox and Steven Pinch

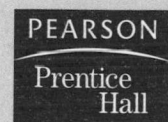

PEARSON

Prentice
Hall

Harlow, England • London • New York • Boston • San Francisco • Toronto • Sydney • Singapore • Hong Kong
Tokyo • Seoul • Taipei • New Delhi • Cape Town • Madrid • Mexico City • Amsterdam • Munich • Paris • Milan

Pearson Education Limited
Edinburgh Gate
Harlow
Essex CM20 2JE
England

and Associated Companies throughout the world

Visit us on the World Wide Web at:
www.pearsoned.co.uk

First published 1982
Second edition 1987
Third edition 1995
Reprinted 1996, 1998
Fourth edition 2000
Fifth edition 2006

© Pearson Education Limited 1982, 2006

ISBN-13: 978-0-13-124944-8
ISBN-10: 0-13-124944-4

British Library Cataloguing-in-Publication Data
A catalogue record for this book is available from the British Library

Library of Congress Cataloging-in-Publication Data
A catalogue record for this book is available from the Library of Congress

10 9 8 7 6 5 4 3 2 1
11 10 09 08 07 06

Typeset in 9.75/13pt Minion by 35
Printed by Ashford Colour Press Ltd., Gosport

The publisher's policy is to use paper manufactured from sustainable forests.

Contents

Contents

6 Structures of building provision and the social production of the urban environment   120

Supporting resources

Visit **www.pearsoned.co.uk/knox** to find valuable online resources

Companion Website for students and instructors
➤ Links to relevant sites on the web
➤ Project assignments for further investigation and practice
➤ Essay questions

For more information please contact your local Pearson Education sales representative or visit **www.pearsoned.co.uk/knox**

A guide to using this book

In order to help you learn about urban social geography, this book incorporates the following features:

➤ At the beginning of each chapter is a list of the key questions that will be addressed.

➤ Each time a new important concept is introduced, it is outlined in bold.

➤ At the end of each chapter there is:

a brief summary of the key points,
a list of the key concepts used in the chapter,
a list of recommended reading.

Together with the discussion in the text, there is a definition of the key concepts in the Glossary at the end of the book. The Glossary also contains extensive cross-referencing to related concepts. It can therefore be used as a memory aid after reading a chapter and also as a way of tracking other related concepts.

Acknowledgements

We are grateful to the following for permission to reproduce copyright material:

Figure 1.1 from *The Guardian*, G2 supplement, 21.01.05 © The Guardian Newspapers Limited; Figure 2.1 from 'Testing the model of the pre-industrial city: the case of ante-bellum Charleston, South Carolina' (Radford J.P., 1979) in *Transactions of the Institute of British Geographers*, published by Blackwell Publishing; Figure 2.2 from *The Limits to Capital* (Harvey, D., 1982) published by Blackwell Publishing; Figures 4.1 and 4.2 from 'Spatial organization of towns at the level of the smallest urban unit: plots and buildings', in *Urban Landscape Dynamics: A Multi-level Innovation Process* (A. Montanari, G., Curdes and L. Forsyth (eds), 1993). Published by Ashgate Publishing; Figures 4.3 and 4.4 from 'The evolution of twentieth-century residential forms: an American case study' (Mouden A.V., 1992), published by Taylor & Francis; Figure 4.6 from 'Liverpool' by Chandler, B. *et al.*, in *Urban Landscape Dynamics* (A. Montanari, G. Curdes and L. Forsyth (eds), 1993), published by Ashgate Publishing; Figure 4.8 from 'Out Changing Cities'(Hart, J.F. (ed.), 1991), pp. 244–245 Figure 12.21 © 1991 Reprinted with permission of The Johns Hopkins University Press; Figure 4.13 from *Factorial Ecology of Metropolitan Toronto* (Murdie R.A., 1969) published with kind permission of The University of Chicago Press; Figure 4.16 from *The Geography of Social Wellbeing in the United States* (Smith, D.M., 1973) published by McGraw-Hill, New York; Figure 4.17 from 'Environments of disadvantage: geographies of persistent poverty in Glasgow' (Pacione, M., 2004), *Scottish Geographical Magazine*, 120, 117–132 published by the Royal Geographical Society; Figure 5.2 from *Environment and Planning D: Society and Space*, 9, 120 (Fitzgerald, 1991), reprinted with permission from Pion Ltd; Figure 6.3 from *Conflict, Politics, and the Urban Scene* (Taylor, P.J. and Hadfield H., 1982), Longman, London; Figures 6.4 and 12.3 from 'Public and private consumption and the city' (Clarke, D.B. and Bradford, M.G. 1998), *Urban Studies* 35, 865–888 published by Taylor & Francis www.tandf.co.uk/journals; Figure 6.6 from Roof, 1, 111 (Weir, 1976), reprinted with permission from Roof Shelter Publication; Figures 6.7 and 6.8 from *The New Urban Frontier: Gentrification and the Revanchist City* (Smith, N. 1996), publisher Taylor & Francis; Figure 7.2 from 'Community, communion, class and community action' (Bell and Newby), *Social Areas in Cities* Vol 2: *Spatial Perspectives on Problems and Policies* (eds Herbert and Johnston, 1976) copyright John Wiley & Sons Limited, reproduced by permission; Figure 7.3 from *The City* (Park R.E., Burgess, E.W. and McKenzie, R.D., 1925), published with kind permission of University of Chicago Press; Figure 8.1 from JPR report No. 5, 'Long-term planning for British Jewry, Final report and recommendations', 2003, p. 54; Figures 9.1 and 9.2 from *Environment and Planning D: Society and Space*, 9, 417–31 (Simonsen, 1991), reprinted with permission from Pion Ltd; Figure 9.3 from 'Areal differentiation and post-modern human geography' by Gregory, D., from *Horizons in Human Geography* Gregory, D. and Walford, R. (eds), 1989), published by Palgrave; Figure 9.4 from *Environment and Planning A*, 18, 231–52 (Moos and Dear, 1986), reprinted with permission from Pion Ltd; Figure 9.6 from *Architecture et comportment*, 2, 17–122 (Knox, 1984), reprinted with permission from Architecture and Behaviour, Ecole Polytechnique Federale de Lausanne; Figure 10.1 from *Society, Action and Space: An alternative Human Geography* (Werlen, B., 1993), published by Routledge, London; Figure 10.2 from 'The Geography of Civility Revisited: New York Blackout

Looting' (Wohlenberg, E.H., 1982), *Economic Geography* 58 (1), Figure 2, p. 36. Reprinted with permission of Clark University; Figure 10.3 from *Policing and the State* (Bottoms, A.E. and Wiles, P., 1992), published by Taylor & Francis; Figure 10.4 from 'Crime delinquency and the urban environment', in *Progress in Human Geography* (Herbert D.R., 1977) © 1977 Edward Arnold (Publishers) Limited; Figure 10.5 from 'Geographical aspects of the incidence of residential burglary in Newcastle-under-Lyme' (Evans D. and Olds G., 1984), in *Tijdschrift voor Economische en Social Geografie*, published by Blackwell Publishing; Figure 10.6 from *Crime, Policing and the State* (Evans, Fyfc and Herbert (eds), 1992), reprinted with permission from Routledge; Figure 11.1 from *Environment and Planning D: Society and Space*, 6, 37–54 (Winchester and White, 1988), reprinted with permission from Pion Ltd; Figure 11.2 from *Sexuality, Communality and Urban Space: An Exploration of Negotiated Senses of Communities Amongst Gay Men in Brighton*, unpublished PhD Thesis (Wright, 1999), reprinted with permission from David Wright; Figure 12.4 from *Urban Social Areas* (Robson, 1975), by permission of Oxford University Press; Figure 13.1 from 'Public Service Provision and the Urban Development'(Smith C.J., 1984), published by Taylor & Francis; Figure 13.4 from 'Managing urban change: the case of the British inner city' by Diamond, D., in *Global Change and Challenge* (R. Bennett and R. Estall (eds), 1991), published by Taylor & Francis; Figure 13.5 from *Unruly Cities* (Pile, Brook and Mooney (eds), 1999), reprinted with permission from Open University Press.

Table 4.1. from 'Spatial organization of towns at the level of the smallest urban unit: plots and buildings', in *Urban Landscape Dynamics: A Multi-level Innovation Process* (A. Montanari, G., Curdes and L. Forsyth (eds), 1993); Table 5.1 from 'Local governance, the crises of Fordism and the changing geographies of regulation' (Goodwin, M. and Painter J., 1996) in *Transactions of the Institute of British Geographers*, published by Blackwell Publishing; Table 9.1 from *The Condition of Postmodernity* (Harvey D.W., 1989), published by Blackwell Publishing; Table 11.1 reprinted with permission from Paul Chapman Publishing from Maddock and Parkin, 'Gender culture: how they affect men and women at work', *Women and Management* (Davidson and Burke (eds), 1994), copyright Paul Chapman Publishing; Table 11.2 from The Guardian, 13 October 1995, reprinted with permission; Table 12.1 from *The Western European City: A Social Geography* (White, 1984), published by Longman and reprinted with permission from Pearson Education; Table 13.1 from 'The geographically uneven development of privatisation: towards a theoretical approach' (Stubbs, J.G. and Barnett, J.R., 1992), *Environment and Planning A*, 24, 1117–1135 published by Pion Limited; Table 13.2 from Environment and Planning A, 21, 905–26 (Pinch, 1989), reprinted with permission from Pion Ltd.

In some instances we have been unable to trace the owners of copyright material, and we would appreciate any information that would enable us to do so.

Chapter 1

Social geography and the sociospatial dialectic

Key questions addressed in this chapter

➤ Why are geographers interested in city structures?

➤ What are the distinctive contributions that geographers can make to understanding these structures?

➤ In what ways do city structures reflect economic, demographic, cultural and political changes?

Why do city populations get sifted out according to race and social class to produce distinctive neighbourhoods? What are the processes responsible for this sifting? Are there any other criteria by which individuals and households become physically segregated within the city? To what extent is territory relevant to the operation of local social systems? How does a person's area of residence affect his or her behaviour? How do people choose where to live, and what are the constraints on their choices? What groups, if any, are able to manipulate the 'geography' of the city, and to whose advantage? These are some of the key questions that we will be examining in this book.

As many writers now acknowledge, the answer to most of these questions is ultimately to be found in the wider context of social, economic and political organization. Cities can be understood only in relation to their historical, cultural and economic settings. It follows that a proper understanding of any city requires a cross-disciplinary approach, whatever the ultimate focus of attention. In the city, everything is connected to everything else, and the deficiencies of one academic specialism must be compensated for by the emphases of others. Within geography as a whole, there are several different approaches to knowledge and understanding. Four main approaches have been identifiable in the recent literature of urban geography.

1

1.1 Different approaches within human geography

The quantitative approach

First there is the **quantitative approach**, which attempts to provide descriptions of the geographical structure of cities using statistical data represented in the form of maps, graphs, tables and mathematical equations. Much of the inspiration for this approach has come from **neoclassical economics** and **functionalist sociology**. These approaches aim to be 'scientific', providing objective descriptions of cities in such a way that the values and attitudes of the observer do not influence the analysis. This attempt to separate the observer from the observed is often termed the **Cartesian approach**, after the philosopher Descartes. However, many have questioned whether such neutrality is possible. The feminist writer on theories of knowledge, Donna Haraway (1991), has argued that the scientific approach represents the desire for a '**god trick**', attempting to see 'everything from nowhere'. She argues that this goal is impossible since the values of the researcher will inevitably be reflected in the data that are chosen and the theoretical frameworks, words and metaphors that are used to represent this data. This external scrutiny of peoples and places by observers is also sometimes termed the '**gaze**' or (since traditionally most urban analysts have been men) the 'male gaze'.

The behavioural approach

Second is the so-called **behavioural approach**, which initially emerged as a reaction to the unrealistic **normative assumptions** (i.e. theories concerning what *ought* to be, rather than what actually exists) of neoclassical-functional description. The emphasis here is on the study of people's activities and decision-making processes (where to live, for example) within their perceived worlds. Many of the explanatory concepts are derived from social psychology although **phenomenology**, with its emphasis on the ways in which people experience the world around them, has also exerted a considerable influence on behavioural research.

The structuralist approach

Third, there is the approach generally known as **structuralism**. Unlike the quantitative and behavioural approaches, structuralists are very suspicious of everyday appearances and people's subjective reactions to, and interpretations of, the world. Instead, they argue that to understand society one needs to probe beneath the obvious external world to apprehend the underlying mechanisms at work. Since these mechanisms cannot be observed directly, one needs to study them through processes of abstract reasoning by constructing theories. This structuralist approach was initially used to study 'primitive' societies. Despite the diversity of cultural forms that can be found throughout the world, it was argued that there were underlying universal cultural structures that govern all human behaviour (such as prohibitions on incest). However, most geographers have allied structuralist approaches with **Marxian** perspectives, rather than anthropology. These attempt to update the ideas devised by Karl Marx in the context the nineteenth century industrial city (sometimes termed **classical Marxism**) in the light of developments in the twentieth century. These updated Marxian theories are also sometimes termed **neo-Marxist approaches**.

Marx argued that the key underlying mechanism in a capitalist society was a conflict between two major classes over the issue of value: first, the class made up of owners of capital and, second, the class of workers who owned little but their labour power. Of course, much has changed since Marx was writing in the nineteenth century; in particular, both the class structure and the role of the state have become much more complex. Nevertheless, at root, Marxian perspectives attempt to relate contemporary societal developments to the class struggle over value. Thus, structuralist approaches stress the constraints that are imposed on the behaviour of individuals by the organization of society as a whole and by the activities of powerful groups and institutions within it. At its broadest level, this approach looks to political science for its explanatory concepts, focusing on the idea of power and conflict as the main determinants of locational behaviour and resource allocation.

Marxist approaches tend to play down the importance of individuals' subjective interpretations (which are sometimes termed 'false consciousness'). But critics have argued that, in playing down the perceptions of people, Marxian theorists ignore the fact that there are many different conflicts in society in addition to those based around class, such as those conflicts based around gender, ethnicity, age, sexuality, religion, disability, nationality, political affiliation, location of neighbourhood and so on. There is growing recognition that there are many different interests in the city, many different 'voices' and different theories that can represent these interests.

It is frequently argued by critics that Marxian theories also have a poor sense of **human agency** (i.e. the capacity of people to make choices and take actions to affect their destinies). It is argued that within Marxian theory people are frequently portrayed as hapless dupes thrown about by broader economic forces. Nevertheless, it is undoubtedly true that many people *are* relatively defenceless in the face of economic forces. Furthermore, it is important to note that Marxian theories are diverse in character and many scholars have tried to overcome these limitations in recent years. As we will see later in this book, the basic principles of structuralist thinking provide us with powerful tools for understanding contemporary social change (see also Box 1.1 on David Harvey – the key exponent of Marxian approaches in human geography).

Poststructuralist approaches

Poststructuralist approaches are strongly opposed to the idea that the world can be explained by a single, hidden, underlying structure, such as class-based conflict. Instead, it is argued that there are numerous shifting and unstable dimensions of inequality in society. In addition, it is argued that these inequalities are reflected in various forms of representation, including language, intellectual theories, advertising, popular music and city landscapes. All of these forms of representation involve sets of shared meanings – what are called **discourses**. This implies that the words and ideas that we use to represent the world are not straightforward reflections of an external reality (the **mimetic approach**). Instead, these words shape and create the world through the underlying assumptions

and discourses that they incorporate. Poststructuralism therefore argues that there is no simple undistilled experience – all our experiences are filtered through particular sets of cultural values. It therefore follows that the method by which we represent reality is as important as the underlying reality itself.

Clearly, then, words are not neutral but have powerful underlying assumptions and meanings. This means that analysis of **culture** is crucial to understanding the language and discourse. The effect of poststructuralist thinking on urban social geography has been substantial, to the point where the subdiscipline has taken a clear '**cultural turn**'. This, it should be emphasized, is not a turn toward the traditional notion of culture within cultural geography (sometimes termed the '**superorganic**' view; see Duncan, 1980), but to the anthropological idea of cultures as systems of shared meanings.

The study of urban social geography

The implications of these differing perspectives will be highlighted in greater detail at various stages throughout this book. For the present we should note that cities are not just physical structures – they are also products of the human imagination. The way in which we use these imaginings to conjure up visions of areas and the people within them may be termed **imaginative** (or **imagined**) **geographies**. The plural geograph*ies* is commonly used to reflect the fact that different people have widely differing notions of geographical areas and, of course, our own visions of these spaces can change over time. This means that there can be no *one* urban social geography. The crucial point is that these imaginative geographies shape the physical structures of cities and the ways in which we are, in turn, shaped by these structures.

For example, within writings on cities, the suburb has often been portrayed as the socially homogeneous, relatively safe, female-dominated domestic and *private* sphere. In sharp contrast, the central city has been portrayed as a socially heterogeneous, male-dominated, relatively dangerous *public* space. As will be described in subsequent chapters, there is undoubtedly a great deal of truth in this characterization. However, there is a growing body of research that highlights the

Box 1.1

Key thinkers in urban social geography – David Harvey

The career of David Harvey illustrates the enormous methodological changes that have taken place in geography since the 1970s. Indeed, it is difficult to think of anyone who has had more influence upon human geography during this time.

Harvey began his career in the late 1960s as an advocate of a 'scientific' approach, positivism and quantitative methods, and he made a strong case for these approaches in his book *Explanation in Geography* (1969). However, during the early 1970s he began to feel that geographers were paying insufficient attention to the numerous social issues that were manifest during those troubled times. His next major text, *Social Justice and the City* (1973), charts his shift from a liberal position (the first half of the book using mainstream social science to analyze problems) towards a Marxian approach (updating classical Marxism to explain recent changes). Cities appear to have played an important part in Harvey's 'conversion'. His move in 1969 from the University of Bristol in the United Kingdom to Johns Hopkins University in Baltimore made him conscious of the enormous scale of poverty endemic in many US cities that were suffering from extensive job losses at that time.

David Harvey has continued to be a staunch advocate of Marxian interpretations of changes in cities, resisting the widespread adoption of poststructuralist perspectives. His defence is expressed in one of the best-selling social science textbooks of recent years *The Condition of Postmodernity* (1989b). Harvey's work has been criticized for concentrating on class-based economic sources of conflict and for downplaying other sources of inequality in cities related to factors such as gender, ethnicity and sexuality (Deutsche, 1991). Harvey (1996) argues that while there are many sources of 'difference' in society, underpinning all of these are basic economic processes that lead to exploitation.

Key concepts associated with David Harvey (see Glossary)

Commodity fetishism, neo-Marxism, 'spatial fix', structured coherence, time–space compression, urban entrepreneurialism.

Further reading

Castree, N. (2004) David Harvey, in P. Hubbard, R. Kitchin and G. Valentine (eds) *Key Thinkers on Space and Place* Sage, London

Deutsche, R. (1991) Boy's town *Environment and Planning D: Society and Space*, **9**, 5–30

Harvey, D. (1996) Justice, *Nature and the Geography of Difference* Blackwell, Oxford

Links with Other Chapters

Chapter 2: Marx and the Industrial City, Fordism and the Industrial City, Neo-Fordism

Chapter 5: Structuralist Interpretations of the Political Economy of Contemporary Cities, The Local State and the Sociospatial Dialectic, The Question of Social Justice in the City

Chapter 6: Builders, Developers and the Search for Profit, Manipulating Social Geographies; Blockbusting and Gentrification

Chapter 9: Constructing Place Through Spatial Practices, Box 9.3 Henri Lefebvre

Chapter 10: Alienation, Structuralist Theory

Chapter 13: Urban Change and Conflict, the Aggregate Effects of Aggregate Patterns

limitations and simplifications of the public/private, male/female distinctions (Bondi, 1998a; Dowling, 1998). Nevertheless, these ideas or stereotypes have formed a powerful ideology (i.e. a dominant set of ideas) that has affected urban planning and design. As we will see in greater detail in Chapter 3, the material dimensions of the city therefore reflect our cultural values and at the same time help to shape them.

Central to these imaginings of cities are the use of various metaphors (see Table 1.1). A metaphor is a way of describing one thing in a figurative sense by reference to another thing that is not literally appropriate (e.g. as in the term *urban jungle*). Many of the metaphors that have been used to describe, analyze and comprehend cities have negative overtones, reflecting the anti-urban feelings that underlie a great deal of Western thinking about cities (the city as 'labyrinth', 'sewer' or 'jungle'). However, the ambiguous role of urban settings is illustrated by those metaphors that portray cities as places of excitement, liberation and enlightenment (the city as

Table 1.1 Metaphors commonly used to describe cities

Metaphor	Connotations/Examples	Metaphor	Connotations/Examples
Arena	A place of contest between differing interest groups	Hell	A nightmarish place of punishment (e.g. the dark satanic mills of industrial capitalism in the nineteenth century)
Babel	A cacophony of discordant, non-communicating voices (after the biblical tower)	Jerusalem	A potentially utopian place of salvation
Babylon	A place of luxury and affluence, but also vice, corruption and tyranny (derived from the ancient Mesopotamian civilization but applied by marginalized groups who feel dispossessed in contemporary cities such as London)	Jungle	A threatening and dangerous place in which the inexperienced and unwary may not survive
		Kaleidoscope	A continually changing mixture of images
		Labyrinth/maze	A confusing place from which there is no escape
Body	A living organism with circulation through various arteries, limbs, bowels and control centres (often portrayed as sick or unhealthy)	Machine/system	A set of interrelated parts that can be analyzed and controlled (as in comprehensive planning models, Fritz Lang's film *Metropolis* and the architecture of Le Corbusier)
Bohemia	A place in which people defy social conventions (especially artists and intellectuals)	Market/bazaar	A place in which goods and services are exchanged, often exotic, mysterious and enticing in character
Cesspool/sewer	A dirty insanitary place of physical and moral decay, squalor and corruption (often used to describe nineteenth-century cities)	Melting pot	A creative place in which diverse ethnic groups mix together producing new cultural forms
Circuit/flow	A place in which money, people, goods, services and ideas continuously circulate and recirculate	Mosaic/patchwork	A diverse set of residential areas and land uses with distinct patterns (unlike fragments)
Fabric	A place with many interwoven elements (that can also be tattered and torn)	Network	A conjunction of many overlapping webs of social and economic interaction
Forum	A democratic place in which people can give expression to many diverse opinions	Nightmare	A disturbing mix of almost surreal images and experiences
Fragments	Diverse, randomly placed and disconnected spaces (as in 'postmodern' cities)	Organism	A living system with a hierarchy of cells
		Text	A mixture of landscapes and images that can be 'read' like a book for cultural meaning
Galaxy	A widely dispersed set of diverse elements (as in large sprawling cities)	Theatre	A city of diverse sets and backdrops in which people play out different roles
		Theme park	A place of fantasy, spectacle and excitement
Game	A place in which economic and social development is like a lottery, casino game or Monopoly	Urban village	A place of many small communities in which people have close personal contacts

'theme park', 'theatre' or 'melting pot'). This complexity is revealed in descriptions of inner-city ghetto areas occupied by ethnic minorities; on the one hand they are often presented as crime-ridden decaying zones but, on the other hand, they are also often envisaged as spaces of cultural resurgence. It follows, therefore, that these metaphors are not just artistic licence; they are used to understand cities and can justify different approaches to urban policy.

Related to the above is a further crucial point – the metaphors, theories, concepts and modes of representation we use to analyze cities cannot be regarded as neutral, objective and value free. Instead, they tend to represent particular theoretical perspectives and interest groups. These interests are not always immediately obvious. What might appear to be a neutral theory depicting inequalities in a city as some natural, inevitable, outcome might serve to support existing conditions in society. For example, few studies have been more influential than that of the Chicago School of Urban Sociology (see Chapter 7), whose theorists used a biological metaphor. Although they sought to produce a 'scientific' view of the city, their choice of metaphor helped to make the existing social order seem natural and inevitable (Sibley, 1995).

It follows that all ideas and theories or 'claims to knowledge' should be viewed critically. The ideas and theories described in this book represent the best efforts of numerous highly observant, aware and intelligent people to understand the complex world around them. However, until comparatively recently, most urban geography, even if focused on the poor and marginalized, has been written by white, middle-class, physically active, mostly heterosexual (or at least not explicitly homosexual) men. This has inevitably affected what has been studied and the way it has been studied. For example, until recently issues such as child-care in the city have been neglected while the perspectives of women, gays and lesbians and people with disabilities have also been ignored. Fortunately, these omissions are being rectified, as some of the new literature in this book will reveal.

1.2 The sociospatial dialectic

Urban spaces are created by people, and they draw their character from the people that inhabit them. As people live and work in urban spaces, they gradually impose themselves on their environment, modifying and adjusting it, as best they can, to suit their needs and express their values. Yet at the same time people themselves gradually accommodate both to their physical environment and to the people around them. There is thus a continuous two-way process, a **sociospatial dialectic** (Soja, 1980), in which people create and modify urban spaces while at the same time being conditioned in various ways by the spaces in which they live and work. Neighbourhoods and communities are created, maintained and modified; the values, attitudes and behaviour of their inhabitants, meanwhile, cannot help but be influenced by their surroundings and by the values, attitudes and behaviour of the people around them. At the same time, the ongoing processes of urbanization make for a context of change in which economic, demographic, social and cultural forces are continuously interacting with these urban spaces (Knox, 1994). It is helpful to follow Dear and Wolch (1989) in recognizing three principal aspects of the sociospatial dialectic:

➤ Processes wherein social relations are *constituted* through space, as when site characteristics influence the arrangements for settlement.

➤ Processes wherein social relations are *constrained* by space, such as the inertia imposed by an obsolete built environment, or the degree to which the physical environment facilitates or hinders human activity.

➤ Processes wherein social relations are *mediated* by space, as when the general action of the 'friction of distance' facilitates the development of a wide variety of social practices, including patterns of everyday life.

Space, then, cannot be regarded simply as a neutral medium in which social, economic and political processes are expressed. It is of importance in its own right in contributing both to the pattern of urban development and to the nature of the relationships between different social groups within the city. Space and distance are undeniably important as determinants of social networks, friendships and marriages. Similarly, territoriality is frequently the basis for the development of distinctive social milieux which, as well as being of interest in themselves, are important because of their capacity to mould the attitudes and shape the behaviour of their inhabitants. Distance also emerges as a significant determinant of the quality of life in different parts of the city because of variations in physical accessibility to opportunities and amenities such as jobs, shops, schools, clinics, parks and sports centres.

Because the benefits conferred by proximity to these amenities contribute so much to people's welfare, locational issues often form the focus of inter-class conflict within the city, thus giving the spatial perspective a key role in the analysis of urban politics. The partitioning of space through the establishment of *de jure* territorial boundaries also represents an important spatial attribute which has direct repercussions on several spheres of urban life. The location of municipal boundaries helps to determine their fiscal standing, for example; while the boundaries of school catchment areas have important implications for community status and welfare; and the configuration of electoral districts is crucial to the outcome of formal political contests in the city.

1.3 The macro-geographical context

This book is concerned with cities in developed countries that have '**postindustrial**' societies (i.e. societies in which industrial employment has been in decline and employment in service industries has been growing). The term postindustrial can be misleading in that there are still substantial manufacturing industries in these cities. However, the 'post' label alludes to the fact that

Box 1.2

Key debates in urban social geography – How to classify cities in different types of nation state?

This book is primarily concerned with cities in what are generally termed the 'advanced', 'industrialized' or 'developed' economies. As we will see later, there is much debate about the different types of cities that can be found within these nations. For example, in recent years a key debate has been over differences and similarities between cities in the United States and Europe, and the differences between Japanese cities and Western cities have also received attention. However, there is also much discussion concerning how to typify cities in other types of country. In this context it is important to remember that over half of the world's population live in some form of urban area and that of all the 13 cities with at least ten million inhabitants, only a few are found in the West (New York, Los Angeles, Tokyo and Osaka).

The term 'Third World' was coined some years ago to distinguish countries that were neither 'First World' capitalist, market-based, economies nor 'Second World' centrally planned communist states. There has been extensive controversy over the characteristics of 'Third World' cities. These cities display wide variations but they

have characteristics that distinguish them from the cities considered in this book; in particular, low levels of economic development and very rapid rates of urban growth with extremely poor living conditions in extensive slum areas on the outskirts of cities.

With the collapse of communist regimes there has been a renewed focus upon the characteristics of 'transitional cities', those that have moved very rapidly from central planning towards market economies. These cities are increasingly characterized by many of the urban forms to be found in developed economies: greater social inequality, residential segregation, suburban shopping malls and inner-city retail developments.

The term 'Third World' has come in for increasing criticism in recent years, with 'developing economies' now finding greater favour. Indeed, whether there is a distinctive 'Third World/developing-economy city' is now seriously questioned. It has been argued that in highlighting the distinctive characteristics of non-Western cities (whether characterized as 'developing' or 'transitional') one may ignore similar processes of globalization affecting most urban areas.

Key concepts related to city classification (see Glossary)

Paradigmatic city, transitional cities, world cities.

Further reading

Andrusz, G.M., Harloe, M. and Szelenyi, I. (eds) (1996) *Cities After Socialism: Urban and Regional Change and Conflict in Post-Socialist Societies* Blackwell, Oxford

Dick, H.W. and Rimmer, P.J. (1998) Beyond the Third World city: the new urban geography of South-east Asia *Urban Studies*, **35**, 2303–2321

Fielding, A.J. (2004) Class and space: social segregation in Japanese cities *Transactions, Institute of British Geographers*, **29**, 64–84

Potter, R.B. and Lloyd-Evans, S. (1998) *The City in the Developing World* Pearson Education, Harlow

Links with Other Chapters

Chapter 2: Globalization

Chapter 10: Box 10.3 The first and second nature of cities

these cities have experienced various changes that distinguish them from the classic industrialized cities of the nineteenth and early twentieth centuries. The societies in which these changes are most advanced are in Europe and North America, where levels of urbanization are among the highest anywhere. References to cities elsewhere are included not to redress this bias but to provide contrasting or complementary examples and to place arguments within a wider setting (see also Box 1.2).

Even within the relatively narrow cultural and geographical realm of Europe and North America, however, there are important differences in the nature of urban environments. These will be elaborated in the body of the text but it is important to guard against **cultural myopia** from the beginning of any discussion of urban geography (i.e. assuming that the arrangements in one own country or culture are the only possible set or arrangements or that these are a superior approach). It is, therefore, important to acknowledge the principal differences between European and North American

cities. For one thing, European cities are generally much older, with a tangible legacy of earlier modes of economic and social organization embedded within their physical structure.

Another contrast is in the composition of urban populations, for in most of Europe the significance of minority groups is generally much less than in North America. The relatively large size of the migrant workforce in US cities combined with limited employment restrictions enables the employment of large numbers of workers in low paid service sectors or manufacturing 'sweatshops'. European nations tend to have lower migration levels, coupled with restrictive labour market barriers and more generous unemployment benefits (resulting in higher levels of unemployment than the United States). However, there are wide variations in this pattern, while Germany is typical, the United Kingdom tends more towards a US approach. A third major difference stems from the way in which urban government has evolved. Whereas North American

European patterns of urban development: co-operative housing at Alt-Erla, Vienna. Photo Credit: Paul Knox.

cities tend to be fragmented into a number of quite separate and independent municipalities, European cities are less so and their public services are funded to a significant level by the central government, making for a potentially more even-handed allocation of resources within the city as a whole. This is not unrelated to yet another important source of contrast – the existence of better-developed welfare states in Europe. This not only affects the size and allocation of the **social wage** within cities but also has profound effects on the social geography of the city through the operation of the housing market. Whereas fewer than 5 per cent of US urban families live in public housing, over 20 per cent of the families in many British cities live in what can be termed 'social housing' (i.e. dwellings rented from public authorities or housing associations operating in the not-for-profit sector).

Finally, it is worth noting that in Europe, where the general ideology of **privatism** is less pronounced and where there has for some time been an acute awareness of the pressures of urban sprawl on prime agricultural land, the power and influence of the city planning machine are much more extensive. As a result, the morphology and social structure of European cities owe much to planning codes and philosophies. Thus, for example, the decentralization of jobs and homes and the proliferation of out-of-town hypermarkets and shopping malls has been much less pronounced in Europe than in North America, mainly because of European planners' policy of urban containment. The corollary of this, of course, is that the central business districts (CBDs) of European cities have tended to retain a greater commercial vitality than many of their North American counterparts. Finally, it should be noted that there are important *regional* and *functional* differences in the social geography of cities. The cities of the American northeast, for example, are significantly different, in some ways, from those of the 'Sunbelt', as are those of Canada and the United States.

A changing context for urban social geography

Cities have become impossible to describe. Their centers are not as central as they used

to be, their edges are ambiguous, they have no beginnings and apparently no end. Neither words, numbers, nor pictures can adequately comprehend their complex forms and social structure.

(Ingersoll, 1992, p. 5)

Just when we'd learned to see, and even love, the peculiar order beneath what earlier generations had dismissed as the chaos of the industrial city . . . along came a tidal wave of look-alike corporate office parks, mansarded all-suite hotels, and stuccoed town houses to throw us for another monstrous, clover-leaf loop.

(Sandweiss, 1992, p. 38)

It is now clear that cities throughout the developed world have recently entered a new phase – or, at least, begun a distinctive transitional phase – with important implications for the trajectory of urbanization and the nature of urban development. This new phase has its roots in the dynamics of capitalism and, in particular, the **globalization** of the capitalist economy, the increasing dominance of big conglomerate corporations, and the steady shift within the world's core economies away from manufacturing industries towards service activities. Yet, as this fundamental economic transition has been gathering momentum, other shifts – in demographic composition, and in cultural and political life – have also begun to crystallize.

Economic change and urban restructuring

Since the 1970s, the economies of Europe and North America have entered a substantially different phase in terms of *what* they produce, *how* they produce it, and *where* they produce it. They are now characterized by **neo-Fordist** regimes of production (see Chapter 2).

In terms of *what* they produce, the dominant trend has been a shift away from agriculture and manufacturing industries towards service activities. There have been, however, substantial differences in the performance of different *types* of services. Contrary to the popular view of retail and consumer services as a driving force in advanced economies, they have not in fact grown

9

...r, it has been producer services ...ces), public sector services and non- ...mainly higher education and certain ...lth-care) that have contributed most to ...n in service-sector employment. As we shall see ... sequent chapters, these economic shifts have been written into the social geography of contemporary cities in a variety of ways as labour markets have been restructured.

In terms of *how* production is organized, there have been two major trends. The first has been towards oligopoly as larger and more efficient corporations have driven out their competitors and sought to diversify their activities. The second has been a shift away from mass production toward flexible production systems. This trend has had much greater significance for urban social geography, since the flexibility of economic activity has imprinted itself onto the social organization and social life of cities, creating new cleavages as well as exploiting old ones.

In terms of *where* production takes place, the major trend has been a redeployment of activity at metropolitan, national and international scales – largely in response to the restructuring of the big conglomerates. As the big new conglomerates have evolved, they have rationalized their operations in a variety of ways, eliminating the duplication of activities among regions and nations, moving routine production and assembly operations to regions with lower labour costs, moving 'back office' operations to suburbs with lower rents and taxes, and consolidating head-office functions and R & D laboratories in key settings. As a result, a complex and contradictory set of processes has begun to recast many of the world's economic landscapes.

One of the major outcomes in relation to urban social geography has been the **deindustrialization** of many of the cities and urban regions of the industrial heartlands in Europe and North America. Another has been the accelerated **decentralization** of both manufacturing and service employment within metropolitan regions. A third has been the transformation of a few of the largest cities into **world cities** (also termed **global cities**) specializing in the production, processing and trading of specialized information and intelligence (see Chapter 2). And a fourth has been the **recentralization** of high-order producer-service employment.

Meanwhile, economic globalization and the influence of new digital telecommunications technologies have drawn individual cities and parts of cities into different – and rapidly changing – roles in the ever-broadening and increasingly complex circuits of economic and technological exchange of neo-Fordism (see Chapter 2). Traditional patterns of urbanization began to be overwritten by a very new dynamic dominated by enclaves of superconnected people, firms and institutions, with their increasingly broadband connections to elsewhere via the Internet, mobile phones and satellite televisions and their easy access to information services. The uneven evolution of networks of information and communications technologies began to forge new urban landscapes of innovation, economic development and cultural transformation while at the same time intensifying social and economic inequalities within cities, resulting in what Stephen Graham and Simon Marvin (2001) have termed **splintering urbanism**. Changes in this fundamental have inevitably led to major changes in the social geography of every city, affecting everything from class structure and community organization to urban service delivery and the structure of urban politics. Meanwhile, economic restructuring and the transition to neo-Fordism has produced some important changes in the composition of urban labour markets, not least of which is a tendency toward economic polarization. One conspicuous outcome has been a decisive increase in unemployment in the cities of the world's industrial core regions. Another significant outcome has been that the shift away from manufacturing has resulted in a substantial decrease in blue-collar employment and a commensurate increase in white-collar employment. White-collar employment itself has been increasingly dichotomized between professional and managerial jobs on the one hand and routine clerical jobs on the other.

Within the manufacturing sector, meanwhile, advances in technology and automation have begun to polarize employment opportunities between those for engineers/technicians and those for unskilled/semi-skilled operatives. Within the service sector, retailing and consumer services have come to be dominated by part-time jobs and 'secondary' jobs (jobs in small firms or in the small shops or offices of large firms, where few skills are required, levels of pay are low and there is little opportunity for advancement. Government services, on the other hand, tend to have increased the pool of 'primary' jobs (jobs with higher levels of pay and security).

One key consequence of these changes, from the point of view of urban social geography, is that a growing proportion of both working- and middle-class families find it increasingly difficult to achieve what they had come to regard as an acceptable level of living on only one income. One response to this has been the expansion of the two-paycheque household; another has been the growth and sophistication of the informal economy, which in turn has begun to create new kinds of household organization, new divisions of domestic and urban space, and new forms of communal relations.

Finally, it is important to bear in mind that many of the changes emanating from economic transformation are taking place simultaneously within most large cities. Thus we see, side by side, the growth of advanced corporate services and the development of sweatshops operated by undocumented workers, the emergence of newly affluent groups of manager-technocrats and the marginalization of newly disadvantaged groups. As a result, the emerging geography of larger cities is complex. It is creating distinct new social spheres, yet it has to link these spheres within the same functional unit.

The imprint of demographic change

In the past 30 or so years, some important demographic changes have occurred that have already begun to be translated into the social geography of the early twenty-

first-century city. The storybook family in Dick-and-Jane readers (with an aproned mother baking cakes for the two children as they await father's return from a day of breadwinning) has by no means disappeared, but it is fast being outnumbered by other kinds of families. In the United States, for instance, most people live in households where there are two wage-earners. The single-parent family is the fastest-growing of all household types, and almost one in every three households consists of a person living alone (Table 1.2). Similar changes are occurring in most other Western societies in response to the same complex of factors.

Central to all these changes is the experience of the generation born after the Second World War (the baby boomers). The baby boom generation – those individuals born between 1946 and 1964 – has been one of the most powerful and enduring demographic influences on European and North American societies. The postwar baby boom was not about women having more children. It was, instead, based on more people marrying overall and having at least two children early in the marriage. During the 1950s and early 1960s, young women also married earlier than the previous generation and had children earlier. The result is that there is a very large cohort of individuals, currently in their late 30s/early 40s to their mid-to-late 50s, who have had, and will continue to have, tremendous impacts on the rest of the population, especially as they enter their 60s, 70s and 80s.

The advent of reliable methods of birth control in the mid-1960s fostered the postponement of childbearing in the baby boom generation just as its first

Table 1.2 Household composition, United States, 1950–2000 (percentages)

	Married couples		Headed by women	Persons living alone	Other	Total
	Home-making wife	Working wife				
1950	59.4	19.6	8.4	10.8	1.8	100
1955	54.2	21.7	8.8	12.8	2.5	100
1960	51.2	23.3	8.5	14.9	2.5	100
1965	47.0	25.6	8.7	16.7	2.0	100
1970	41.6	28.9	8.8	18.8	1.9	100
1975	36.6	29.2	10.0	21.9	2.3	100
1980	30.3	30.6	10.8	26.1	2.2	100
1990	23.0	32.2	11.6	29.8	3.4	100
2000	17.5	34.2	12.2	31.9	4.2	100

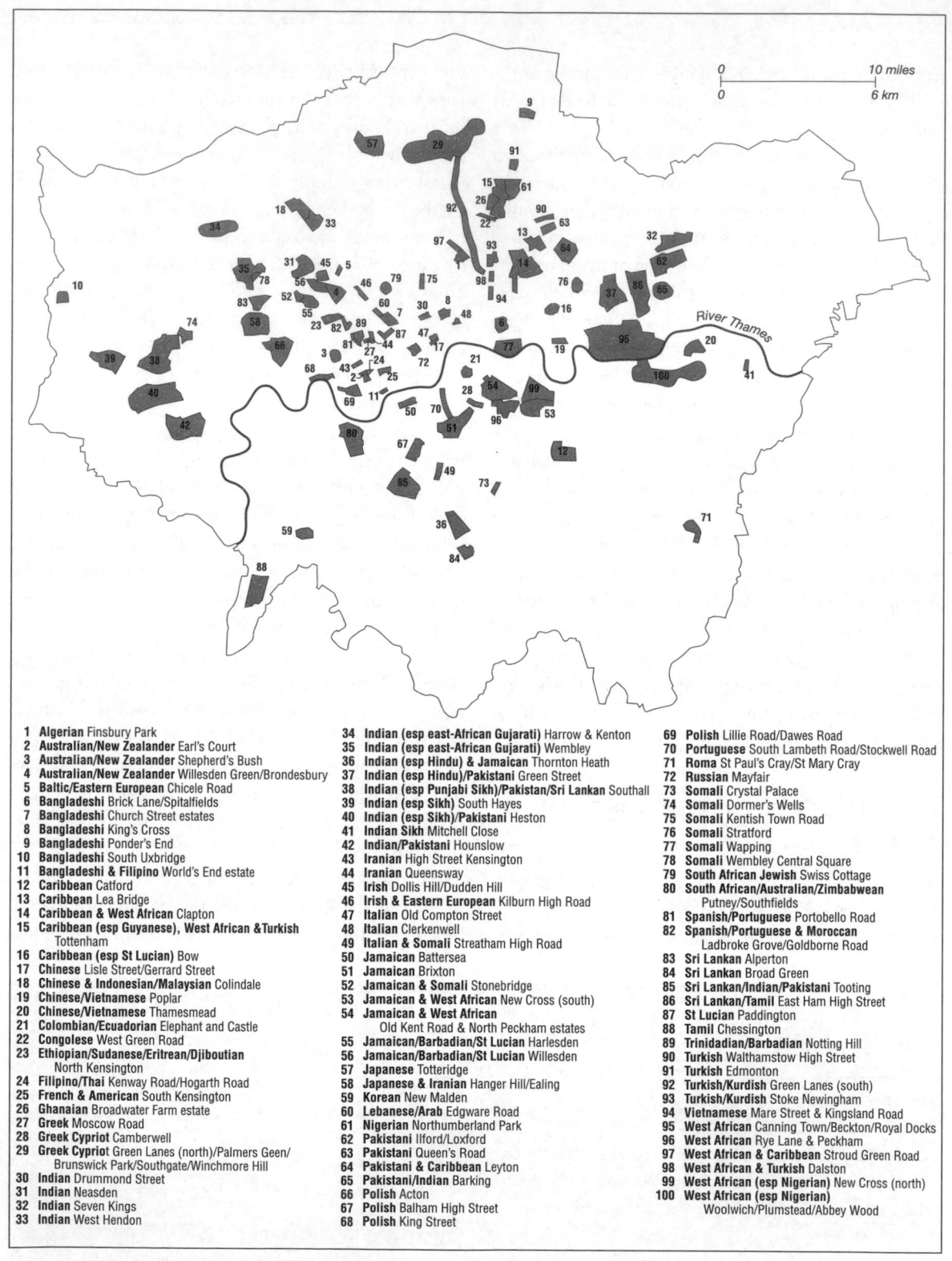

Figure 1.1 Cosmopolitanism in London: the distribution of ethnic minority clusters in 2004.
Source: *Guardian*, G2 Supplement, 21st January, 2005.

1 **Algerian** Finsbury Park
2 **Australian/New Zealander** Earl's Court
3 **Australian/New Zealander** Shepherd's Bush
4 **Australian/New Zealander** Willesden Green/Brondesbury
5 **Baltic/Eastern European** Chicele Road
6 **Bangladeshi** Brick Lane/Spitalfields
7 **Bangladeshi** Church Street estates
8 **Bangladeshi** King's Cross
9 **Bangladeshi** Ponder's End
10 **Bangladeshi** South Uxbridge
11 **Bangladeshi & Filipino** World's End estate
12 **Caribbean** Catford
13 **Caribbean** Lea Bridge
14 **Caribbean & West African** Clapton
15 **Caribbean (esp Guyanese), West African &Turkish** Tottenham
16 **Caribbean (esp St Lucian)** Bow
17 **Chinese** Lisle Street/Gerrard Street
18 **Chinese & Indonesian/Malaysian** Colindale
19 **Chinese/Vietnamese** Poplar
20 **Chinese/Vietnamese** Thamesmead
21 **Colombian/Ecuadorian** Elephant and Castle
22 **Congolese** West Green Road
23 **Ethiopian/Sudanese/Eritrean/Djiboutian** North Kensington
24 **Filipino/Thai** Kenway Road/Hogarth Road
25 **French & American** South Kensington
26 **Ghanaian** Broadwater Farm estate
27 **Greek** Moscow Road
28 **Greek Cypriot** Camberwell
29 **Greek Cypriot** Green Lanes (north)/Palmers Geen/ Brunswick Park/Southgate/Winchmore Hill
30 **Indian** Drummond Street
31 **Indian** Neasden
32 **Indian** Seven Kings
33 **Indian** West Hendon

34 **Indian (esp east-African Gujarati)** Harrow & Kenton
35 **Indian (esp east-African Gujarati)** Wembley
36 **Indian (esp Hindu) & Jamaican** Thornton Heath
37 **Indian (esp Hindu)/Pakistani** Green Street
38 **Indian (esp Punjabi Sikh)/Pakistan/Sri Lankan** Southall
39 **Indian (esp Sikh)** South Hayes
40 **Indian (esp Sikh)/Pakistani** Heston
41 **Indian Sikh** Mitchell Close
42 **Indian/Pakistani** Hounslow
43 **Iranian** High Street Kensington
44 **Iranian** Queensway
45 **Irish** Dollis Hill/Dudden Hill
46 **Irish & Eastern European** Kilburn High Road
47 **Italian** Old Compton Street
48 **Italian** Clerkenwell
49 **Italian & Somali** Streatham High Road
50 **Jamaican** Battersea
51 **Jamaican** Brixton
52 **Jamaican & Somali** Stonebridge
53 **Jamaican & West African** New Cross (south)
54 **Jamaican & West African** Old Kent Road & North Peckham estates
55 **Jamaican/Barbadian/St Lucian** Harlesden
56 **Jamaican/Barbadian/St Lucian** Willesden
57 **Japanese** Totteridge
58 **Japanese & Iranian** Hanger Hill/Ealing
59 **Korean** New Malden
60 **Lebanese/Arab** Edgware Road
61 **Nigerian** Northumberland Park
62 **Pakistani** Ilford/Loxford
63 **Pakistani** Queen's Road
64 **Pakistani & Caribbean** Leyton
65 **Pakistani/Indian** Barking
66 **Polish** Acton
67 **Polish** Balham High Street
68 **Polish** King Street

69 **Polish** Lillie Road/Dawes Road
70 **Portuguese** South Lambeth Road/Stockwell Road
71 **Roma** St Paul's Cray/St Mary Cray
72 **Russian** Mayfair
73 **Somali** Crystal Palace
74 **Somali** Dormer's Wells
75 **Somali** Kentish Town Road
76 **Somali** Stratford
77 **Somali** Wapping
78 **Somali** Wembley Central Square
79 **South African Jewish** Swiss Cottage
80 **South African/Australian/Zimbabwean** Putney/Southfields
81 **Spanish/Portuguese** Portobello Road
82 **Spanish/Portuguese & Moroccan** Ladbroke Grove/Goldborne Road
83 **Sri Lankan** Alperton
84 **Sri Lankan** Broad Green
85 **Sri Lankan/Indian/Pakistani** Tooting
86 **Sri Lankan/Tamil** East Ham High Street
87 **St Lucian** Paddington
88 **Tamil** Chessington
89 **Trinidadian/Barbadian** Notting Hill
90 **Turkish** Walthamstow High Street
91 **Turkish** Edmonton
92 **Turkish/Kurdish** Green Lanes (south)
93 **Turkish/Kurdish** Stoke Newingham
94 **Vietnamese** Mare Street & Kingsland Road
95 **West African** Canning Town/Beckton/Royal Docks
96 **West African** Rye Lane & Peckham
97 **West African & Caribbean** Stroud Green Road
98 **West African & Turkish** Dalston
99 **West African (esp Nigerian)** New Cross (north)
100 **West African (esp Nigerian)** Woolwich/Plumstead/Abbey Wood

cohort reached marriageable age. Meanwhile, the demise of the 'living wage', noted above, prompted still more women to take up full-time employment and to postpone childbearing or to return to work soon after childbirth. The reduced birth rates resulted in a 'baby bust' generation. Meanwhile, as their boomer parents began to reach their peak earning years, they were instrumental in an important cultural shift: a change in lifestyle preferences away from familism towards consumerism (Clarke, 2003).

These trends have some important consequences for urban social geography. In addition to the implications of an increasingly consumerist urban lifestyle, the effects of reduced birth rates on many aspects of collective **consumption**, and the effects of higher proportions of working women on the demand for child-care facilities, there are the implications of the ageing of the boomers. The baby bust generation will, of course, be the generation that has to 'mop up' after the flood of boomers moves up their career ladders towards retirement and beyond. The tremendous size of the baby boom cohort is likely to affect the career and job mobility of the busters, and as more and more of them reach retirement age, so the burden of financing pension funds will fall increasingly on the busters. Perhaps most important change of all is the general, if gradual and incomplete, change in attitudes towards women that has accompanied consumerism, birth control and increased female participation in the labour market. Already, changing attitudes about the status of women have come to be reflected in improved educational opportunities and a wider choice of employment, both of which have fostered the development of non-traditional family structures and lifestyles.

In addition, once the proposition that sex need not be aimed primarily or solely at procreation had become generally accepted, further trends were set in motion. The social value of marriage decreased, with a consequent decline in the rate of marriage, an increase in divorce and an increase in cohabitation without marriage – all conspiring to depress the fertility rate still further and to create large numbers of non-traditional households (single-parent households, in particular) with non-traditional housing needs, non-traditional residential behaviour and non-traditional demands on urban services. Meanwhile, the effects of economic restructuring,

combined with the increased number of female-headed households and the generally inferior role allocated to women in the labour market, has precipitated yet another set of changes with important implications for urban social geography: the economic marginalization of women and the **feminization of poverty**.

One final and important demographic change in Western cities in recent years has been their growing cosmopolitanism, i.e. increasing cultural and ethnic diversity. This is especially noticeable in large cities such as London (Fig. 1.1), Paris, New York and Los Angeles (see also Box 1.3). Such cities have been reception areas for peoples from many nations for many years and much of their contemporary diversity emanates from the descendants of earlier waves of immigration. This is especially true of US cities with their typical patchwork of diverse ethnic neighbourhoods. However, in recent years new waves of immigration have changed the character of cities. The scale of movement may be smaller than in the past but the range of peoples is far more diverse. No longer is the predominant pattern one of migration from Europe to the United States, or from former colonies into European cities, but the movement of peoples from all around the globe into all the main Western cities (Sassen, 1999b). Whether temporary or permanent, legal or illegal, these migrants have been attracted by the economic opportunities offered in Western cities. Some are an elite group of highly skilled workers operating in a global labour market, while others are less skilled workers prepared to undertake relatively low-paid work. Yet other migrants are refugees seeking asylum from political oppression. Others attempt to infiltrate under the guise of political refugees when they are in reality economic migrants.

The city and cultural change

The rise of consumerism and materialistic values has been one of the dominant cultural trends since 1980. One reason for this is the lifestyle shift associated with the baby boom generation, described above. Another is that people have been made more materialistic as capitalism has, in its search for profits, had to turn away

Box 1.3

Key trends in urban social geography – Growing cosmopolitanism in Western cities: the example of London

A major trend in Western cities in recent years has been their growing cosmopolitanism. This cultural diversity reflects not only the descendants of previous generations of immigrants, but also recent migrants such as temporary workers, students, refugees, asylum seekers and long-term tourists.

Few cities illustrate this growing cosmopolitanism better than London. As Hamnett (2003) notes, in the 1960s ethnic minorities in London were small in scale and extent. Yet, as illustrated in the table opposite, by 2001 over a third of the population in London, as measured by the Census, was other than 'White British'. Some cities have larger proportions of individual groups, such as Latinos and African-Americans in Los Angeles. Nevertheless, few can match the sheer range of ethnic origins to be found in London. This diversity reflects not only the colonial past of Britain but also London's status as a centre of international finance, business services and creative and cultural industries.

The resulting atmosphere in London was well captured by the distinguished social commentator Anthony Sampson:

The capital [London] has become the most cosmopolitan in the world, from top to bottom, teeming with Americans,

Ethnic group	Numbers (000)	Percentage of total population
White: British	4287	59.8
White: Irish	220	3.1
White: other	595	8.3
Mixed	226	3.2
Indian	437	6.1
Pakistani	143	2.0
Bangladeshi	154	2.1
Other Asian	133	1.9
Black: Caribbean	344	4.8
Black: African	379	5.3
Other Black	60	0.8
Chinese	80	1.1
Other ethnic group	113	1.6
All ethnic groups	7192	

Europeans, Australians, Asians, Africans and Arabs. . . . The streets and buses are loud with exotic languages, full of Muslim veils and beards and African Robes. The high street has restaurants from 30 countries including Iraq, Iran and Sudan.

It is perhaps hardly surprising therefore that in 1999 *Newsweek* ran a cover story headlined 'London: coolest city on the planet'.

Further reading

Hamnett, C. (2003) *Unequal City; London in the Global Arena* Routledge, London

White, P. (1998) The settlement patterns of developed world migrants in London *Urban Studies* **35**, 1725–1744.

Links with Other Chapters

Chapter 3: Postcolonial Theory and the City

Chapter 8: The Spatial Segregation of Minority Groups, Box 8.1 The Latinization of US cities

Chapter 12: Box 12.3 The Rise of 'Transnational Urbanism'

Chapter 13: Box 13.1 The Emergence of Clusters of Asylum Seekers and Refugees

from the increasingly regulated realm of production towards the more easily exploited realm of consumption. Meanwhile, the relative affluence of the postwar period allowed many households to be more attentive to consumerism. For one reason or another, people soon came to be schooled in the sophistry of conspicuous consumption. One of the pivotal aspects of this trend, from the point of view of urban social geography,

was the demand for home ownership and the consequent emphasis on the home and its accessories as an expression of self and social identity.

Against the background of this overall trend towards consumerism there emerged in the 1960s a distinctive middle-class youth counterculture based on a reaction against materialism, scale and high technology. These ideas can be seen, for example, in the politicization of liberal/ecological values in relation to urban development and collective consumption, and in the realm of postmodernist, neo-romantic architecture and urban design. It should be noted, however, that the spread of these values has not, for the most part, displaced materialism. Rather, they have grown up alongside. The middle classes have come to have their (organic) cake and eat it too, facilitated by the commercial development of products and services geared to liberal/ecological tastes.

Meanwhile, digital telecommunications, innovative forms of electronic representation, and economic globalization have fostered a homogenized cultural mainstream in which the meaning and distinctiveness of individual places and communities has become attenuated. The corollary of *this*, in turn, is an increased concern, among some consumers, with conserving and developing the urban sense of place.

Political change and the sociospatial dialectic

The trends outlined above are pregnant with problems and predicaments that inevitably figure among the dominant political issues. Economic restructuring has led to a realignment of class relations with differences being drawn increasingly along geographical (as well as structural) lines. As labour markets become more segmented, as differential processes of growth and decline work themselves out, and as shifts in the balance of economic and social power reshape the political landscape, there is a constant stream of political tensions.

Especially significant in this context was the spread in the 1980s and 1990s of the idea that welfare states had not only generated unreasonably high levels of taxation, budget deficits, disincentives to work and save, and a bloated class of unproductive workers, but also that they may have fostered 'soft' attitudes towards 'problem' groups in society. Ironically, the electoral appeal of this ideology can be attributed to the very success of welfare states in erasing from the minds of the electorate the immediate spectre of material deprivation. Consequently, the priority accorded to welfare expenditures receded (though the logic and, critically, the costs of maintaining them did not), displaced by **neoliberal policies** that are predicated on a minimalist role for the state, assuming the desirability of free markets as the ideal condition not only for economic organization, but also for political and social life (Brenner, 2002). The retrenchment of the public sector has already brought some important changes to the urban scene as, for example, in the privatization of housing and public services (Pinch, 1997). In broader perspective, these changes can be interpreted as part of the shift in emphasis from collective consumption to capital accumulation that is in turn one of the mechanisms through which the lead economies of the world-system have attempted to steer themselves out of an episode of 'stagflation'.

Meanwhile, deindustrialization and economic recession mean that sociospatial disparities have been reinforced. This has changed the dynamics of inner-city politics, while 'traditional' working-class politics, having lost much of its momentum and even more of its appeal, is being displaced by a 'new wave' of local politics (Chapter 5).

Chapter summary

1.1 Several different approaches – quantitative, behavioural, structural and poststructural – are relevant to an understanding of contemporary cities.

1.2 Cities reflect a *sociospatial dialectic*, a two-way process in which people modify urban spaces while at the same time are conditioned by the spaces in which they live and work.

1.3 City structures reflect their surrounding economic, demographic, cultural and political backgrounds. Consequently, North American cities display somewhat different characteristics to European cities.

Key concepts and terms

(Note: these will be developed in greater depth later in this book.)

behavioural approach	global cities	polarization
Cartesian approach	globalization	postindustrial society
classical Marxism	'god trick'	poststructuralist approach
consumption	human agency	privatism
cultural myopia	imaginative geographies	quantitative approach
cultural turn	Marxian theory	recentralization
culture	mimetic approach	regulation theory
decentralization	neoclassical economics	social wage
deindustrialization	neo-Fordism	sociospatial dialectic
discourse	neoliberal policies	splintering urbanism
feminization of poverty	neo-Marxist approaches	structuralism
functionalist sociology	normative assumptions	superorganic (culture)
gaze	phenomenology	world cities

Suggested reading

The concept of the sociospatial dialectic that under-pins this book can be found in Ed Soja's article 'The socio-spatial dialectic', *Annals of the Association of American Geographers*, 1980: **70**, 207–225. This concept is further developed in Chapter 3 of Soja *Postmodern Geographies* (1989: Verso, London, although this is an advanced book and not for those new to urban studies). Three readers reproduce key articles in the sphere of urban studies and are a useful companion to this book: see Richard Le Gates and Frederic Stout (eds) *The City Reader* (1996: Routledge, London); and the two edited readings by Gary Bridge and Sophie Watson *The Blackwell City Reader* (2002: Blackwell, Oxford) and *A Companion to the City* (2003: Blackwell, Oxford). An invaluable general reference for urban geographers is the fourth edition of *The Dictionary of Human Geography* (2000: Blackwell, Oxford) edited by Ron Johnston, Derek Gregory, Geraldine Pratt and Michael Watts. In addition, *A Glossary of Feminist Geography* (1998: Arnold, London) edited by Linda McDowell and Joanne P. Sharp considers a wide range of concepts relevant to urban social geo-graphy. In recent years there has been a resurgence of interest in urban geography and this has led to a spate of books on this topic. These provide background material complementing the focus of this volume on urban social aspects of urban areas. The following books are well worth tracking down: Blair Badcock *Making Sense of Cities; A geographical survey* (2002: Arnold, London); Michael Dear *The Postmodern Urban Condition* (2000: Blackwell, Oxford); Tim Hall *Urban Geography* (2001: Routledge, London); Kris Olds *Globalization and Urban Change* (2001: Arnold, London); Michael Pacione *Urban Geography: A Global Perspective* (2001: Routledge, London); Ed Soja *Postmetropolis: Critical Studies of Cities and Regions* (2000: Blackwell, Oxford). Two key texts on cities are the volume edited by Ronan Paddison *Handbook of Urban Studies* (2000: Sage, London) and Peter Hall's *Cities in Civilization* (1999: Phoenix Giant, London). An excellent review of planning ideologies and urban developments is Peter Hall's *Cities of Tomorrow* (second edition, 1997: Blackwell, Oxford). Useful texts on social geography are the volume edited by Rachel Pain and others *Introducing Social Geographies* (2001: Arnold, Oxford), Gill Valentine's *Social Geographies: Space and Society* (Prentice Hall, Harlow) and Ruth Panelli's *Social Geographies: From Difference to*

Action (2004: Sage, London). For analysis of recent developments in cultural geography see Holdsworth and Mitchell *Cultural Geography: A Reader* (1998: Arnold, London) and Mike Crang's excellent *Cultural Geography* (second edition, 2003: Routledge, London). A useful overview of developments in British cities is Michael Pacione (ed.) *Britain's Cities: Geographies of Division in Urban Britain* (1997: Routledge, London). The distinctiveness of European cities in comparison with those in North America is discussed on pages 102–106 in Stanley Brunn and Jack Williams' edited volume *Cities of the World* (second edition, 1993: HarperCollins, New York). Another accessible introduction to European cities is Donald McNeill's *New Europe: Imagined Spaces* (2004: Arnold, London). For the changing economic context see the introductory essay in Paul Knox *The Restless Urban Landscape* (1993: Prentice Hall, New York). Mike Savage and Alan Warde's *Urban Sociology, Capitalism and Modernity* (1993: Macmillan, London) provides useful background from the perspective of urban sociology and Simon Parker's *Urban Theory and the Urban Experience* (2004: Routledge, London) provides insights from political science. See also Jan Lin and Christopher Mele (eds) *The Urban Sociology Reader* (2004: Routledge, London). Finally, an extremely useful accompaniment to this book is *Key Thinkers on Space and Place* edited by Phil Hubbard, Rob Kitchin and Gill Valentine (2004: Sage, London); many of those discussed in this volume have influenced urban social geography.

International journals

In addition to the above, there are a number of journals that you should keep an eye on for the very latest research on urban social geography (listed in alphabetical order):

- *Annals of the Association of American Geographers**
- *Antipode**
- *Area**
- *Australian Geographical Studies**
- *Built Environment**
- *Canadian Geographer**
- *Capital and Class**
- *Children's Geographies**
- *Cities*
- *City; analysis of urban trends, culture, theory, policy, action*
- *Cultural Geographies**
- *Economic Geography**
- *Ecumene**
- *Environment and Planning A**
- *Environment and Planning C: (Government and Policy)**
- *Environment and Planning D: (Society and Space)**
- *Ethics, Place and Environment**
- *European Planning Studies**
- *Gender, Place and Culture**
- *Geoforum**
- *Geografiska Annaler**
- *Housing Studies**
- *International Journal of Urban and Regional Research**
- *International Planning Studies**
- *Journal of the American Institute of Planners**
- *Journal of the American Planning Association**
- *Journal of Property Research**
- *Journal of Historical Geography**
- *Journal of Social and Cultural Geography**
- *Journal of Urban Affairs*
- *Journal of Urban Design*
- *Local Economy**
- *Local Government Studies**
- *New Community**
- *New Zealand Geographer**
- *The Planner**
- *Planning Practice and Research**
- *Policy and Politics**
- *Political Geography**
- *Population, Space and Place**
- *Professional Geographer**
- *Progress in Human Geography**†
- *Progress in Planning**
- *Regional Studies**
- *Scottish Geographical Magazine**
- *Service Industries Journal**
- *Social and Cultural Geography* (www.tandf.co.uk/journals)
- *Space and Polity**
- *Tijdschrift voor Economische en Sociale Geografie**

- *Town and Country Planning**
- *Tourism Geographies**
- *Transactions of the Institute of British Geographers** (www.rgs.org/eTransactions) (pre-1994)
- *Urban Affairs Quarterly*
- *Urban Geography*
- *Urban Policy and Research*
- *Urban Studies*

* Only partially concerned with urban issues.

† This journal is especially useful since it contains many review articles covering all aspects of human geography.

Note: this list is by no means exhaustive – there are many other journals containing urban-related material published in English throughout the world.

Chapter 2

The changing economic context of city life

Key questions addressed in this chapter

➤ What were the main characteristics of the preindustrial city?

➤ In what ways did industrial capitalism affect the structure of cities?

➤ What is meant by the Fordist city?

➤ How has neo-Fordism influenced city structures?

➤ What has been the impact of globalization upon the social geography of cities?

➤ What are the likely effects of new technologies upon urban form?

The social geography of cities has been shaped by processes operating over many years. An appreciation of urban history is therefore vital to the student of urban social geography. These processes are diverse in character – economic, political, technological, social and cultural – but scholars have often stressed the importance of *one* of these factors in isolation.

For example, the changing role of technology is often cited as crucial in determining city structures (as with the shift from horse-drawn to steam-powered and later petrol-driven transport, or more recently the rise of new telecommunications systems). However, such approaches can fall into the trap of **technological determinism** – assuming that technology is some independent 'external' force acting 'upon' cities. Instead, technology is best conceptualized as an integral part of the sociospatial dialectic.

There are in fact many competing technologies at any given time, and which ones are ultimately influential usually depends not so much upon their inherent utility but upon economic, political and social influences. For example, in the nineteenth century the 'barons' who controlled the major railway companies had tremendous political power to affect city developments. In the twentieth century, however, the railroad companies in the United States came up against an even more powerful political lobby, the automotive manufacturers, who were able to suppress initiatives for public transport systems in some of the newer urban developments, thereby ensuring greater demand for cars.

Another big controversy has surrounded the relative importance of economic and cultural factors in determining the form of cities. At one extreme, some studies have fallen into the trap of **economic determinism**, assuming that the changing economic structure of cities determines the accompanying social and cultural forms. The other extreme, which has come to the fore more recently, might be seen as a form of 'cultural determinism', since it assumes that cultural issues have become crucial in contemporary economies. Thus, it is argued that culture affects *what* is produced (increasingly videos, films, music and software) as well as *how* it is produced (in firms that take issues of 'work culture' much more seriously than in the past). These processes are sometimes termed the **culturalization of the economy**. Most scholars adopt positions between these two extremes of economic and cultural determinism, acknowledging that issues of culture and economy are 'mutually constitutive'. This means that each realm affects the other but they have room for independent manoeuvre – no one sphere is totally determined by the other.

Bearing the above comments in mind, this chapter examines the influence of changing economic structures upon the structure of Western cities. The economic context is a useful starting point because there can be no doubt that economic systems have a crucial impact on city forms and their social geography. The discussion traces the impact of the shift from the early preindustrial economy, through to the rise of the capitalist economy, as manifest in the classic industrial city, and then considers contemporary developments in the so-called 'postindustrial' city. At various times issues of culture will be considered but these are given extended treatment in Chapter 3.

2.1 The precapitalist, preindustrial city

Before the full emergence of capitalist economies in the eighteenth century and the advent of the Industrial Revolution in the nineteenth century, cities were essentially small-scale settlements based on a mercantile economy and a rigid social order stemming from the tradition of medieval feudalism. Our knowledge of these early settlements is fairly patchy but it seems likely that they varied considerably in structure. Sjoberg (1960) has provided us with an idealized model of the social geography of the **preindustrial city** (see Fig. 2.1). In essence, this is the spatial expression of the division of the preindustrial city into a small elite and larger groups of lower classes and outcasts. The elite lived in a (by the standards of the times) pleasant and exclusive central core while the lower classes and outcasts lived in a surrounding poorly built and garbage-strewn periphery.

According to Sjoberg, the elite group consisted of those in control of the religious, political, administrative and social functions of the city. Merchants – even

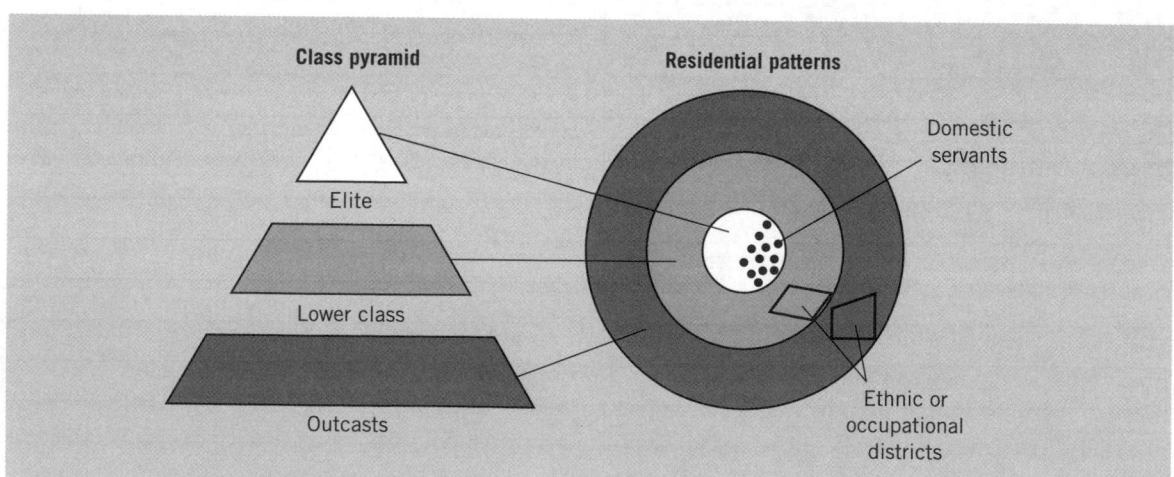

Figure 2.1 Sjoberg's idealized model of the social and geographical structure of the preindustrial city.
Source: Radford (1979) p. 394.

Relic features of the pre-industrial city: Canale Cannereggio, Venice. Photo Credit: Paul Knox.

the wealthy ones – were generally excluded from the elite because of a 'pre-occupation with money and other mundane pursuits ran counter to the religious-philosophical value systems of the dominant group' (Sjoberg, 1960, p. 83). Responding to these values, the elite tended to favour a residential location close to the administrative, political and religious institutions that were typically located in the centre of the city, thus producing an exclusive, high-status core. In time, the elite came to be increasingly segregated from the rest of urban society, partly because of the repulsiveness of the rest of the city and its inhabitants, and partly because of a clustering reinforced by bonds of kinship and intermarriage among the elite.

Beyond this core area lived the lower classes, although not in an undifferentiated mass. Distinct socioeconomic clusters developed as a result of the spatial association of craftsmen of different kinds, reinforced by social organizations such as guilds, which fostered group cohesion and spatial clustering of their members. Less well-organized groups, including the poor, members of ethnic and religious minority groups, people engaged in particularly malodorous jobs (such as tanning) and people who could find only menial employment (carting, sweeping or peddling) found themselves pushed to the outskirts of the city in extensive but densely inhabited tracts of the very worst housing. Table 2.1 summarizes the key features of Sjoberg's idealized model of the preindustrial city.

The idea of a preindustrial social geography characterized by an exclusive central core surrounded by a wider area over which status and wealth steadily diminished with distance from the city centre was questioned by Vance (1971), who attached much greater significance to the occupational clusterings arising from the inter-relationships between social and economic organization

Table 2.1 Characteristics of Sjoberg's idealized model of the
preindustrial city

- Inanimate power base
- Dependence upon slave labour
- City has small population
- City has a small proportion of the total population of the society
- City is surrounded by a wall
- City has important internal walls
- Domination by a feudal elite
- Elite is isolated by traditional values
- Elite is isolated by lifestyle
- Elite is impenetrable
- Elite has large households
- Elite privileges leisure and despises industry and commerce
- Elite women are idle
- Education is confined to the elite
- The merchant class are excluded from the elite
- Merchants are regarded as foreign and suspected of spreading heretical ideas
- Successful merchants use wealth to acquire elite symbols
- City has three classes
- Society has a sovereign ruler
- Rigid class structure
- Existence of an outcast group
- Manners, dress and speech reinforce class divisions
- Craft and merchant guilds
- Time is unregulated
- Credit is poorly developed
- Residential status declines with distance from the core
- Elite occupies city centre
- City centre has symbolic sites
- Outcasts are located on the urban periphery
- Part-time farmers on the periphery
- Ethnic quarters within the city
- Residential areas are differentiated by occupation
- Lack of functional specialization of land use

Sources: Sjoberg (1960); Radford (1979).

promoted by the craft guilds. For Vance, the early city was 'many centred' in distinct craft quarters – metal working, woodworking, weaving and so on – each with its own shops, workplaces and wide spectrum of inhabitants. The political, social and economic advantages conferred by guild membership reinforced external economies derived from spatial association, creating tight clusters of population living under a patriarchal social system headed by the master craftsmen.

Within each of the occupational districts, dwellings, workshops and store rooms were arranged with a vertical rather than horizontal structuring of space, with workshops on the ground floor, the master's family quarters on the floor above and, higher still, the store rooms and rooms of the journeymen, apprentices and

servants. Beyond the specialized craft quarters Vance recognized, like Sjoberg, the existence of a fringe population of the very poorest of the proletariat and a central core inhabited by the city's elite. Unlike Sjoberg, however, Vance interpreted these groups as having only a minor impact on the social geography of the city. The result was a model of the city in which spatial differentiation is dominated by a mosaic of occupational districts, with class and status stratification contributing a secondary dimension that is more important vertically than horizontally.

As with much social history, we do not yet have enough evidence from comparative studies to judge which of these two interpretations of the preindustrial city is more accurate. For present purposes, however, it is probably more helpful to stress the points of common agreement. Both writers portray a city in which everything physical was at a human scale, a 'walking city' in which the distances between home and work were even more tightly constrained by the organization of work into patriarchal and familial groupings. Both portray an immutable social order based on a traditional and essentially non-materialistic value system; and both recognize the existence (though with different emphasis) of a patrician elite residing in the core of the city, a number of occupationally distinctive but socially mixed 'quarters' in intermediate locations, and a residual population of the very poor living on the outskirts of the city.

2.2 The growth of the industrial city

The industrial city inherited few of the social or morphological characteristics of the preindustrial city. Some, such as Bologna, Bruges, Norwich and Stirling, were fortunate enough to retain their castles, cathedrals, palaces and other institutional buildings, together, perhaps, with fragments of the preindustrial residential fabric. Others, such as Aigues Mortes, Bernkastel and Ludlow, were bypassed by change and have consequently retained much of the appearance of the preindustrial city, albeit in a sanitized, renovated and picture postcard way. Many more sprang up with virtually no antecedents, products of a new economic logic that turned urban structure inside out from the preindustrial model, with the rich exchanging their central location for the peripheral location of the poor. Occupational clustering has given way to residential differentiation in terms of status, family structure, ethnicity and lifestyle; power and status in the city are no longer determined by traditional values but by wealth; ownership of land has become divorced from its use; workplace and home have become separated; and family structures have been transformed.

The cause of this profound realignment was primarily economic, rooted in the emergence of capitalism as the dominant means of production and exchange and buttressed by the technologies that subsequently emerged during the Industrial Revolution. Probably the most fundamental change to emerge with the rise of capitalism and its new system of production – the factory – was the creation of two 'new' social groups: the industrial capitalists and the unskilled factory workers. These two groups respectively formed the basis of a new elite and a new proletariat that replaced the old order. As the accumulation of capital by individuals became not only morally acceptable but the dominant criterion of status and power, entrepreneurs introduced a new, materialistic value system to urban affairs.

Meanwhile, competition for the best and most accessible sites for the new factories and the warehouses, shops and offices that depended on them brought about the first crucial changes in land use. Land was given over to the uses that could justify the highest rents, rather than being held by a traditional group of users. The factory and commercial sites secured, there sprang up around them large tracts of housing to accommodate the workers and their families. The new urban structure became increasingly differentiated, with homes no longer used as workplaces, and residential areas graded according to the rents that different sites could command. Social status, newly ascribed in terms of money, became synonymous with rent-paying ability, so that neighbourhoods were, in effect, created along status divisions.

Inevitably, since the size and quality of buildings was positively linked with price, and price with builders' profits, housing built for the lowest-paid, lowest-status groups was of the lowest quality, crammed in at high densities in order to cover the costs of the ground rent. At the same time, the wealthy moved to new locations on the urban fringe. Edged out of the inner city by factories and warehouses, the wealthy were in any case anxious to add physical distance to the social distance between themselves and the bleak misery of the growing working-class neighbourhoods adjacent to the factories. Encouraged by the introduction of new transport services in the early nineteenth century, they were easily lured to the fashionable new dwellings being built in the suburbs by speculators with an eye towards this lucrative new market.

Later, as the full effects of a dramatic excess of births over deaths (which largely resulted from improvements in medical practice and public health) were reinforced by massive immigration (in response to the cities' increased range and number of opportunities), the rate of urban growth surged. Changes in building technology made it possible for cities to grow upwards as well as outwards, and the cyclical growth of the capitalist economy, with successive improvements in urban transport systems, produced a sequence of growth phases that endowed the industrial city with a series of patchy but distinctive suburban zones.

Early models of the spatial structure of industrial cities

For observers in the nineteenth century, one of the most perplexing aspects of the cities was the spatial separation of the classes. Nowhere was this more apparent than in Manchester, the first 'shock city' of the age. Probably the best known and most succinct description of this segregation is Engels's (1844) work on Manchester in *The Condition of the Working Class in England* (a book that provided much of the documentation for the theories of his friend Karl Marx). Engels wrote that the commercial centre of Manchester, largely devoid of urban dwellers, was surrounded by:

> . . . unmixed working peoples' quarters, stretching like a girdle, averaging a mile and a half in breadth . . . Outside, beyond this girdle, lives the upper and middle bourgeoisie, the middle bourgeoisie in regularly laid out streets in the vicinity of the working quarters . . . the upper bourgeoisie in remoter villas with gardens . . . in free, wholesome country air, in fine comfortable homes, passed once every half or quarter hour by omnibuses going into the city.
>
> (Engels, 1844, p. 80)

This pattern of concentric zones, with the working class concentrated near the centre, was to become typical of many Victorian cities. By 1900, London exhibited four distinct zones around the largely unpopulated commercial core of the city. Charles Booth, in his *Life and Labour of the People of London*, provided a series of 'social maps' that showed these zones clearly (Booth, 1903). The innermost zone was characterized by the most severe crowding and extreme poverty, except in the west where there was a sector of extreme affluence.

The second zone was slightly less wealthy in this western sector and rather less crowded and impoverished elsewhere, while the third zone was inhabited by the 'short distance commuter' belonging mainly to the lower-middle class. The fourth zone belonged exclusively to the wealthy. The overall pattern of zones was, however, modified by a series of linear features. As competition for central space drove the price of land up and up, industry began to edge outward from the commercial core, following the route of canals, rivers and railways, and so structuring the city into a series of wedges or sectors.

North American cities also exhibited a spatial structure with a predominantly zonal pattern, but with important sectoral components. In the United States, Chicago was the archetypal example. Indeed, the idealized version of the social geography of Chicago in the early decades of the twentieth century became, for urban geographers, 'the seed bed of theory, the norm, the source of urban fact and urban fiction' (Robson, 1975, p. 4). In Chicago, both sectors and zones were particularly pronounced because of the effects of the massive inflows of immigrant workers and the radial development of the railroads that fanned outwards from the centre of the city, drawing with them corridors of manufacturing industry. The residential communities that developed between the radial corridors during successive phases of urban growth were graphically documented by the Chicago 'School' of urban sociology (Burgess, 1926; Park *et al.*, 1925), providing the basis for ecological ideas that have influenced urban studies ever since (see Chapter 7).

Marx and the industrial city

It was in this rapidly changing context of the nineteenth century industrial city that Karl Marx formed his ideas about capitalist society. The key concept in his analysis was the underlying economic base of society – the system of industrial capitalism – which he termed the **mode of production**. This consisted of two further elements: the **forces of production**, the technology underpinning the production process, and the **social relations of production**, the legal system of property rights and trade union legislation that governed the system of production.

Marx argued that the forces of production tend to be held back by the social relations of production – so that a new set of social relations is required in order to release the productive potential of new technologies when they come along. Preindustrial social relations, Marx pointed out, could not cope with the new forces unleashed by the growth of new trading patterns and industrial technologies. From this perspective, therefore, the economic base of society becomes a key driver of a much wider set of social arrangements (sometimes termed the **superstructure**).

Marx used his (and Engels's) observations to develop a critique of capitalism. Because of the ample supplies of labour in Victorian cities, wages were typically very low. The prevailing economic orthodoxy of the day argued that these wages were just and fair because they represented the most efficient outcome – the intersection of supply and demand in an equilibrium solution. However, Marx argued that price of commodities should not be determined by their **exchange value** (the amount they could command on the market) but by their **use value** (their capacity to satisfy human needs). The difference between what workers were paid for producing goods and the price that goods could command on the market, Marx termed the **surplus value**. According to his **labour theory of value**, instead of market exchange values, the prices of commodities should reflect the amount of 'socially necessary labour' that went into their production. This perspective enabled Marx to argue that this surplus value was being wrongly taken away from the new industrial proletariat in the form of huge profits by factory owners. Rather than a just reward for taking risks with their investments, these profits were seen as an immoral appropriation of the wealth generated by workers.

This exploitative relationship is usually portrayed as a continuous **circuit of production**. This begins with the investment of capital or money (M) in commodities (C) in the form of labour power (LP) raw materials and the means of production (MP), which is used to produce more commodities (C'), which are then sold to acquire more money (M'). A key principle of a Marxist approach, therefore, is the observation that capitalism is not just the ownership of wealth, it is a set of social relations, or institutional arrangements, that affect the relationships between two classes that are

inevitably in conflict – enabling the owners of capital to command labour to produce further wealth.

The labour theory of value has been extremely controversial and has proven extremely difficult to operationalize empirically. Furthermore, much has changed since Marx was writing in the nineteenth century; the interests of capital are now less easily identifiable with a class of individuals (despite the ostentatious presence of the 'super-rich') since wealth has become more diffused among banks, pension funds and investments trusts. Companies are now much less likely to be owned by single individuals; skilled labour and knowledge has become much more important than unskilled manual labour in the production system; and the state has taken on an increasing role in regulating economies. In addition, the collapse of communist regimes in the late twentieth century means that Marxist notions are widely perceived to have failed, both as a political ideology and as a system for promoting economic efficiency. Nevertheless, as we will see in this and subsequent chapters, some of the basic concepts underlying Marx's ideas have proved to be a rich source of inspiration for scholars of Western cities.

Fordism and the industrial city

A key concept used to analyze changes in cities from the 1920s through to the mid-1970s is **Fordism**. The origins of this concept can be traced back to the Italian communist Gramsci (1973) but the notion has been most extensively developed by a group of French scholars in what is known as **regulation theory** (Aglietta, 1979; Lipietz, 1986; Dunford, 1990). This approach attempts to understand why it is that, despite all their inherent tensions and contradictions, capitalist economies manage to survive. Regulation theorists argue that such tensions and problems are overcome by various regulatory mechanisms, such as those embodied in legislation surrounding commerce, trade and labour relations, together with the activities of various institutions that govern these spheres. From time to time these various regulatory mechanisms show some stability, at which point a **mode of regulation** gets established. A crucial feature of the regulation approach is recognition of the fact that regulatory mechanisms vary considerably from nation to nation. Nevertheless, over time they tend to show certain similarities in different places. Furthermore, if we view economic systems from a broader perspective, it is argued that much more general sets of arrangements can be seen which serve to link production and consumption. These broader structures form what is termed a **regime of accumulation**, and Fordism represents one such regime. Fordism is a very wide-ranging concept that can be used to analyze changes in at least three different ways: first, changes in the way people work; second, changes in the way industrial production is structured; and third, changes in the organization of society as whole (in particular the ways in production and consumption are coordinated). Fordism as a way of working and a way of organizing industry is associated with the factory system developed in the early part of the twentieth century by Henry Ford in Detroit to mass-produce automobiles. Ford was an early advocate of **Taylorism** (named after an American engineer called Frederick Taylor), a system of production in which the planning and control of work in manufacturing industries are allocated entirely to management, leaving production workers to be assigned specialized tasks that are subject to careful analysis – 'scientific management' – using techniques such as time-and-motion studies.

The genius of Henry Ford's approach was to integrate these ideas with the moving assembly line – an idea he got from the method used to transport carcasses of meat around the slaughter-houses of Chicago. Each worker on the assembly line did a relatively simple task, assisted by specialized machines. This approach enhanced productivity to such an extent that Henry Ford was able to cut the cost of his cars by a half, while at the same time paying his workers $5 a day, a sum that was twice the average industrial wage at the time. This highly efficient system, combined with the widespread availability of credit, led to a revolution in production. On the one hand, the product, the Model T car, was just what consumers wanted, being reliable and simple to drive and maintain. On the other hand, the system of production suited the labour market of American cities, which at the time were crammed with migrants from many European nations. The relatively simple jobs on the assembly line could be undertaken by immigrants since they required limited training or knowledge of English. Henry Ford's factory system resulted in a productive linking of the technical division of labour (the work tasks that need to be done) with the social division of labour (the skills of the people available to do the work). The result was an increase in both supply of, and demand for, the product and the development of mass production.

Keynesianism and the 'long boom'

Although Fordism brought a capacity for vastly increased outputs of consumer goods in the 1920s and 1930s, the system faltered during this time because of a lack of demand, which resulted in the onset of a huge economic slump known as the Depression. After the Second World War, however, there emerged a system that, for a quarter of a century, seemed to create a relatively harmonious relationship between production and consumption. This period is often called the '**long boom**' of Fordism. Underpinning this time period was a government policy known as **Keynesianism**, based on the economic principles of the economist John Maynard Keynes. He argued that governments should intervene to regulate the booms and slumps that characterize capitalist economies. In particular, governments should spend in times of recession to create more effective demand for private goods and services.

In the United States after the Second World War the economy was greatly stimulated by government spending on the interstate and intra-urban highway systems. These new roads enabled unprecedented numbers of households to decentralize out of inner-city areas into surrounding low-density suburban areas. This resulted in greater distances between home, work and shops and therefore greatly boosted the automobile industry. The construction industry was also kept busy building new suburban dwellings as well as roads and there was also a huge demand for domestic consumer products such as televisions, cookers and refrigerators. Drawing heavily upon the work of the famous French urban analyst Henri Lefebvre, David Harvey formulated a theory linking these new geographical arrangements with the needs of capitalism. Harvey (1978) argues that this massive process of suburbanization represented a shift from the 'primary

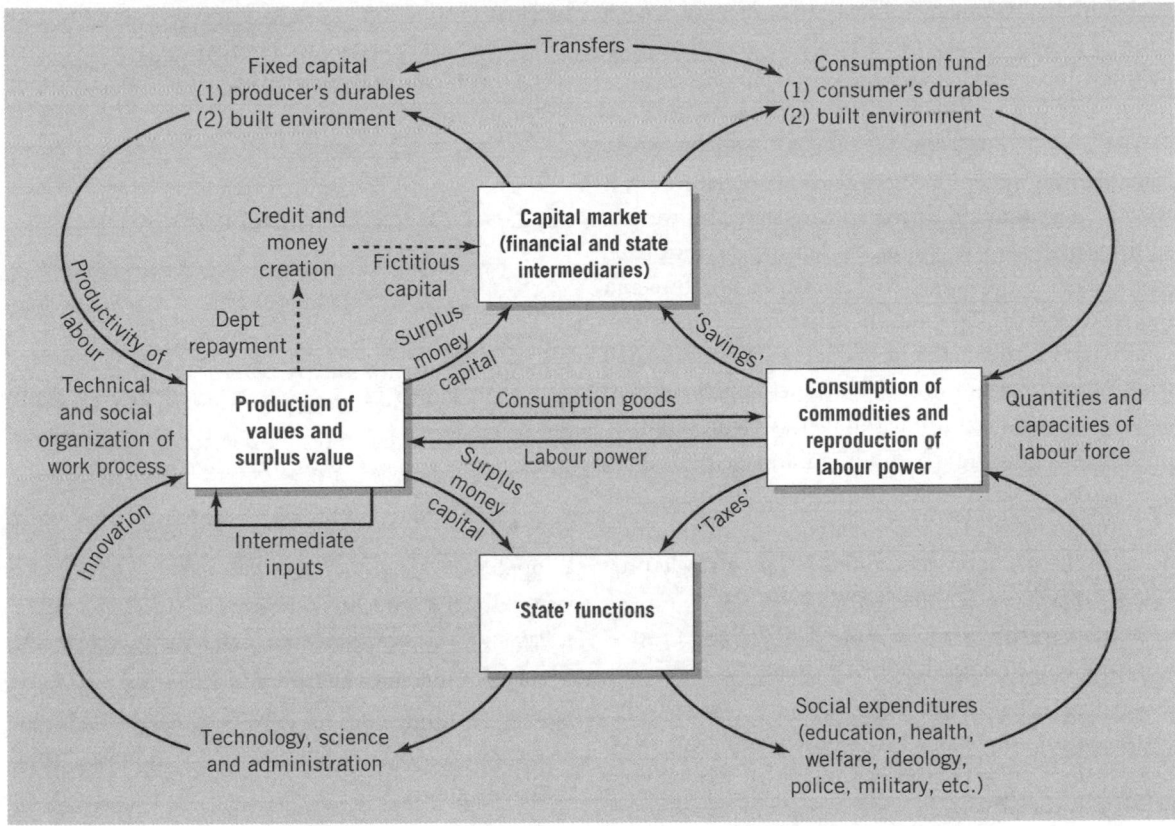

Figure 2.2 David Harvey's model of the circuits of capital involved in suburbanization.
Source: Harvey (1982), Fig. 21.1, p. 408.

circuit' of capital (investment in the production system) into the 'secondary circuit' (various consumption funds including the built environment) (see Fig. 2.2). This was extremely useful for the capitalist system at this time in a variety of ways (amounting to what he calls a '**spatial fix**'). Harvey argues that suburbanization stimulated a '**commodity fetishism**' – an obsessional tendency for households to compete with one another and display their wealth through consumer products. In addition, since most families needed to raise a mortgage to purchase their properties, it was argued that this tended to stabilize the entire socioeconomic system, producing a class of debt-encumbered persons who were unlikely to petition for radical change.

Some have argued that this interpretation of decentralization portrays suburban families as hapless ciphers in an economic system. They argue that Harvey's interpretation therefore smacks of **functionalism**, attempting to 'read off' the causes of suburbanization through its assumed effects. Even if these effects were in evidence (and this is not always clear), it cannot be assumed that policies were developed with these particular consequences in mind. Others would therefore stress the role of human agency and the ways in which capitalism satisfies the needs of people for material goods together with their desire to escape from overcrowded inner-city environments. The main problem with this focus upon human choice is that it falls into the opposite trap: of **voluntarism** – assuming that people have complete freedom to do what they wish, free of all economic constraints. It is perhaps best to steer a path between the extremes of functionalism and voluntarism here, acknowledging that people display human agency but in a context in which their desires and wishes are actively framed by powerful economic interests.

In European cities, as noted in Chapter 1, there was much less suburbanization during the Fordist era. Nevertheless, relatively inexpensive European cars such as the original German Volkswagen Beetle, the French Citroen 2CV, the Italian Fiat Topolino and the British Morris Minor, together with the various autobahn, autoroute, autostrada and motorway systems of Europe, played a similar role to the Model T and the interstate highway system in the United States. However, it was the development of the **welfare state** and **welfare statism** that helped stimulate demand in European cities. Welfare states vary enormously in structure and scope (see Pinch, 1997) but they shared a common goal of attempting to ameliorate the inequalities associated with market mechanisms. In British cities this resulted in state-provided housing (known as local authority or 'council' housing) in suburban areas as well as in inner-city renewal areas. In continental European cities less housing was provided directly by the state, with a greater reliance upon state-funded, but privately provided, forms of social housing.

Welfare statism involves more than the direct provision of goods and services by the public sector; it is a broader set of arrangements to ensure full employment, minimum wages, safe working conditions and income transfers to the less well-off – what is sometimes called the **social wage**. As we will see in Chapter 13, Western governments have in recent years tended to renege on these arrangements but in the 'long boom' of Fordism they helped to boost consumer demand. Relative affluence was also boosted by the industrial relations systems prominent during this time – trade unions and collective bargaining ensured that workers were relatively well rewarded for their efforts.

It is important at this stage not to paint an idealized picture of the 'long boom' of Fordism. For example, while some of the large industrial sectors such as car assembly provided relatively good wages, other 'sweatshop' industries did not. Furthermore, women, ethnic minorities and other marginalized groups were effectively excluded from certain sectors of the economy (see, for example, Hanson and Pratt, 1995). In addition, there were still periodic booms and slumps in the economy. In the United States, for example, car sales boomed in the early 1950s, reaching a peak in 1955, dipping somewhat towards the end of the decade before soaring again in the 1960s. Nevertheless, within the conventions of the day, unemployment was relatively low and there was a 'virtuous circle' of arrangements that brought together production and consumption. From the mid-1970s onwards, however, the Fordist system encountered severe problems and these eventually impacted upon city structures.

2.3 The contemporary city

There is a great deal of controversy over how the Fordist system ran into trouble. A core problem was declining productivity, which has been linked to a variety of factors. Among the most important of these factors are the following:

➤ A failure to invest sufficiently in research and development (a particular problem in the United States and the United Kingdom).

➤ The increasing costs of raw materials (intensified by oil price increases in 1974 and 1979–80).

➤ Market saturation of mass-produced goods and increasing consumer hostility to uniform, poor-quality goods.

➤ System rigidity stemming from the high capital costs of establishing production lines under Fordism.

➤ Repetitive, boring, physically demanding assembly line work leading to alienation among the workforce and poorly assembled low-quality products.

➤ Adversarial industrial relations and widespread labour unrest.

➤ The increasing costs associated with safety and environmental legislation.

As a result of these problems, Fordism and Keynesianism have imploded. In some ways, Fordism was a victim of its own success, saturated markets for mass consumption having pushed producers towards niche markets, packaging, novelty and design in the search for profit. Keynesianism lost its effectiveness when the influence of organized labour and the authority of national governments were short-circuited by the global reach of transnational corporations. Fordism and Keynesianism have been replaced by a 'new economy' – a neo-Fordist system underpinned by information technologies and networked around the globe – and **neoliberalism** – the view that the state should have a minimal role. National, state and local governments have found it difficult to regulate and control the new economy and have shed many of their traditional roles as mediators and regulators.

Meanwhile, the proponents of neoliberal policies have advocated free markets as the ideal condition not only for economic organization, but also for political and social life. Ideal for some, of course. Free markets have generated uneven relationships among places and regions, the inevitable result being an intensification of economic inequality at every scale, from the neighbourhood to the nation state. The pursuit of neoliberal policies and free market ideals has also dismantled a great deal of the framework for city building and community development that Western societies used to take for granted: everything from broad concepts such as the public good to the nuts and bolts of the regulatory environment (Filion, 1996; Brenner, 2002). Globalization has meanwhile contributed to the emergence of a postmodern culture in which the symbolic properties of places and material possessions have assumed unprecedented importance, with places becoming important objects of consumption.

Neo-Fordism

A crucial problem with the Fordist system was its rigidity in the face of increasing market and technological change. *Flexibility* is therefore the key factor underlying the numerous changes that have modified the Fordist system to the point where we now have to think of a **neo-Fordist** (or post-Fordist) system (see Table 2.2). In particular, this involves the capacity of firms to adjust the levels and types of their output in response to varying market conditions. One method of increasing flexibility is through increased use of technology, such as computer-aided design and manufacture (CAD/CAM). Also of crucial importance has been flexible use of labour. Workers are now much more likely to be multiskilled (often termed **functional flexibility** or **polyvalency**), rather than committed to just one task as under Fordism. In addition, the labour force increasingly exhibits **numerical flexibility**, the capacity to be hired and laid-off when necessary, as is the case with part-time, temporary, agency or subcontract

workers (also referred to as **contingent workers**). It has been suggested that this is leading to a dual labour market, as shown in Fig. 2.3.

At the core of the labour market are various workers who are rewarded for their functional flexibility by relatively secure well-paid jobs with good working conditions and company benefits. Surrounding this core, however, are various types of secondary, contingent or peripheral workers who exhibit numerical flexibility with limited rewards, job insecurity and relatively poor working conditions. Such increased use of 'non-core' workers such as part-timers, agency and temporary workers is sometimes termed **casualization**. This is an idealized model and there are many variations from this structure; some core workers' jobs are relatively insecure while many peripheral workers have job stability if they wish.

Crucial to profitability in the Fordist system was control of costs, especially those of labour. However, it is argued that firms compete less on the basis of cost and increasingly on the basis of factors such as reliability, style and innovation. Research and development, product innovation and marketing have therefore become much more important in a neo-Fordist society. Under Fordism, costs were also reduced by taking advantage of **internal economies of scale**. These are factors that decrease the average cost of manufacturing a product as the level of output increases. Because of the high start-up costs of specialized machinery, it took high volumes of output to make big profits under Fordism. In neo-Fordist society, however, market volatility means that there are much smaller product runs and therefore decreasing internal economies of scale. One response has been what is termed the **externalization of production**, the tendency for firms to subcontract functions to other firms and agencies. This enables firms to reduce costs by getting a variety of organizations to compete to provide goods and services. In addition, it can help to offload the risks associated with new joint ventures with other companies (known as strategic alliances).

A major consequence of these developments is that firms are becoming much smaller. This has arisen because existing firms have been subdivided (sometimes known as **vertical disintegration**) as well as through the rapid growth of new small firms. Rather than the internal economies of scale dominant under Fordism, neo-Fordist

Table 2.2 Differences between the ideal types of Fordism and neo-Fordism

Fordism	Neo-Fordism
The labour process	
Unskilled and semi-skilled workers	Multiskilled workers
Single tasks	Multiple tasks
Job specialization	Job demarcation
Limited training	Extensive on-the-job training
Labour relations	
General or industrial unions	Absence of unions, 'company' unions
	'No-strike deals'
Taylorism	'Human relations management'
Centralized national pay bargaining	Decentralized local plant-level bargaining
Industrial organization	
Vertically integrated large companies	Quasi-vertical integration, i.e. subcontracting
	Decentralization, strategic alliances, growth of small businesses
Technology	
Machinery dedicated to production of single products	Flexible production systems, CAD/CAM robotics, information technology
Organizing principles	
Mass production of standardized products	Small batch production
Economies of scale, resource driven	Economies of scope, market driven
Large buffer stocks of parts produced just-in-case	Small stocks delivered 'just-in-time'
Quality testing after assembly	Quality built into production process
Defective parts concealed in stocks	Immediate rejection of poor-quality components
Cost reductions primarily through wage control	Competitiveness through innovation
Modes of consumption	
Mass production of consumer goods	Fragmented niche marketing
Uniformity and standardization	Diversity
Locational characteristics	
Dispersed manufacturing plants in spatial division of labour	Geographical clustering of industries in flexible industrial districts
Regional functional specialization	Agglomeration
World-wide sourcing of components	Components obtained from spatially
Growth of large industrial conurbations	Proximate quasi-integrated firms
	Growth of 'new industrial spaces' in rural semi-peripheral areas
Role of the state	
Keynesian Welfare State	The 'Workfare State'
Demand management of economy	Encouragement of innovation and competition
Provision of state funded and supplied	Privatization, deregulation
Protection of the 'social wage'	Encouragement of self-reliance
Problems	
Inflation	High rates of unemployment
Market saturation	Labour market dualism
Poor-quality products	Social polarization, exclusion and associated social tensions
Inflexibility	
Alienated workforce	Instability of consumer confidence through economic insecurity
Divergence between rising wages and declining productivity growth	Market volatility
Fiscal crisis of state	

Source: adapted from Pinch (1997).

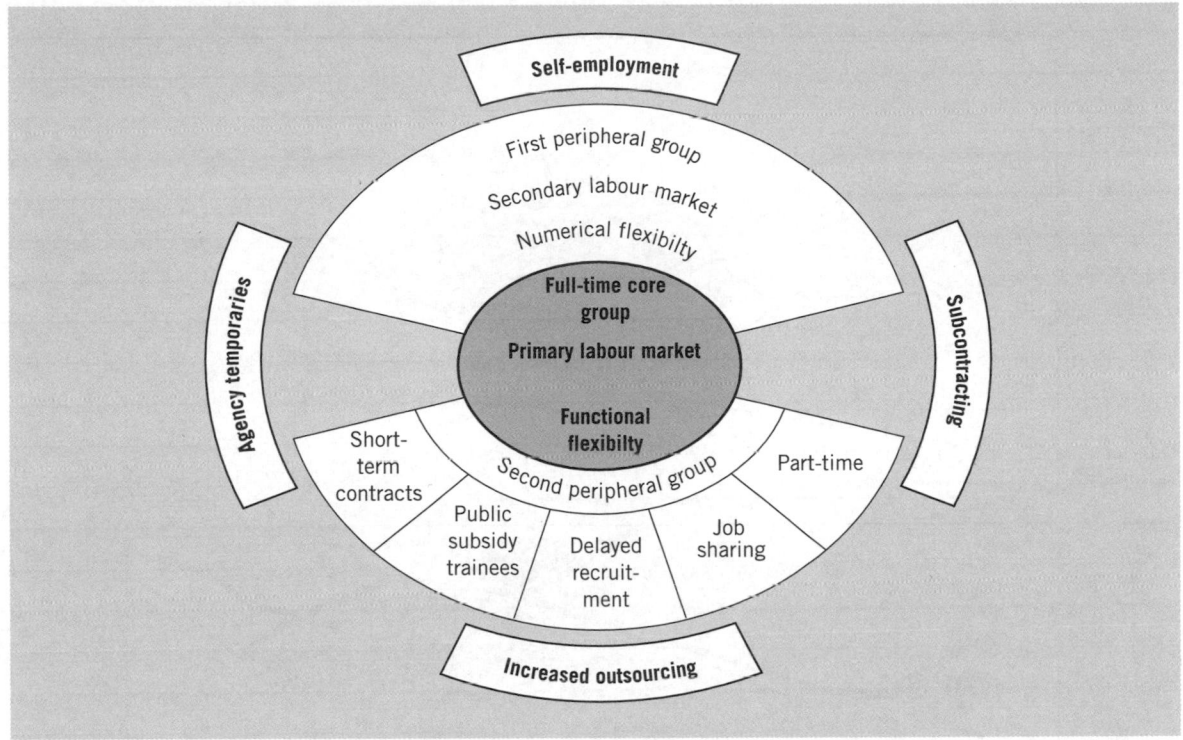

Figure 2.3 An idealized model of the flexible firm in a neo-Fordist economy.
Source: Atkinson (1985), Fig. 2.1, p. 19.

firms are therefore based on external economies of scope. **Economies of scope** are factors that make it possible to produce a *range* of products rather than to produce an individual item on its own. **External economies of scope** arise when the industry to which the firm belongs to is large (i.e. there are a large number of producers). Some have argued that these changes are such that they constitute a new regime of accumulation – often termed **flexible accumulation** (Harvey, 1989b).

Economic and urban change

It is at this stage that we can finally begin to link up neo-Fordism with the changing structure of urban areas. One of the main consequences of neo-Fordist technologies and working practices is that far fewer people are needed to manufacture things. In addition, the production of well-established 'mature' products has often been shifted to low-cost locations outside the Western nations. The result has been massive **deindus-**trialization and the consequent transformation of the classic industrial city. The decline of traditional heavy manufacturing industry has been especially pronounced in the industrial heartlands of Britain – the Midlands, the North, Wales and Scotland – and in the 'rustbelt' of the United States – including classic industrial cities such as Chicago, Cleveland, Detroit and Pittsburgh.

Parallel with this decline has been the creation of *new* industrial clusters, often termed **new industrial spaces**. Despite the increased availability of advanced telecommunications systems there is growing evidence that many interactions are best undertaken on a face-to-face basis. This is especially the case when complex items of knowledge have to be exchanged and where transactions are dependent upon trust. One way of reducing the costs of transactions and facilitating face-to-face interactions is for these interacting organizations to cluster together. There are many examples of this clustering: Silicon Valley and Orange County in California and Route 128 around Boston in the United States, the M4 Corridor and 'Motor Sport Valley' in the United Kingdom, the

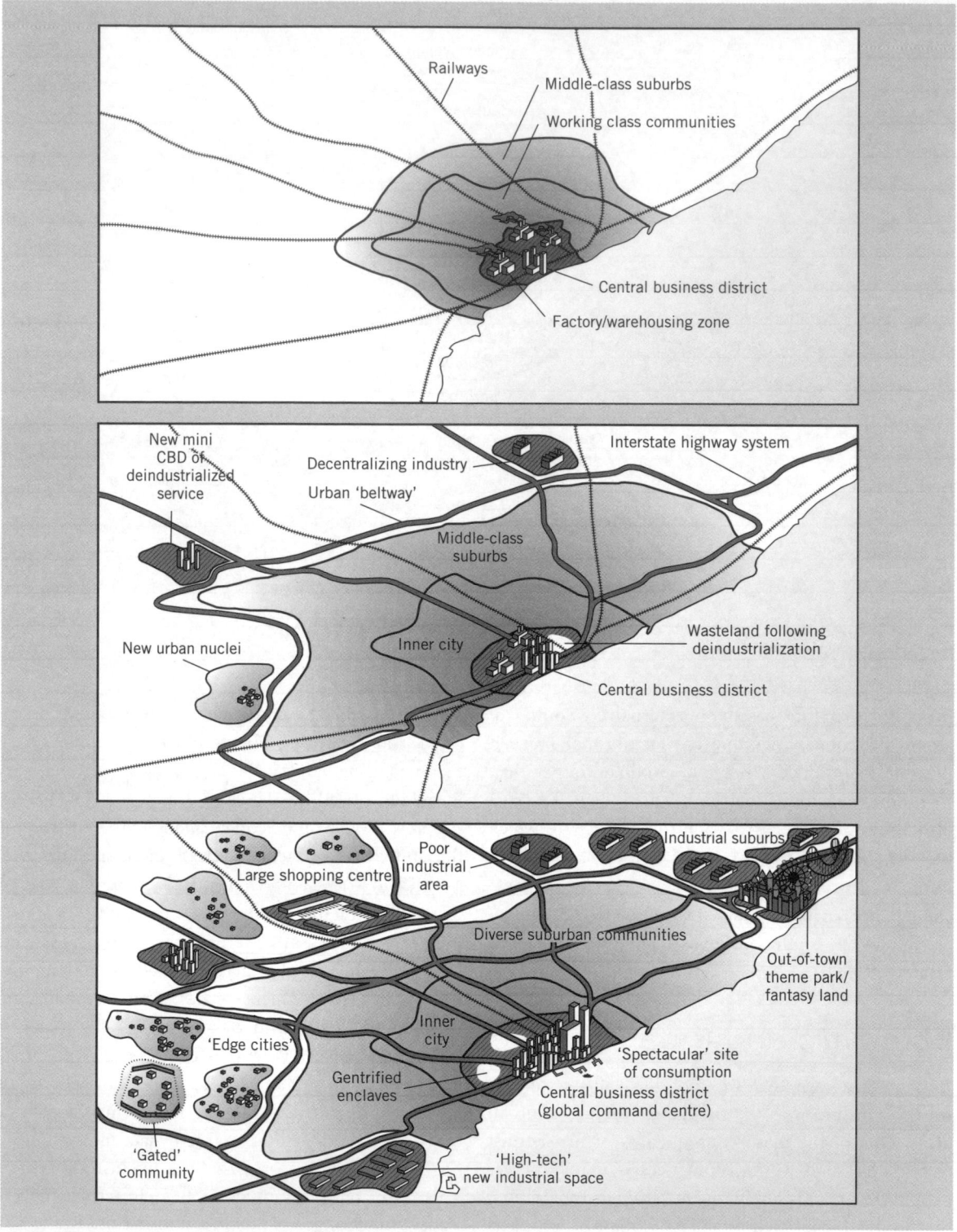

Figure 2.4 The transition from the classic industrial city, *circa* 1850–1845 (upper) to the Fordist city, *circa* 1945–1975 (middle) and neo-Fordist metropolis, *circa* 1975– (lower).

'Third Italy' (Bologna, Emilia and Arezzo), Grenoble in France and Baden-Württemberg in Germany.

The urbanization associated with these clusters differs considerably in character depending upon the industries involved. On the one hand, there are the rapidly growing 'sunbelt' areas of California and the South West of the United States, characterized by extensive suburban development and conservative political regimes. Then there are clusters of industrial sectors such as finance, design and marketing in the large capital cities of the world such as New York, Paris and London. Finally, in Europe there are regions specializing in products such as shoes, ceramics and textiles that have left-wing local governments.

Los Angeles is often cited as the exemplar of the product of urbanization under the new regime of flexible accumulation and the harbinger of future urban form. Soja (1989) has described variously the **postmodern global metropolis, cosmopolis** and **postmetropolis** (1997) – a physically and socially fragmented metropolis (see Fig. 2.4). Los Angeles has experienced a strong process of recentralization in the form of the command centres linked into the new global economy together with a strong process of decentralization in the form of numerous subcentres and **edge cities** (Garreau, 1992). These are not the exclusively affluent suburbs of an earlier era but show enormous variations in character, some being industrial and commercial, and others being relatively poor and/or with distinctive ethnic minorities (see also Box 2.1). Soja (1992) develops this theme into the concept of **exopolis** – a city that has been turned inside-out. In such an environment it is difficult for individuals to have a sense of belonging to a coherent single entity.

Another important theme associated with urbanization under a regime of flexible accumulation is increasing social polarization and the development of protective measures by the more affluent sections of society in an environment in which violence and crime are routine. One manifestation of these defensive measures is the growth of so-called **bunker architecture** (also termed 'citadel', 'fortified' and 'paranoid' architecture): urban developments with gates, barriers and walls, security guards, infra-red sensors, motion detectors, rapid response links with police departments and surveillance equipment such as CCTV. These form what Davis (1990) terms a '**scanscape**', all designed to exclude those regarded as undesirable. These systems are often established in residential areas but exclusionary measures may also be undertaken in shopping malls and city centres. In some central parts of Los Angeles park benches are curved in such a way as to inhibit people from sleeping on them over night. It is little wonder that Davis (1990) talks of the 'militarization' of city life. However, it is not just the marginalized who are kept under surveillance in the contemporary city. New technologies centred around credit and loyalty cards, computers and pay-for-service facilities mean that corporations and governments can access vast amounts of information about people's travel and consumption habits.

Postindustrial society

A key weakness of the regulationist concepts of Fordism and neo-Fordism is their neglect of services. Ironically the **postindustrial society** thesis emerged from experience in the 1950s and 1960s when Fordism was at its zenith. Bell (1973) pointed out that heavy industries were beginning to decrease in importance while employment in service industries was growing, especially in sectors such as finance, business services, retailing, leisure and entertainment industries. The implications of this shift, Bell suggested, pointed to profound shifts in social structure and social relations.

The geographical patterns of this service growth have been complex but a key trend has once again been development away from traditional manufacturing centres. In the United States, service growth has been especially pronounced in the 'sunbelt' areas, while in the UK service growth has been rapid in the South East. In addition, corporate headquarters have tended to decline in what were previously heavy industrial areas and have grown rapidly in regional service economies (Knox *et al.*, 2003). The growth of the service economy has had important consequences for the social geography of cities. One of the most important developments has been the tendency to intensify social polarization. Whereas the traditional manufacturing industries tended to have substantial proportions of relatively well-paid, blue-collar, middle-income jobs, services tend to be characterized by both relatively high-paying and relatively low-paying jobs.

Box 2.1

Key trends in urban social geography – 'Edge cities'

A major change in Western cities in recent decades has been the creation of urban development on the fringes of existing suburbs. These developments reflect the decentralization, not only of people, but also employment, services and retailing from inner-city areas. In some cases these urban areas have become functionally independent of existing central city areas, effectively creating new settlements. However, such cities often lack autonomy, being on the fringes of a number of existing political and administrative jurisdictions. Typically, such developments have taken place on the intersections of major highway systems, as in the much cited case of Tyson's Corner on the edge of Washington, DC.

In the US public attention was draw to the phenomenon of these 'outer cities' in a book written by the journalist Joel Garreau (1992). He coined the term 'edge city', although various other neologisms have been used (e.g. 'stealth city', 'technoburb', 'suburban downtown', 'perimeter city' and 'cyberbia'). Garreau defined edge cities with the following characteristics:

➤ Has 460 000 square metres (5 million square feet) or more of leasable office space.

➤ Has 56 000 square metres (600 000 square feet) or more of leasable retail space.

➤ Has more jobs than bedrooms (to indicate that these are more than dormitory suburbs of commuters).

➤ Is perceived by the population as being one place.

➤ Was nothing like a 'city' as recently as 30 years ago.

Garreau's book celebrated the lifestyles of residents in edge cities whom he portrayed as innovative pioneers of new ways of living, combining the best of urban and rural lifestyles. This might account for the good sales of the book, since it gave suburban dwellers (long derided by many urban commentators as conservative and dull!) a flattering view of themselves. More recently, researchers at Virginia Tech's Metropolitan Institute have shown how metropolitan decentralization has gone well beyond 'edge cities' to produce 'boomburbs' – especially fast-growing suburban municipalities that lack a dense business core – and 'edgeless cities' – sprawling swaths of low-density commercial and residential development that lack a clear physical boundary.

Recognizing a different context, researchers have been cautious in applying notions of 'edge cities' in Europe, referring instead to use the terms 'edge urban areas' or 'peripheral urban areas'. These terms indicate that European edge cities are typically less autonomous than their US counterparts. There is much greater planning of such urban forms in Europe, with local governments and the public sector being much more active partners in their creation (see Phelps and Parsons, 2003). Nevertheless, edge cities are the subject of intense interest throughout the Western world, especially since some of them are the source of innovation in high-technology industries (and even if they have sometimes been derided as 'Nerdistans'!).

Key concepts associated with edge cities (see Glossary)

Decentralization, exopolis, metropolitan fragmentation, multiple nuclei model, new industrial spaces.

Further reading

Beauregard, R.A. (2003) Edge cities: peripheralizing the center *Urban Geography*, **16**, 708–721

Garreau, J. (1992) *Edge City: Life on the New Frontier* Random House, New York

Lang, R.E. (2003) *Edgeless Cities: Exploring the Elusive Metropolis* Brookings Institution Press, Washington, DC.

Phelps, N.A. and Parsons, N. (2003) Edge urban geographies: notes from the margins of Europe's capital cities *Urban Studies*, **40**, 1725–1749

Links with Other Chapters

Chapter 9: Suburban Neighbourhoods: Community Transformed?, Splintering Urbanism and the Diversity of Suburbia

Chapter 13: Accessibility to Services and Amenities, Decentralization and Accessibility to Services, Urban Social Sustainability

Another consequence of service growth has been increasing competition among cities for employment. It is argued that whereas heavy manufacturing industries tended to be firmly rooted in particular places (because of their proximity to certain raw materials, their dependence upon large amounts of capital investment in buildings, machinery, equipment and specialized skilled labour), service industries are much more mobile. The reason for this mobility is that the basic ingredients for many routine service industries – suitable office properties and large supplies of female workers – are much more geographically dispersed. Consequently, there is much more locational freedom on the part of service companies. The result has been vigorous campaigns by city authorities to attract major service employers.

Globalization

A key focus of regulation theory as it was originally developed is the changing relationships between national states. Thus, it is argued that the Fordist system was based around the dominance of the United States over the world economy in the period immediately after the Second World War, as structured by the Bretton Woods financial system (Leyshon and Thrift, 1997). This system created fixed exchange rates between currencies, based around the US dollar as the medium of exchange. Gradually, however, as other countries, especially Japan and countries in Europe, gained in economic strength, the Bretton Woods system began to break down and this led to a new financial regime based on floating exchange rates. This new financial regime opened up national economies to the forces of **globalization** – a concept that has a number of dimensions.

Globalization is a complex and controversial topic. Nevertheless one can identify a number of elements within concepts of globalization. The term is usually associated with the growing importance of **multinational** (or **transnational**) corporations operating in more than one country. In fact, large companies such as Ford have long manufactured outside their home nation, producing goods for local consumption in distant markets. Globalization is thus a more recent process in which the operations of transnationals, both in the spheres of production and marketing, are increasingly *integrated* on a global scale. Thus, products are made in multiple locations from components manufactured in many different places.

Globalization is also associated with the development of a broader global *culture*. This is a controversial

The command centre in a global city: looking across the financial area of the City of London.
Photo Credit: Paul Knox.

idea but essentially involves the widespread diffusion of Western values of materialism. Globalization can also be seen in the popularity of Hollywood films throughout the world and the increased popularity of 'world music' (Taylor, 1997). It is therefore argued that globalization involves the homogenization of the culture – the development of cultural interrelatedness throughout the globe. This process has been encouraged by new telecommunications systems that facilitate rapid transmission of information and images around the world (Graham and Marvin, 2001). However, there has been a resistance to these global forces through the assertion of local cultural identities – most notably in the form of Islamic fundamentalism and various popular movements for regional autonomy, but also at the level of urban systems (King, 1997) and the individual city (Luke, 1994; Oncu and Weyland, 1997; Nijman, 1999).

Globalization has had a number of profound effects upon urban social geography. Most notably, it has led to the emergence of so-called **world cities** – command centres such as New York, London and Tokyo which are key players in the new concentrated world financial system (also sometimes called **global cities**: Sassen, 2001; Taylor, 2004). Sassen argues that one of the main features of global cities is social polarization. In large measure, this inequality stems from the characteristics of financial services; they are dependent upon a narrow stratum of relatively well-paid workers who require many consumer services such as restaurants, shops and cleaners, which in turn utilize large numbers of low-paid workers. This social inequality is also manifest in the social geography of global cities. Housing for affluent workers in financial services may be built in close proximity to poor-quality housing, as in the revitalized London Docklands. The social tensions associated with such inequality mean that the affluent may need to resort to many protective strategies (see Chapter 3).

While few cities can claim true global status as command centres in the world economy, there is a sense in which all urban centres are now global for they are all affected by events and decisions outside of their boundaries. Furthermore, they are all engaged in a fierce competition to attract mobile capital into their areas. There is therefore a close interaction between global and local forces – a process that has been dubbed **glocalization** and which is also referred to as the **global–local nexus**.

Knowledge economies and the informational city

Some of the newest ideas relating to the changing economic context of city development point to the growing importance of *knowledge* in contemporary economies. Various terms or have been used to summarize these developments including *knowledge-based capitalism* (Florida, 1995); the *network society* (Castells, 1996); *reflexive accumulation* (Lash and Urry, 1994); *soft capitalism* (Thrift, 1998) and the *weightless world* (Coyle, 1997) (see also Table 2.3). All these theories point in

Table 2.3 Grand metaphors used to predict the impact of telematics and knowledge economies upon city life and society

➤ The 'city in the electronic age'	➤ Communities without boundaries
➤ 'Cyberspace'	➤ Cyberville
➤ The 'dematerialized economy'	➤ Electronic communities
➤ Electronic cottage	➤ Electronic spaces
➤ The 'flexicity'	➤ The 'information city'
➤ The 'informational' city'	➤ The 'intelligent city'
➤ The 'invisible' city'	➤ The 'knowledge-based city'
➤ Knowledge economies	➤ 'Netscape'
➤ 'Networld'	➤ 'Reflexive accumulation'
➤ The 'overexposed city'	➤ 'Soft capitalism'
➤ 'Space of flows'	➤ The 'telecity'
➤ 'Teletopia'	➤ The 'virtual city'
➤ The 'virtual community'	➤ The 'weightless world'
➤ The 'wired city'/'wired society'	

Box 2.2

Key thinkers in urban social geography – Manuel Castells

One of the key thinkers in urban social geography in the past three decades has been Manuel Castells. Born in Spain, Castells' radical activities as a student forced him to flee the dictatorship of General Franco and move to France. However, his participation in the riots of May 1968 led to expulsion by the French Government (although he was subsequently pardoned) (Hubbard, 2004). Castells' ideas on public service prevision (what he termed 'collective consumption') as developed in his book *The Urban Question*, were enormously influential in the 1970s (see also Chapter 5). However, his move to the University of California, Berkeley, was associated with less radical work on local protest groups (*The City and the Grassroots*, 1983) and information technology (*The Informational City*, 1989). He also undertook some of the earliest work on gay spaces (Castells and Murphy, 1982).

Castells was one of the first scholars to highlight the role of new forms information technology in enhancing inequalities between social groups in cities. This he did in three volumes with the general title *The Information Age*: Volume 1 was subtitled *The Rise of the Network Society* (1996), Volume 2 was subtitled *The Power of Identity* (1997) and Volume 3 was subtitled *End of Millennium* (1998).

In these writings Castells argued that we should not envisage cities as static places defined by fixed boundaries, but instead as the sites of what he calls 'spaces of flows' – ever-increasing circulations of peoples, ideas, images and consumer goods. Castells' approach is one of a number of recent 'fluidity-type' conceptualizations in social science that envisage social life as consisting of movement and flows. Castells links these ideas with those who suggest that globalization is eroding the distinctiveness of particular places. A good example of this erosion is the bland 'international style' that pervades many 'postmodern' city skylines as well as the interiors of many public buildings such as airports, hotels and shopping malls.

According to Castells, the global 'network society' consists of three levels. First, there is infrastructural level, the information technology that permits rapid knowledge flows around the world. Second, there are the global translation centres, the 'world cities' that provide the structures for the flow of knowledge. Finally, there is the managerial elite that operationalizes this global economy.

As with many original, influential, thinkers, Castells' writings can be challenging at times. Furthermore, some of his earlier work on issues such as collective consumption, though highly influential in their day, have been heavily criticized and found wanting. His recent ideas on the network society have also been heavily criticized (see Crang, 2002; Smart, 2000). Nevertheless, Castells' ideas remain highly influential and are essential reading for any serious student of the city.

Key concepts associated with Manuel Castells (see Glossary)

Collective consumption, informational city, megacities, network society, spaces of flows.

Further reading

Crang, M. (2002) Between places: producing hubs, flows and networks *Environment and Planning A*, **34**, 569–574

Hubbard, P. (2004) Manuel Castells, in P. Hubbard, R. Kitchin and G. Valentine (eds) *Key Thinkers on Space and Place* Sage, London

Smart, B. (2000) A political economy of the new times? Critical reflections on the network society and the ethos of informational capitalism *European Journal of Social Theory*, **3**, 51–65

Susser, I. (ed.) *The Castells Reader on Cities and Social Theory* Blackwell, Oxford

Links with Other Chapters

Chapter 11: Homosexuality and the City

Chapter 13: Service Sector Restructuring

various ways to the increasing importance of knowledge in economic development.

First, there is the ever-increasing and rapidly changing technological sophistication of both goods and services. Thus, products such as computers and cameras become superseded by more complex models in a very short period of time. Keeping up with this rapid pace of change puts a great premium on knowledge acquisition by manufacturers. Second, the design or fashion element of products is becoming increasingly important,

Box 2.3

Key films related to urban social geography – Chapter 2

Blue Collar (1978) One of the very few films to show the problems facing workers on a Fordist assembly line in Detroit.

The Deer Hunter (1978) The middle section of this film is an extremely harrowing portrayal of the Vietnam War (although the Russian roulette theme is apparently entirely fictional). Nevertheless, the first and last sections, set in a Pennsylvanian steel town, illustrate the powerful links forged between industrial workers and their local milieu.

The Full Monty (1997) An extremely funny, but also at times touching, movie about the trials and tribulations facing a group of redundant male steel workers seeking new employment as strippers! Illuminates in a not too serious manner some of the challenges to male identities brought about by deindustrialization and the rise of the new service economy.

Matewan (1987) This film is not set in a major city, but instead is located in a West Virginian coal mining com-

munity in the 1920s. Nevertheless, a complex portrayal of class conflict in the United States.

Roger and Me (1989) Controversial because of what some regard as director Michael Moore's 'fast and loose' attitude to the truth, nevertheless, a documentary portrayal of the social consequences of deindustrialization in Flint, Michigan that is at times extremely funny but also frequently sad and troubling.

especially in a world in which there are an increasing number of specialized niche markets. As in the case of technological change, these fashion changes seem to be taking place at an ever-increasing rate. This tendency to make products outdated by changes in fashion, even though they may function adequately, is sometimes called **semiotic redundancy** (after the term semiotics – the study of signs – see Chapter 3). Finally, there is a growing recognition of the need to integrate design with the engineering and production end of manufacturing. All of these processes mean that firms no longer tend to gain competitive advantage on the basis of price alone but increasingly rely upon factors such as quality, performance, design and marketing.

Closely linked to the above developments has been the evolution of new technologies in the sphere of telecommunications – teleconferencing, faxes, e-mail, the Internet and various other specialized computer network systems. These new technologies mean that it is now possible for many routine and 'back-office' functions to be undertaken in low-cost labour areas in peripheral regions of the Western economies and in less-developed and newly industrializing countries. However, there is growing evidence that digital telecommunications technologies are creating a division between favoured cities linked into global command centres and less-favoured cities in older, previously

industrialized areas (Sassen, 1997, 1999a). Within cities, too, there is evidence of a 'digital divide' that intensifies and reinforces social polarization. Some of the most influential ideas relating to these new social processes and their effect on cities have come from Manuel Castells (see Box 2.2).

Conclusion

It should be clear by now that the contemporary processes discussed in this chapter, neo-Fordism, service sector growth, globalization and new telecommunications systems, are all highly interrelated. None of these perspectives in isolation provides a comprehensive explanation for the changing economic context of city growth. However, taken together they begin to illuminate some of the factors that have so radically altered city structures in recent years. Cities are no longer single entities, unified around production and consumption, linked into elaborate national hierarchies. Instead, they are increasingly multi-centred phenomena, based around both producer and consumer services, and are linked into global networks underpinned by new telecommunications technologies. These developments have impacted upon, and have been influenced by, the cultures of cities which are considered in the next chapter.

Chapter summary

2.1 Preindustrial cities were essentially small-scale 'walking' cities. Although they displayed an element of vertical differentiation based on social divisions within the districts of the occupational guilds, their main division appears to have been that between the elite who lived in the exclusive central core and the mass of population who lived around the periphery of the city.

2.2 Industrial capitalism inverted the structure of the preindustrial city by forcing the poor into poor-quality inner-city districts while the middle and upper classes retreated to the urban periphery. The polarized class structure of the early industrial cities was gradually replaced by more complex social divisions which made the capitalist class less easily identifiable.

2.3 The 'long boom' of Fordism brought about a relatively harmonious linking of mass production and consumption that was manifest in extensive suburbanization. The numerous problems associated with the Fordist economic system led to various neo-Fordist developments that have been manifest in new urban forms based around agglomerating industries.

2.4 Globalization has had profound impacts upon cities, leading to the emergence of world cities, centres of corporate and financial control. It has also fostered increased competition between cities and intensified social polarization.

2.5 New telecommunications systems have allowed the exchange of ever more complex information over greater distances. However, as yet, they have not been associated with a decline in the strength of cities as centres for information production and exchange.

Key concepts and terms

bunker architecture
casualization
circuit of production
commodity fetishism
contingent workers
cosmopolis
culturalization of the economy
deindustrialization
economic determinism
economies of scale
economies of scope
edge city
exchange value
exopolis
external economies of scope
externalization of production
flexible accumulation
forces of production
Fordism
functional flexibility
functionalism

global cities
globalization
global–local nexus
glocalization
internal economies of scale
Keynesianism
labour theory of value
'long boom'
mode of production
mode of regulation
multinational
neo-Fordism
neoliberalism
new industrial space
numerical flexibility
polyvalency
postindustrial society
postmetropolis
postmodern global metropolis
preindustrial city
reflexive accumulation

regime of accumulation
regulation theory
'scanscape'
semiotic redundancy
social division of labour
social relations of production
social wage
spatial fix
superstructure
surplus value
Taylorism
technical division of labour
technological determinism
transnational
use value
vertical disintegration
vertical integration
voluntarism
welfare state
welfare statism
world cities

Suggested reading

Sir Peter Hall's *Cities in Civilization* (1997: HarperCollins, London) is a masterful survey of the intersection of social history and urban history. The classic study of the preindustrial city is G. Sjoberg *The Pre-industrial City* (1960: Free Press, Chicago). This should be contrasted with J.E. Vance, Jr, 'Land assignment in pre-capitalist and post-capitalist cities', *Economic Geography*, 47, 1971, 101–120. A clear review of the issues surrounding the preindustrial city is also to be found in J.P. Radford 'Testing the model of the pre-industrial city: the case on ante-bellum Charleston, South Carolina', *Transactions, Institute of British Geographers* (New Series) 1979: 4, 392–410. Those wishing to study the preindustrial city in greater depth would do well to examine two books that are part of the Open University *Cities and Technology* course: first, *Pre-industrial Cities and Technology* edited by Colin Chant and David Goodman (1999: Routledge, London) and second, *The Pre-industrial Cities and Technology Reader* edited by Colin Chant (1999: Routledge, London). David Harvey's work on the historical geography of Paris, *The Capital of Modernity: Paris in the Second Empire* (2003: Routledge, New York) provides detailed examples of the transition to Modernity, while his *The Condition of Postmodernity* (1989: Blackwell, Oxford), is the single most important reference for this chapter. A good general introduction to changing economic geography is Paul Knox, John Agnew and Linda McCarthy, *Geography of the World Economy*, 4th edition (2003: Arnold, London) which can be usefully supplemented by Eric Sheppard and Trevor Barnes' *A Companion to Economic Geography* (2002: Blackwell, Oxford) and Diane Perrons' *Globalization and Social Change* (2003: Routledge, London). The relations between economy and culture are also discussed at length in an excellent set of readings in Roger Lee and Jane Wills (eds) *Geographies of Economies* (1997: Arnold, London). On the concepts of Fordism and neo-Fordism see Richard Meegan 'A crisis of mass production?', in John Allen and Doreen Massey (eds) *The Economy in Question* (1988: The Open University Press, London); Ash Amin (ed.) *Post-Fordism: A Reader* (1994: Blackwell, Oxford); and John Allen 'Fragmented firms, disorganized labour', in Allen and Massey *op. cit.* plus the article by John Allen 'Post-industrialism and post-Fordism', in S. Hall, D. Held and T. McGrew, (eds) *Modernity and Its Futures* (1988: Polity Press, Cambridge). For changes in financial services see Andrew Leyshon and Nigel Thrift *Money/Space* (1996: Routledge, London) and Ron Martin (ed.) *Money and the Space Economy* (1998: John Wiley, Chichester) while services are examined in Neil Marshall and Peter Woods *Services and Space* (1995: Longman, London) and Colin C. Williams *Consumer Services and Economic Development* (1997: Routledge, London). Relevant books on world (or global) cities are Saskia Sassen's *Cities in a World Economy* (2000: Pine Forge, London), and her *The Global City* (2000: Princeton University Press, Princeton, second edition), Peter Taylor *World City Network* (2004: Routledge, London) and Neil Brenner and Roger Keil (eds) *The Global Cities Reader* (2005: Routledge, London). The development of cultural industries in cities is discussed in Allen Scott's book *The Cultural Economy of Cities* (2000: Sage, London). The emergence of megacities is discussed by Edward Soja in *Postmetropolis* (1999: Blackwell, Oxford). On globalization see UN Centre for Human Settlements *Cities in a Globalizing World* (2001: Earthscan, London). An excellent review of place marketing is J.R. Gold and S.V. Ward (eds) *Place Promotion: The Use of Publicity and Marketing to Sell Towns and Regions* (1994: Wiley, Chichester). Labour market change is discussed in Susan Hanson and Geraldine Pratt *Gender, Work and Space* (1995: Routledge, London). The question of sociospatial polarization in world cities is explored by Peter Marcuse and Ronald Van Kempen (eds) *Global Cities: An International Comparative Perspective* (1999: Blackwell, Oxford). There is a huge literature on so-called 'new industrial spaces' and Manuel Castells and Peter Hall's *Technolpoles of the World: The Making of 21st Century Industrial Complexes* (1994: Routledge, London) provides an overview. The crucial book on the telematics and digital technologies is Stephen Graham and Simon Marvin *Splintering Urbanism* (2001: Routledge, London), which can be supplemented with the *Cybercities Reader* also edited by Stephen Graham (2003: Routledge, London). It is also worth examining Manuel Castells' *The Rise of the Network Society* (1996: Blackwell, Oxford) and *The Information Age. Economy, Society and Culture: Vol. 2 The Power of Identity* (1997: Blackwell, Oxford).

Chapter 3

The cultures of cities

Key questions addressed in this chapter

➤ What is culture?

➤ Why have the concerns of cultural studies and postcolonial theory become so important in the study of cities?

➤ What are the roles of space and place in the formation of culture?

➤ What is meant by the postmodern city?

Without doubt the most important change in urban studies in recent years has been the increased attention given to issues of *culture*. This development reflects a broader trend in the social sciences that has become known as the '**cultural turn**'. As suggested by the sociospatial dialectic, cities have long had a crucial impact upon, and have in turn been influenced by, cultures. What makes cities especially interesting in the contemporary context is that they bring together many different cultures in relatively confined spaces. Appadurai (1996) calls the diverse landscape of immigrants, tourists, refugees, exiles, guest workers and other moving groups to be found in many contemporary cities an **ethnoscape**. This juxtaposition of peoples often leads to innovation and new cultural forms as cultures interact. But it can also lead to tensions and conflict, especially if cultural groups retreat into particular areas of the city. It is therefore not surprising that there has accumulated over the years a substantial body of work on city cultures. In the last decade the 'cultural turn' has radically influenced the study of city cultures, affecting both *what* is studied and the *way* cities are examined.

Understanding the so-called 'cultural turn' poses a number of difficulties, however, for a number of reasons. There has been a bewildering flood of new terms to describe a large set of new concepts that are all highly interrelated and build up to form radically new ways of looking at society. Taking a few 'cultural' concepts in isolation therefore tends to neglect their wider significance. In recognition of these difficulties, this chapter attempts to provide a guide to understanding the 'cultural turn' and to explain why issues of culture have so radically altered our ways of looking at cities. This will equip us with a series of concepts and perspectives that we can use to understand the many issues that will be considered in following chapters including:

citizenship (Chapter 5); ethnic segregation (Chapter 8); neighbourhood formation (Chapter 9); and sexuality (Chapter 11).

3.1 What is culture?

Culture is popularly thought of as 'high art' in the form of paintings, sculpture, drama and classical music, as found in museums, art galleries, concert halls and theatres. However, in the social sciences culture is usually interpreted in a much broader sense. Culture is a complex phenomenon, and therefore difficult to summarize briefly, but may best be thought of as consisting of 'ways of life' (Giddens, 1989). These ways of life involve three important elements:

➤ Values that people hold (i.e. their ideals and aspirations). These values could include: the desire to acquire wealth and material goods; seeking out risk and danger; attaining spirituality through abstinence from worldly pursuits; caring for a family; help friends and neighbours; or campaigning for political change.

➤ Norms that people follow (i.e. the rules and principles that govern their lives). These include legal issues such as whether to exceed speed limits when driving or whether to engage in illegal forms of tax evasion. Norms also involve issues of personal conduct such as whether to be faithful to one's partner or whether to put self-advancement over the needs of others.

➤ Material objects that people use. For most people in relatively affluent Western societies this is an enormous category, ranging from everyday consumer goods through to transportation systems, buildings and urban facilities.

The above statements might seem like common sense and hardly the basis for a fundamental shift in the ways we think about cities. However, a number of important insights flow from these points.

The materiality of cultures

It should be clear that these elements of culture – values, norms and objects – are all highly interrelated. Culture is *not* just about ideas: the material objects that we use also provide clues about our value systems. The widespread use of the automobile, for example, says something about the value placed upon personal mobility. The same applies to the urban landscape. The structures of cities – whether they display extensive motorway networks or integrated public transport systems, uncontrolled urban sprawl or tightly regulated high-density development – provide us with indications of the wider set of values held by the society.

City landscapes therefore raise numerous questions. For example, is the society based on principles of privatism and independent wealth acquisition, or are social inequalities ameliorated by wider notions of a social consensus? Does the society attempt to impose some over-arching set of social norms or does it allow for the expression of cultural diversity? We can therefore attempt to 'read off' these values from the landscape. Consequently, the landscape can regarded as a '**text**' that can be read for layers of inner meaning, in an analogous manner to a book. The relationship between material objects and culture may be encapsulated by the notion of **intentionality**. This concept draws attention to the fact that objects have no meaning in themselves but only acquire meaning through the uses that people put them to. To sum up, culture involves much more than high art; indeed, in cultural studies a text comprises any form of **representation** with meaning and can include advertising, popular television programmes, films, popular music and even food (Bell and Valentine, 1997).

Shared meanings

Central to culture are shared sets of understandings – what are often called **discourses** or narratives. The study of the signs that give clues about these meanings is known as **semiology** (or semiotics). The things that point to these wider meanings are termed **signifiers** while the cultural meaning is called the **signified**. Activities that are full of cultural symbolism (which includes virtually all social activity!) are known as **signifying practices**. The study of meanings behind urban landscapes is more generally known as **iconography**. A good example of iconography is the downtown corporate headquarters. The imposing and expensive architectural forms of such buildings may be seen as signifiers of a discourse of corporate power and influence and are therefore said to express **symbolic capital**. Large architectural projects that are full of such symbolism are often called **monumental architecture** (Sennett, 1990). Monumental architecture involves imposing buildings and monuments in which an attempt is made to symbolize particular sets of values. Restricted access to these buildings is often a key component of their monumental and imposing status. For example, Jane M. Jacobs (1996) has documented the imperialist aspirations underpinning many of the proposals to redevelop buildings in the financial centre of London (known as the City of London). Preserving the architectural heritage of buildings that dated from the time when the British Empire was at its height is seen as a way of reinforcing the global status of the City of London in an era of financial deregulation. These issues will be taken up again in Chapter 9.

Diversity and difference

While societies do have dominant value systems, these are often resisted by many groups. For example, a large downtown office block may for some be a symbol of financial strength and influence but others may regard this as a symbol of unfair working practices and corporate greed. One of the most striking and best documented clashes of urban cultures in recent years comes from London – in the Docklands – a previously run-down inner-city area that has been transformed by an urban development corporation. These corporations provided infrastructure to encourage private sector investment (see Chapter 5 for more details). The huge monumental buildings of Canary Wharf in the London Docklands are therefore highly symbolic of this new role for private capital in urban regeneration (even if the original investors, Olympia

Monumental architecture at Canary Wharf in London's redeveloped Docklands.
Photo Credit: Paul Knox.

and York, went bankrupt in 1992). However, the Docklands office developments have juxtaposed affluence and relative poverty as highly paid professional groups have moved into newly constructed residential areas. Furthermore, the development of the Docklands has involved bypassing traditional forms of democratic accountability. While public meetings are held, giving an aura of consultation and inclusion, they have typically been used to legitimate planning decisions that have already been taken behind closed doors. The result has been a deep-rooted hostility on the part of many local working-class communities.

Such antagonisms have prompted attempts in many cities to write urban history from the perspective of local people. These accounts contrast with the sanitized versions of local history produced by development corporations as they attempt to sell inner-cities as sites of potential profit for speculators. It follows from the above discussion that the values embodied in landscapes and other material objects are not natural and inevitable, but have been created and are actively fought over by many contesting groups in society. Nevertheless, the dominant sets of values will tend to reflect the most powerful forces in society through advertising, politics, the media, education systems and so on. Thus Sharon Zukin (1995) has called the restructured areas of major Western cities **landscapes of power**.

A key theme of cultural studies, then, is *diversity* and *difference*. There are, within the dominant values of a society, many smaller subgroups with their own distinctive cultures, known as **subcultures** (and sometimes termed **'deviant' subcultures** if the norms are significantly at variance with the majority). The term **alterity** is sometimes used to denote a culture that is radically different from and totally outside that to which it is opposed. However, exponents of cultural studies are keen to emphasize the limitations of assuming these subgroups are homogeneous, for even within these subcultures there are usually many distinctions and divisions. In the United Kingdom, for example, what is often referred to in the local press as *the* 'Muslim community' contains different sects and groupings. As with landscapes, our definitions of these groupings are often determined by the most powerful in society. These issues are developed further in Chapter 8.

Identities

The diversity of cultural values in society raises issues of **identity** (i.e. the view that people take of themselves). There is a perspective, stretching back to the philosopher Descartes, that assumes that people's identities are single, rational and stable. However, this assumption is disputed by a cultural studies perspective. To begin with, it is argued that our identities are shaped by many factors such as class, age, occupation, gender, sexuality, nationality, religious affiliation, region of origin and so on. Surrounding these factors are various discourses about characteristics and abilities – what are called **subject positions** – that affect the way we behave. These subject positions are described by Dowling (1993, p. 299):

> . . . certain ways of acting, thinking and being are implicit in specific discourses and in recognising themselves in that discourse a person takes on certain characteristics, thus producing a person with specific attributes and capabilities.

Identity may therefore be thought of as being shaped by the intersection of many subject positions. Of course, to a large extent these views are also shaped by people's individual personalities and life experiences. All of these things combine to form *subjectivity* – or more accurately, since they are continually changing, **subjectivities**. One consequence is that our identities are not fixed, but vary over time and space. The crucial point is that these unstable identities and subjectivities depend upon who it is we are comparing ourselves with. For example, we are often made aware of a sense of national identity when visiting another country, when differences in habits and customs may be thrown sharply into focus.

Urban environments also have a crucial impact upon subjectivities because they tend to bring together in close juxtaposition many different types of people. This mingling requires a response on the part of the city dweller, whether this is indifference, fear, loathing, incomprehension, admiration or envy. Sometimes these comparisons are based on stereotypes – exaggerated, simplified or distorted interpretations – of the group in question. Indeed, the use of **binaries** – two-fold divisions – is often central to the creation of these differences (e.g. male/female, healthy/sick, sane/mad, heterosexual/homosexual, non-foreigner/foreigner, authentic/fake). Furthermore, issues of power lie behind many of these comparisons; we often feel either superior or inferior to the group we are comparing ourselves with.

The process whereby a group is analyzed in a way that constructs them as being inferior is sometimes termed **objectification**. Identities and cultural values do not therefore evolve in isolation but require opposition from other sets of identities and values that are excluded and demoted in some sort of hierarchy of value. These processes leading to identity formation can have a crucial impact on the social geography of the city for, as we will see in many following chapters, they help to create social exclusion and residential differentiation. The processes of identity formulation may be multiple and unstable but they are anchored in power relations and the allocation of material, political and psychological resources in cities.

3.2 Postcolonial theory and the city

Some of the most important insights into this process of identity formation have come from a field of study closely related to cultural studies: **postcolonial theory**. Before discussing this work we need to make a distinction between colonialism and imperialism. **Colonialism** may be defined as the direct rule of one country by another through the imposition of new settlements. **Imperialism,** in contrast, is defined by the attitudes and actions of a domineering centre upon a distant territory. **Neocolonialism** refers to the economic and political domination of previously colonized nations by Western powers after the ending of formal colonialism. The crucial point is that although colonialism has largely ended, imperialist attitudes still persist in many forms in a neocolonial era.

Postcolonial theory therefore attempts to examine the **imperialist discourses** running through Western representations of non-Western societies, both in the colonial period and in contemporary contexts. As such, it attempts to undermine **ethnocentrism** – the notion that Western thought is superior. The insights from the approach are valuable in studying Western cities because many immigrants have come from what were previously colonies of the West. Indeed, many would argue that the distinction between the West and 'the Rest' has been subverted by this infusion of ideas from throughout the world into the heart of the major Western cities.

Imperialist discourses whereby subject peoples have been constructed as inferior are often encapsulated in the term '**othering**'. In addition, those subject to these hegemonic (i.e. dominant) discourses are often termed the **subaltern** classes or groups. Edward Said (1978) wrote about the ways in which European thought constructed views of Oriental peoples in his highly influential book *Orientalism*. Said argued that the notion of the Orient is a Western invention, conjuring up visions of exotic and sensual peoples. The crucial point is that this conception of the Orient was defined in relation to perceived notions of the Occident (the West) as the opposite – rational, civilized and safe. In other words, the Orient helped to define the West's view of itself. However, Said's work has been criticized for creating a binary division between colonizer and colonized and for assuming that colonial discourse is primarily the product of the colonizer. Others – including the leading postcolonial theorist Homi Bhabha (1994) – have argued that there is a mutual interaction between the colonizer and the colonized. Said's work has also been criticized for neglecting the gendered nature of colonial discourses (i.e. they were mostly constructed by men).

Hybridity

The idea that one culture is superior to another is therefore undermined by another key concept in postcolonial theory – **hybridity** – the idea that all cultures are *mixtures*. A torrent of new terms has emerged to capture these complex processes of cultural hybridization. For example, this process whereby dominant and subordinate cultures intermix is also known as **transculturation**. Somewhat similar is the notion of **creolization** – a term originally used to denote miscegenation (i.e. interbreeding) between colonizers and indigenous people but now used in a much more general sense. The terms **nomadization** and **deterritorialization** are also used to refer to the destabilization of identities, either in a metaphorical sense, or as a result of geographical migration. Many new terms have been devised to refer to geographical areas in which this intermixing takes place including: **liminal space, heterotopia, borderlands** and **third space**.

Postcolonial studies therefore dispute the notion of **authenticity** – the idea that there is some basic, pure, underlying culture. Cultures are inevitably hybrid mixtures and yet the mistaken notion of authenticity is one that has fuelled (and continues to fuel) many nationalist movements. In a different realm, it can be suggested that the mistaken desire for 'authentic', 'ethnic' music reflecting some pure, underlying, culture, untainted by the commercialism of Western society,

explains much of the popularity of so-called 'world' music (see also Box 3.1).

Often, the hybrid mixing of cultures – especially in a colonial context – led to **ambivalence**, a complex combination of attraction and repulsion on the part of the colonizers and the colonized. The colonizers, for example, were often flattered to have their culture copied by the colonized but at the same time did not want to be replicated entirely because this would undermine their feeling of superiority. The colonized often admired the colonial culture and copied this (a process called **appropriation**) but at the same frequently resented their subjugated position.

Bhabha (1994) has argued that the colonial culture is never copied exactly, hence the growth of hybridity.

Box 3.1

Key trends in urban social geography – Hybridization: the case of popular music

A good example of hybridization is the evolution of popular music for, as Shuker (1998, p. 228) notes, 'Essentially, all popular music consists of a hybrid of musical styles and influences'. The reason for this is that in contemporary societies music is highly transportable, transmissible and reproducible. Musical forms are highly vulnerable to manipulation and they can easily be appropriated, copied and transformed. As Negus notes, 'once in circulation, music and other cultural forms cannot remain "bounded" in any one group and interpreted simply as an expression that speaks to or reflects the lives of that exclusive group of people' (Negus, 1992, p. 121). Thus the emergence of rock and roll was a popularized amalgam of white working class country music and African American rhythm and blues (which was also a mixture of various black styles of music – especially blues, jazz and soul). Soul itself was a secularized form of gospel music blended with jazz and funk. More recent forms of

hybridization are Tex Mex (a mixture of rock, country, blues, R & B and traditional Spanish and Mexican music) and salsa (music from Cuba popularized by Puerto Ricans in New York and incorporating elements of big band, jazz, soul, rock and funk).

Popular music is today one of the most widely disseminated of consumer products. Furthermore, it is an important part of the culture, lifestyle and identity of many groups – especially youth cultures upon which it is so heavily targeted (despite the existence of 'Dad rock' for older generations). For example, surf rock in the early 1960s celebrated a particular creation of southern Californian lifestyle based around sun, sand, surf, sex, hot rods and drag racing.

This hybridity of musical forms is a problem for cultural theorists who would wish to pin down distinctive musical styles as emerging from particular localities. Nevertheless, the intermixing of musical styles does take place at particular locations

(usually cities). There is much debate about the characteristics that give rise to innovation, including some of the following:

➤ the geographical juxtaposition of differing cultural groups;

➤ the presence of proactive local record companies;

➤ the involvement of an active student community;

➤ an active club scene.

Musical innovation often takes place in 'marginal' locations where global mainstream trends interact with local subcultures. For example, the Seattle Sound characterized by *grunge* is partly attributed to its distance from Los Angeles; the Dunedin Sound was perhaps distinctive because of its relative isolation in New Zealand and new hybrid musical forms including Celtic elements have emerged on the East Coast of Canada.

Sources of popular music creativity are shown opposite:

continued

Location	Approx. time period	Genres	Examples of groups/artists
New York	1945–1955	Doo-wop	Ravens, Oriels
Chicago	1940s–1960s	Electric blues	Muddy Waters, BB King, John Lee Hooker
Memphis	1950s and 1960s	Rock and roll	Elvis Presley, Chuck Berry, Jerry Lee Lewis
Liverpool, UK	Late 1950s–1960s	Britbeat	Beatles, Gerry and the Pacemakers
Los Angeles	Early 1960s	Surf rock	Beach Boys, Dale and the Del-Tones, Surfaris
San Francisco	1960s	Psychedelic rock	Jefferson Airplane, The Grateful Dead, Moby Grape
Bronx, New York	Late 1970s, 1980s	Early rap	
Athens, Georgia	1980s	Alternative	B-52s, Love Tractor, Pylon, REM
Boston/Amherst	1980s	Alternative	Dinosaur r Jnr., The Pixies, Throwing Muses, The Lemonheads
Dunedin, New Zealand	1980s	Alternative	The Chills, The Verlaines, The Clean, Toy Love
Ireland	1980s, 1990s	Celtic-rock/punk/ grunge/new age	Van Morrison, Clannad, The Chieftains, Enya, Corrs, Pogues
East Canadian Coast	1980s, 1990s	As above	Rawlins Cross, Ashley MacIsaac, The Barra McNeils, The Rankin Family
Bristol, UK	Mid-1990s	Trip-hop	Massive Attack, Portishead, Tricky
Seattle	1990s	Alternative/grunge	Nirvana, Pearl Jam, Soundgarden
Manchester, UK	1990s	Alternative/Britpop	Oasis, Happy Mondays, Stone Roses

Key terms associated with popular music (see Glossary)

Appropriation, cultural industries, hybridity.

Further Reading

Leyshon, A., Matless, D. and Revill, G. (1995) The place of music *Transactions, Institute of British Geographers*, **20**, 430–433

Negus, K. (1992) *Producing Pop* Edward Arnold, London

Shuker, R. (1998) *Key Concepts in Popular Music* Routledge, London

Links with Other Chapters

Chapter 8: Box 8.4 Rap as Cultural Expression (and Commercialization)

Indeed, appropriation can sometimes lead to **mimicry** – a parody or pastiche of the colonial culture. The classic example of the mimic is the Indian civil servant aping the manners of the English bureaucrat. This is potentially threatening for the colonizers since mimicry is not far removed from mockery and it has the potential to undermine colonial authority. The general point to note here is that these processes continue in a postcolonial era and that all cultures are the product of appropriation in some form. Hybridity is not therefore just a mixing of cultures – it involves a

destabilization of many of the symbols of authority in the dominant culture.

The social construction of culture

Both cultural studies and postcolonial theory draw attention to the fact that cultures are, above all, *social constructs*. In the case of national identities, for example, certain aspects are usually isolated as being distinct, authentic, elements. Benedict Anderson (1983) used the term **imagined communities** to draw attention to the socially constructed nature of national identities. He argued that the formation of national identities involves the use of a great deal of imagination because, although it is physically impossible to know everyone in a country, strong patriotic bonds are forged between large numbers of people through imaginative projections resulting from the influence of books, newspapers and television. However, in previous eras, as in the feudal period in Europe, for example, national identities were much weaker.

Identification with the nation state developed rather later, following the French and American Revolutions. National identities were subsequently constructed by comparisons with 'others', especially colonized peoples. Thus, British identity was seen as based on 'reason', 'democracy' and 'civilization' compared with the supposed 'uncivilized' cultures of the British Empire. The communities to be found within contemporary cities also have a socially constructed and imagined character, for it is only in the smallest of settlements that we have face-to-face contacts with all the members of a group. Senses of community within other units such as neighbourhoods, towns and cities all involve imaginative elements shaped by many factors such as mass media and elements of popular culture.

3.3 Space, power and culture

Another insight that emerges from the cultural studies movement – and one we have already signalled above – is the crucial role of *space* in the formation of culture. The reason for this connection is that space, like culture, is a social construct and is therefore intimately bound up with power and authority.

Foucault and the carceral city

Michel Foucault, one of the key figures underpinning contemporary cultural studies, has been highly influential in drawing attention to these issues (see also Box 3.2). Rather like Gramsci (the originator of the notion of Fordism discussed in Chapter 2), Foucault was concerned with understanding the ways in which consent is achieved in society (i.e. the processes through which people agree to have their lives determined by others).

Foucault was opposed to the idea that such consent could be explained by any single, over-arching, theory. Instead, he argued that consent was achieved by various types of discourse. These discourses are a crucial component in the exercise of power, since they help to shape the view that people take of themselves. Foucault thought of power as a crucial component in daily life that helps to construct the ordinary, everyday, actions of people. According to Foucault, therefore, power is not something that some people have and others do not; what makes people powerful is not some individual characteristic, or position in society, but the recognition by others of their capacity to exercise that power. Power, then, is a process rather than a thing that is exercised. Foucault also argued that power was like a network of relations in a state of tension. The term **micropowers** was used to encapsulate these processes. In addition, Foucault coined the term **carceral city** to indicate an urban area in which power was decentred and in which people were controlled by these micropowers (from the Latin term *carcer* meaning prison – hence the English term *incarceration*).

In this sense people may be envisaged as agents of their own domination. Foucault used the metaphor of the Panopticon to describe these processes in what he termed the **disciplinary society**. The **Panopticon** was a model prison devised by the nineteenth-century thinker Jeremy Bentham in which inmates could be kept under observation from a central point. Although the design was never directly implemented, Foucault's metaphor of the Panopticon (panoptic meaning 'all embracing in a single view') has been used to describe the surveillance practices that take place in contemporary city spaces such as shopping malls through the use of close circuit television (CCTV) and private security guards.

Some have argued that Foucault's concept of power is too passive and says too little about the capacity of people to resist disciplinary forces. For example, Warren (1996) highlights some of the tactics used by people to subvert the surveillance and control exercised in Disney theme parks, some of the most intensively controlled spaces of the contemporary world. Control measures even include guards dressed up in comic costumes as Keystone Cops or as 'tourists' in order that they may

Box 3.2

Key thinkers in urban social geography – Michel Foucault (1926–1984)

In the late twentieth and early twenty-first centuries French scholars have had a profound influence upon Western intellectual thought – Barthes, Baudrillard, Bourdieu, Camus, Deleuze, Derrida, Fanon, Lacan, Lefebvre, Levi-Strauss, Lyotard and Sartre – to mention but a few! The attitudes of Anglo-Saxon scholars towards these thinkers is often polarized – some seem bewitched and dazzled by their complex and enigmatic ideas, others regard them as intellectual charlatans and impostors.

Whatever stance is adopted, is seems clear that of all these gurus, *the* cult intellectual underpinning the cultural studies movements in contemporary social science is Michel Foucault. Both his life and his work have a complex, enigmatic, mysterious and colourful character, and both have been the subject of numerous books and biographies (e.g. Macey, 1993; Jones and Porter, 1998; Falzon, 1998). His style is certainly difficult at times but his ideas have been extremely influential. Indeed, a large 'Foucault industry' seems to have evolved, reinterpreting, criticizing and extending his ideas.

To do them justice, Foucault's ideas need a book to themselves. Nevertheless, we should note here that his ideas have generated great interest among geographers through their concern with the role of space in the exercise of power and the formation of various types of knowledge.

Key concepts associated with Michel Foucault (see Glossary)

Heterotopia, micropowers, othering, Panopticon, queer theory.

Further reading

Barry, A. Osborne, T. and Rose, N. (1996) *Foucault and Political Reason* UCL Press, London

Falzon, C. (1998) *Foucault and Social Dialogue* Routledge, London

Jones, C. and Porter, R. (1998) *Reassessing Foucault: Power, Medicine and the Body* Routledge, London

Macey, D. (1993) *The Lives of Michel Foucault* Vintage, London

McHoul, A. and Grace, W. (1995) *A Foucault Primer* UCL Press, London

Links with Other Chapters

Chapter 11: Sexuality and the City

---

watch both visitors and employees. Nevertheless, such measures are not able to prevent some tourists and employees smuggling in drugs and alcohol (considered by some to be essential to get the best out of the rides or maybe to endure the boredom of the queues).

The social construction of space

Space is something that we move through and often take for granted. It is therefore not surprising that some are sceptical about the idea that space is something more than an 'empty container', since this notion seems to run against common sense. However, cities provide many examples of the relationships between culture, space and power that help to clarify what is meant by space as a 'social construct'. For example, some people are excluded from **public spaces**. This was most starkly revealed by the very earliest public spaces in cities – the *agora* of ancient classical Greek cities – the open market

in which democracy was supposed to flourish. However, this space was based around a particular notion of the democratic citizen that excluded women, slaves and foreigners (see also Chapter 5).

In contemporary cities measures are also taken to exclude certain groups from 'public' spaces including gangs of youths, drunks, the homeless and 'deviants' such as those who appear to be mentally ill. Lees (1997), for example, describes how the public space of Vancouver's new public library is policed by security guards to ensure standards of hygiene among the homeless as well as providing a sense of security for women and children. The reason why such groups are excluded is that they disrupt certain codes of behaviour: sobriety, cleanliness and the like. In many cases it is the *perception* that these groups are likely to disrupt these codes of behaviour that is important.

Spaces therefore reinforce cultures because the patterns of behaviour expected within them reflect particular cultural values. As Sibley (1995) notes,

Box 3.3

Key thinkers in urban social geography – David Sibley

David Sibley is an example of a geographer whose ideas have had a big influence upon urban social geography, albeit somewhat late in his academic career. This belated recognition would seem to reflect the fact that his work was ahead of its time (and this led him to experience problems in getting it published earlier in his career). Although he trained as a quantitative geographer at the University of Cambridge, Sibley became interested in psychoanalytic theories relating the attitudes of adults to their early child-rearing experiences, theories that have only recently been taken up on any scale in human geography.

Significantly, Sibley and his partner spent some time outside academia developing a school for gypsy children. Based on these experiences, Sibley developed a series of ideas about processes of exclusion and their relationship with space, ideas that were formulated in his highly influential book *Geographies of Exclusion* (1995). He argued that powerful groups in society have a tendency to 'purify' space by attempting to exclude what are perceived to be the 'polluting' and 'dirty' influences of 'outsiders'. Such exclusionary practices help people to preserve their sense of identity but also

reflect subconscious desires, developed in early childhood experiences, to maintain cleanliness and purity.

The difficulties Sibley experienced in getting his ideas published led to him becoming interested in the processes that lead to certain groups and their associated intellectual perspectives being excluded by the types of people who govern academic life (i.e. mostly white, middle aged, middle-class, physically able men). He therefore highlighted the factors that led to the denigration of the ideas developed by the (mostly female) Chicago School of Social Service Administration who were overlooked and overshadowed by the Chicago School of human ecology (Sibley, 1990; see also Chapter 7). Although many are sceptical about theories related to psychoanalysis, Sibley's ideas have helped to foster the recent interest in geographies of youth and childhood (see Skelton and Valentine, 1998; Holloway and Valentine, 2000).

Key concepts associated with David Sibley (see Glossary)

Exclusion, exclusionary zoning, othering, purified communities, purified space.

Further reading

Holloway, S.L. and Valentine, G. (eds) (2000) *Children's Geographies: Playing, Living, Learning* Routledge, London

Mahtani, M. (2004) David Sibley, in P. Hubbard, R. Kitchin and G. Valentine (eds) *Key Thinkers on Space and Place* Sage, London

Sibley, D. (1988) Survey 13: Purification of space *Environment and Planning D*, **6**, 409–421

Sibley, D. (1990) Invisible women? The contribution of the Chicago School of Social Service Administration to urban analysis *Environment and Planning A*, **22**, 733–745

Skelton, T. and Valentine, G. (eds) (1998) *Cool Places: Geographies of Youth Cultures* Routledge, London

Links with Other Chapters

Chapter 10: Box 10.1 How to Understand Geographies of Childhood and Youth Culture

---

segregation is therefore crucial to the creation of landscape and space, creating what can be termed **spaces of exclusion**. Power is expressed through the monopolization of spaces by some groups and the exclusion of certain weaker groups to other spaces. However, Sibley goes on to point out that these power relations are often taken for granted as a 'natural' part of the routine of everyday life (see also Box 3.3). This leads to what Young (1990) calls **cultural imperialism**, whereby the dominant power relations in society become 'invisible' while less powerful groups are marked out as 'other'.

Cultural imperialism is quite common in discussions of ethnic identities. The focus is frequently upon the distinctive characteristics of a minority ethnic group and not upon the wider society or what 'whiteness' means, for example (Hesse, 1997).

Space and identity

Space is therefore crucial to all the processes of identity formation, stereotype construction, othering, objectification and binary construction noted above. The term

spatialized subjectivities is often used to describe the processes leading to identity formation. And again, cities have played a crucial part in the formation of such identities. Most notably, the perception of the working classes as dirty, disease-ridden and dangerous was fostered by the increased spatial separation of classes that emerged with the early cities of the industrial revolution. Richard Sennett drew attention to these issues in another of his highly influential books *The Uses of Disorder* (Sennett, 1971), in which he used the term '**purified communities**' to draw attention to the ways in which groups build walls around themselves to exclude others.

Once again, we can see the sociospatial dialectic at work here. On the one hand, an area of a city may serve as a social setting in which particular cultural values can be expressed; on the other hand, the neighbourhood can serve to form and shape those distinctive cultural values. However, it is crucial to remember at this point that the cultures of the city do not emerge in these spaces in isolation. Not only are they defined in relation to cultures in other areas but they also involve a hybrid mixing of various elements from elsewhere. For example, even something as traditional as the (once?) staple fare of English working class culture – fish and chips – is a remarkable demonstration of cultural hybridity. The large English chip (French fry) is a direct descendant of the *pomme frite* first introduced into England by the Huguenots from France, while battered fried fish were brought by Russo-European Jews (Keith, cited in Parker, 2004). In fact, recent surveys show that Britain's most popular dish is no longer fish and chips but a 'curry-style' dish called chicken tikka massala, an entirely hybrid concoction that has only a passing relationship to its assumed Indian heartland. Indeed, yet another irony is that the popular 'Indian' cuisine of the UK is largely the product of chefs from Bangladesh.

The history of a particular space is therefore intimately connected with events outside that space. As Massey notes:

> We need to conceptualise space as constructed out of sets of interrelations, as the simultaneous coexistence of social interrelations and interactions at all spatial scales, from the most local to the most global.

(Massey, 1992, p. 80)

All of the above considerations point to the utter folly of attempts to argue for some superior, pure, culture that is rooted in a particular geographically bounded space and that may be 'tainted' or 'corrupted' by mixture with other 'alien' cultures. Yet, such assumptions of authenticity have been the basis for feelings of nationalism – one of the most powerful and dangerous forces of the twentieth century. Furthermore, within nations, racist notions continue to be the basis for urban conflict. Some writers have begun to analyze the imperialist discourses that underpin the racism that is found in some parts of British cities. As Hall points out, those who object to the influx of 'Asians' forget the long history of empire that has led to the evolution of these groups in British cities (cited in Hesse, 1997).

The socially constructed nature of space means that cities are, in a sense, a text that is rewritten over time. The term **scripting** may therefore be used to describe this process whereby we 'produce' or 'construct' cities through our representations. This does not mean that they can be anything we chose to make them. Cities have obvious physical attributes that constrain and influence how they can be presented but, nevertheless, these representations are quite malleable. This is well illustrated by the changing representations over time of the impoverished East End of London (Jacobs, 1996; Cohen, 1997). In the nineteenth century the East End was seen as a dangerous place but from the 1890s onwards, under the influence of many factors – urban reform movements, the Labour Party, the church, state education, housing redevelopment and representations in music halls – the East Enders became transformed into cheerful, patriotic Cockneys. With the influx of ethnic minorities, especially from Bangladesh, and the growth of Thatcherism in the 1980s, the area became an 'imagined community' of self-reliant entrepreneurs.

Other examples of changing city images can be seen in the attempts by public agencies to 'rebrand' cities and make them attractive to investors. In the United States in particular, **place promotion** (or **place marketing**) has become a multi-billion dollar industry as consultants and public relations firms specialize in the packaging, advertising and selling of cities. Understandably, some of the most vigorous re-representations occur in cities that have been badly affected by deindustrialization, as illustrated by the attempts to reconstruct the image of the city of Syracuse (Short *et al.*, 1993). The city's logo, dating from the nineteenth century, symbolized a classic industrial city with factories and smokestacks whereas the new logo reflects the lakefront and environmental amenity. The crucial point is that logos are not simply reflections of a changing environment, they are attempts to consciously engineer a view about that environment that will affect future development.

Robins (1991) has argued that processes of globalization have broken some of these links between culture and territory. The reason for this is that new technologies of mass media and telecommunications have enabled transnational corporations to impose what he terms an 'abstract electronic space' across pre-existing cultural forms. This is especially noticeable in the film and music industries whose stars are viewed simultaneously throughout the world. Audiences are constructed around common shared experiences on a global scale and culture is less dependent upon local forms of knowledge.

This process whereby local cultures are eroded by the processes of globalization is sometimes called **delocalization** (see also Chapter 2). However, many would argue that this tendency towards homogenization of culture can be overstated. Indeed, we have recently seen a reaffirmation of local forms of identity through various nationalist movements and distinctive cultural expressions in spaces within cities – perhaps largely as a response to the perceived threat of some external mass culture.

Culture, then, is not a preserve of elite groups in society; it is something that is all around us in consumer goods, landscapes, buildings and places. Furthermore, it is not a static thing but is a continually evolving and disputed realm that is alive in language and everyday social practices.

3.4 Postmodernism

One way of summarizing recent cultural shifts in cities is through the concepts of **modernism** and **postmodernism**. As with the Fordist/neo-Fordist division noted in Chapter 2, the concepts of modernism and postmodernism can be used in many complex ways: as a particular *cultural style*, as a *method of analysis*, or as an *epoch in history* (Dear, 1999). To understand how these concepts are used we first need to examine the concept of modernism.

Modernism is usually regarded as a broad cultural and philosophical movement that emerged with the Renaissance, coming to full fruition in the late nineteenth and early twentieth centuries. Above all, it was characterized by the belief that the application of rational thought and scientific analysis could lead to universal progress (see Harvey, 1989b). This eventually led to **social engineering**, the notion that society could be improved by rational comprehensive planning and the application of scientific principles. This philosophy was implicit in comprehensive urban renewal (i.e. slum clearance) and traffic management schemes within cities in the 1960s. Modernism may similarly be regarded as a basic foundation for the quantitative and behavioural approaches discussed in Chapter 1. As the twentieth century progressed, however, wars, famines, political repression and disillusionment with the social and ecological costs of advanced technologies and attempts at comprehensive planning led to a significant degree of disillusionment with the concept of modernism.

Modernism was also undermined by what might be termed the '**linguistic turn**', the forerunner to the 'cultural turn'. A key influence here was a Swiss linguist called Ferdinand de Saussure. He argued that the meanings of words are derived not directly from the world to which they refer, but from common sets of understandings among people. These commonly accepted meanings are created internally *within* the language. To understand the meanings of words we therefore need to understand the broader cultural context from which the language emerges. What might at first glance seem like an obscure and abstract insight has a crucially important consequence: it means that our language is not a direct reflection (or mirror) of the world around us. Instead, we very largely *create* the world through the words we use and the concepts to which they refer. A key feature of postmodern thought, therefore, is that the meanings attached to words are always defined in relation to other things and there is no one single meaning to a word.

It follows from this that there is not just one version of the truth embodied in words but many possible versions depending upon the common understandings being represented. These common understandings will vary within societies as well as between them. The crucial point is that our understandings of the world are always filtered through particular theoretical perspectives. What claims to be a superior way of understanding the world is therefore often an attempt by one group to impose their understandings on others.

A central feature of postmodernism, therefore, is that *there is no pure, immediate, undistilled experience – all our experiences are mediated through sets of cultural values that are embodied in language and other forms of representation.* Because of this, postmodernism focuses upon the methods and systems of representation rather than the reality itself. Postmodernists therefore argue that theories or 'knowledge claims' are bound up with power – they represent an attempt to impose views on others. Postmodernism therefore marks a break with the belief in a single path of knowledge leading to universal progress. It argued that there is not just one effective way of analyzing the world (what is sometimes called a **metanarrative** or **totalizing discourse**). Instead, there are many differing ways of representing truth, depending upon the power relations involved.

We should note at this stage that the validity of postmodernism is much questioned. Many have objected to postmodernism because of its implications of moral relativism – suggesting at times a complacency towards the activities, experiences and living conditions of other people. Giddens, meanwhile, argues that rather than being postmodern, contemporary societies reflect a **late modernism**, characterized by an intense degree of individual **reflexivity**, whereby people are increasingly aware of the attitudes they adopt and the choices they make in their everyday life and are also self-conscious about their reflexivity.

Postmodernism in the city

A key feature of postmodernist thinking, which is in keeping with the cultural studies movement, is a recognition of the diversity of different groups in society. This diversity of groups and their aspirations is reflected not only in academic writing but also in popular forms of representation such as music, advertising and literature. In terms of the landscape of cities, for example, postmodernism is reflected in a diversity of architectural styles rather than the rectilinear appearance of Modernist styles. Postmodern styles may range from 'high-tech' to neo-classical and frequently involve an eclectic blend of many different motifs. Postmodern design often attempts to be playful or to allude to layers of meaning. Whereas the architecture and urban planning of the modernist city reflected a striving for progress, contemporary buildings represent consumption, hedonism and the creation of profit with little regard for the social consequences. This issue is developed further in Chapter 9.

Another key theme of postmodernist interpretations of the city therefore is the increasing importance of signs and images in everyday life. The leading postmodernist writer Baudrillard (1988) argued that postmodern culture is based on images or copies of real world (known as **simulacra**) that take on a life of their own and are difficult, if not impossible, to distinguish from the reality they imitate.

A good example of simulcra in popular culture is the music of Bruce Springsteen. This music is marketed as the 'authentic' voice of the blue-collar American male with references to manual work, deindustrialized urban environments, riding motorcycles of freeways and, of course, an all-consuming desire for the opposite sex. Sales of his records across the globe have thus created an American urban male mythology. This music has been criticized for its lack of 'authenticity' and yet such myths and legends can influence the ways people live their lives (Negus, 1992). The crucial point is that such simulacra are neither true or false; instead they are based on a series of signs and shared understandings (often called codes) that create an imagined tradition of experience that connect an artist with their audience. These traditions are in part based on the experiences of the audience but at the same time they serve to shape that experience. Similar comments apply to the creation of authenticity within 'country' music based on an imagined community of largely white, rural working class (Negus, 1999).

Another feature of postmodern culture is that advertising and mass media produce signs and images that have their own internal meanings, producing what is termed a **hyperreality**. An environment dominated by hyperreality may be termed a **hyperspace**. The theme

A 'space of consumption': the Potsdamer Platz, Berlin. Photo Credit: Alan Latham.

parks of Disneyland are perhaps the most extreme and obvious example of this hyperreality since they present a non-threatening, sanitized view of the world that extols patriotism, traditional family values and free enterprise and that glosses over issues of violence, exploitation and conflict (Warren, 1996). This conscious creation of environments has been termed **imagineering**. Archer (1997) points out that in Orlando, Florida, the creation of a particular Disneyesque view of community and history has required an extreme degree of control over a privatized space. The reason for this is that a less con-trolled public space might lead to dissent, disruption and alternative visions of reality. It might also lead to people becoming more aware of the imagineered nature of the space and lead to demands for participation in the formation of such spaces.

The ultimate extension of the Disney philosophy is the creation of Celebration City in Orlando, an integrated, privatized, residential community in which the tensions associated with social polarization can be excluded. Sorkin (1992) argues that in a postmodern era the city has a whole is becoming one big 'theme park' in which a variety of simulations present a highly distorted view of the world (a space he terms **ageographia**). The result is that the diverse postmodern buildings of the contemporary city present a shallow facade of culture (sometimes termed **Disneyfication**) (see also Box 3.4).

The aestheticization of consumption

The concept of a **positional good** – one that displays the superiority of the consumer – is well established. However, it is argued that in postmodern society the act of consumption has assumed much greater significance. No longer is position ascribed by birth: rather, people

Box 3.4

Key debates in urban social geography – To what extent does the film *Blade Runner* represent the dystopian postmodern city?

Few films have received more analysis in recent years than the cult science fiction classic *Blade Runner*. Quite apart from the content, the history of the film is in itself fascinating. Based on the novel by the science fiction writer Philip K. Dick, entitled *Do Androids Dream of Electric Sheep?*, the film was something of a commercial and critical flop on its release in 1982. A mixture of 1940s detective *film noir* and science fiction, cinema-goers expecting an action movie were confronted with a downbeat vision underpinned by a mournful and haunting soundtrack by the Greek composer Vangelis. Early test audiences found the film incomprehensible and melancholic and (contrary to the director's wishes) a crude voice-over was hastily added to explain the plot, while film clips of pastoral mountain scenery from Stanley Kubrick's film *The Shining* were used to give a rather creaky happy ending. However, *Blade Runner* was in many respects a film 'ahead of its time' and it attracted something of a cult following among devoted fans who purchased videos and attended regular late-night screenings in art-house cinemas.

A key turning point in the film's fortunes was a screening of director Ridley Scott's original version (minus the voice-over and happy ending) at a film festival in Venice California. It is alleged that spontaneous and rapturous applause greeted the film as it became evident how emasculated the original released version had been. Sensing commercial success at last, Warner Brother released a 'Director's Cut' in 1991. Since then, the film

has spawned a virtual '*Blade Runner* industry' of books, articles and films. The visually stunning character of the film has also had a crucial impact upon subsequent science fiction movies. One of the reasons for this extraordinary amount of attention is that the film (like Dick's original paranoid vision) seems to highlight with remarkable prescience some of the key elements of the evolving postmodern city:

➤ *Environmental disaster*: set in Los Angeles in 2019, instead of a sun-bathed land of opportunity, the city is portrayed as dark, polluted, cloud-ridden and rain-soaked (cynics have joked that this vision more closely corresponds with that of a bad night in the industrial city of Gateshead in the North East of England where the film's director Ridley Scott grew up!). The crucial point about the film is that, rather than portray the future as a place of technological sophistication and ease, it was one of the first science fiction films to portray the future as problematic.

➤ *Urban underclass*: Los Angeles is a socially polarized city with a large, powerless, multi-ethnic underclass – although curiously without African-Americans. (In *Bladerunner* this underclass has a strong Japanese influence – perhaps reflecting the paranoia about the growing influence of Japanese economic might in the early 1980s.)

➤ *Corporate domination*: *Blade Runner* portrays a world in which people are dominated by large, faceless, corporations.

➤ *Genetic engineering*: new technologies are utilized to produce androids – 'replicants' – that are virtually indistinguishable from humans. (It is surely no coincidence that the chief protagonist, Deckard, seems to be a pun on Decartes who gave us the famous phrase 'I think therefore I am'!)

➤ *Postmodern culture*: the dominant discourses in the city are those based around commercial interests. Consequently, the urban underclass lacks any sense of collective vision. The prevailing ethos is one of selfishness, the main priority simply survival in a hostile world.

Key concepts related to *Blade Runner* and postmodernism (see Glossary)

Hyperreality, postmodernism, simulacra

Further reading

Doel, M.A. and Clarke, D.B. (1997) From Ramble City to the Screening of the Eye: *Blade Runner*, death and symbolic exchange, in D.B. Clarke (ed.) *The Cinematic City* Routledge, London

Links with Other Chapters

Chapter 14: Cinema and the City

are able to choose various types of identity through the goods they consume. This process has been described as the **aestheticization** of consumption:

> What is increasingly being produced are not material objects but signs. . . . This is occurring not just in the proliferation on non-material objects that comprise a substantial aesthetic component (such as pop music, cinema, video etc.), but also in the increasing component of sign value or image in material objects. The aestheticization of material objects can take place either in the production or the circulation and consumption of such goods.
>
> (Lash and Urry, 1994, p. 15)

A classic example of an item that became a branded fashion icon is the Nike athletic shoe. Initially developed for playing sports, the shoe became a symbol of rejection of the mainstream when adopted by young people in the Bronx. The Nike marketers were not slow in exploiting this symbolic cultural value through the addition of innovations such as air-cushioned soles, coloured laces and 'pump-action' fittings. Together with a high-profile advertising campaign, this led to large profits and frequently considerable hardship to relatively poor families as they struggled to purchase these shoes. Of course, once it became widespread in society, the product lost much of its association with transgression and subversion.

A key element in recent postmodern theorizing about cities, therefore, is the crucial role of consumption in the shaping of the identities. As Jackson and Thrift (1995, p. 227) note, 'identities are affirmed and contested through specific acts of consumption'. It is argued that in purchasing particular products, people not only differentiate themselves from others but they also find a means of self-expression in which they can adopt and experiment with new subject positions. Thus it is argued (somewhat controversially) that people are increasingly defined by what they consume rather than by traditional factors such as their income, class or ethnic background.

One of the most striking manifestations of this increased role of consumption in contemporary cities is the increased amount of space devoted to shopping – not only in vast suburban malls but also in revitalized city centres and festival marketplaces. The internal architecture of these new spaces of consumption are carefully constructed to encourage people to spend their money (Goss, 1993). In an attempt to attract people to spend in these new shopping malls, special events, street theatre and ever more dramatic architectural forms are employed. One of the most cited examples is the West Edmonton Mall in Canada which proved to be a model for many other retail developments in Europe and North America (Shields, 1989).

Within these new spaces of consumption, postmodern culture has also a big impact through advertising. To begin with, as in the sphere of postmodern architecture, irony, pastiche and internal jokes abound. One strategy is to construct commodities as exotic and 'other'. This is an approach adopted by Banana Republic, a clothing store that sells its products by portraying the Third World and colonial Africa as some desirable, distant place (Crang, 1998). Irony is another key feature of postmodern culture. Thus, what was once seen as the *kitsch* culture of the 1970s has recently become regarded as *chic* in some quarters and used to sell a variety of products including records, films and clothing.

Harvey (1989b) and Jameson (1984) both argue that postmodernism is the logical cultural partner to the regime of flexible accumulation. Not only does postmodern culture help to produce many diverse niche markets but it tends to produce a fragmented populace supposedly bewitched by the glamour of consumption and lacking the collective institutions to mount a challenge to the dominant powers in society. This explanation has great appeal, but as with all such overarching theories, it has overtones of functionalism and should be treated with some scepticism. The cultures of cities are diverse and it is by no means certain that the structures evolving in cities will be sustainable in the longer run. For example, the social fabric of many cities is being stretched almost to breaking point through increasing social polarization.

3.5 Conclusions

This chapter has illustrated some of the numerous complex changes in the cultures of Western cities in recent years. Such is the complexity of these changes that many have argued it is becoming increasingly difficult to 'read' and understand the city as a single cultural landscape. There is no longer *one* urban geography (not that there ever was) but a whole set of urban geograph*ies*. It is argued that the city has become decentred, not only spatially and structurally, but also socially and conceptually (these elements all being closely bound together). However, a major criticism of this culturally inspired work is that it has led to a focus upon numerous 'local' forms of knowledge and identity while losing sight of the broader structure of political economy in which these cultures operate.

One important consequence of the concern to avoid essentializing, objectifying, 'othering' and the creating binary dualisms, is that many cultural geographers are suspicious of, and frequently hostile towards, the sorts of maps and tables that have traditionally been the staple fare of urban social geography. These established forms of representation are distrusted because they are seen as masking power relations and, indeed, contributing towards the very stereotypes that cultural geographers seek to subvert. Instead, the 'new' cultural geography places much greater emphasis upon ethnographic

methods and, in particular, in-depth interviews to reflect the complexity and diversity of people's views.

Students of the city vary enormously in their attitudes towards these issues. At the extremes are those wedded exclusively to either quantitative or qualitative methods. However, Anthony Giddens, one of the world's leading social commentators, has claimed 'All social research, in my view, no matter how mathematical or quantitative, presumes ethnography' (Giddens, 1991, p. 219). In this vein it is suggested that a judicious mixture of both quantitative and qualitative methods is appropriate in studying urban social geography. Qualitative methods can reveal the diversity of voices in the city but maps, graphs and tables, if viewed with sufficient awareness of their limitations, are crucial to reveal a broader picture. Crucially, the latter – together with insights from the 'inside' voices of the city – are needed to effect collective, coordinated policies for social improvement. Both these methods are therefore utilized in the next chapter, in which we consider some of the basic patterns of social differentiation in cities.

Box 3.5

Key films related to urban social geography – Chapter 3

Blade Runner (1984) This is *the* cult movie for all fans of science fiction, *film noir* and debates concerning the nature of postmodernism (see also Box 3.4). Essential viewing for any student of cities.

Lone Star (1995) A detective story set on the Tex-Mex border that deals in a subtle way with issues of history, identity and cultural hybridity. Directed by John Sayles whose other films are also worth seeking out (e.g. *Matewan, Passion Fish, Sunshine State*).

Chapter summary

3.1 Cities play a crucial role in the formation of cultures. These cultures involve 'ways of life' including the values that people hold, the norms that they follow and the material objects that they use.

3.2 All cultures are hybrid mixtures of various influences that change over time and so the notion that there is some pure authentic culture is a myth.

3.3 Space plays a crucial role in the evolution of cultural values since, like culture, it is a social construct intimately bound up with power and authority.

3.4 Although postmodernism is a much disputed concept, many of the recent changes in cities such as the focus upon consumption and growing fragmentation and diversity can be interpreted as a manifestation of the postmodern condition.

Key concepts and terms

aestheticization	hybridity	positional good
ageographia	hyperreality	postcolonial theory
alterity	hyperspace	postmodernism
ambivalence	iconography	public spaces
appropriation	identity(ies)	purified communities
authenticity	imagined communities	reflexivity
binaries	imagineering	representation scripting
borderlands	imperialism	semiology
carceral city	imperialist discourse	signified
colonial discourse	intentionality	signifier
colonialism	landscape of power	signifying practices
constitutive otherness	late modernism	simulacra
creative cities	liminal space	social engineering
creolization	linguistic turn	spaces of exclusion
cultural imperialism	metanarrative	spatialized subjectivities
'cultural turn'	micropowers	subaltern
culture	mimicry	subculture
delocalization	modernism	subject formation
deterritorialization	monumental architecture	subjectivities
deviant subculture	neocolonialism	subject positions
disciplinary society	nomadization	symbolic capital
discourse	objectification	text
Disneyfication	othering	third space
ethnocentrism	Panopticon	totalizing discourse
ethnoscape	place marketing	transculturation
heterotopia	place promotion	

Suggested reading

Cultural studies are immensely popular and have spawned a huge amount of literature in recent years. A key text is Don Mitchell's *Cultural Geography: A Critical Introduction* (2000: Blackwell, Oxford). In addition, an accessible introduction to cultural geography is Mike Crang's *Cultural Geography* (second edition, 2002: Routledge, London). See also Peter Jackson's *Maps of Meaning* (1989: Unwin Hyman, London), Kay Anderson and Fay Gales' *Cultural Geography* (second edition, 1996: Longman, London) and Deborah Stevenson *Cities and Urban Cultures* (2003: Open University Press, Maidenhead). Other collections of essays on cultural geography are James S. Duncan and David Ley (eds) *Place/Culture/Representation* (1993: Routledge, London), Steve Pile and Nigel Thrift (eds) *Mapping the Subject: Geographies of Cultural Transformation* (1995: Routledge, London), Kay Anderson and Fay Gale (eds) *Inventing Places: Studies in Cultural Geography* (1992: Longman Cheshire, Melbourne), Ruth Fincher and Jane M. Jacobs (eds) *Cities of Difference* (1998: The Guildford Press, New York), M. Featherstone and S. Lash *Spaces of Culture: City, Nation, World* (1999: Sage, London), Pamela Shurmer-Smith (ed.) *Doing Cultural Geography* (2002: Sage, London), Sallie Westwood and John Williams (eds) *Imagining Cities* (1996: Routledge, London), James S. Duncan *et al.*

A Companion to Cultural Geography (2003: Blackwell, Oxford), Kay Anderson *et al. Handbook of Cultural Geography* (2002: Sage, London) and Alison Blunt *et al. Cultural Geography in Practice* (2003: Hodder Arnold, London). A series of innovative chapters on identity formation in public spaces can be found in Nicholas Fyfe (ed.) *Images of the Street* (1998: Routledge, London). Also worth seeking out are the series of texts linked to the Open University course 'Understanding Cities': Doreen Massey, John Allen and Steve Pile (eds) *City Worlds* (1999: Routledge London), John Allen, Doreen Massey and Michael Pryke (eds) *Unsettling Cities? Movement/Settlement* (1999: Routledge, London) and *Unruly Cities?: Order/ Disorder* (1999: Routledge, London). On the role of consumption see Neil Wrigley and Michelle Lowe (eds) *Retailing, Consumption and Capital: Towards the New Retail Geography* (1996: Longman, London), Daniel Miller (ed.) *Acknowledging Consumption* (1995: Routledge, London), Daniel Miller, Peter Jackson, Nigel Thrift, Beverly Holbrook and Michael Rowlands *Shopping, Place and Identity* (1998: Routledge, London) and David B. Clarke *Consumer Society and the Post-modern City* (2003: Routledge, London). A clear introduction to culture is C. Jenks *Culture* (1993: Routledge, London) which can be usefully supplemented by Simon During (ed.) *The Cultural Studies Reader* (1998: Routledge, London). Two useful readers are Malcolm Miles *et al.* (eds) *The City Cultures Reader* (2000: Routledge, London) and Ash Amin and Nigel Thrift (eds) *The Blackwell Cultural Economy Reader* (2004: Blackwell, London). A comprehensive guide to issues of identity is the volume edited by Paul Du Gay, Jessica Evans and Peter Redman *Identity: A Reader* (2000: Sage, London). A useful introduction to postcolonial theory is Bill Ashcroft, Gareth Griffiths and Helen Tiffin (eds) *Key Concepts in Post-colonial Theory* (1997: Routledge, London). This can be supplemented by *The Post-Colonial Studies Reader* (1993: Routledge, London) by the same authors. Also of use in this context is Ania Loomba *Colonialism/Postcolonialism* (1998: Routledge, London). An influential extension of postcolonial ideas to the study of cities is Jane M. Jacobs *Edge of Empire: Postcolonialism and the City* (1996: Routledge, London). A thorough examination of the implications of postmodernity can be found in *The Postmodern Urban Condition* (1999: Blackwell, Oxford) by Michael Dear. See also in this context Sophie Watson and Katherine Gibson (eds) *Postmodern Cities and Spaces* (1995: Basil Blackwell, Oxford). Central to the themes of this chapter is Sharon Zukin's *The Cultures of Cities* (1995: Basil Blackwell, Oxford). A fascinating introduction to the idea of the city as a theme park is John Hannigan's *Fantasy City: Pleasure and Profit in the Postmodern Metropolis* (1998: Routledge, London). The phenomenon of cities 'selling' themselves to investors and tourists is discussed in Gerry Kearns and Chris Philo (eds) *Selling Places: The City as Cultural Capital, Past and Present* (1993: Pergamon Press, Oxford).

Patterns of sociospatial differentiation

In parallel with the tradition of regionalization within the discipline as a whole, urban social geographers have sought to 'regionalize' towns and cities in attempts to produce high-level generalizations about urban form and structure. These generalizations can be thought of as capturing the outcome, at a particular point in time, of the sociospatial dialectic. They provide useful models with which to generate and test hypotheses and theories concerning urban growth processes and patterns of social interaction in cities. Moreover, many geographers regard the analysis of **areal differentiation** within cities as a basic task of geographical analysis (like regionalizations of countries or continents), providing an overall descriptive synthesis that is of fundamental utility in its own right. Whatever the perspective, the initial objective is to identify areas within cities that exhibit distinctive characteristics and that can be shown to be relatively homogeneous. The word *relatively* is crucial here for, as Chapter 3 indicated, we need to be careful, when labelling such areas, not to assume that all the inhabitants share similar characteristics, for there is diversity and difference in even the most homogeneous-looking urban areas. Nevertheless, cities do display distinctive spatial patterns. In this chapter, we shall establish the fundamental patterns that occur in both the physical and the socioeconomic dimensions of contemporary cities and describe them from a variety of perspectives.

4.1 Urban morphology and the physical structure of cities

The study of the physical qualities of the urban environment is one of the longest-established branches of urban geography, especially in Europe, where the study of 'townscapes' and 'morphological regions' has occupied a prominent place in urban studies.

House types, building lots and street layouts

To a large extent, morphological patterns are based on two fundamental elements: the size and shape of plots of land, and the layout of streets. Both vary according to historical period, economics and socio-cultural ideals. Where there is a shortage of building land, or where as many buildings as possible have to be accommodated along a given frontage (as on a waterfront or around a market square), small, deep plots tend to result. Elsewhere, the size and form of the plot tend to be determined by the predominant house type (for example, the towns of England, the Netherlands and the north German coast were historically characterized by small, rational housetypes that required small plots with only a narrow (5-metre) frontage; the standard nineteenth-century tenement building of American cities required only a 9-metre frontage; whereas the standard apartment house needed 32 metres of frontage; Curdes, 1993).

In general, just a few major forms have come to be dominant (Moudon, 1992):

➤ The fundamental morphological framework provided by plots of land has traditionally been *rectilinear*, because of the ease of surveying, the efficiency of use of space, and the ease of laying out streets (Fig. 4.1). It follows that city blocks and street layouts have also tended to be rectilinear (Fig. 4.2).

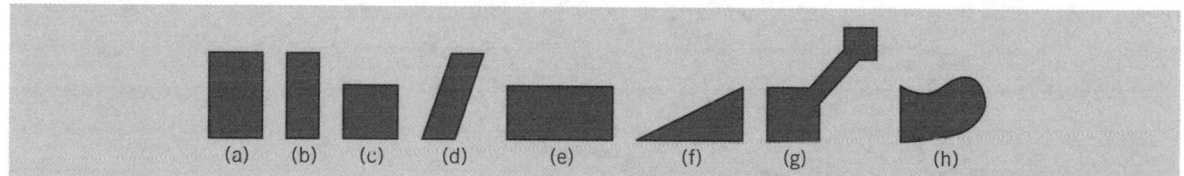

Figure 4.1 Plot forms in European cities, in order of frequency: (a) rectangular-deep, (b) narrow-deep, (c) square, (d) parallelogram-deep, (e) rectangular-wide, (f) triangular, (g) many-sided, different combinations, (h) irregular.
Source: Curdes (1993), Fig. 14.1, p. 282.

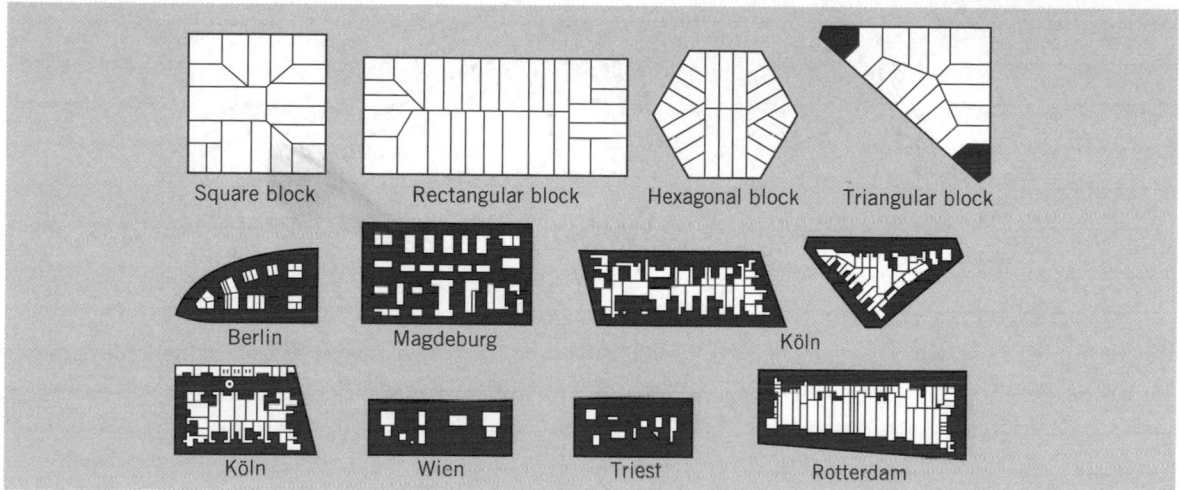

Figure 4.2 Block forms in European cities.
Source: Curdes (1993), Figs 14.2 and 14.3, p. 282.

The dominance of this form eventually provoked a reaction.

➤ The continuous *curvilinear* form, with wide, shallow plots that reinforce the impression of spaciousness (Fig. 4.3). The origins of this form can be traced to the Romantic suburbs of the mid- to late nineteenth century (e.g. the Chicago suburb of Riverside, designed by F.L. Olmsted); it became widespread in the United States in the 1930s (as the newly created Federal Housing Administration promoted its virtues) and in Europe in the 1950s (as developers of suburbia copied the American model to which consumers aspired).

➤ A refinement of the curvilinear form is the *loop road with culs-de-sac* (Fig. 4.4), an attempt to retain the aesthetics of curvilinear layouts while mitigating the nuisance and dangers of automobile traffic. Such subdivisions can be traced to the Radburn garden city model in New Jersey that was sponsored in the 1930s by the Regional Planning Association of America. Since the 1970s, this has been the dominant form of suburban morphology in most developed countries. The loop road provides access to arterial through-streets, while the houses (typically semi-detached dwellings on zero-lot-line plots or garden apartments with no internal plot boundaries) face away from the loop, served by culs-de-sac.

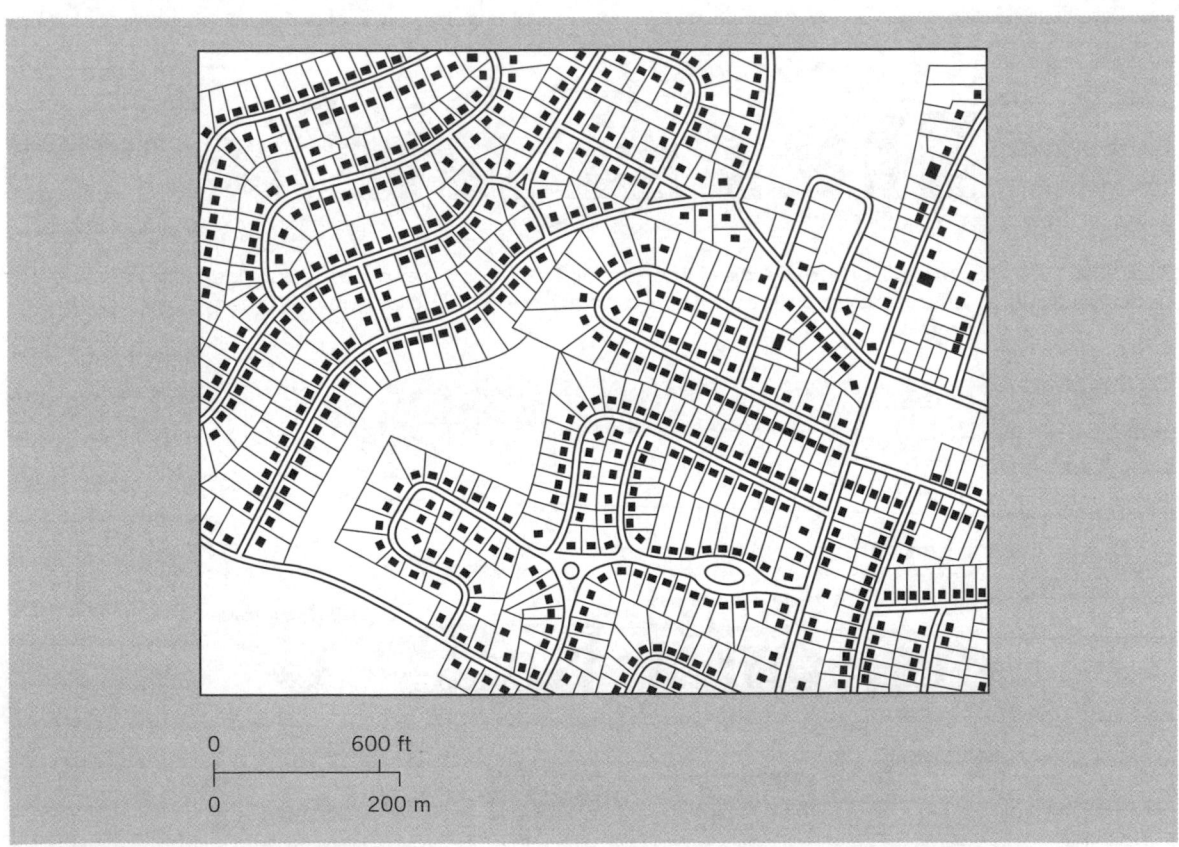

Figure 4.3 An example of the continuous curvilinear form of street network, with wide, shallow plots.
Source: Moudon (1992), Fig. 7.4, p. 176.

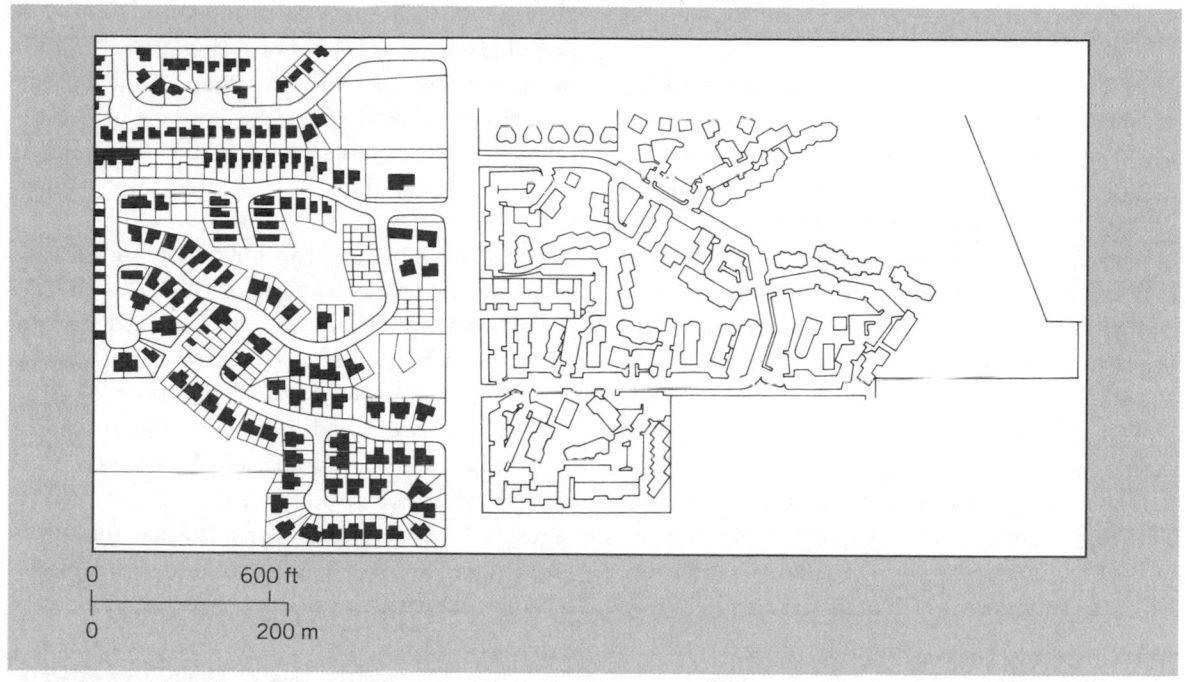

Figure 4.4 An example of the loop road street network, with (left) zero lot-line plots and (right) garden apartments.
Source: Moudon (1992), Fig. 7.5, p. 177.

The ever-changing physical fabric of cities: abandoned housing in Dundee. Photo Credit: Paul Knox.

Morphogenesis

Morphogenesis refers to the processes that create and reshape the physical fabric of urban form. Over time, urban morphology changes, not only as new urban fabric is added but also as existing fabric is modified. Basic forms, consisting of house, plot and street types of a given period, become hybridized as new buildings replace old, plots are amalgamated or subdivided, and street layouts are modified. Each successive phase of urban growth is subject to the influence of different social, economic and cultural forces, while the growth of every town is a twin process of outward extension and internal reorganization. Each phase adds new fabric in the form of accretions and replacements. The process of outward extension typically results in the kind of annular patterns of accretion shown in Fig. 4.5. The process of reorganization is illustrated

by Fig. 4.6, which shows the changes that occurred in part of Liverpool as institutional land users (including the University, the Roman Catholic Cathedral and hospitals and clinics) encroached into the nineteenth-century street pattern, replacing the fine grain of residential streets with a much coarser fabric of towers and slab blocks.

The oldest, innermost zones of the city are especially subject to internal reorganization, with the result that a distinctive morphological element is created, containing a mixture of residential, commercial and industrial functions, often within physically deteriorating structures. Small factories and workshops make an important contribution to the ambience of such areas. Some of these factories may be residual, having resisted the centrifugal tendency to move out to new sites, but a majority are 'invaders' that have colonized sites vacated by earlier industries or residents. Typically, they occupy

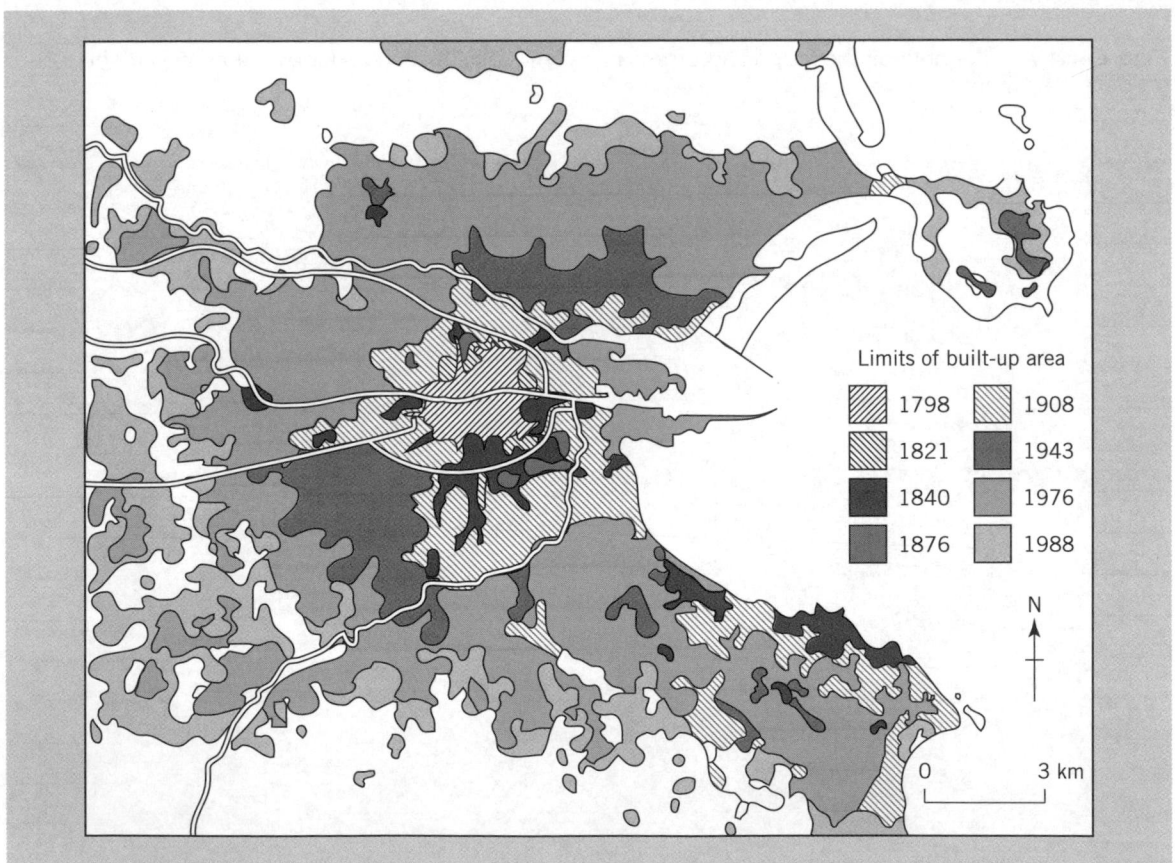

Figure 4.5 Growth phases in Dublin.
Source: MacLaran (1993), Fig. 2.9, p. 42.

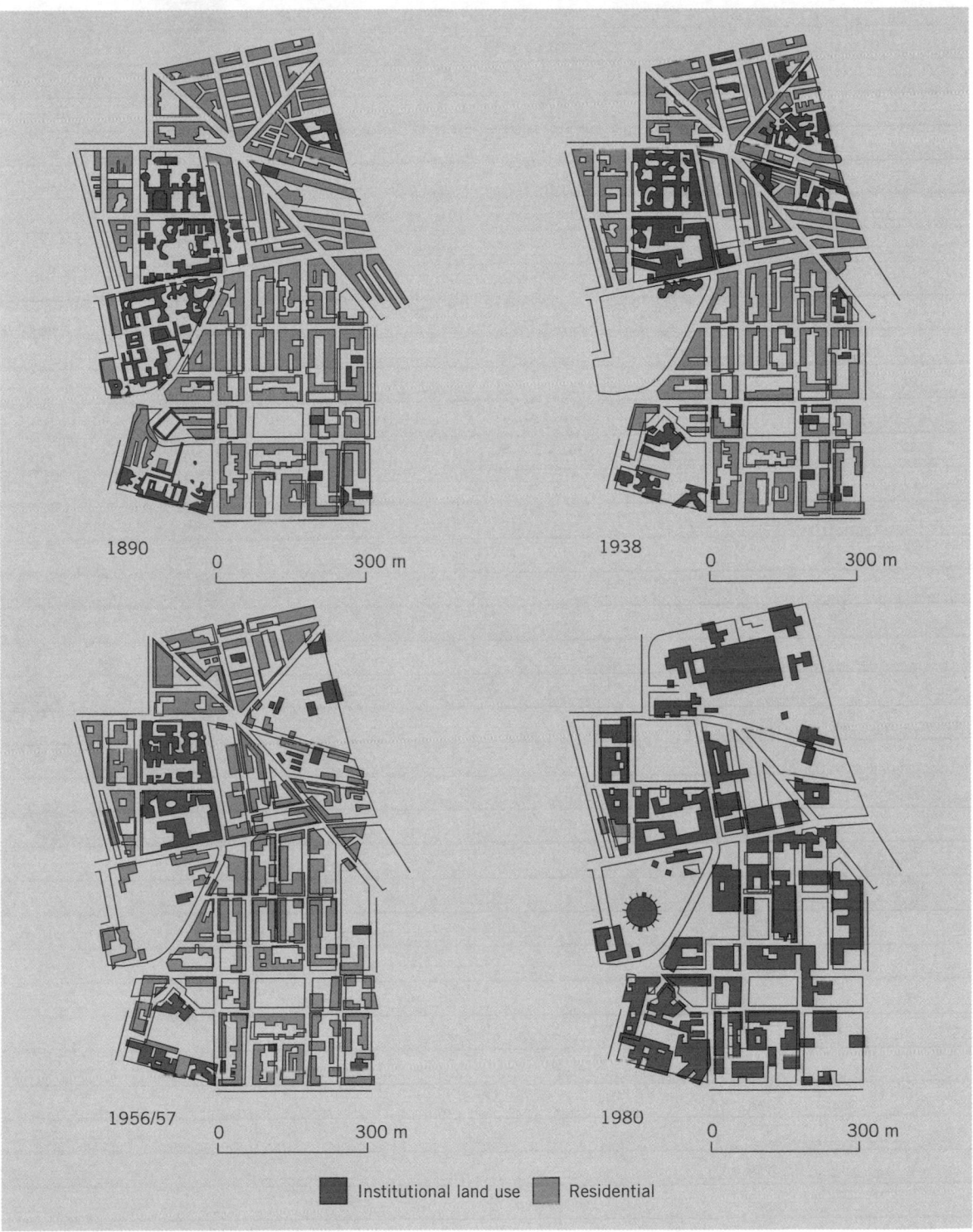

1890 0   300 m

1938 0   300 m

1956/57 0   300 m

1980 0   300 m

■ Institutional land use   ■ Residential

Figure 4.6 Morphological reorganization of the university district in Liverpool, 1890–1980.
Source: Chandler *et al* (1993), Fig. 5.3, p. 112.

old property that has become available in side streets off the shopping thoroughfares in the crowded but decaying residential zone surrounding the CBD.

Beyond this inner zone, industrial, commercial and residential morphological elements tend to be clearly differentiated, although typically arranged in an imperfect zonation, interrupted by radial arteries of commercial and industrial development and by major roads and railway tracks, and distorted by the peculiarities of site and situation. In addition, a good deal of urban development is characterized by the persistence of enclaves of relict morphological units (e.g. castles, cathedrals, university precincts, boulevards, public parks and common lands, all of which tend to resist the logic of market forces and so survive as vestigial features amid newly developed or redeveloped neighbourhoods). These relict units tend to impair the symmetrical pattern that may otherwise emerge.

Even in relatively new and homogeneous suburban residential areas a good deal of morphological reorganization can take place. Figure 4.7 shows the extent of morphological reorganization (between the mid-1950s and the late 1980s) in sample sites taken from both the inner and the outer reaches of suburban neighbourhoods of towns in southeast England. In this case, reorganization resulted from pressure for more intensive residential development that arose because of a combination of a reduction in the average size of households and increases in population, employment opportunities and incomes in southeast England (Whitehand, 1992). The result, reflected in all of the sample sites shown in Fig. 4.7, is a considerable degree of plot subdivision (the truncation of corner plots, the subdivision of original parcels, or both) and redevelopment (including the residential development of previously non-residential land). Such patterns represent

Figure 4.7 Plot subdivision and redevelopment in low-density residential areas in sample towns in southeast England, *c.* 1995–1986.
Source: Whitehand (1992), Fig. 4.4, p. 141.

an important dimension of change – piecemeal infill and redevelopment – that is central to the sociospatial dialectic of the ageing suburbs of cities throughout the developed world.

The sociospatial dialectic is indeed an important aspect of morphogenesis. Over the broader sweep of time, morphogenesis is caught up in the continual evolution of norms and aesthetics of power, space and design. Successive innovations in urban design (Table 4.1) are not only written into the landscape in the form of extensions and reorganizations but also come to be symbolic of particular values and attitudes that can be evoked or manipulated by subsequent revivals or modifications. Within this context, innovations in transport technology are of particular importance, since they not only contribute to the evolution of the norms and aesthetics of power, space and design (as in the development of subdivisions based on culs-de-sac and loop roads in

response to the intrusiveness of automobiles) but also exert a direct influence on the overall physical structure of urban areas (Goodman and Chant, 1999). Major innovations in transport technology (the railway, the streetcar, rapid transit, automobiles, trucks and buses) have the effect of revolutionizing urban structure because they allow for radical changes in patterns of relative accessibility. The result is a pattern of physical development that is the product of successive epochs of transport technology. These have been identified in US cities (Fig. 4.8) as:

➤ the pre-rail (before 1830) and 'iron horse' (1830–1870) epochs;

➤ the streetcar epoch (1870–1920); and

➤ the auto–air–cheap oil epoch (1920–1970) and jet propulsion–electronic communication epoch (1970–).

Table 4.1 Innovations in urban design

Time	Innovation	Location
1100–1500	Medieval irregular towns	Middle Europe
1200–1400	Medieval regular towns	France, south-west Germany, Baltic Sea, east of Elbe
1500–1700	Renaissance town concepts	Italy, France, Germany, USA
1600–1900	Baroque town concepts	Rome, Paris
1800–1830	Classical grid/block reverting to renaissance principles	Krefeld, Prussia
1800–1880	Geometric town design	Middle Europe
1850–1900	Haussmann: axis concept, circus, triangle, boulevard, point de vue	Paris
1857	Ring concept	Vienna, Cologne
1889–1930	Sitte, Henrici, Unwin: artistic movement	Austria, Germany, UK
1898–1903	Howard, Parker and Unwin: Garden city	Letchworth
1902–1970	Garden City movement	World-wide
1900–1930	Modern blocks	Netherlands, Germany
1920–1930	Corbusier, Taut, May, Gropius: Rationalism and 'Neues Bauen'	France, Germany
1930–1945	Fascist neoclassicism	Italy, Germany
1945–1975	Flowing space and free	
1975 to date	Reurbanization: reverting to block systems	Europe
1975 to date	Postmodernism	World-wide
1985 to date	Deconstructivism	Western world

Source: Curdes (1993), Table 14.2, p. 287.

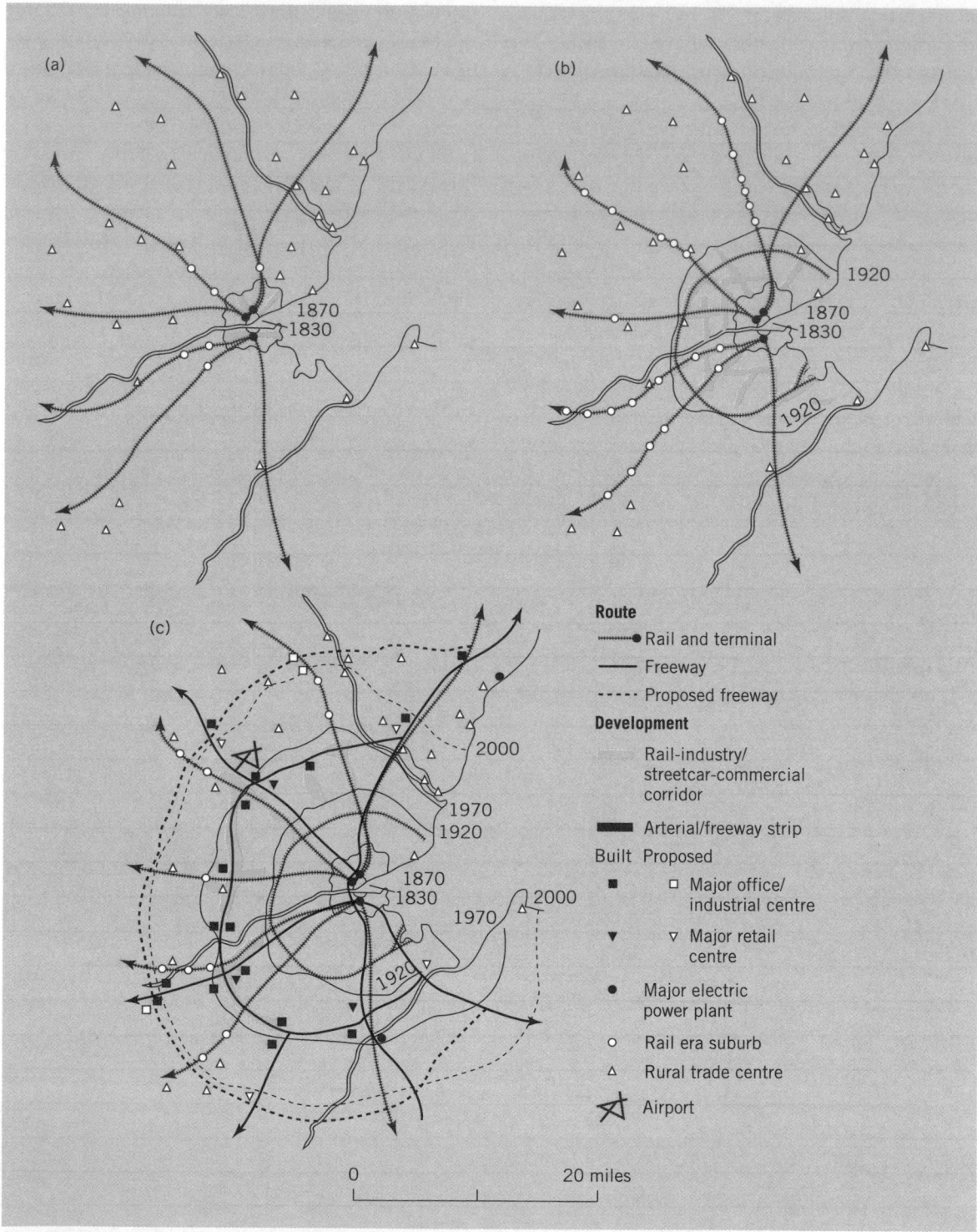

Figure 4.8 Schematic maps of development patterns and age rings in a 'generic' high-order metropolitan area: (a) to 1870; (b) 1870–1920; (c) 1970– .

Source: Borchert (1991), Fig. 12.21.

Environmental quality

One specific dimension of the built environment that is worth special consideration from a social perspective is that of spatial variations in environmental quality. Implicit in the previous sections is the fact that not only are different urban sub-areas built to different levels of quality and with different aesthetics, but that at any given moment some will be physically deteriorating while others are being renovated and upgraded. Because they are closely tied in to the sociospatial dialectic through patterns and processes of investment and disinvestment (Chapter 6) and of social segregation (Chapter 8), the qualitative dimension of the built environment tends to exhibit a considerable degree of spatial cohesion.

Take, for example, the patterns of physical upgrading and downgrading in Amsterdam (Fig. 4.9), where the stability of the outermost sub-areas contrasts with the renewal and upgrading of much of the inner, nineteenth-century residential districts and in most of the central neighbourhoods along the canals, where private housing predominates. In this example, downgrading is very limited in extent, being restricted to four small sub-areas adjacent to older industrial works. In some cities and metropolitan areas, however, physical decay and substandard housing is a serious problem. Figure 4.10 shows the extent to which the northeastern sector of inner Paris is riven with substandard housing; while Fig. 4.11 shows the highly localized impact of urban decay in New York City, where in parts of the Bronx (Fig. 4.11b) some sub-areas lost between 50 and 80 per cent of their occupied housing units within a brief (ten-year) but devastating period.

The actual condition of streets and buildings is an aspect of the built environment that has been of increasing interest to town planners and community

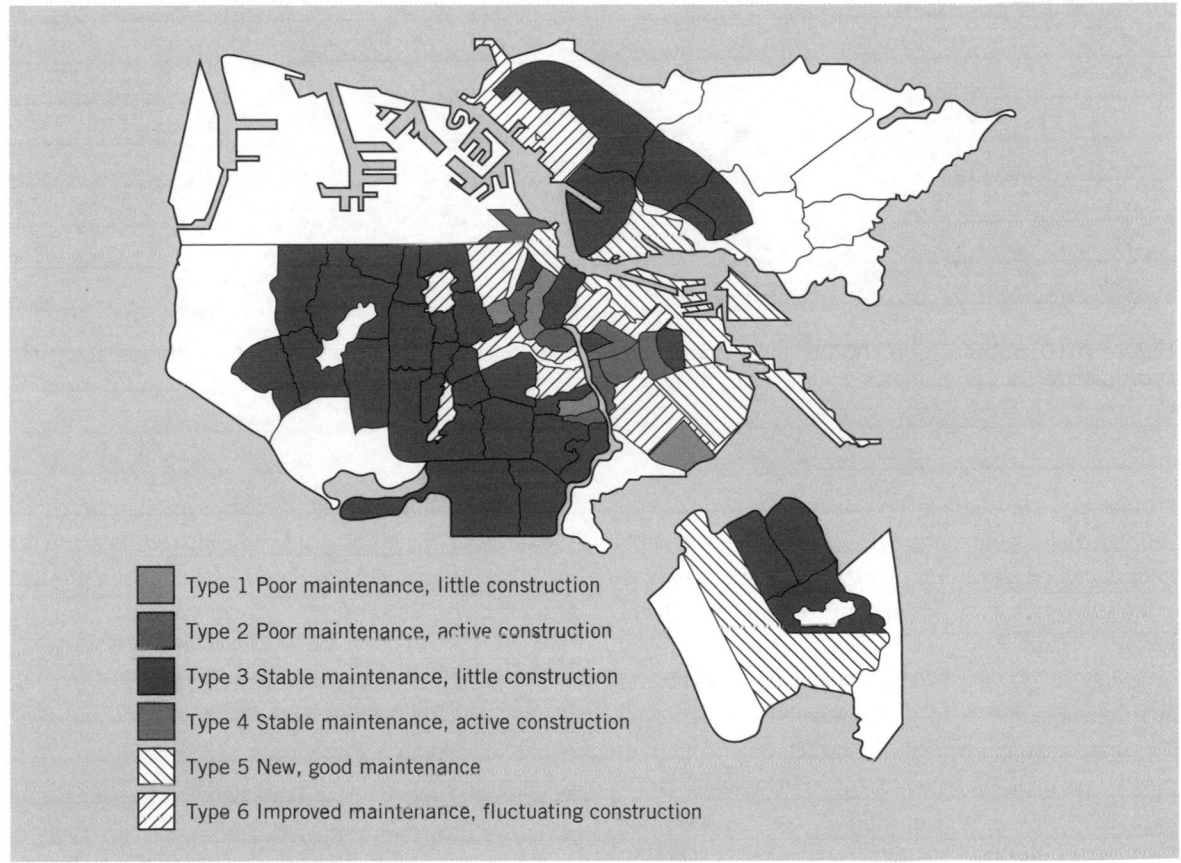

Type 1 Poor maintenance, little construction
Type 2 Poor maintenance, active construction
Type 3 Stable maintenance, little construction
Type 4 Stable maintenance, active construction
Type 5 New, good maintenance
Type 6 Improved maintenance, fluctuating construction

Figure 4.9 Physical upgrading and downgrading in Amsterdam.
Source: Musterd (1991), Fig. 5, p. 37.

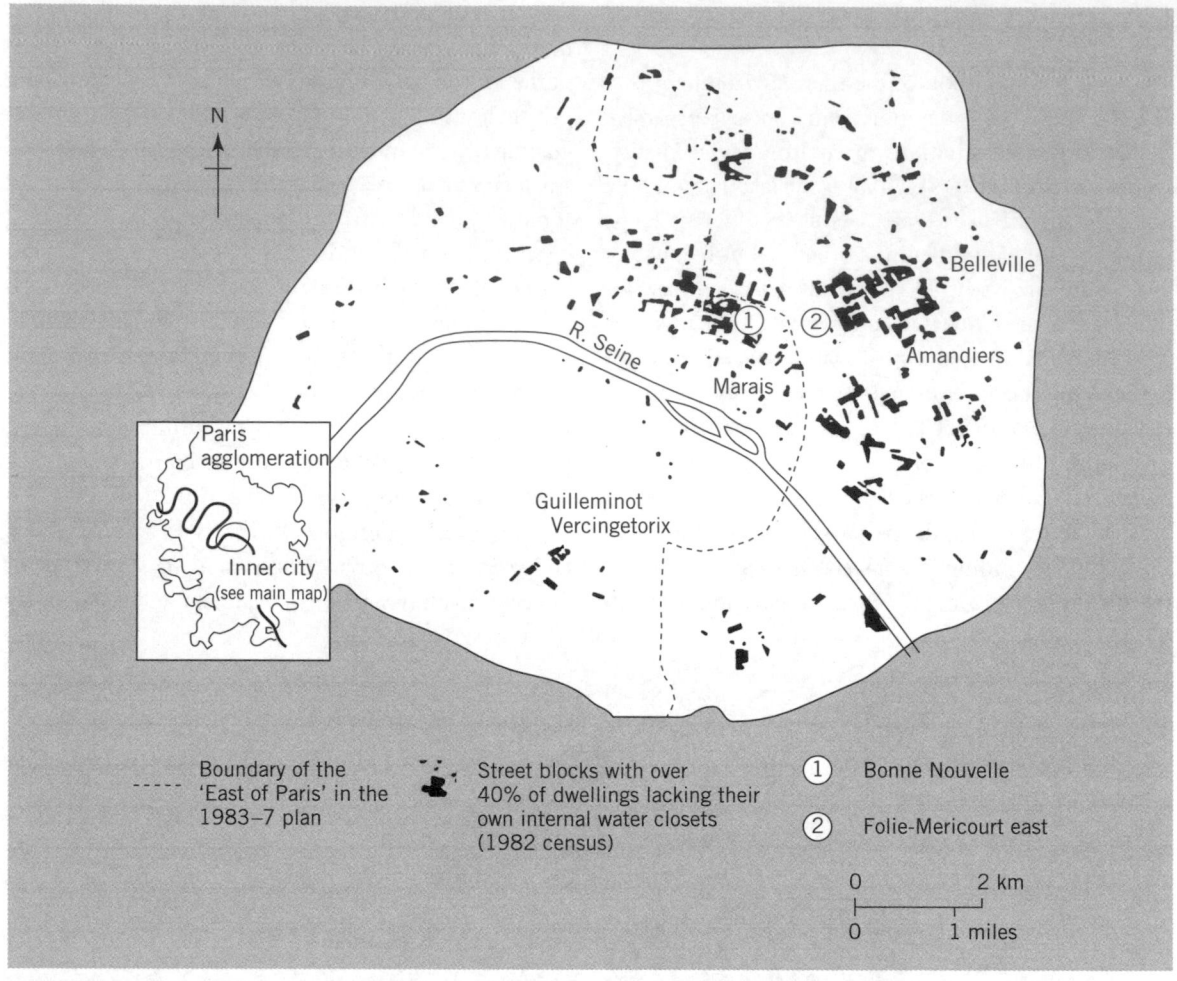

Figure 4.10 Substandard housing in Paris: street blocks with over 40 per cent of dwellings lacking their own WCs.
Source: White and Winchester (1991), Fig. 1, p. 41.

groups as well as to geographers. We should note, though, that while certain aspects of environmental quality can be measured objectively, other aspects are quite subjective. Also, environmental quality is highly income-elastic. The less well-off, with more urgent needs to satisfy, may well be relatively unconcerned about many aspects of environmental quality; while the rich, having satisfied their own material needs, may be particularly sensitive to environmental factors such as the appearance of houses, streets and gardens. Despite such problems, some researchers have attempted to analyze the determinants and patterns of environmental quality using questionnaire data. A study of a

low-status public housing district in Glasgow (Pacione, 1982) concluded that residents' satisfaction with the quality of their neighbourhood was largely a product of their perceptions of:

➤ traffic problems,

➤ street cleanliness and maintenance,

➤ accessibility to open space,

➤ antisocial activity (e.g. vandalism, roving dogs),

➤ accessibility within the city as a whole,

➤ social interaction, and

➤ landscaping.

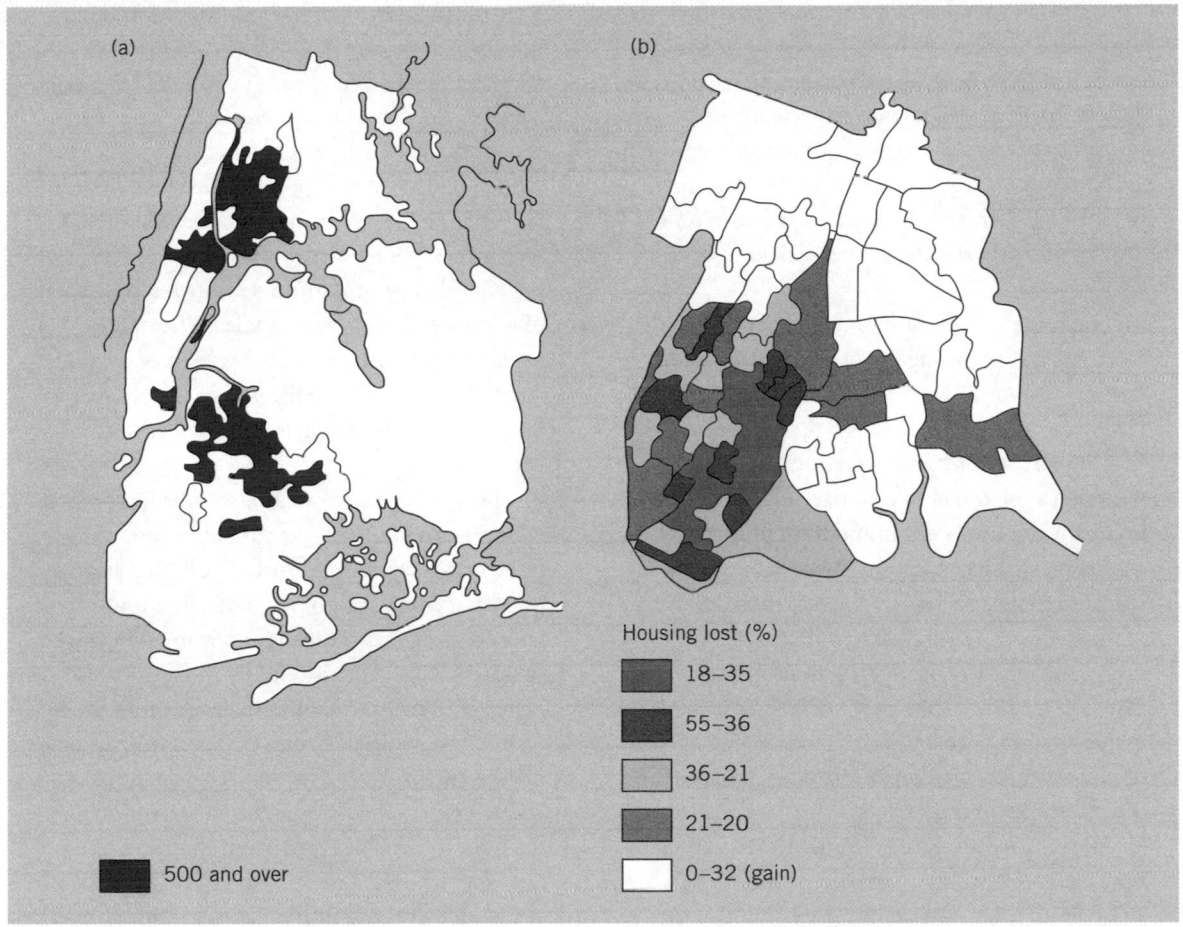

Figure 4.11 Urban decay: (a) census tracts in New York City that each lost 500 or more occupied housing units between 1970 and 1980; (b) percentage loss of occupied housing units in the Bronx, 1970–1980.
Source: Wallace (1989), Fig. 4 (part), p. 1590; Wallace and Fullilove (1991), Fig. 2, p. 1703.

Meanwhile, an analysis of data from the US Annual Housing Survey has shown overall satisfaction with neighbourhood environmental quality to be inversely related to centrality and to city size, with low levels of satisfaction being a function, in particular, of the presence of rundown housing units, abandoned buildings, litter, street crime, air pollution, street noise and heavy traffic (Dahmann, 1985). These, of course, are aggregate findings: different types of household, with different backgrounds and values, react in different ways to the same kind of environment. Younger, married female heads, blacks and people with many children, for example, tend to express significantly lower levels of satisfaction in any residential context (Galster and Hester, 1981).

Townscapes and the *genius loci* of the built environment

Morphological regions are not simply the sum of the attributes of building style and function, plot and street layout, and environmental quality: it is the form in which these morphological elements and attributes are set in relation to each other and to unbuilt spaces that creates the *genius loci* which lends distinctiveness to one morphological region in comparison with another. In many ways, this distinctiveness can only be captured subjectively, articulated by written descriptions rather than formal analysis.

71

The strength of written description in conveying the flavour and character of particular parts of the city is illustrated in these brief extracts from Richard Sennett (1990), who emphasizes the subjectivity of his own responses to places. The following extracts are from his description of his regular walk from his apartment, in New York's Greenwich Village, to the restaurants in midtown on the East Side, where he likes to eat:

> To reach the French restaurants I have to pass from my house through a drug preserve just to the east of Washington Square. Ten years ago junkie used to sell to junkie in the square and these blocks east to Third Avenue. In the morning stoned men lay on park benches, or in doorways; they slept immobile under the influence of the drugs, sometimes having spread newspapers out on the pavement as mattresses. . . . The dulled heroin addicts are now gone, replaced by addict-dealers in cocaine. The cocaine dealers are never still, their arms are jerky, they pace and pace; in their electric nervousness, they radiate more danger than the old stoned men. . . .
>
> Along Third Avenue, abruptly above Fourteenth Street, there appear six blocks or so of white brick apartment houses built in the 1950s and 1960s on the edges of the Grammercy Park area; the people who live here are buyers for department stores, women who began in New York as secretaries and may or may not have become something more but kept their jobs. Until very recently, seldom would one see in an American city, drinking casually in bars alone or dining quietly with one another, these women of a certain age, women who do not attempt to disguise the crowsfeet at the edge of their eyes; for generations the blocks here have been their shelter. It is a neighborhood also of single bald men, in commerce and sales, not at the top but walking confidently through to the delis and tobacco stands lining Third Avenue. All the food sold in shops here is sold in small cans and single portions; it is possible in the Korean groceries to buy half a lettuce. . . .
>
> Unfortunately, in a few minutes of walking this scene too has disappeared, and my walk now takes an unexpected turn. The middle Twenties between Third and Lexington is the equestrian center of New York, where several stores sell saddles and Western apparel. The clientele is varied: polo players from the lusher suburbs, Argentines, people who ride in Central Park, and then another group, more delicate connoisseurs of harnesses, crops, and saddles. The middle twenties play host as well to a group of bars that cater to these leather fetishists, bars in run-down townhouses with no signs and blacked-out windows. . . .
>
> The last lap of my walk passes through Murray Hill. The townhouses here are dirty limestone or brownstone; the apartment buildings have no imposing entrance lobbies. There is a uniform of fashion in Murray Hill: elderly women in black silk dresses and equally elderly men sporting pencil-thin mustaches and malacca canes, their clothes visibly decades old. This is a quarter of the old elite in New York. . . . The center of Murray Hill is the Morgan Library, housed in the mansion at Thirtysixth and Madison of the capitalist whose vigor appalled old New York at the turn of the century. . . . Near the Morgan Library is B. Altman's, an enormous store recently closed which was regularly open in the evenings so that people could shop after work. One often saw women, of the sort who live nearby in Grammercy Park, shopping for sheets there; the sheet-shoppers had clipped the advertisement for a white sale out of the newspaper and still carried it in their unscuffed calf handbags; they were hardworking, thrifty.
>
> (Sennett, 1990, pp. 123–4, 125, 130)

What is most striking about these extracts is the importance of *people* in giving character to place. To Sennett's eye, the squares, streets, apartment houses, townhouses, bars, stores and institutions along his walk are given character and meaning by people – the addict-dealers, the 'single bald men' and the 'women of a certain age' – and vice versa. This is a theme that we shall explore systematically in Chapters 7 and 9.

Qualitative methods and observational fieldwork in urban areas

The growing influence of cultural studies (see Chapter 3) has led to an ever-increasing use of qualitative methods to assess the nature of urbanism and the distinctive 'spirit' of place. Table 4.2 shows a typical example of the sorts of data that might be collected via such a qualitative approach. These qualitative methods are often – somewhat mistakenly – thought of as less useful than quantitative methods. However, if used rigorously, qualitative methods can enable the researcher to take a serious look at the assumptions underpinning the methods being used in the analysis (Burgess and Jackson, 1992). In particular, qualitative methods raise questions about the values inherent in the choice of data being collected together with the values of those being observed *and* those undertaking the observation – what has been termed by Anthony Giddens the **double hermeneutic**. Key questions include:

➤ *Why* is this particular method being used and does it raise ethical issues?

➤ *Who* is being observed?

➤ *Where* will be observation take place, in one locale or several?

➤ *What* aspects of behaviour will be observed?

➤ *When* will this observation take place?

➤ *How* are the subjects of the study to be engaged, covertly (in the role of a 'spy' or 'voyeur') or overtly (in the role as a 'member' or 'fan' of the group)?

Qualitative observation schedules such as that listed in Table 4.2 can provide valuable insights into the dominant assumptions underpinning the behaviour expected in particular spaces, the role of surveillance, the complex links between a local space and other more distant spaces and temporal variations in behaviour patterns.

Table 4.2 An example of an ethnographic data-gathering schedule

Southampton 'Streetwork' Study

Student name: Location of study: Times of study:

Use the sections listed below to focus your ethnographic enquiry

1. The built environment
(i) What kinds of buildings are located in this area and what types of cultural values do they convey?
(ii) Are there any monuments in your space and, if so, what kinds of cultural values do they convey?
(iii) Are there any signs of resistance to these values (e.g. graffiti)?

2. A global 'sense of place'?
(i) In what ways does the physical environment indicate wider networks and connections beyond the city?
(ii) What effects do these wider networks and connections appear to have on the identity of your study area?

3. Spaces of power
(i) What are the sources and symbols of power and authority in your space?
(ii) What sorts of behaviour might be 'out of place' here?
(iii) What sort of surveillance/policing is present in this space?

4. People and place
(i) What kinds of people occupy your space, and what are they doing?
(ii) How are these people dressed?
(iii) What do the appearances and practices of these people suggest about their social and cultural lives?

5. Sensing place
(i) What is the mood/atmosphere of your space (pay attention to sights, sounds, smells, etc.)?
(ii) What are your feelings as you observe the area and its cultural life?
(iii) How do you feel as you participate in the cultural life of the area?
(iv) Do you feel that you are altering the place as you study it?

Source: University of Southampton Wessex Fieldcourse, March 1999 (authors Alex Hughes and Steven Pinch).

Box 4.1

Key debates in urban social geography – The role of non-representational theory

The continually changing nature of methodological approaches in human geography is illustrated by one of the most recent developments – non-representational theory. This approach has a very different view as to what constitutes knowledge compared with many of the well-established approaches in the social sciences. In essence, non-representational theory argues that much of the knowledge that we use as we participate in everyday city life cannot be directly written down and interpreted with formal logic. This type of knowledge is often termed tacit knowledge, and contrasts with knowledge that can be formally expressed, known as codifiable knowledge. Everyday tasks that are full of tacit knowledge include driving, cycling, cooking and dancing. In addition to that which cannot be formally expressed, tacit knowledge is also envisaged as consisting skills of which the exponent is often not fully conscious.

Non-representational theories therefore focus upon the 'how' rather than the 'why' of city life. Social life is conceptualized as an ongoing process, a set of 'performances', based on both codifiable and tacit knowledge, undertaken in a similar fashion every day, yet always with the possibility of experimentation and change. This type of work is relatively new but is part of a reconceptualization of many parts of the social sciences that attempts to understand social life through a series of metaphors stressing fluidity (such as flow, circulation, networks and movement). It follows from this approach that phenomenon such as cities, firms and organizations are conceived, not in static terms, but as the centres of networks held together by continuous flows of money, people, resources and knowledge. This type of thinking has been applied to cities in Amin and Thrift's path-breaking book *Reimagining the Urban* (2002):

> The city has no completeness, no centre, no fixed parts. Instead, it is an amalgam of often disjointed processes and social heterogeneity, a place of near and far connections, a concatenation of rhythms, always ending in new directions.
> (Amin and Thrift, 2002, p. 8)

Key concepts related to non-representational theory (see Glossary)

Performativity, subjectivities.

Further reading

Environment and Planning A (2003) Vol. 35, Special Volume edited by Latham, A. and Conradson, D. on Making Place, Performance, Practice and Space

McCormack, D. (2002) A paper with an interest in rhythm *Geoforum*, **33**, 469–485

Nash, N. (2000) Performativity in practice: recent work in cultural geography *Progress in Human Geography*, **24**, 653–664

Smith, S. (2000) Performing the (sound)world *Environment and Planning D*, **18**, 615–637

Thrift, N. (2000) Afterwords *Environment and Planning D*, **18**, 213–255

Links with Other Chapters

Chapter 1: Different Approaches within Human Geography

Chapter 2: Box 2.2 Manuel Castells

Chapter 9: Box 9.4 Nigel Thrift

Chapter 13: Box 13.2 Automobility

4.2 Difference and inequality: socioeconomic and sociocultural patterns

As might be expected given the increasing complexity and social polarization of Western cities discussed in Chapters 2 and 3, a major theme of urban social geography is the spatial patterning of difference and inequality. In detail, such patterns can present a kaleidoscope of segregation, juxtaposition and polarization:

> At one end of Canon St. Road, London E1, you can pay £4 for a two-course meal. At the other end of the street, less than 500 metres away, the same amount of money will buy a single cocktail in Henry's wine bar in a postmodern shopping mall come upmarket residential development. The very urban fabric here, as in so many other

cities across the globe, has altered at a feverish rate in the past decade.

The street runs south from the heartland of the rag trade and clutter of manufacturing, retail and wholesale garment showrooms on Commercial Road. Residentially, the north end is occupied almost exclusively by the Bengali community in one of the poorest parts of any British city. Three hundred yards south, the road crosses Cable Street, a short distance away from a mural commemorating a defiant Jewish community confronting Moseley's fascist Blackshirts in 1926, the caption 'they shall not pass' now addressed to the adjacent gentrified terrace. A few hundred yards further and the microcosm is completed by Tobacco Dock, cast as the 'Covent Garden of the East End,' although suffering badly in the depression of the early 1990s.

The leitmotif of social polarization is unavoidable. Golf GTIs share the streets uneasily with untaxed Ford Cortinas. Poverty is manifest, affluence is ostentatious. Gentrification sits beside the devalorization of old property. The appeals for information in the police posters tell of yet another racist attack, just as the graffiti with which they are decorated demonstrate the credence given locally to the powers of police investigation.

(Keith and Cross, 1992, p. 1)

Seen in broader perspective, patterns of inequality and spatial differentiation exhibit a certain regularity that is often consistent from one city to another. In societies based on the competition and rewards of the marketplace, personal income is probably the single most significant indicator, implicated as it is with people's education, occupation, purchasing power (especially of housing), and with their values and attitudes towards others. It has long been recognized that the geography of income within cities is characterized not only by steep gradients and fragmented juxtapositions at the microlevel but also by clear sectors dominated by high-income households and by sinks of inner-city poverty. Consider, for example, the map of incomes in the Tucson metropolitan area (Figure 4.12), where in

1999 the median family income in the affluent northern suburbs was between four and six times the median family income of inner-city census tracts.

While socioeconomic differentiation is arguably the most important cleavage within contemporary cities, it is by no means the only one. Demographic attributes such as age and family structure are also of central importance to social life, yet are only loosely related, if at all, to differences in socioeconomic status. There are, however, clear patterns to the geodemographics of cities – in large part because of the tendency for certain household types to occupy particular niches within the urban fabric. Thus, for example, families with preschoolchildren are typically found in disproportionately high numbers in new, peripheral suburban subdivisions and apartment complexes; the elderly, on the other hand, typically tend to be concentrated as a residual population in older, inner-city residential neighbourhoods.

Embedded in the sociospatial framework delineated by the major stratifications of money and demographics are the marginalized subgroups of contemporary society. The idea of marginality is of course a relative concept, and depends on some perceived norm or standard. Hilary Winchester and Paul White (1988) suggested that these norms and standards can be economic, social and/or legal.

They identify four groups of the *economically marginal*:

➤ the unemployed, particularly the long-term unemployed;

➤ the impoverished elderly;

➤ students; and

➤ single-parent families.

In addition, they identify another three groups that can be categorized as *both economically and socially marginal*, the two dimensions generally reinforcing one another:

➤ ethnic minorities;

➤ refugees; and

➤ the handicapped (either mentally or physically), and the chronically sick (notably including people with AIDS).

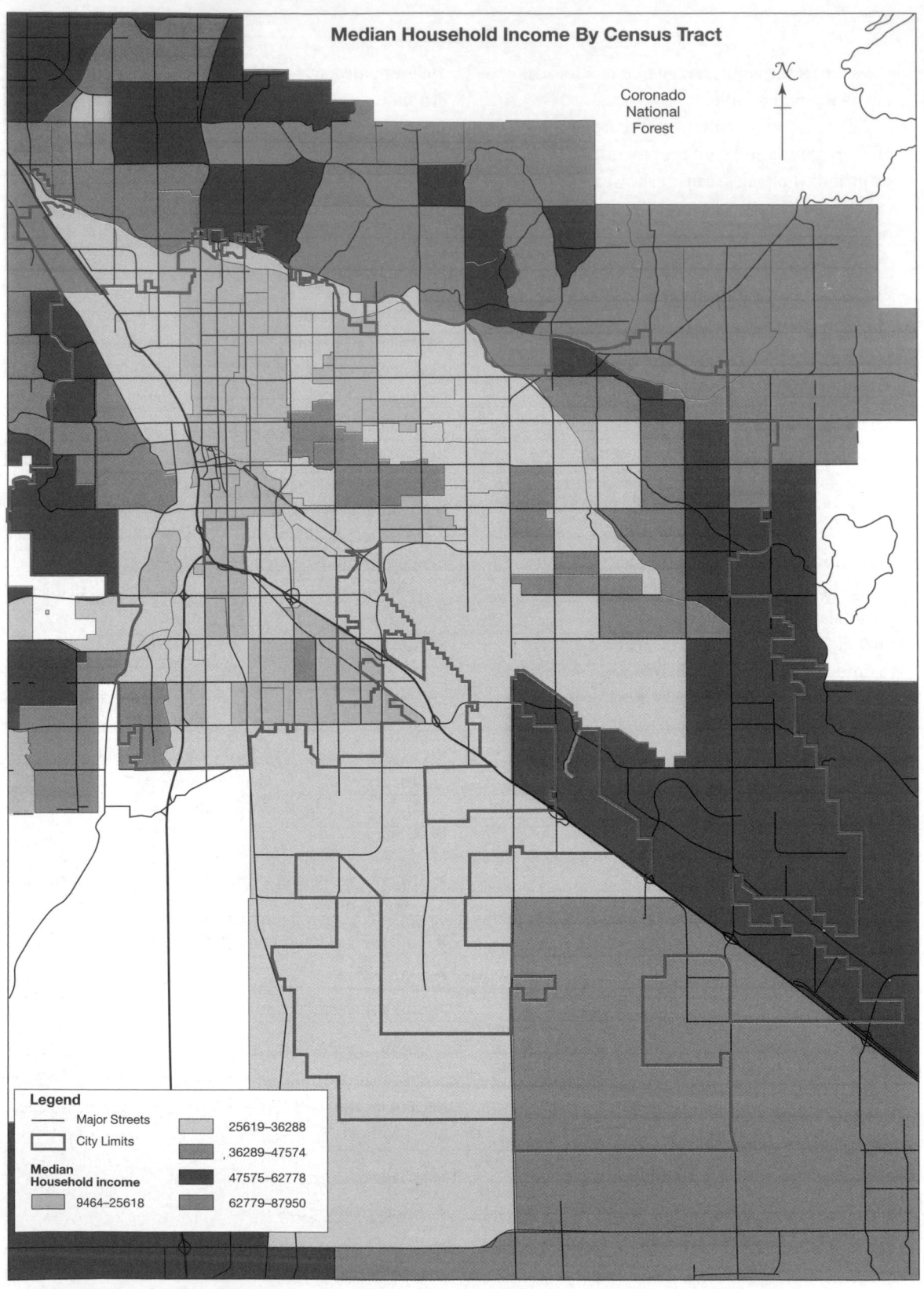

Figure 4.12 Persons aged 65 and over living in poverty, 2000, Atlanta.
Source: McCullen and Smith (2004) Census Issue 7, p. 30.

The remaining marginalized groups are marked by elements of *legal* as well as economic and/or social marginality:

➤ illegal immigrants;

➤ down-and-outs;

➤ participants in drug cultures;

➤ petty criminals;

➤ prostitutes; and

➤ homosexuals (both male and female).

Not surprisingly, these groups also tend to be marginalized spatially, both in terms of their residential locations and in terms of their activity spaces. In general, this translates into fractured, isolated and localized clusters – though with some numerically larger groups, such as the impoverished lone elderly,

the clustering tends to be somewhat less pronounced. With the possible exception of some criminals, prostitutes and homosexuals, this localization is determined by niches of the most economically and socially marginal housing: older, residual inner-city blocks, blighted and abandoned spaces, and lower-grade social (public) housing.

The high degree of localization of female-headed households within the inner-city areas is largely a product of their concentration in inner-city public housing projects. In many cases, such clusters are in fact shared spaces for several marginal subgroups (the subgroups listed above being by no means mutually exclusive, in any case), so that specific sub-areas can take on a very definite character – bohemia, ghetto, slum, drug market – according to the mix of inhabitants (see also Box 4.2). This, of course, begs the more general question of how

Twilight years or the Third Age?: senior citizens in Todi, Italy. Neighbourhoods are not only distinguished by persons of differing class, ethnicity, sexuality and nationality, but also by age. Photo Credit: Paul Knox.

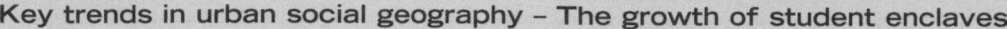

Box 4.2

Key trends in urban social geography – The growth of student enclaves

An important trend in many UK cities in recent years, but one that has so far received little research, is the development of residential areas dominated by young adults (i.e. those aged between 18 and 24 years). There is growing evidence that the geographical distribution of these age groups is becoming more concentrated in British cities. To a large extent this development reflects the increasing proportion of people entering higher education in the UK. In contrast to many nations, a high proportion of UK students move out of their parental homes and away from their immediate localities to universities and colleges in other towns and cities. This migration leads to a demand for accommodation that cannot be met by purpose-built university properties, and so about three-quarters of all students move to the privately rented sector. Such property tends to be overwhelmingly concentrated in inner-city areas, although sometimes these can be relatively affluent neighbourhoods adjoining university campuses (such as Headingly, a suburb of northwest Leeds).

These patterns have a number of important consequences. Most notorious is the clash of lifestyles between older inhabitants and the transient student population. Late-night parties, loud music and rowdy behaviour can lead to bad relations and the exodus of older people. Indeed, as in the case of certain ethnic minorities, sometimes a 'tipping point' is reached whereby the proportion of students reaches such a level that not only is out-migration of the established population intensified, but non-student households are inhibited from moving into such areas. Another impact is the development of city quarters catering for the needs of students with pubs, clubs, restaurants, discos and bars. These areas may cater for all young adults in a city, although sometimes hostilities between students and the indigenous young adult population can lead to areas that tend to cater exclusively for students.

There is a growing recognition of the enormous impact that higher education can have on urban economies. Universities and colleges are often major employers, injecting vast sums into local economies through the purchase of local services and spending power of staff salaries. In addition, universities can be the catalyst for local economic development through the exploitation of scientific and technological assets in the form of new firm 'spin-offs'. In more general terms, it seems that the intellectual and cultural facilities provided by universities forms an attractive milieu for the talented, 'creative' sections of the population who appear to be increasingly important in ensuring local economic growth.

Key terms associated with student enclaves (see Glossary)

Community action, externality, subculture, 'tipping point', 'turf' politics.

Further reading

Chatterton, P. (1999) University students and city centres – the formation of exclusive geographies. The case of Bristol UK *Geoforum*, **30**, 117–133

Chatterton, P. and Hollands, R. (2003) *Urban Nightscapes: Youth Cultures, Pleasure Spaces and Power* Routledge, London

Links with Other Chapters

Chapter 6: The Decline of Private Renting

Chapter 10: Box 10.2 The Development of 'Urban Nightscapes'

Chapter 11: Box 11.2 Richard Florida

Chapter 13: Externality Effects

cities are patterned according to the attributes and relative homogeneity of their neighbourhoods. It is a question that is most effectively addressed empirically through the study of factorial ecology.

Studies of factorial ecology

Factor analysis, together with the associated family of multivariate statistical techniques that includes **principal components analysis**, has become one of the most widely used techniques in social research of all kinds; and it is now generally the preferred approach for dealing with the complex question of measuring urban sociospatial differentiation. In this context, factor analysis is used primarily as an inductive device with which to analyze the relationships between a wide range of social, economic, demographic and housing characteristics, with the objective of establishing what common patterns, if any, exist in the data. The 'quantitative revolution' in geography in the 1970s led to factorial ecology studies of a wide range of cities, thus forming the basis for reliable high-level generalizations about urban sociospatial structure in Western, twentieth-century cities.

It is not within the scope of this book to discuss details of the methodology of factor analysis and related techniques. Essentially, they can be regarded as summarizing or synthesizing techniques that are able to identify groups of variables with similar patterns of variation. These are expressed in terms of new, hybrid variables called factors or components. Each factor accounts for measurable amounts of the variance in the input data and, like 'ordinary' variables, can be mapped or used as input data for other statistical analysis. The relationships and spatial patterns that the factors describe are known collectively as a **factorial ecology**. The usual statistical procedure produces a series of hybrid variables (factors), each representing a major dimension of co-variance in the data, each statistically independent of one another and each successively accounting for a smaller proportion of the total variance in the input data.

By far the major finding of factorial ecology studies has been that residential differentiation in the great majority of cities of the developed, industrial world have been dominated by a socioeconomic status dimension, with a second dimension characterized by **family status**/life-cycle characteristics and a third dimension relating to segregation and **ethnic status**. Moreover, these dimensions appear to have been consistent even in the face of variations in input variables and in the statistical solution employed; and evidence from the limited number of studies of factorial ecology change that have been undertaken shows that these major dimensions have tended to persist over periods of two or three decades at least. There also appears to be a consistent pattern in the *spatial expression* of these dimensions, both from city to city and from one census year to the next.

This consistency suggests that socioeconomic status, family status and ethnicity can be regarded as representing major dimensions of social space that, when superimposed on the physical space of the city, serve to isolate areas of social homogeneity 'in cells defined by the spider's web of the sectoral-zonal lattice'. The resultant idealized model of urban ecological structure is shown in Fig. 4.13. Yet it should be acknowledged that

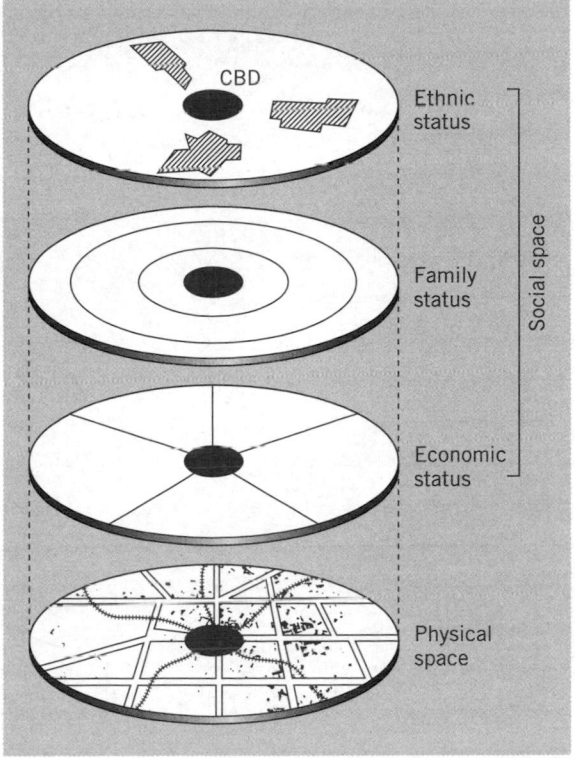

Figure 4.13 Idealized model of urban ecological structure.
Source: Murdie (1969), p. 8.

these sectors and zones are not simply superimposed on the city's morphology: they result from detailed interactions with it. Radial transport routes, for example, are likely to govern the positioning of sectors and to distort zonal patterns. Similarly, the configuration of both sectors and zones is likely to be influenced by specific patterns of land use and by patterns of urban growth. By introducing such features to the idealized model it is possible to provide a closer approximation to the real world.

It is important to emphasize that this classic model represents a high level of generalization and that the results of some studies are ambiguous or even contradictory. An analysis of Montréal, for example, found that the socioeconomic status dimension was not 'pure', for it contained some 'ethnic' elements (Foggin and Polese, 1977). Nevertheless, many geographers have suggested that the classic three-factor model has substantial generality throughout the Western culture

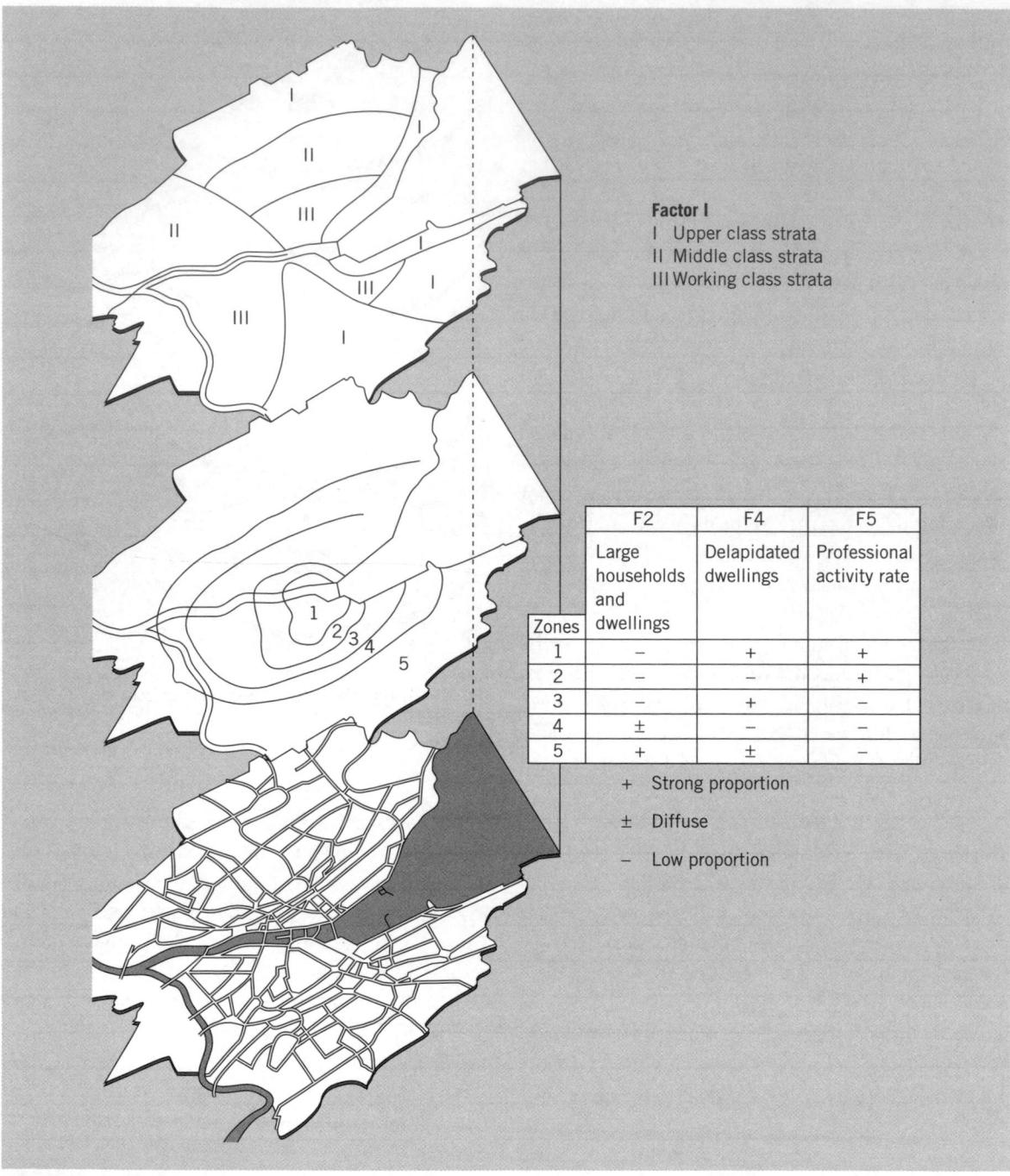

Factor I
I Upper class strata
II Middle class strata
III Working class strata

Zones	F2 Large households and dwellings	F4 Delapidated dwellings	F5 Professional activity rate
1	–	+	+
2	–	–	+
3	–	+	–
4	±	–	–
5	+	±	–

+   Strong proportion

±   Diffuse

–   Low proportion

Figure 4.14 The factorial ecology of Geneva.
Source: Bassand (1990), Map 1, p. 73.

area. This is certainly borne out by factorial ecologies of cities in Canada, Australia and New Zealand, but evidence from studies of European cities tends to be less conclusive. Overall, residential differentiation in continental European cities does tend to be dominated by a socioeconomic status dimension (as in the example of Geneva: Fig. 4.14), though it is often associated with housing status and the localization of self-employed workers. Continental cities such as Geneva also tend to conform to the 'classical' ecological model in that family status figures prominently (though often in a complex manner) in the factor structure. Ethnicity, however, does not generally occur as an independent dimension, partly because of the absence of substantial ethnic minorities, and partly because those that do exist appear to be more integrated – at census tract level – with the indigenous population. British cities, however, do not conform so closely to the general Western model. Indeed, British cities exhibit a somewhat distinctive ecological structure, with the principal dimensions of the classical model being modified by the construction and letting policies associated with the large social housing sector. Figure 4.15 shows the typical outcome.

Factorial ecologies as a product of social structure

If there *is* such a thing as a general model of Western city structure, then these modifications must be viewed as the product of special conditions, or of the absence or attenuation of the basic conditions necessary for the classic dimensions to emerge. But what are these necessary conditions? Janet Abu-Lughod (1969) attempted to answer this question in relation to the socioeconomic status and family status dimensions. She suggested that residential differentiation in terms of socioeconomic status will only occur:

➤ where there is an effective ranking system in society as a whole that differentiates population groups according to status or prestige; and

➤ where this ranking system is matched by corresponding subdivisions of the housing market.

Similarly, she suggested that a family status dimension will occur where families at different stages of the family life cycle exhibit different residential needs and where the nature and spatial arrangement of the housing stock are able to fulfil these needs. Implicit in these conditions is the important assumption that the population is sufficiently mobile to match social status and life-cycle needs to existing housing opportunities. Abu-Lughod points out that these conditions are characteristic of contemporary North American society:

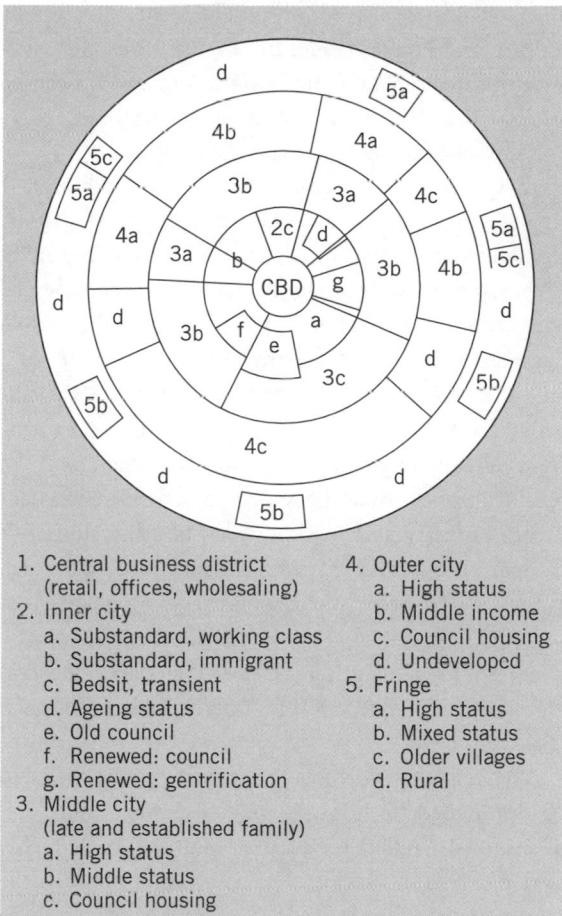

1. Central business district
   (retail, offices, wholesaling)
2. Inner city
   a. Substandard, working class
   b. Substandard, immigrant
   c. Bedsit, transient
   d. Ageing status
   e. Old council
   f. Renewed: council
   g. Renewed: gentrification
3. Middle city
   (late and established family)
   a. High status
   b. Middle status
   c. Council housing
4. Outer city
   a. High status
   b. Middle income
   c. Council housing
   d. Undeveloped
5. Fringe
   a. High status
   b. Mixed status
   c. Older villages
   d. Rural

Figure 4.15 A model of social regions for British cities.
Source: Davies (1984), Fig. 9.8, p. 341.

a prewelfare state in which people are geographically very mobile, and where social status is ascribed principally by occupation and income.

Accepting the validity of these ideas, it is clearly possible to relate factorial ecologies to a wider view of society and to begin to build a body of theory around the generalized model of the Western city. Abu-Lughod's work provides a useful framework against which deviations from the general model can be explained. In Montréal, for example, where the socioeconomic status dimension overlapped with ethnicity, the explanation can be found in the unusually large minority population of French-speakers that occupies most of the lower part of the social ladder, with the result that ethnicity and social status are not independent phenomena (Foggin and Polese, 1977). In Swedish cities, the existence of three separate family status dimensions can be attributed to the relative immobility of Swedes, compared with American norms (Janson, 1971).

In Geneva, the fact that three factors exhibited a strong concentric zonal pattern (Fig. 4.14) can be explained by the two main stages of the city's growth. The first, preindustrial phase, established a commercial core (zone 1) surrounded by fortifications that, when demolished, formed the template for zone 2; zones 3 and 4, meanwhile, emerged as the result of early suburban growth. In the second stage, the innermost zone was reaffirmed by commercial redevelopment and the recentralization of tertiary activities, while manufacturing industry and working-class housing shifted to the periphery (zone 5). At the same time, the zonal structure of the city was consolidated by the pattern of investment in road and public transport circuits and by a series of town planning decisions concerning the legal regulation of construction in different zones of the city (Bassand, 1990).

The tendency for the ecology of British cities to be dominated by housing market characteristics can be seen as a reflection of the country's more highly developed public sector. The association between the family status dimension and measures of crowding found in most British studies, for example, can be related to the letting policies of local authority housing departments, many of which allocate public housing on the basis of family size, among other things, as an indicator of housing need. Similarly, the use of economic criteria of housing need in determining people's eligibility for council houses ensures that there is a close association between socioeconomic status and housing tenure.

A historical perspective

Davies provided a rather different framework for explaining variations in urban structure (Davies, 1984). He suggested that, historically, four major dimensions of social differentiation have dominated cities everywhere – social rank, family status, ethnicity and migration status – and that *these are combined in different ways in different types of society* to produce varying urban structures. In traditional or feudal societies, family-related considerations dominated the social structure, since prestige and status were based primarily on kinship. In the 'feudal' city, therefore, a single axis of differentiation can be expected, combining social rank and family status as well as the limited amount of ethnic and migrant variation.

With economic specialization and the development of external economic linkages, division of labour intensifies, a merchant class is added to the political elite, and selective migration streams add to the social and ethnic complexity of cities. Davies postulates that these changes led to the creation of three very different types of urban structure, each composed of two dominant axes of differentiation that combined the four basic dimensions in different ways. In 'preindustrial' cities, the perpetuation of family kinship patterns and the continued importance of the established elite combined to produce a single axis of differentiation (social rank/family status), while arrival of migrant groups of different ethnic origins created a second major axis of differentiation. In 'colonial' cities that were located in previously settled areas, immigrants would be politically and socially dominant, so that social rank, ethnicity and migration status would be collapsed into a single dimension. Meanwhile, family status characteristics would represent an independent dimension of differentiation. In 'immigrant' cities, the indigenous political elite remained dominant, while the age-, ethnic- and sex-selective process of in-migration tended to overwhelm residential variations in family status. As a result, social rank and migration/ethnicity/family status emerge as the major dimensions of residential structure.

The onset of industrialization brought a great increase in specialization, while income and wealth became more important as yardsticks of social prestige. As transport technologies made large-scale suburbanization possible, these changes produced a transformation in the characteristics of social prestige and in family organization, eventually leading to quite distinct patterns of differentiation in terms of social rank and family status. Similarly, processes of segregation led to the separation of various ethnic and migrant groups in different parts of cities, thus completing the 'classic' structure of the modern industrial city.

Now, these patterns are dissolving under the 'splintering urbanism' associated with the evolution of Western economies to a postindustrial basis within a global economy. As a result, the classical model of factorial ecology is in the process of being overwritten. As we saw in Chapters 1, 2 and 3, cities of the developed world have entered a new phase as a fundamental economic transition has gathered momentum, accompanied by demographic, cultural, political and technological changes that underpin a 'splintering urbanism'. Advances in telecommunications have already begun to remove many of the traditional frictions of space for households as well as for economic activities, opening up the possibility of the dissolution of traditional urban spaces and the irruption of a diversity of new ones. This is not to suggest that residential differentiation and segregation will disappear, but that they will be manifest *in more complex ways and at a finer level of resolution* than the sectors, zones and clusters that have been associated with socioeconomic status, family status and ethnicity.

The implications of economic, technological, demographic and social change for ecological structure suggest that the increasing complexity of society is likely to result in more dimensions of differentiation. Among these are the following:

➤ The emergence of migrant status as a potent source of differentiation.

➤ The reinforcement of ethnic differentiation with the arrival of new immigrant groups.

➤ The emergence of new dimensions of occupational differentiation related to the expansion of service jobs.

➤ The appearance of significant distinctions in the degree of welfare dependency.

➤ A relative increase in the importance of poverty and substandard housing as a result of the consolidation of the urban underclass.

➤ The increased sociospatial differentiation of young adults and of the elderly as a result of changes in household organization.

➤ The emergence of a distinctive social ecology in the urban fringe.

At the same time, longstanding differences between suburban and central city areas have already begun to disappear as the suburbs have become the hub of daily economic activity and sections of central cities have been renewed, upgraded or gentrified, bringing some 'suburban' socioeconomic and demographic profiles to some inner-city neighbourhoods.

Patterns of social well-being

One of the major shortcomings of traditional factorial ecology studies is that the mix of input variables overlooks many important aspects of urban life, including environmental quality, accessibility to facilities such as hospitals, shopping centres, libraries and parks, and the local incidence of social pathologies such as crime, delinquency and drug addiction. The emergence of 'quality of life' and 'territorial justice' as important concerns within human geography has meant that much more attention has been given to such issues, demanding a rather different perspective on patterns of socioeconomic differentiation.

Renewed interest in patterns of social well-being reflects a number of factors. First, there is the growing social inequality in Western societies that we noted in Chapter 2. Second, following the growing influence of continental European intellectual traditions, there has been a focus upon ideas of *social exclusion* (Room, 1995; Lister, 1998). This approach defines poverty not only in terms of access to material resources, but also in terms of issues such as social participation and belonging. This approach is embodied in a redefinition of the concept of *citizenship* which is taken up in Chapter 5. Third, in the wake of the Rio Summit of 1992 and the Istanbul Habitat II Conference of 1996 there has been a resurgence of interest in environmental issues. This has encouraged the search for measures of environmental impact and broader quality-of-life factors in addition to measures of economic growth.

As David Smith (1977) demonstrated, **territorial social indicators** provide a very useful descriptive device in the context of geographical analysis. Smith made the case for a 'welfare approach' to human geography, with the central concern being 'who gets what, where and how?' Following this approach, territorial social indicators are seen as fundamental to 'the major and immediate research task' of describing the geography of social well-being at different spatial scales. This, it is argued, will not only provide the context for empirical research concerned with *explaining* the mechanisms and processes that create and sustain territorial disparities in well-being but will also facilitate the *evaluation* of these disparities in the light of prevailing societal values and, if necessary, the *prescription* of remedial policies. Two kinds of study are of particular interest here: those that attempt to describe variations in the overall level of local social well-being – 'quality of life' studies – and those that attempt to identify particular sub-areas whose residents are relatively disadvantaged – studies of 'deprivation' (see also Box 4.3).

Box 4.3

Key debates in urban social geography – Are Western cities becoming socially polarized?

A key debate in urban studies centres around the changing class and occupational structures of Western cities. Particularly contentious is the issue of whether these cities are becoming increasingly polarized. The so-called 'polarization thesis' argues that the decline in relatively well-paid manufacturing occupations since 1980, combined with the rapid expansion of both low-paying and high-paying service sector jobs, is leading towards the decline of middle incomes. In other words, the social structure of Western societies is moving towards an 'hour glass' structure. It has been argued that this trend towards a 'disappearing middle' is most pronounced in 'world' (Friedmann and Woolff, 1982) or 'global' (Sassen, 1991) cities with their high-level business and financial services which tend to be especially remunerative for senior employees.

The conspicuous consumption of new urban elite groups, combined with the growth of destitution on the streets of major cities in the form of 'skid rows', makes this polarization thesis seem intuitively appealing and in line with common sense. However, proving that social polarization is taking place raises both conceptual problems and methodological difficulties (see Pinch, 1993, and Hamnett, 2003, for more details). For example, the polarization thesis does not imply simply that the rich are getting richer and the poor are getting poorer. Instead, it refers to the *proportions* in the occupational structure. It is therefore possible for polarization to be leading to increasing inequality of incomes but it could also mean that, although there are more poor, they are doing somewhat better than in the past in relative terms. Furthermore, there are two further theses in opposition to the polarization theory. One argues that there is an upward reskilling of occupations in the West (the professionalization thesis) and the other that there is progressive deskilling (the proletarianization thesis).

Taking the evidence as a whole, there seems little doubt that the expansion of financial and business services sectors, tax cuts for the more affluent, the decline of relatively well-paid manufacturing jobs and the decline in the welfare state, have in combination all led to growing inequality of incomes in Western cities. These processes are most developed in US cities where larger immigrant labour forces and less restrictive employment legislation has encouraged the rapid expansion of low-paid service sector jobs. In Europe, relatively smaller immigrant workforces, more restrictive employment legislation and stronger (though threatened) welfare states have led to less inequality but much more unemployment. Nevertheless, there is also a general pattern of an upward shift in the occupational structures of Western cities.

Key concepts related to social polarization (see Glossary)

Neoliberal policies, professionalization thesis, proletarianization thesis, residualization, underclass, world cities.

Further reading

Friedmann, J. and Woolff, K. (1982) World city formation: an agenda for research and action *International Journal of Urban and Regional Research*, **6**, 309–344

Hamnett, C. (2003) *Unequal City: London in the Global Arena* Routledge, London

Pinch, S. (1993) Social polarization: a comparison of evidence from Britain and the United States *Environment and Planning A*, **25**, 779–795

Sassen, S. (2000) *The Global City* (second edition) Princeton University Press, Princeton, NJ

Links with Other Chapters

Chapter 2: Postindustrial Society

Chapter 13: Social Polarization

Chapter 14: Los Angeles and the California School

Intra-urban variations in the quality of life

Quality-of-life studies are of interest here because they offer the possibility of portraying the essential socio-geographical expression of urban communities on a conceptual scale that ranges along a continuum from 'good' to 'bad', thus providing a potent index with which to regionalize the city. The construction of such an index presents a number of difficulties, however. The first task is to set out a definition of social well-being that can be translated into a composite statistical measure: something that has taxed social scientists a great deal. The range of factors that potentially influence people's well-being for better or worse is enormous. Moreover, opinions about the importance of different factors often vary between sociogeographical groups; and factors that might be important at one geographical scale can be completely irrelevant at another. Any search for conclusive or universal definitions of social well-being is therefore futile. Nevertheless, as Smith himself argued, 'the imperative of empirical analysis in welfare geography means that we must be prepared to move in where the angels fear to tread' (Smith, 1973, p. 47).

Smith's original analysis of Tampa, Florida, provides a good case study of intra-urban variations in the quality of life. In operationalizing the concept Smith drew on measures of welfare dependency, air pollution, recreational facilities, drug offences, family stability and public participation in local affairs – a marked contrast to the conventional spectrum of variables deployed in studies of factorial ecology. Smith acknowledged that his selection of variables was 'a compromise between the ideal and what was possible given the constraints of time and resources', but maintained that 'the data assembled provide a satisfactory reflection of the general concept of social well-being and embody many important conditions which have a bearing on the quality of individual life' (Smith, 1973, p. 125). An overall measure of social well-being was derived from these data using the relatively simple procedure of aggregating, for each census tract, the standardized scores on all the variables. The resultant index is mapped in Fig. 4.16. Despite the rather peculiar shape of the city, with its CBD close to the bay and its southern suburbs surrounded on three sides by water, there is a clear pattern: a sink of ill-being occupies the inner city area, with relatively poor areas extending towards the city limits in a north-easterly direction. The best areas occupy the opposite sector of the city, although most suburban neighbourhoods enjoy a quality of life that is well above the average. Similar results have emerged from more recent quality of life studies: they typically describe a sharply bipolar society, in which the geography of social well-being exhibits both sectoral and zonal elements. In addition, most studies have revealed a close association between race and the quality of life.

One of the major potential weaknesses of this kind of approach is the implicit assumption that the simple aggregation of a series of measures of different aspects of social well-being will produce a meaningful statistic. Although this procedure may be an acceptable expedient in many circumstances, it is clear that social well-being should in fact be regarded as the product of a series of contributory factors that are *weighted* according to their relative importance to the people whose well-being is under consideration. It is evident from social surveys, for example, that British and American people do not regard housing conditions as being as important to their well-being as their health, whereas both factors are felt to be much more important than accessibility to recreational facilities.

Moreover, these values tend to vary significantly among sociogeographical groups: in Britain, for example, intra-urban variations in attitudes to education have become part of the conventional wisdom of a whole generation of educationalists. There are plenty of reasons for such variations. To begin with, some aspects of social well-being (leisure and material consumption, for example) are highly income-elastic, so that successive increases in expendable income will bring about marked increases in the intensity with which they are valued.

This conforms neatly with Maslow's suggestion that human motivation is related to a hierarchy of human needs, so that as people's basic needs – for nutrition, shelter and personal safety – are satisfied, motivation turns towards higher goals such as the attainment of

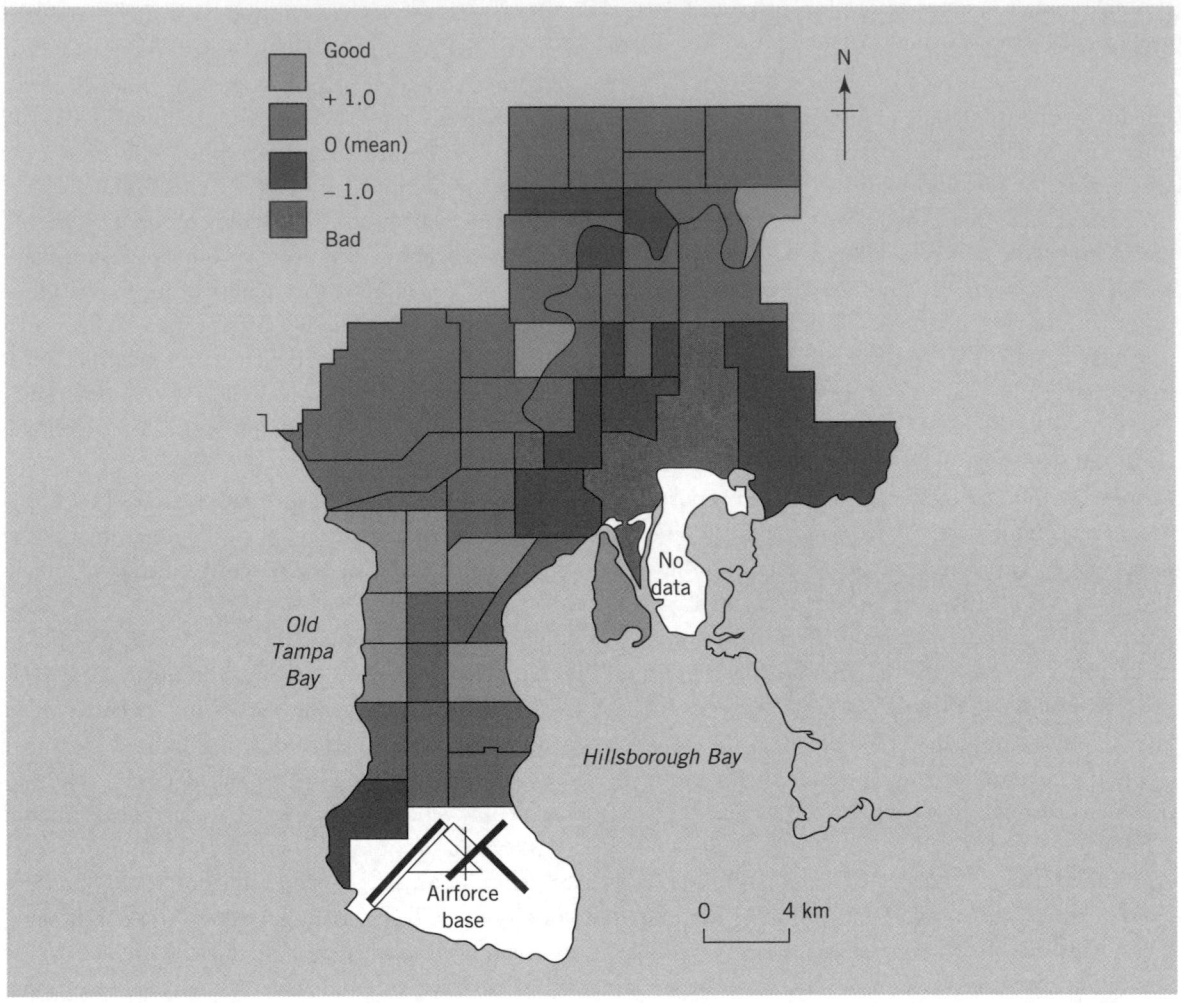

Figure 4.16 Standard scores on a general indicator of social well-being for Tampa, Florida.
Source: Smith (1973), p. 126.

social status, prestige and self-expression (Maslow, 1970). It follows that people with low levels of material well-being will attach more importance to material-istic than to aesthetic, spiritual or cultural aspects of life. People's values also vary according to their stage in the family life cycle, and to their membership of particular religious or cultural groups. Moreover, the social geography of the city is itself likely to generate or reinforce differences in values from one neighbour-hood to another, for the sociodemographic composition of different neighbourhoods creates distinctive local reference groups that contribute significantly to people's attitudes to life (see Chapter 7).

The geography of deprivation and disadvantage

Patterns of deprivation represent a particularly import-ant facet of the social geography of the city. In this con-text, it is useful to regard deprivation as multidimensional (hence the term **multiple deprivation**), directing atten-tion to the spatial configuration and interrelationships of different aspects of deprivation. The tendency within many cities is for the accumulative distribution of depriva-tions, with low-status neighbourhoods tending to fare badly on most dimensions of deprivation and prosperity. In this context, it seems fair to aggregate indicators to produce a single index of 'multiple deprivation'. A good example of this approach is provided by Fig. 4.17, which shows the worst 1 per cent and 5 per cent of enumeration districts in Glasgow in 2001, based on a composite measure of disadvantage derived from a principal components analysis of ten census indicators of deprivation covering access to transportation, health, housing tenure, household composition, overcrowding, social status and unemployment. In Glasgow, pockets of deprivation are found throughout the city, not only in central areas of older, private, tenement housing, but also – and indeed predominantly – in some of the newer, peripheral public housing estates.

This kind of approach can be criticized on several grounds, however, including the desirability of aggregating indicators of several different aspects of deprivation and the validity of assigning them equal weight in the overall index (recall the discussion of quality-of-life indicators). It is also necessary to guard against the dangers of the **ecological fallacy** (i.e. making inferences about individuals with data based on aggregates of people). Thus not everyone in a deprived area is necessarily deprived and not every deprived person in an area of 'multiple deprivation' is necessarily multiply deprived. Consider Figure 4.18, for example, which shows the results of a special tabulation of census data that allowed

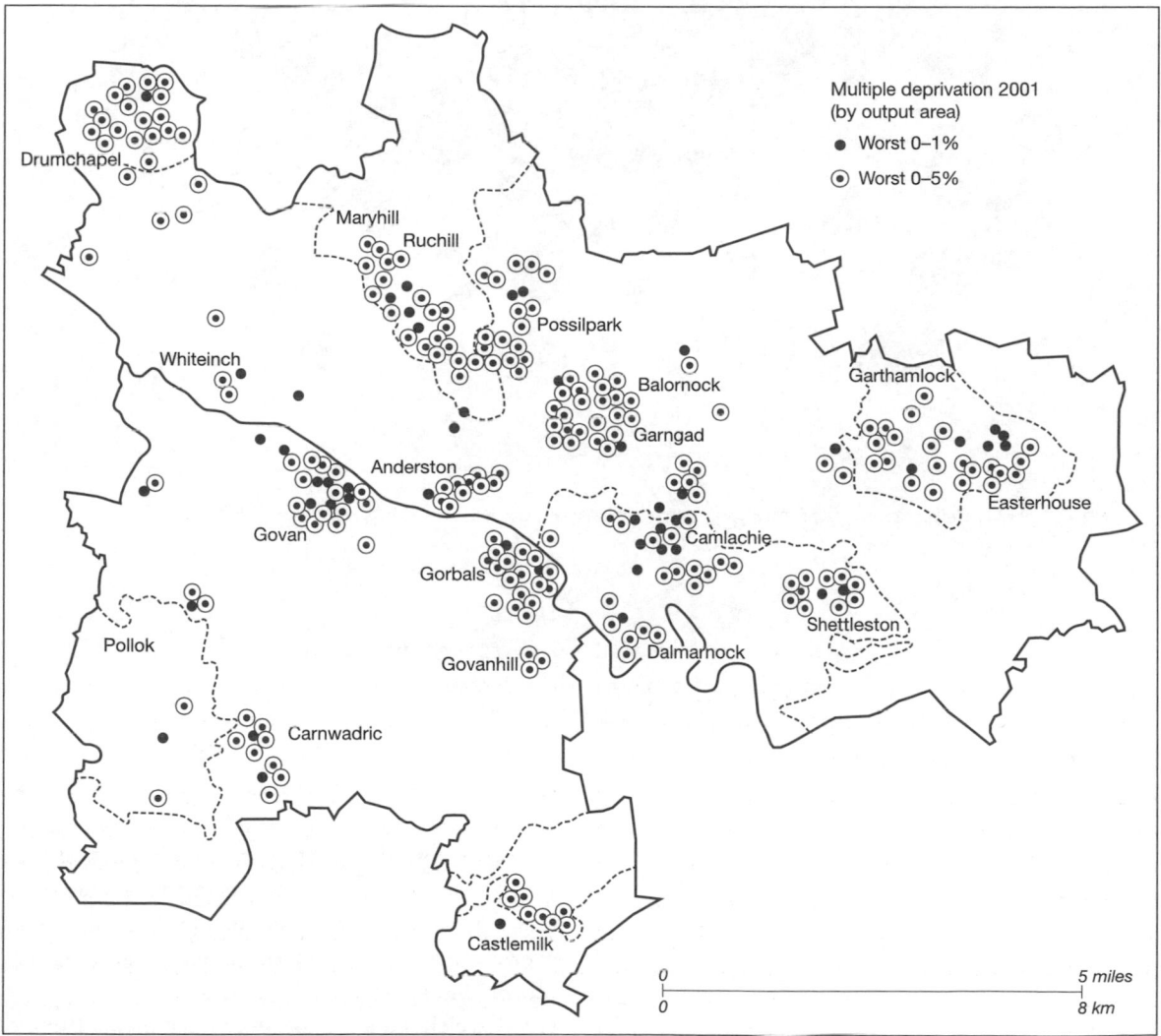

Figure 4.17 Social deprivation in Glasgow, 2001.
Source: Pacione (2004).

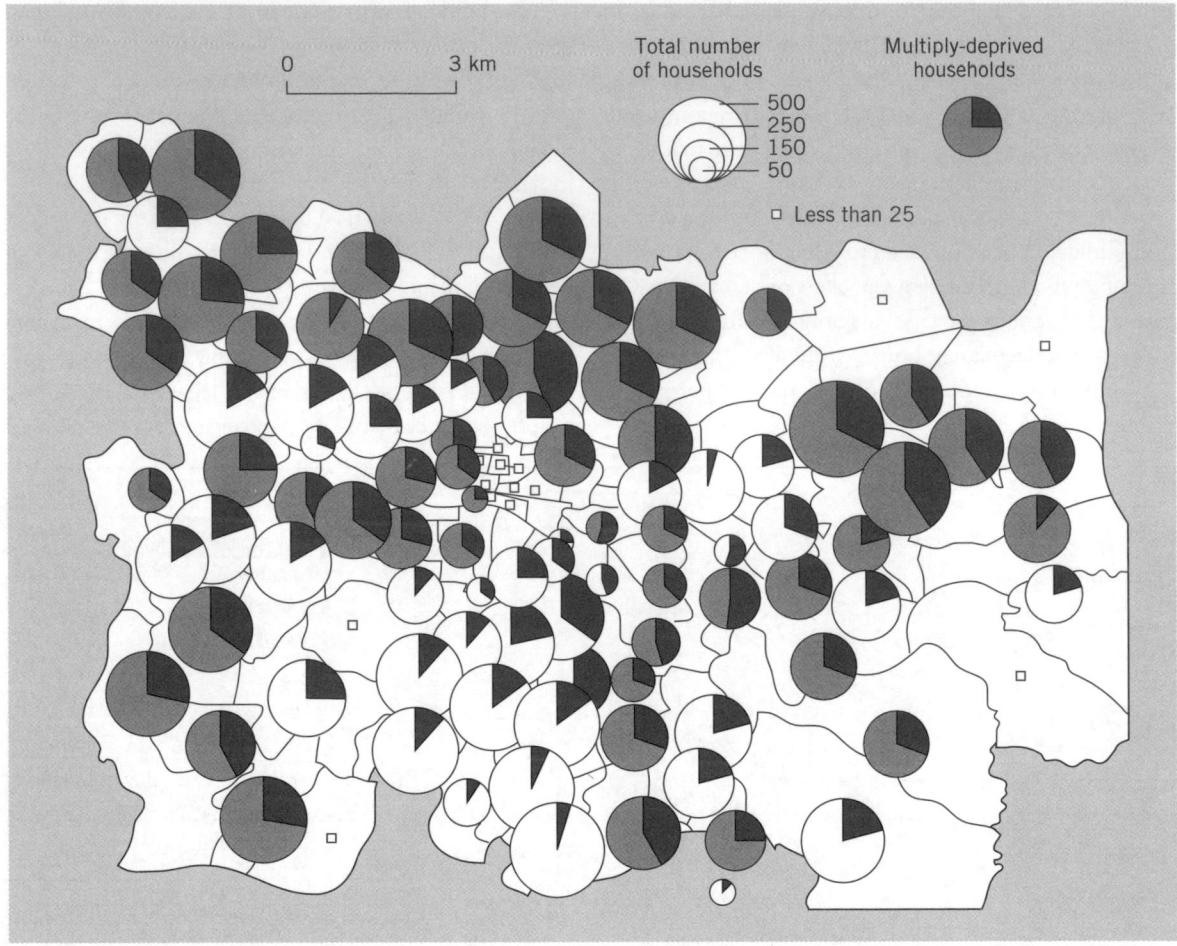

Figure 4.18 The distribution of multiply deprived households in Glasgow.
Source: Knox (1986), Fig. 19-2, p. 261.

the overlap of 'multiple deprivation' to be identified at the level of individual households, thus avoiding the dangers of the ecological fallacy. Multiple deprivation is shown to be a very widespread phenomenon, with significant levels in nearly every part of the city (although it should be acknowledged that Glasgow is a city with an unusually high overall level of multiple deprivation).

Microsimulation of disadvantage

The development of territorial social indicators was for some years inhibited by the absence of geographically disaggregated information on issues related to household income, wealth, taxation and welfare benefits. In the UK for example, little information on income is published below the level of relatively large regions (Green, 1998). Thus most analysis of data on incomes has been undertaken by economists, who are seldom concerned with the geographical implications of their studies. However, in recent years this impasse has been overcome by microsimulation techniques that take advantage of recent developments in **geographical information systems (GIS)**.

These techniques enable the simulation of estimates of data for small areas based on detailed data published at larger spatial scales. To illustrate, in the United Kingdom there is a wealth of complex data relating to households in the Sample of Anonymized Records (SARs). These show the interrelationships between many variables (known as **cross-tabulations**); but to maintain confidentiality this information is published only at large spatial scales. However, certain statistical techniques (linear programming and iterative proportional fitting) can simulate the interrelationships between variables for small areas such as enumeration districts. Furthermore, these techniques facilitate the linking of data from a number of different data sources at new small spatial scales. The major drawback of such approaches is the difficulty of validating the newly created data sets. Two solutions to this problem are; first, to undertake some detailed fieldwork for small areas to compare the results with the microsimulations; or second, see how the simulations work by reaggregating the new data to spatial scales for which there is detailed information.

Microsimulation holds out prospect of reassessing our views of **area-based urban policy** (i.e. urban policy that is targeted upon specific high-need areas in cities). Edwards (1995, p. 710) has argued that these policies are largely cosmetic since:

> by far the greatest part of the social and economic needs of inner city residents will not be met by urban specific policies but by mainline housing, health, income support and education provision.

However:

> we know precious little about the effectiveness of such programmes either in targeting the deprived, spatially or otherwise, or in providing for the (sometime) multiplicity of needs or the different additional needs that may be found within individual households.

> (Edwards, 1995, p. 711)

Recent work has begun to redress this gap in our knowledge. For example, the New Policy Institute and the Joseph Rowntree Foundation have recently devised a new set of deprivation indicators based on household poverty, wealth and social exclusion (Howarth *et al.*, 1998). In addition, the UK's Office of the Deputy Prime Minister (ODPM) has commissioned a comprehensive measure of multiple deprivation that enables measure of disadvantage to be determined for small-scale areas (wards) in England (see Table 4.3). Although highly influential, this complex index has been criticized, not only for the nature of the data, but the ways in which they are combined (see Deas *et al.*, 2003). Urban analysts have begun to develop similar indicators for small areas within cities. Thus, Bramley and Lancaster (1998) developed a model that generated income data for small areas based on the numbers of workers in household, economic activity rates and household tenure. Caldwell *et al.* (1998) have also used microsimulation techniques in the US context to show the geography of wealth, taking into account factors such as financial assets as well as debts.

Chapter summary

4.1 The physical structures of cities display complex patterns that reflect many processes superimposed over the years. There are wide variations in environmental quality in cities. Urban areas also have distinctive characteristics that can often best be captured through subjective written descriptions that reflect the 'feel' of a neighbourhood.

4.2 Studies of Western cities reveal interrelationships between groups of variables reflecting three main dimensions: economic inequalities, family status and ethnicity, which are often reflected respectively in sectors, concentric rings and clusters. There are, however, many variations in residential structure reflecting the distinctive processes to be found within cities. Territorial social indicators also reveal wide variations in the quality of life in cities. Often these indices overlap to reveal multiple deprivation.

Table 4.3 The UK Office of the Deputy Prime Minister's (ODPM) Index of Multiple Deprivation 2004

Income Deprivation Domain (22.5% weighting)
➤ Adults and children in income support households (2001)
➤ Adults and children in income-based job-seekers households (2001)
➤ Adults and children in Working Families Tax Credit households whose equivalized income (excluding housing benefits) is below 60% of median before housing costs (2001)
➤ Adults and children in Disabled Person's Tax Credit households whose equivalized income (excluding housing benefits) is below 60% of median before housing costs (2001)
➤ National Asylum Support Services asylum seekers in England in receipt of subsistence only and accommodation support (2002)

Employment Deprivation Domain (22.5% weighting)
➤ Unemployment claimant count of women aged 18–59 and men aged 18–64 (2001)
➤ Incapacity benefit claimants women aged 18–59 and men aged 18–64 (2001)
➤ Participants in the New Deal for the 18–24s who are not included in the claimant count (2001)
➤ Participants in the New Deal for 25+ who are not included in the claimant count (2001)
➤ Participants in New Deal for Lone Parents aged 18 and over (2001)

Health Deprivation and Disability Domain (13.5% weighting)
➤ Years of potential life lost (1997–2001)
➤ Comparative illness and disability ratio (2001)
➤ Measures of emergency admissions to hospital (1999–2002)
➤ Adults under 60 suffering from mood or anxiety disorders (1997–2002)

Education Skills and Training Deprivation Domain (13.5% weighting)
Subdomain: children/young people
➤ Average point score of children at Key Stage 2 (2002)
➤ Average point score of children at Key Stage 3 (2002)
➤ Average point score of children at Key Stage 4 (2002)
➤ Proportion of young children not staying on in school or school level education above 16 (2001)
➤ Proportion of those aged under 21 not entering Higher Education (1999–2002)
➤ Secondary school absence rate (2001–2002)
Subdomain: Skills
➤ Proportions of working age adults (aged 25–54) in the area with no or low qualifications (2001)

Barriers to Housing and Services Domain (9.3% weighting)
Subdomain: wider barriers
➤ Household overcrowding (2001)
➤ LA percentage receiving assistance for homelessness
➤ Difficulty of access to owner occupation
Subdomain: geographical barriers
➤ Road distance to GP premises (2003)
➤ Road distance to a supermarket or convenience store (2002)
➤ Road distance to a primary school (2001–2002)
➤ Road distance to a Post Office (2003)

Crime Domain (9.3% weighting)
➤ Burglary (4 recorded crime offence types, April 2002 to March 2003)
➤ Theft (5 recorded crime offence levels, April 2002 to March 2003)
➤ Criminal damage (10 recorded crime offence levels, April 2002 to March 2003)
➤ Violence (10 recorded crime offence levels, April 2002 to March 2003)

The Living Environment Deprivation Domain (9.3% weighting)
Subdomain: the 'indoors' living environment
➤ Social and private housing in poor condition (2001)
➤ Houses without central heating (2001)
Subdomain: the 'outdoors' living environment
➤ Air quality (2001)
➤ Road traffic accidents involving injury to pedestrians and cyclists (2000–2002)

Source: http://www.odpm.gov.uk/stellent/groups/odpm_urban_policy/documents/page/odpm

Key concepts and terms

area-based urban policy	factor analysis	multiple deprivation
areal differentiation	factorial ecology	principal components analysis
cross-tabulations	family status	quality-of-life indices
double hermeneutic	geographical information	Social Area Analysis
ecological fallacy	systems	social well-being
economic status	morphogenesis	territorial social indicators
ethnic status	morphological regions	urban social areas

Suggested reading

A thorough and comprehensive treatment of morphogenesis and the physical structure of British cities can be found in Jeremy Whitehand's book on *The Making of the Urban Landscape* (1992: Blackwell, Oxford). The evolution of twentieth-century residential forms in America is reviewed and analyzed in an essay by Anne Moudon in *Urban Landscapes: International Perspectives*, edited by Jeremy Whitehand and Peter Larkham (1992: Routledge, London), while both the physical landscapes and the social ecology of the 'postmodern urban matrix' in American cities are described in Chapter 8 (pp. 207–236) of *The Restless Urban Landscape* (Paul Knox, 1993: Prentice Hall, New York). *A Glossary of Urban Form*, edited by Peter Larkham and Andrew Jones (1991: Historical Geography Research Series No. 26, Urban Morphology Research Group, University of Birmingham), is a very useful resource that contains examples and commentaries as well as dictionary-type entries. Every student will also find *The Dictionary of Human Geography* (Ron Johnston, Derek Gregory and David Smith (eds); 4th edition, 2000: Blackwell, Oxford) an invaluable resource, not just for the material in this chapter but also for the entire subject area of the book. A thorough review of geographical approaches to urban residential patterns is provided by Ron Johnston in his extended essay (pp. 193–236) in *Social Areas in Cities*, edited by Ron Johnston and David Herbert (1976: Wiley, London). More detailed and specific treatments of factorial ecology and patterns of residential differentiation are provided by Wayne Davies in *Factorial Ecology* (1984: Gower, Aldershot) and by Michael White in *American Neighborhoods and Residential Differentiation* (1987: Russell Sage Foundation, New York). Qualitative methods for studying cities are discussed in the special edition of the *Journal of Geography in Higher Education* (1992, Vol. 16). The key reference to patterns of social well-being is David Smith's *The Geography of Social Well-being in the United States* (1973: McGraw-Hill, New York), while the geography of deprivation and disadvantaged is introduced in Paul Knox's review essay (pp. 32–47) in *Social Problems and the City: New Perspectives* (David Herbert and David Smith, editors; 1989: Oxford University Press, Oxford). A recent useful review of the field is Michael Pacione 'Quality-of-life research in urban geography' (*Urban Geography*, 2003: **24**, 314–339) and for a recent application see James E. Randall and Peter H. Morton 'Quality of life in Saskatoon 1991 and 1996; a geographical perspective' (*Urban Geography*, 2003: **24**, 691–722). Issues of method in geographical research are discussed in Robin Flowerdew and David Martin (eds) *Methods in Human Geography* (1997: Longman, London). One of the best reviews of measure of multiple deprivation is Martin Senior's chapter in P. Rees, D. Martin and P. Williamson (eds) *The Census Data System* (2002: Wiley, Chichester). For extensive documentation on the Indices of Deprivation used by the UK's Office of the Deputy Prime Minister (ODPM) see: http://www.odpm.gov.uk/stellent/groups/odpm

Spatial and institutional frameworks: citizens, the state and civil society

Key questions addressed in this chapter

➤ In what ways are the structures of cities influenced by legal, governmental and political structures?

➤ What are the consequences of metropolitan fragmentation?

➤ In what ways does the institutional structure of cities affect the functioning of democracy?

➤ How is power distributed in cities?

In this chapter, we explore some fundamental components of the sociospatial dialectic: the social, legal and political structures surrounding citizenship, democracy and civil society. The physical and socioeconomic patterns described in Chapter 4 are all outcomes of complex, interlayered processes in which social and spatial phenomena are intermeshed – the sociospatial dialectic described in Chapter 1. These processes are all

played out, moreover, within spatial and institutional frameworks – electoral districts, school catchment areas, legal codes, homeowner association deeds and so on – that are themselves both outcome and medium of social action. Individually and collectively, we act out our lives and pursue our interests both *in* and *through* these institutional and spatial frameworks. Our lives and our lifeworlds are facilitated, shaped and constrained by these frameworks but we also, consciously and unconsciously, contribute to their shape and character.

5.1 The interdependence of public institutions and private life

It was the emergence of capitalist democracies that forged the basis for modern urban society. The scale, rhythm and fragmentation of life required by the new logic of industrial capitalism meant that traditional societies had to be completely restructured. Local and informal practices had to be increasingly standardized and codified in order to sustain the unprecedentedly extensive, complex and rapidly changing economic and social structures of an urbanized and industrialized system. At the heart of this process was the growth and transformation of public institutions in order to be able to facilitate and regulate the new political economy. This was the era when many new nation states were established and most of the old ones were recast with modern institutions of governance, democracy and judicial process.

Yet these institutions did not simply emerge, autonomous, from the flux of change in the eighteenth and nineteenth centuries. The **public sphere** that came to encapsulate the realms of private life derived its *raison d'être* from the changing needs (and demands) of citizens who 'freely join[ed] together to create a public, which forms the critical functional element of the political realm' (Marston, 1990). According to social theorist Jurgen Habermas, the public sphere and the citizens who populate it can be seen as one of four fundamental categories of social organization characteristic of modern societies. The others are the *economy, civil society* and *the state* (Habermas, 1989). The meaning of **civil society** has changed over time but is generally understood to involve all the main elements of society outside government. The emergence of these categories, Habermas points out, requires the working-through of an established relationship of the public to the private spheres of life. He has suggested that in most instances this relationship has come to rest on the recognition of three sets of common rights (Habermas, cited in Marston, 1990, p. 83):

➤ Those related to rational critical public debate (freedom of speech and opinion, freedom of the press, freedom of assembly and association, etc.).

➤ Those related to individual freedoms, 'grounded in the sphere of the patriarchal conjugal family' (personal freedom, inviolability of the home, etc.).

➤ Those related to the transactions of private owners of property in the sphere of civil society (equality before the law, protection of private property, etc.).

The way these rights are articulated and upheld in particular locales determines, among other things, the nature of access to economic and political power and to social and cultural legitimacy. It follows that issues of citizenship, legal codes and the roles claimed by (or given to) urban governments have a great deal to do with the unfolding of the sociospatial dialectic.

Citizenship, patriarchy and racism

The idea of **citizenship** 'refers to relationships between individuals and the community (or State) which impinges on their lives because of who they are and where they live' (Smith, 1989b, p. 147). In contrast to the premodern hierarchies of rights and privileges tied to the notion of the allegiance of subjects to a monarch (see Chapter 2), citizenship implies a rationality that is accompanied by mutual obligations. The citizenship that emerged with the onset of modernity (see Chapter 3) was tied to the territorial boundaries of new and reconstituted national states rather than to the divine authority of nobility. It was the construct through which political and civil rights were embedded in national constitutions. Later, there developed in most of the economically more-developed countries an ideal of citizenship that embraced social as well as political and civil rights – the right to a minimum level of personal security and of economic welfare, for example.

The process of constructing this modern idea of citizenship inevitably provoked a running debate over who is and who is not a citizen, especially in countries such as Australia and the United States, which drew demographic and economic strength on the basis of immigration. The result was that the social construction of citizenship has been mediated through deep-seated prejudices and entrenched cultural practices, as introduced in Chapter 3. **Sexism** and **racism**, in short, found their way into conceptions of citizenship and from there into the relationship between the public sphere and private life and to the very heart of the sociospatial dialectic through which contemporary cities have been forged.

In the first instance, of course, citizenship was available only to white, property-owning males. Women and minority populations 'in essence retained their subject status' (Marston, 1990, p. 450). The exclusion of women can be traced, in large measure, to the **paternalism** of

Western culture: in particular, to naturalistic assumptions about the social roles of men and women. The basic assumptions are (1) that the dominance of husband over wife is a 'law of nature' and (2) that men by nature are more suited to the aggressive pursuits of economic and public life while women by nature are more suited to the nurturing activities of the domestic sphere.

In addition, the social philosophers of the transition to modernity cultivated the assumption that women are by nature sentimental rather than rational, and so incapable of developing the proper sense of justice required for participation in civil society. The idea of 'Public Man' (whose corollary was 'Private Woman') persisted even after the franchise was extended to women and indeed still persists, well after the 'women's liberation' of the 1960s. Even in Australia, a comparatively progressive country in terms of incorporating women's issues into mainstream policymaking, gender inequalities persist in the form of:

> discriminatory controls on the availability of social citizenship rights which determine how women are defined in Australian social policy. . . . State policies, from social security and income tax to industrial and family law, construct women as wives, mothers and carers, regulating their social role and reinforcing women's dependency on men.
> (Smith, 1989b, p. 149)

The exclusion of minority populations has in general been more explicit, not least in antebellum America, where black slavery represented the very antithesis of citizenship. The inherent racism of 'mainstream' society overtly circumscribed the participation of native American, Chinese and black populations in the full rights of citizenship all through the 'melting-pot' of American urbanization in the late nineteenth and early twentieth centuries, until the Civil Rights legislation of the late 1960s. In Europe, racism was focused on Jews and gypsies until after the Second World War, when immigration brought large numbers of Asians and Africans to the cities of Britain, France and Germany. In addition to the overt and formal limits imposed on these immigrants in terms of the civil and political rights of citizenship, systematic discrimination has circumscribed their social rights of citizenship, so that, for example:

> despite the theoretical eligibility of black Britons for welfare benefits, and despite their disproportionate contribution to the welfare state (through labour and taxation), their ability to secure State-subsidized services and resources may actually be deteriorating relative to that of whites.
> (Smith, 1989b, p. 149)

We shall see in Chapter 8 how these consequences of racism come into play in the social production of space and the maintenance of sociospatial segregation.

The law and civil society

The law stands as an important link between the public and private spheres, and between the state and the economy. As a key component of the sociospatial dialectic, the law must be seen as both a product of social forces and spatial settings and as an agent of sociospatial production and reproduction. There are several specific elements to the law in this context. It is *formulated* (usually in quite abstract and general ways) by elected legislatures that in turn draw on citizens' conceptions of justice, equity, etc. It is subsequently *applied* in specific places and circumstances by a variety of agencies (such as the police, social workers, housing authorities, etc.) to whom responsibility is delegated by the national state. Where problems and disputes emerge as to the specific meaning of law, it is *interpreted* through other mechanisms of civil society, principally the courts (Blomley, 1989).

It is now acknowledged that each of these elements is deeply geographical in that they involve the interpenetration of place and power (Clark, 1990). Indeed, recent research on law and geography has been explicitly framed within the concept of the sociospatial dialectic through which the spatiality of social life is reproduced, reinforced or transformed (e.g. Blomley and Clark, 1990; Delaney, 1993). Among the best-documented examples of the interpenetration of law, civil society and urban geography are the decisions of the US Supreme Court in cases involving voting rights, school desegregation, open housing, and land use zoning. To take just a few examples, these include decisions on *Brown v. Board of Education* (1954), which declared school segregation unconstitutional; *Shelley v. Kraemer* (1948), which ruled that racially restricted covenants on property sales are illegal; *Euclid v. Ambler* (1926), which established the right of municipalities to zone land use in order to protect the public interest; and *NAACP v. Mt. Laurel* (1972), which struck down an exclusionary zoning ordinance (Clark, 1991).

Through such cases, particular social values and moral judgements are mapped onto the urban landscape while others are deflected or eradicated. This, of course, is by no means straightforward. Apart from anything else, the formulation, application and interpretation of law take place not only at the national scale but also at the level of the municipality (or '**local state**'), making for a complex and sometimes contradictory framework of legal spaces that are superimposed on, and interpenetrated with, the social spaces of the city. At the same time, the continual evolution and reorganization of society introduces elements of change that cumulatively modify the tenor of civil society itself, alter the relationships between central and local states, and raise new challenges for law and urban governance. Legal practice and discourse, observes Nicholas Blomley:

> contain multiple representations of the spaces of social and political life. . . . Spatial representations play a vital role in legal reasoning: legal categories – property, the public and the private, the individual, the municipal corporation – are all spatially conceived and defined. The spaces of local discretion and autonomy, for example, are frequently cast as problematic within formal legal practice by virtue of their contextuality and 'communality.' Indeed, in much the way that geography as a discipline is unsure how to map theoretically the sociospatial world, so legal actors struggle and quarrel over the spatiality of the law. The city, for example, can be cast as a dangerously collectivized agency that, by virtue of its semi-autonomous location within the interstices of the state, poses a threat to individual liberty, or it can be hailed as the epitome of democratic participation and political life.
>
> (Blomley, 1993, p. 6)

The changing nature of urban governance

As the economic base of cities has shifted, the fortunes of different groups have changed, cities themselves have thrown up new problems and challenges, and urban government has attracted different types of people with different motivations and objectives. The ethos and orientations of urban government, reflecting these changes, have in turn provided the catalyst for further changes in the nature and direction of urban development. Today, the scope of urban governance has broadened to the point where it now includes the regulation and provision of all kinds of goods and services, from roads, storm drainage channels, street lighting, water supplies and sewage systems to law enforcement, fire prevention, schools, clinics, transport systems and housing. All these activities have a direct and often fundamental effect on the social geography as well as on the physical morphology of cities, as we shall see in Chapter 13. Moreover, the economic and legislative power of modern local authorities makes them a potent factor in moulding and recasting the urban environment. In general terms, it is useful to distinguish five principal phases in the evolution of urban governance.

1. The earliest phase, dating to the first half of the nineteenth century, was a phase of virtual non-government, based on the doctrine of utilitarianism. This **laissez-faire** philosophy rested on the assumption that the maximum public benefit will arise from unfettered market forces. In practice, an oligarchy of merchants and patricians presided over urban affairs but did little to modify the organic growth of cities.

2. The second phase, dating between 1850 and 1910, saw the introduction of '**municipal socialism**' by social leaders in response to the epidemics, urban disorder and congestion of the Victorian city. The law and urban governance in this period were based on a strong ethos of public service and paternalism, and the result was a wide range of liberal reforms. At the same time, the increasing power and responsibility of political office-holders facilitated the widespread development of corruption in urban affairs.

3. Between 1910 and 1940 there occurred a critical event – the Depression – that finally swung public opinion in favour of a permanent and more fundamental municipal role in shaping many aspects of social life and well-being. Cities everywhere expanded their activities in health, welfare, housing, education, security and leisure. At the same time,

the composition and character of city councils shifted once more. In Britain, the 75 per cent de-rating of industry by the Local Government Act (1929) and the central government's policy of industrial protection in the 1930s combined to remove from many businessmen the incentive to participate in local affairs. In contrast, members of the working and lower-middle classes found a new rationale for being on the council: to speak for the city's growing number of salaried officials and blue-collar employees. These developments led to the replacement of paternalistic businessmen and social leaders by 'public persons' drawn from a wider social spectrum. In addition, representatives of the working class were installed on city councils through the agency of the Labour party, and *party politics* soon became an important new facet of urban governance.

4. Between 1940 and 1975, the many roles of urban government generated large, vertically segregated bureaucracies of professional administrators geared to managing the city and its environment. The professional and the party politician came to rule as a duumvirate, the balance of power between the two being variable from function to function and from city to city. By this time, however, a deep paradox had clearly emerged to confront all those concerned with urban affairs. The paradox was this. *Although urbanization was the vehicle that capitalism needed in order to marshal goods and labour efficiently, it created dangerous conditions under which the losers and the exploited could organize themselves and consolidate.* Urban governance and management, facing this paradox, became hybrid creatures, dedicated on the one hand to humanistic and democratic reform, but charged on the other with the management of cities according to a particular kind of economic and social organization. Inevitably, the demands of this task led to an escalation in the number of professional personnel employed to assist councillors in their decision-making. At the same time, however, the effective power of councillors to formulate policy initiatives decreased. As the technical complexities of municipal finance, public health, educational administration and city planning increased, councillors became more and more dependent on the expertise of professional personnel and their staff. Consequently, most cities have become permanently dependent on large bureaucracies staffed by specialist professionals.

5. In the most recent phase, from the mid-1970s, radical economic transformation at the national and international scale has set in motion a pronounced bout of metropolitan restructuring that has been accompanied, in most large cities, by chronic problems of physical deterioration and **fiscal stress**. Faced with the rapid decentralization of jobs and residents, and with an increasingly externalized and distant control of their economies, local governments lost a good deal of power and autonomy to central governments. In their weakened and somewhat desperate position, they began to 'privatize' many of the functions and responsibilities that they had acquired in previous phases. In the vacuum left by the retreat of the local state, **voluntarism** has become a principal means of providing for the needs of the indigent, while in more affluent communities various forms of 'stealthy', 'private' governments, such as homeowner associations, have proliferated. Local governments have turned increasingly to the private sector for capital for economic and social investment through public–private partnerships of various kinds, and now give much greater priority to economic development than to the traditional service-providing and regulating functions of the local state. This '**civic entrepreneurialism**' has fostered a speculative and piecemeal approach to the management of cities, with a good deal of emphasis on set-piece projects, such as downtown shopping centres, festival market places, conference and exhibition centres and the like, that are seen as having the greatest capacity to enhance property values (and so revivify the local tax base) and generate retail turnover and employment growth (Hall and Hubbard, 1996). Meanwhile, economic restructuring and the decentralization of manufacturing have changed the complexion of urban politics, undermining the former strength of working class constituencies. At the same time, the growth and recentralization of producer-service jobs have created (in some cities, at least) a new bourgeoisie with a distinctively materialistic sort of liberal ideology that has come into play in urban politics and policy-making (see also Box 5.1).

Box 5.1

Key trends in urban social geography – The growth of 'urban entrepreneurship'

In 1989 David Harvey drew attention to 'urban entrepreneurialism' – an emerging set of changes in the way cities are governed. Since then, the trends that he highlighted have intensified and others have undertaken extensive analyses of the 'entrepreneurial city' (e.g. Hall and Hubbard, 1996: Jessop, 1997).

Entrepreneurs are people who introduce new goods and services. In a city context, entrepreneurialism also refers to both new ways of providing traditional services and also completely new policies. Such innovation has been necessary in response to the 'hollowing out' of the central state, the devolution of responsibilities down towards smaller units and greater involvement of the private sector in the provision of services.

This has involved more than the public sector simply behaving in a more efficient 'business-like' manner, adopting the private sector language of targets, markets and customers. It has also involved engaging in public–private partnerships, attracting private sector capital and utilizing private sector expertise to provide local services and solve local problems. As a result, cities have increasingly come into competition with one another as they attempt to attract inward investment from both central governments and the private sector. This has led to a concern with the external image of cities, 'city branding' and 'place marketing'. Cities have also become involved in new schemes to transform their neighbourhoods and encourage indigenous innovation and entrepreneurship.

Despite the emphasis upon cities behaving in new innovative ways, in many respects urban entrepreneurialism involves a new form of orthodoxy as cities attempt to adopt the most successful policies of their rivals. Common themes are the promotion of high-technology clusters, gentrified inner cities, waterfront redevelopment, the promotion of cultural quarters and retail-led regeneration. It has also been argued that such policies circumvent traditional forms of democratic representation, giving greater powers to private capital.

Key concepts associated with urban entrepreneurship (see Glossary)

Governance, 'hollowing-out', place marketing, pro-growth coalitions, property-led development.

Further reading

Harvey, D. (1989a) From managerialism to entrepreneurialism: the transformation of urban governance in late capitalism *Geografiska Annaler*, **71B**, 3–17.

Hall, T. and Hubbard, P. (1996) The entrepreneurial city: new urban politics, new urban geographies *Progress in Human Geography*, **20**, 153–174.

Hall, T. and Hubbard, P. (1998) *The Entrepreneurial City: Geographies of Politics, Regime and Representation* Wiley, Chichester

Jessop, B. (1997) The entrepreneurial city: re-imagining localities, redesigning economic governance or restructuring capital, in N. Jewson and S. MacGregor (eds) *Transforming Cities: Contested Governance and New Spatial Divisions* Routledge, London

Links with Other Chapters

Chapter 11: Box 11.2 Richard Florida

Chapter 13: Box 13.3 The Rise of Social Entrepreneurship

In the early part of the twenty-first century it is perhaps possible to see an emerging phase of local government centred around issues of economic **sustainability**. This follows increasing public concern over environmental issues and the 1992 Rio Earth Summit that produced Agenda 21, a framework for sustainable global development. This was followed up by Habitat II, a summit in Istanbul in 1996 that brought to the fore issues of cities and sustainability. In the United Kingdom, for example, although local authorities have had their powers reduced in the spheres of water and transport services, and they have been undermined by the contracting-out of services to the private sector, they have assumed an increasing role in the sphere of environmental regulation and recycling (Patterson and Theobold, 1995). However, critics point to the fact that local policies for urban sustainability have been patchy and limited in scope, since in their existing form local authorities lack power to influence many of the realms affecting the environment. In addition, it is intended that Agenda 21 be implemented through the notion of **subsidiarity** – the devolution of decision-making down to the most appropriate level. However, this raises the issue of just what is the most appropriate level for decision-making; sustainable policies require local level empowerment and democratic participation, yet they also need central coordination across administrative boundaries.

5.2 *De jure* urban spaces

The geopolitical organization of metropolitan areas is an important element in the sociospatial dialectic. *De jure* **territories** are geographical areas as enshrined in law (i.e. with legal powers as in political and administrative regions) and can be seen as both outcome and continuing framework for the sociospatial dialectic. In this section we examine the evolution of *de jure* spaces at the intra-metropolitan level and discuss some of the major implications of the way in which urban space has been partitioned for political and administrative purposes.

Metropolitan fragmentation and its spatial consequences

Modern metropolitan areas are characterized by a complex partitioning of space into multipurpose local government jurisdictions and a wide variety of special administrative districts responsible for single functions such as the provision of schools, hospitals, water and sewage facilities (hence the term **jurisdictional partitioning**). This complexity is greatest in Australia and North America, where the ethic of local autonomy is stronger, and it reaches a peak in the United States, where the largest metropolitan areas each have hundreds of separate jurisdictions. While never reaching these levels of complexity, the same phenomenon can be found in Europe. In Britain, for example, the government of London is still fragmented among 32 boroughs; and in every city there are special district authorities that are responsible for the provision of health services and water supplies.

Much of this complexity can be seen as the response of political and administrative systems to the changing economic and social structure of the metropolis. In short, the decentralization of jobs and residences from the urban core has brought about a corresponding decentralization and proliferation of local jurisdictions. New local governments have been created to service the populations of new suburban and exurban dormitory communities, resulting in the 'balkanization' of metropolitan areas into competing jurisdictions. In the United States, this process has been accelerated by policies that, guided by the principle of local autonomy, made the annexation of territory by existing cities more difficult while keeping incorporation procedures very easy.

New single-function special districts, on the other hand, have proliferated throughout metropolitan areas, largely in response to the failure of existing political and administrative systems to cope with the changing needs and demands of the population. Between 1942 and 1972, the number of non-school special districts in the US increased from 6299 to 23 885. By 1992, Cook County, Illinois, contained 516 separate jurisdictions, one for every 10 000 residents. Special districts are an attractive solution to a wide range of problems because they are able to avoid the statutory limitations on financial and legal powers that apply to local governments. In particular, a community can increase its debt or tax revenue by creating an additional layer of government for a specific purpose. Special districts also have the advantage of corresponding more closely to functional areas and, therefore, of being more finely tuned to local social organization and participation. Another reason for their proliferation has been the influence of special interest groups, including (a) citizen groups concerned with a particular function or issue and (b) business enterprises that stand to benefit economically from the creation of a special district.

Yet although spatial fragmentation can be defended on the grounds of fostering the sensitivity of politicians and administrators to local preferences, it can also be shown to have spawned administrative complexity, political disorganization and an inefficient distribution of public goods and services. Not least of these problems is the sheer confusion resulting from the functional and spatial overlapping of different jurisdictions. Decentralized decision-making leads to the growth of costly bureaucracies, the duplication of services and the pursuit of conflicting policies. Of course, not all public services require metropolitan-wide organization: some urban problems are of a purely local nature. But for many services – such as water supply, planning, transport, health-care, housing and welfare – economies of scale make large areal units with large populations a more efficient and equitable base.

The balkanization of general-purpose government in the United States has also led to the suppression of political conflict between social groups:

> Social groups can confront each other when they are in the same political arena, but this possibility is reduced when they are separated into different arenas. Political differences are easier to express when groups occupy the same political system and share the same political institutions, but this is more difficult when the groups are divided by political boundaries and do not contest the same elections, do not fight for control of the same elected offices, do not contest public polities for the same political units, or do not argue about the same municipal budgets.
>
> (Newton, 1978, p. 84)

This subversion of democracy means in turn that community politics tends to be low key, while the politics of the whole metropolitan area are often notable for their absence. The balkanization of the city means that it is difficult to make, or even think about, area-wide decisions for area-wide problems. 'The result is a series of parish-pump and parochial politics in which small issues rule the day for want of a political structure which could handle anything larger' (Newton, 1978, p. 86).

Fiscal imbalance and sociospatial inequality

One of the most detrimental consequences of **metropolitan fragmentation** is the **fiscal imbalance** that leaves central city governments with insufficient funds and resources relative to the demands for the services for which they are responsible (Orfield, 2002). The decentralization of jobs and homes, the inevitable ageing of inner-city environments and the concentration of a residuum of elderly and low-income households

in inner-city neighbourhoods has led to a narrowing tax base accompanied by rising demands for public services. The ageing, high-density housing typical of inner-city areas, for example, requires high levels of fire protection; high crime rates mean higher policing costs; and high levels of unemployment and ill-health mean high levels of need for welfare services and health-care facilities. As a result of these pressures, many central cities in the United States have experienced a *fiscal squeeze* of the type that led to the near-bankruptcy of New York City in 1975 (and again in the early 1990s). Some have suggested that such problems are aggravated by additional demands for public services in central city areas that stem from suburbanites working or shopping there. This is the so-called **suburban exploitation thesis.** There is no question that the presence of suburban commuters and shoppers precipitates higher expenditures on roads, parking space, public utilities, policing and so on; on the other hand, it is equally clear that the patronage of downtown businesses by suburbanites enhances the central city tax base while their own suburban governments have to bear the cost of educating their children. The extent to which these costs and benefits balance out has not yet been conclusively demonstrated.

A more compelling argument has interpreted fiscal squeeze as a product of the nature of economic change. In this interpretation, it has been the growth of new kinds of *private* economic activity that has imposed high costs on the *public* sector. In general, the growth of new kinds of urban economic activity has been expensive because it has failed to provide employment and income for central city residents, and it has made demands on the public sector for infrastructure expenditures that were not self-financing:

> On the one hand, new economic growth in the central cities did not provide sufficient employment and income benefits to the central city's residents. Industrial jobs were taken by suburbanized union workers. Construction work was dominated by restrictive craft unions. And the new office economy was drawing on the better educated, better heeled, suburban workforce. Industrial investments were now part of vast multilocational networks of plants, thus weakening

the local multipliers from local plant investments. This export of the income benefits of local economic growth meant a continuous reservoir of poor, structurally unemployed people who turned to city governments for jobs and services.

On the other hand, the rising office economy of the central city required a restructuring of urban space to move people and information most efficiently. This required a massive investment in public capital for mass transit, parking, urban renewal, and the more traditional forms of infrastructure.

(Friedland, 1981, pp. 370–1)

These infrastructural investments were insulated from conflict through the exploitation of new forms of administration and financing: autonomous special districts, banker committees, and new forms of revenue and tax increment bonding. As a result there emerged two worlds of local expenditure: one oriented to providing services and public employment for the city's residents, the other to constructing the infrastructure necessary to profitable private development:

These two worlds – of social wage and social capital – were structurally segregated. The former was governed by electoral politics and the excesses of patronage. The latter was housed in bureaucratic agencies, dominated by men in business who survived by their efficiency.

(Friedland, 1981, p. 371)

Fiscal crises of the sort epitomized by the plight of New York City in the mid-1970s and early 1990s are seen by some commentators as an important catalyst for change in the political economy of central cities. Few, however, are optimistic as to the eventual outcomes.

Fiscal mercantilism

In a classic economic interpretation of urban public economies, Tiebout, noting the different 'bundles' of public goods provided by different metropolitan jurisdictions, suggested that households will tend to sort themselves naturally along municipal lines according to their ability to pay for them (Tiebout, 1956). It is now increasingly recognized, however, that a good deal of

sociospatial sorting is deliberately engineered by local governments. This unfortunate aspect of metropolitan political fragmentation arises from the competition between neighbouring governments seeking to increase revenue by attracting lucrative taxable land users. The phenomenon has been called **fiscal mercantilism**. Its outcome has important implications for residential segregation as well as the geography of public service provision.

In a fiscal context, desirable households include those owning large amount of taxable capital (in the form of housing) relative to the size of the household and the extent of its need for public services. Low-income households are seen as imposing a fiscal burden, since they not only possess relatively little taxable capital but also tend to be in greatest need of public services. Moreover, their presence in an area inevitably lowers the social status of the community, thus making it less attractive to high-income households. In competing for desirable residents, therefore, jurisdictions must offer low tax rates while providing good schools and high levels of public safety and environmental quality, and pursuing policies that somehow keep out the socially and fiscally undesirable.

The most widespread strategy in the United States involves the manipulation of land use **zoning** powers, which can be employed to exclude the fiscally undesirable in several ways. Perhaps the most common is 'large lot zoning', whereby land within a jurisdiction is set aside for housing standing on individual plots of a minimum size – usually at least half an acre (0.2 ha) – which precludes all but the more expensive housing developments and so keeps out the fiscally and socially undesirable. It is not at all uncommon, in fact, for American suburban subdivisions to be zoned for occupation at not more than one acre per dwelling. Other exclusionary tactics include zoning out apartments, the imposition of moratoria on sewage hook-ups, and the introduction of building codes calling for expensive construction techniques.

The existence of large tracts of undeveloped land within a jurisdiction represents a major asset, since it can be zoned to keep out the poor and attract either affluent households or fiscally lucrative commercial activities such as offices and shopping centres. Inner metropolitan jurisdictions, lacking developable land, have to turn to other, more expensive strategies in order to enhance their tax base. These include the encouragement of gentrification and/or urban redevelopment projects designed to replace low-yielding slum dwellings with high-yielding office developments – both of which also have the effect of displacing low-income families to other parts of the city, often to other jurisdictions. The end result is that central city populations are left the privilege of voting to impose disproportionate costs of social maintenance and control upon themselves.

Municipal service delivery and sociospatial inequality

The conflicts over resources that are embodied in the issues surrounding metropolitan fragmentation, fiscal imbalance and fiscal mercantilism are at once the cause and the effect of significant inter-jurisdictional disparities in public service provision. Here, then, we see another facet of the sociospatial dialectic: spatial inequalities that stem from legal and institutional frameworks and sociopolitical processes, inequalities that in turn are constitutive of the relations of power and status (Smith, 1994).

The extent of these disparities in a fragmented metropolitan area can be illustrated with reference to the example of social services for the elderly in Greater London. From a geographical point of view, the 'ideal' distribution of such resources might be one that is in direct proportion to the levels of need in each of the 32 Greater London boroughs: a situation that represents 'territorial justice'. An examination of the provision of home help, meals-on-wheels, home nurses and residential accommodation for the elderly, however, found evidence of considerable variability in the extent to which territorial justice is achieved (Pinch, 1979). In order to quantify local levels of need for social services for the elderly, the study used an index of social conditions based on a mixture of variables measuring local levels of health, housing conditions, unemployment and socioeconomic status, as well as the incidence of pensioners living alone. Levels of provision were measured both in terms of *financial input* committed to each service by the local authorities and in terms of the *extensiveness* and *intensity* of the services provided by this expenditure. The extensiveness of service provision is taken to be the proportion of those eligible for a service who

actually receive it (e.g. the percentage of the elderly who receive home helps or meals-on-wheels), while the intensity of service provision is evaluated in terms of the average amount of monetary or physical resources provided per recipient of the service.

Correlations between the index of need and a series of measures of service provision showed that although the overall trend was for a positive relationship between need and provision, the situation fell a long way short of the criterion of territorial justice. Indeed, there were some aspects of the home nursing and health visiting services for the elderly for which the overall spatial distribution was regressive. Most of the domiciliary services for the aged were distributed more equitably, although the correlation between needs and provision is far from perfect. This is illustrated by Fig. 5.1, which shows the relationship between the index of social conditions and a variety of indices of social service provision. Not surprisingly, several of the boroughs with poor social conditions (i.e. high levels of need) and relatively low levels of provision were inner-city jurisdictions – the likes of Islington, Lambeth and Lewisham, where fiscal problems were most severe. On the other hand, there were several needy inner-city boroughs that provided a relatively high standard of service – Camden, Hammersmith and Southwark, for example. The explanation for this seems to be rooted in local political disparities which, like the local resource base, are very closely influenced by the spatial configuration of local government boundaries.

5.3 The democratic base and its spatial framework

In all Western cities, the political framework is structured around the democratic idea of power resting, ultimately, with an electorate in which all citizens have equal status. A distinctive feature of urban politics in practice, however, is the low turnout of voters at election time. Seldom do more than 50 per cent of the registered voters go to the polls in municipal elections: the more likely figure is 30 per cent, and it is not uncommon for the vote to drop to less than 25 per cent

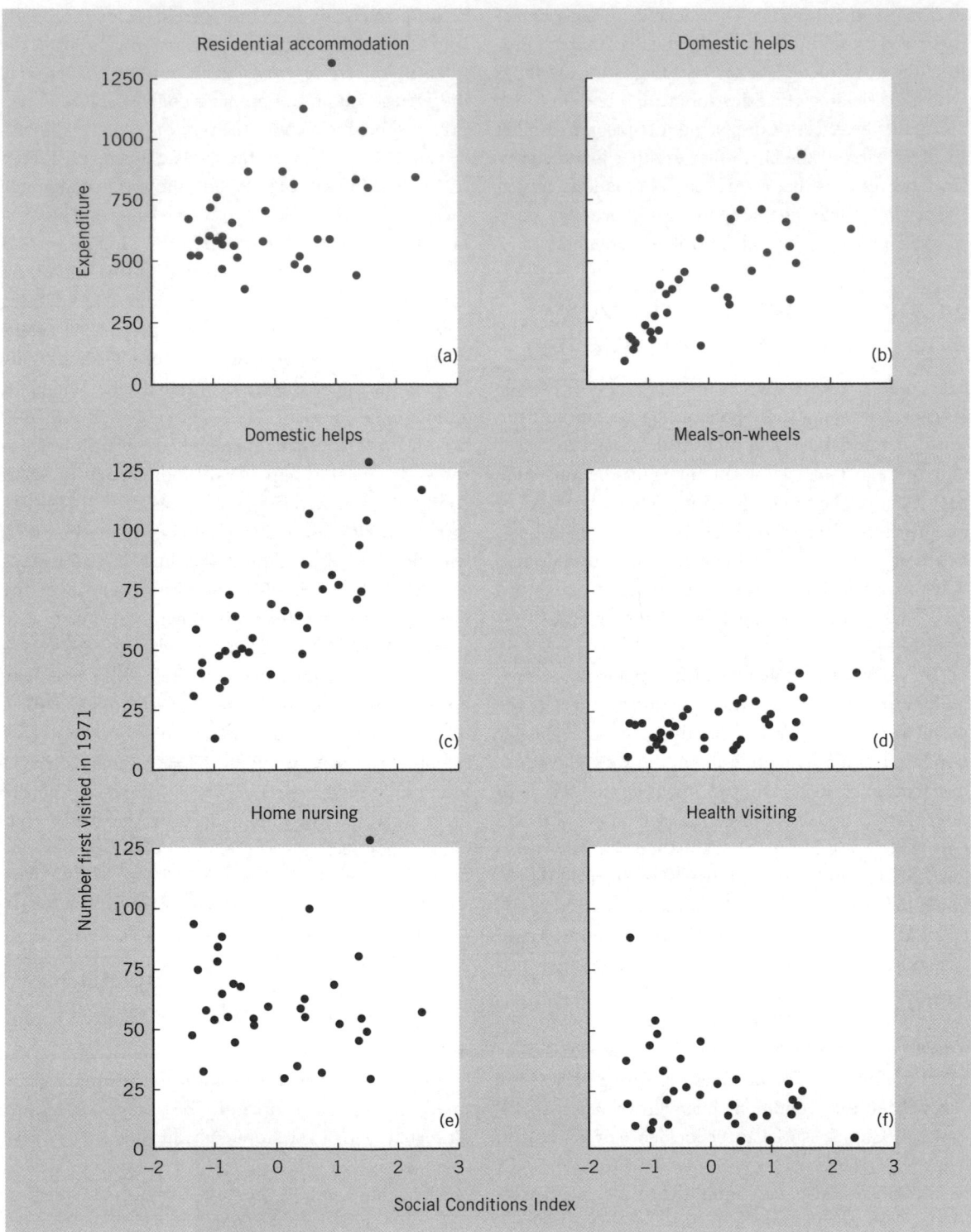

Figure 5.1 Degrees of 'territorial justice': correlations between indices of resource allocation and service provision for selected social services in the Greater London boroughs.
Source: Pinch (1985), Fig. 2.4, p. 55.

of the electorate. Moreover, although data on voting behaviour at municipal elections is rather fragmentary, it seems probable that about a fifth of the electorate never vote at all. This passivity can be attributed to two very different perspectives on life. On the one hand there are those who feel that their interests are well-served by the existing power structure and who therefore do not feel the need to act; and on the other there are those who feel that their interests are consistently neglected or sacrificed by government but who feel they can do little about it.

Such passivity is clearly undesirable from the standpoint of civic vitality; even more serious is the consequent lack of sensitivity of the political system to the interests of all sectors of society, since non-voters are by no means distributed randomly throughout the population. People living in rented accommodation tend to vote less than home owners; women tend to vote less than men; young people and retired people are less likely to go to the polls than are people in intervening age groups; people with lower incomes and lower educational qualifications tend to vote less than the rich and the well educated; and recent in-migrants are less inclined to vote than long-term residents. This represents another important facet of the sociospatial dialectic. It amounts to a distortion of the democratic base that inevitably leads to a bias in the political complexion of elected representatives; and, given the overall composition of voters compared with non-voters, it is logical to expect that this bias will find expression in municipal policies that are conservative rather than liberal.

The spatial organization of elections

At the same time, the *spatial organization* of the electoral process is itself also a source and an outcome of conflict. Electoral results, in other words, can be influenced by the size and shape of electoral districts in relation to the distribution of the electorate. Put another way, a vote can be regarded as a resource whose value varies according to the degree to which it permits the voter to secure preferred policies. This value tends to vary from one neighbourhood to another according to the individual voter's relative numerical

importance within his own constituency. At the same time, it is a function of the marginality of his or her constituency in terms of the balance between the major conflicting social groups and political parties.

Evaluating the effects of geopolitical organization on urban affairs in this context is not easy, since systems of electoral representation and their associated spatial frameworks can be very complex. This complexity is compounded by the existence, in most countries, of a hierarchy of governments responsible for a variety of different functions. The Greater London boroughs discussed above, for example, have been both independent multipurpose jurisdictions and constituencies for the Greater London Council, which was (until its abolition by the Thatcher government) responsible for certain aspects of strategic planning, housing management and slum clearance. They also happen to be constituencies for the House of Commons and they are themselves divided into wards for the election of their own political office-holders. A person may therefore find his or her vote much more effective in influencing policy at one level of government than at another.

A further complicating factor is the electoral system itself which, for any given set of constituencies may operate on the basis of: (1) a single-member plurality system; (2) a multimember plurality system; (3) a weighted plurality system; (4) preferential voting in single-member constituencies; (5) preferential voting in multimember constituencies; or (6) a list system in multimember constituencies. It is not possible to do justice here to the potential effect of each of these systems on the sociospatial dialectic. Rather, attention is directed towards two of the more widespread ways in which the spatial organization of electoral districts has been engineered in favour of particular communities, social groups and political parties: malapportionment and gerrymandering.

Malapportionment and gerrymandering

Malapportionment refers to the unequal population sizes of electoral subdivisions. Quite simply, the electorate in smaller constituencies will be over-represented in most electoral systems, while voters in larger-than-average constituencies will be under-represented.

Deliberate malapportionment involves creating larger-than-average constituencies in the areas where opposing groups have an electoral majority. In the United States, malapportionment of congressional, state senatorial and state assembly districts was ended by the Supreme Court in a series of decisions between 1962 and 1965 that began the so-called 'reapportionment revolution'. But malapportionment continues to exist at the city council level, and deviations in constituency size as large as 30 per cent are not unusual. Deviations of this magnitude also exist at the intrametropolitan level in Britain, effectively disenfranchising large numbers of citizens. If, as is often the case, the malapportioned group involves the inner-city poor, the problem assumes an even more serious nature. Policies such as rent control and garbage collection, and questions such as the location of noxious facilities or the imposition of a commuter tax will be decided in favour of the outer city (O'Loughlin, 1976, p. 540).

Gerrymandering occurs where a specific group or political party gains an electoral advantage through the spatial configuration of constituency boundaries – by drawing irregular-shaped boundaries so as to encompass known pockets of support and exclude opposition supporters, for example; or by drawing boundaries that cut through areas of opposition supporters, leaving them in a minority in each constituency. Not all gerrymanders are deliberate, however: some groups may suffer as a result of any system of spatial partitioning because of their geographical concentration or dispersal. Gerrymandering by a party for its own ends is usually termed 'partisan gerrymandering' and it occurs most frequently where – as in the United States – the power to redraw constituency boundaries lies in the hands of incumbent political parties.

It is very difficult to prove that gerrymandering has taken place: all that is usually possible is to produce strong circumstantial evidence, although there is considerable evidence of gerrymandering against African-Americans in US cities. Finally, it should be noted that although the definition of constituency boundaries in Britain and some other countries is left to non-partisan Boundary Commissions, their efforts often result in an *unintentional* gerrymander that favours the majority party, thus reinforcing its dominance in the electoral assembly.

The spatiality of key actors in urban governance: elected officials and city bureaucrats

The formal democracy of urban affairs is also subject to imperfections in the behaviour of the elected holders of political office. Although city councillors are ostensibly representative of their local communities, there are several reasons for doubting their effectiveness in pursuing their constituents' interests within the corridors of power. Apart from anything else, councillors are by no means representative in the sense that their personal attributes, characteristics and attitudes reflect those of the electorate at large. Even in large cities the number of people who engage actively in local politics is small and they tend to form a community of interest of their own. Moreover, those who end up as councillors tend to be markedly more middle-class and older than the electorate as a whole, and a large majority are men.

Notwithstanding these differences between councillors and their constituents, it is in any case doubtful whether many councillors are able – or, indeed, willing – to act in the best interests of their constituents. Councillors with party political affiliations, for example, may often find that official party policy conflicts with constituency feelings. Alternatively, some councillors' behaviour may be influenced by the urge for personal gain or political glory.

Another doubt about the effectiveness of councillors as local representatives stems from the conflicting demands of public office. In particular, it is evident that many councillors soon come to view their public role mainly in terms of responsibility for the city as a whole or in terms of their duties as committee members rather than general spokespersons for specific communities. It is therefore not surprising to find that empirical evidence suggests that there is a marked discrepancy between the priorities and preoccupations of local electorates and those of their representatives. At least part of this gap between councillors' perceptions and those of their constituents must be attributable to the dearth of mutual contact. When there is contact between local councillors and their constituents it tends to take place in the rather uneasy atmosphere of councillors' advice bureaux and clinics, where discussion is focused almost exclusively on personal grievances of one sort of another – usually connected with housing.

Bureaucracy and sociospatial (re)production

In theory, the expert professional is 'on tap' but not 'on top', but there are many who believe, like Max Weber, that the sheer complexity of governmental procedure has brought about a 'dictatorship of the official' (Weber, 1947). Lineberry, for instance, argued that the influence of professional personnel can be so great and the **decision rules** by which they operate so complex as to effectively remove the allocation of most public services from the control of even the strongest political power groups (Lineberry, 1977). The crucial point here is that the objectives and motivations of professional officers are not always coincident with the best interests of the public nor in accordance with the views of their elected representatives. Although it would be unfair to suggest that bureaucrats do not have the 'public interest' at heart, it is clear that *they are all subject to distinctive professional ideologies and conventions*; and their success in conforming to these may be more valuable to them in career terms than how they accomplish their tasks as defined by their clientele.

There are several techniques that bureaucrats are able to use in getting their own way. Among the more widely recognized are: (1) 'swamping' councillors with a large number of long reports; (2) 'blinding councillors with science' – mainly by writing reports that are full of technicalities and statistics; (3) presenting reports that make only one conclusion possible; (4) withholding information or bringing it forward too late to affect decisions; (5) rewriting but not changing a rejected plan, and submitting it after a decent interval; and (6) introducing deliberate errors in the first few paragraphs of a report in the hope that councillors will be so pleased at finding them that they let the rest go through.

But there is by no means common agreement as to the degree of autonomy enjoyed by professional officers. There are broad economic and social forces that are completely beyond the control of any bureaucrat, as well as strong constraints on their activities that derive from central government directives. In addition, it can be argued that the highest stakes in urban politics are won and lost in the budgetary process, to which few professional officers are privy. Thus, having set the 'rules of the game', politicians 'can leave the calling of the plays to the bureaucratic referees' (Rich, 1979).

Key actors such as city officials should not simply be seen, however, in terms of resource allocation and their direct interventions in the democratic process. As David Wilson points out, they are

> complex carriers of spatiality, mediating past and present sociospatial configurations through the lens of evolving biographies. They . . . proceed through life paths, producing value orientations rooted in encounters with distinctive sociospatial landscapes. These values, imported into the current organization, inform role interpretation, worker regulation, and resource allocation.
>
> (Wilson, 1992)

Like everyone else, city officials are caught up in the sociospatial dialectic; they are different only in the degree to which their experience and interpretation of space and place carries over into values and attitudes (as well as resource-allocating decisions and direct interventions in civic affairs) that (re)produce the spatiality of urban life.

The parapolitical structure

Bureaucrats as well as politicians may in turn be influenced by elements of what has been called the **parapolitical structure** – informal groups that serve as mediating agencies between the individual household and the machinery of institutional politics. These include business organizations, trade unions and voluntary groups of all kinds, such as tenants' associations and conservation societies. Although relatively few such organizations are explicitly 'political' in nature, many of them are 'politicized' inasmuch as they occasionally pursue group activities through the medium of government.

Indeed, there is a school of thought among political scientists that argues that, in American cities at least, private groups are highly influential in raising and defining issues for public debate (see Banfield and Wilson, 1963, for the origins of this interpretation). According to this school of thought, politicians and officials tend to back off until it is clear what the alignment of groups on any particular issue will be and whether any official decision-making will be required. In essence, this gives urban government the role of umpiring the struggle among private and partial interests, leaving these outsiders to decide the outcome of major issues in all but a formal sense.

Business

Business leaders and business organizations have of course long been active in urban affairs. One of the more active and influential business organizations in most towns is the Chamber of Commerce, but it is by no means the only vehicle for private business interests. Business itself typically engages in **coalition building**. Business executives take a leading part in forming and guiding a number of civic organizations, they often play a major role in fund raising for cultural and charitable activities, and they hold many of the board positions of educational, medical and religious institutions. Because of its contribution to the city's economic health in the form of employment and tax revenues, the business community is in an extremely strong bargaining position and, as a result, its interests are often not so much directly expressed as *anticipated* by politicians and senior bureaucrats, many of whom seek the prestige, legitimacy and patronage that the business elite is able to confer.

The basic reason for business organizations' interest in urban affairs is clearly related to their desire to influence the allocation of public resources in favour of their localized investments. In general, the most influential nexus of interests is often the 'downtown business elite': directors of real estate companies, department stores and banks, together with retail merchants and the owners and directors of local newspapers who rely heavily on central city business fortunes for the maintenance of their advertising revenue.

The policies for which this group lobbies are those that can be expected to sustain and increase the commercial vitality of the central city. Given the widespread trend towards the decentralization of jobs and residences, one of their chief objectives has been to increase the accessibility and attractiveness of the CBD as a place in which to work and shop, and this has led business interests to support urban motorway programmes, improvements to public transport systems, urban renewal schemes and the construction of major amenities such as convention centres and theatres from public funds. In Washington, DC, for example, merchants formed a Downtown Retail Merchants Association to fight for what they described as the 'survival of small retailers'.

Labour

Organized labour, in the form of trade unions, represents the obvious counterbalance to the influence of the business elite in urban affairs. But, while organized labour is a major component of the parapolitical structure at the national level, it has traditionally exercised little influence on urban affairs. It is true that union representation on civic organizations has been widespread, and many union officials have been actively engaged in local party political activities; but organized labour in general has been unwilling to use its *power* (the withdrawal of labour) over issues that are not directly related to members' wages and conditions. In Britain, Trade Councils have provided a more community-based forum for trade unionists and have taken a direct interest in housing and broader social problems, but they are concerned primarily with bread-and-butter industrial issues rather than those related to the size and allocation of the 'social wage'.

The point is that organized labour in most countries (France and Italy being the important exceptions) is essentially and inherently reformist. Occasionally, however, union activity does have direct repercussions on the urban environment. In Australia, for example, the Builders' Labourers Federation has organized 'Green Bans' that have held up development projects on the grounds that they were environmentally undesirable; construction unions in the United States have resisted changes to building regulations that threatened to reduce the job potential of their members; and the pressures of local government fiscal retrenchment have drawn public-employee unions directly into the local political arena.

Citizen organizations and special interest groups

It is commonly claimed in the literature of political sociology that voluntary associations are an essential

Box 5.2

Key trends in urban social geography – The resurgence of Barcelona

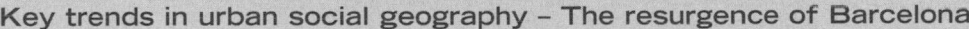

One of the most celebrated examples of *urban entrepreneurialism* (see also Box 5.1) is the case of Barcelona, a city in the Catalan region of northeast Spain. In the 1970s Barcelona was a decaying, dirty, traffic-congested port, yet within a few decades it had become a prime example of economic growth based around cultural regeneration.

Under the leadership of mayor Pasqual Maragall, a coalition of liberal middle classes and working-class immigrants has enabled Catalonian identity and regional autonomy to find expression in a redeveloped and cosmopolitan city. Not least important in encouraging the renaissance was the hosting of the Olympic Games in 1992. This brought enormous publicity for the city and encouraged considerable inward investment from the private sector. Just as Glasgow has exploited the architectural heritage left by Mackintosh, so Barcelona has made the most of the remarkable buildings created by the unique and enigmatic architect Gaudi. These have been supplemented by a staggering array of new architectural forms catering for museums, art galleries, leisure centres and offices.

However, as in other resurgent cities, the regeneration of Barcelona has not been without social costs. Many poorer groups have been dislocated following the gentrification of inner-city districts. In addition, the famous tree-lined boulevard popularly known as La Rambla, has become so popular that crowds of tourists have begun to undermine the charms of strolling past the shops bars and cafes (McNeill, 2004a). Finally, Port Vell, part of the gentrified waterfront, has ceased to be a venue sought out by the affluent. Instead, a mixture of young adult tourists, immigrant groups and working-class citizens have led to social tensions, sometimes resulting in crime and violence. Things came to a head following the drowning of an Ecuadorian immigrant who was beaten by door staff after his ejection from a prominent night-club (Chatterton and Hollands, 2003). City authorities have subsequently developed plans to restrict alcohol consumption and promote greater social diversity and family-oriented activities.

Key concepts associated with Barcelona (see Glossary)

Civic boosterism, gentrification, growth coalitions, property-led development, regime theory, spectacle, urban entrepreneurialism.

Further reading

Chatterton, P. and Hollands, R. (2003) *Urban Nightscapes: Youth Cultures, Pleasure Spaces and Power* Routledge, London.

Marshall, T. (ed.) (2005) *Transforming Barcelona* Routledge, London

McNeill, D. (2004a) Barcelona as imagined community; Pasqual Maragall's spaces of engagement *Transactions, Institute of British Geographers*, **26**, 340–352

McNeill, D. (2004c) *New Europe: Imagined Spaces* Arnold, London

Links with Other Chapters

Chapter 10: Box 10.1 The Development of 'Urban Nightscapes'

Chapter 11: Box 11.3 The Resurgence of Dublin

---

component of the democratic infrastructure, helping to articulate and direct the feelings of individuals into the relevant government channel. The most graphic examples can be seen where grand plans for urban change have failed to take account of the complexity of local feelings. Ravetz (1980), for example, recounted the case of Chesterfield market, where local opposition was mounted against the local authority's scheme to demolish listed buildings and an open air market to make way for a speculative developer's shopping complex. The proposal successfully found its way through the various statutory stages, only to be shelved after a protest march of several thousand citizens and a petitions signed by over 30 000 persons. A similar example is provided by the proposed redevelopment of Covent Garden in London (Christensen, 1982).

But relatively little is actually known about the number of citizen organizations of different kinds in cities, whose interests they represent, and how many of them are ever politically active. There have been numerous case studies of pressure group activity over controversial issues such as urban renewal, transportation and school organization, but these represent only the tip of the pressure-group iceberg, leaving the remaining nine-tenths unexplored. This other nine-tenths encompasses a vast range of organizations, including work-based clubs and associations, church clubs, welfare organizations, community groups such as tenants' associations and parents' associations, sports clubs, social clubs, cause-oriented groups such as Shelter, Help the Aged and the Child Poverty Action Group, groups that emerge over particular local issues (e.g. the 'Save Covent Garden' campaign), as well as political organizations per se.

Given the nature of local government decision-making, many of these are able to influence policy and resource allocation on the 'squeaky wheel' principle. This need not necessarily involve vociferous and demonstrative campaigns. The Los Angeles system of landscaped parkways, for example, is widely recognized as the result of steady lobbying by *Sunset* magazine, the official organ of obsessive gardening and planting in southern California. But not all organizations are politically active in any sense of the word. Their passivity is of course a reflection of the passivity of the community at large. It should not be confused with neutrality, however, since passivity is effectively conservative, serving to reinforce the status quo of urban affairs.

Homeowners' associations: private governments

In contrast to this passivity, affluent homeowners have come to represent an increasingly important element within the parapolitical structure. This has been achieved through homeowners' associations (also known as residential community associations and property owners' associations). Legally, these are simply private organizations that are established to regulate or manage a residential subdivision or condominium development. In practice, they constitute a form of private government whose rules, financial practices and other decisions can be a powerful force within the sociospatial dialectic. Through boards of directors elected by a group of homeowners, they levy taxes (through assessments), control and regulate the physical environment (through covenants, controls and restrictions attached to each home's deed), enact development controls, maintain commonly owned amenities (such as meeting rooms, exercise centres, squash courts and picnic areas) and organize service delivery (such as garbage collection, water and sewer services, street maintenance, snow removal and neighbourhood security).

The private nature of these associations means that they are an unusually 'stealthy' element of the parapolitical structure. In most of the developed industrial nations, homeowners' associations have proliferated, but their numbers and activities remain largely undocumented (Teaford, 1997). In the United States, where the phenomenon is probably most pronounced, homeowners' associations range in size and composition from a few homes on a single city street to thousands of homes and condominiums covering hundreds of acres. Altogether, it is estimated that there are over 250 000 homeowners' associations in the United States (compared with fewer than 500 in the early 1960s and around 20 000 in the mid-1970s), together covering more than 20 per cent of the nation's households and 80 million people. At least half of all housing currently on the market in the 50 largest metropolitan areas and nearly all new residential development in California, Florida, New York, Texas and suburban Washington, DC, is subject to mandatory governance by a homeowners' association.

The earliest homeowners' associations, from the first examples in the 1920s to the point in the mid-1960s when a new wave of suburbanization provided the platform for the proliferation of a new breed of associations, were chiefly directed towards exclusionary segregation. They were, as Mike Davis put it, 'overwhelmingly concerned with the *establishment* of what Robert Fishman has called "bourgeois utopia": that is, with the creation of racially and economically homogeneous residential enclaves glorifying the single-family home' (Davis, 1990, pp. 169–170). Their activities involved crude and straightforward legal-spatial tactics. At first, the most popular instrument was the *racially restrictive covenant*. This was a response to the Supreme Court's judgment against segregation ordinances enacted by public municipalities (*Buchanan v. Warley*, 1917); it was, in turn, outlawed by a Supreme Court case (*Shelley v. Kraemer*, 1948) (Delaney, 1993). Later, they turned to campaigns for *incorporation* that would enable them, in their metamorphosis to a public government, to deploy 'fiscal zoning' (e.g. limiting the construction of multifamily dwellings, raising the minimum lot size of new housing) as a means of enhancing residential exclusivity.

The explosive growth of homeowners' associations in recent years has been driven by the logic of the

Box 5.3

Key thinkers in urban social geography – Mike Davis

A writer who has had a big influence on urban social geography in recent years and our view of one city in particular – Los Angeles – is Mike Davis. In a similar vein to David Harvey (see Box 1.1), Davis's approach is staunchly in the Marxian tradition and underpinning the diversity of urban forms in Los Angeles he sees evidence of a continuing class struggle. However, in contrast to Harvey's somewhat abstract economic theorizing, Davis's writings are full of empirical details about Los Angeles, in particular, the rapacious behaviour of developers, the bigoted conservatism of suburban homeowners and the struggles of exploited Latino and African-American workers.

The popularity and influence of Davis's work has been enhanced by a highly accessible, almost 'journalistic' style. Not that Davis's work is a pleasant read: he presents a rather doom-laden view of Los Angeles that has led some to label him a latter-day Jeremiah (McNeill, 2004b; Merrifield, 2002). Nevertheless, the dystopian vision

outlined in Davis's book *City of Quartz: Excavating the Future in Los Angeles* (1990) seemed to chime with the mood of the times, as manifest in the subsequent urban riots in Los Angeles in 1992.

In his follow-up volume *Ecology of Fear: Los Angeles and the Imagination of Disaster* (1998) Davis played much greater attention to issues of nature than previous urban scholars. He portrayed Los Angeles as a city in a overstretched ecosystem at the mercy of various catastrophes of earthquake, fire and drought. In this book Davis also outlined the impact of growing crime, drug trafficking and violence through measures such as increased surveillance, private armies and gated communities.

Contrary to his reputation for gloomy prognostications on urban development, his third major book, *Magical Urbanism: Latinos Reinvent the American City* (2000) is a celebration of the diversity and vitality brought to US inner cities by new Latino communities.

Key concepts associated with Mike Davis (see Glossary)

Bunker architecture, Los Angeles School, 'scanscape'.

Further reading

McNeill, D. (2004b) Mike Davis, in P. Hubbard, R. Kitchin and G. Valentine (eds) *Key Thinkers on Space and Place* Sage, London

Merrifield, A. (2002) *MetroMarxism: A Marxist Tale of the City* Routledge, New York

Links with Other Chapters

Chapter 2: Neo-Fordism

Chapter 8: Box 8.1 The Latinization of US Cities

Chapter 14: Los Angeles and the California School

---

real-estate industry, which saw mandatory membership in pre-established homeowners' associations as the best way to ensure that ever-larger and more elaborately packaged subdivisions and residential complexes would maintain their character until 'build-out' and beyond. Initially concerned chiefly with the preservation of the aesthetics and overall design vision of 'high end' developments, these common-interest associations soon moved to defend their residential niches against unwanted development (such as industry, apartments and offices) and then, as environmental quality became an increasingly important social value, against any kind of development.

This 'Sunbelt Bolshevism', as Davis calls it in the context of southern California, became an important element in the no-growth/slow-growth politics of

American suburbs: 'the latest incarnation of a middle-class political subjectivity that fitfully constitutes and reconstitutes itself every few years around the defence of household equity and residential privilege' (Davis, 1990, p. 159). At the same time, homeowners' associations established themselves as regular participants at public meetings of city councils, school districts and planning boards. Complaining about encroachment and undesirable development, they represent the vanguard of the **NIMBY** (Not In My Back Yard) movement (see also Box 5.3).

Urban social movements

The impact of affluent homeowner groups raises the question of whether it is possible for disadvantaged groups to sidestep the traditional institutions of urban affairs so that they too can somehow achieve greater power. The conventional answer has been negative, but a distinctive new form of urban **social movement** was precipitated in the 1970s by fiscal stress that in turn led to crises in the provision of various elements of collective consumption. These crises had a serious impact on many of the skilled working classes and the lower-middle and younger middle classes as well as on disadvantaged and marginalized groups. As a result, concern over access to hospitals, public transport (especially commuter rail links), schools and so on, together with frustration at the growing power of technocratic bureaucracies and disillusionment with the formal institutions of civil society, gave rise to a new kind of urban social movement that was based on a broad alliance of anti-establishment interests.

Nevertheless, urban social movements have generally been sporadic and isolated, and some observers have questioned the initial expectation that community consciousness, activated by issues related to collective consumption, can in fact lead to the kind of class consciousness that is assumed to be a prerequisite to achieving a degree of power on a more permanent basis. The answer here seems to be contingent: it depends on whether a community can mobilize an awareness of the structural causes of local problems. This caveat is the basis of the typology of political movements shown in Fig. 5.2. Two of these categories represent urban social movements. *Community-defined* movements are those that are purely local: issue-oriented movements that are bounded by particular context and circumstance.

Community-based movements are those that transcend the initial issue, context and circumstance to form the basis of alliances that are able to achieve a broader and more permanent measure of power (Fitzgerald, 1991).

5.4 Community power structures and the role of the local state

How are the relationships between these various groups and decision-makers structured? Who really runs the community and what difference does it make to the local quality of life? These are questions that have concerned political scientists and urban sociologists for some time (see, for example, Judge *et al.*, 1995) and that have now caught the attention of urban social geographers because of their implications for the sociospatial dialectic. There are two 'classic' types of urban power structure – monolithic and pluralistic – each of which has been identified in a wide range of cities since their 'discovery' by Hunter (1953) and Dahl (1961) respectively. In his study of 'Regional City' (Atlanta), Hunter found that nearly all decisions were made by a handful of individuals who stood at the top of a stable power hierarchy. These people, drawn largely from business and industrial circles, constituted a strongly entrenched and select group: with their blessing, projects could move ahead, but without their express or tacit consent little of significance was ever accomplished.

In contrast, the classic **pluralistic model** of community power advanced by Dahl in the light of his analysis of decision-making in New Haven, Connecticut, holds that power tends to be dispersed, with different elites dominant at different times over different issues. Thus, if the issue involves public housing, one set of participants will control the outcome; if it involves the construction of a new health centre, a different coalition of leaders will dominate. In Dahl's model, therefore, business elites of the kind Hunter found to be in control of Atlanta are only one among many influential 'power clusters'. As Dahl put it:

> the Economic Notables, far from being a ruling group, are simply one of many groups out of which individuals sporadically emerge to influence

		Community consciousness	
		low	high
Class consciousness	low	competitive individualism	community-defined movement
	high	class struggle	community-based movement

Figure 5.2 A typology of political movements.
Source: Fitzgerald (1991), Fig. 1, p. 120.

the politics and acts of city officials. Almost anything one might say about the influence of the Economic Notables could be said with equal justice about half a dozen other groups in the New Haven community.

(Dahl, 1961, p. 72)

Dahl argued that the system as a whole is democratic, drawing on a wide spectrum of the parapolitical structure and ensuring political freedom through the competition of elites for mass loyalty. When the policies and activities of existing power structures depart from the values of the electorate, he suggested, people will be motivated to voice their concerns and a new power cluster will emerge. According to the pluralist model, therefore, we may expect the interplay of views and interests within a city to produce, in the long run, an allocation of resources that satisfies, to a degree, the needs of all interest groups and neighbourhoods: problems such as neighbourhood decay are seen merely as short-term failures of participation in the political process. On the other hand, we might expect monolithic power structures to lead to polarization of well-being, with few concessions to the long-term interests of the controlling elite.

Regime theory

More recently, it has been suggested that urban politics should be seen not in terms of monolithic versus pluralistic power structures but in terms of the evolution and succession of a series of regimes. **Regime theory** attempts to examine how various coalitions of interest come together to achieve outcomes in cities. Frequently these are the interests of '**pro-growth coalitions**' put together by political entrepreneurs in order to achieve concrete solutions to particular problems (Lauria, 1996; Rutheiser, 1996; Feagin, 1997; Jonas and Wilson, 1999). The crucial point about regime theory is that power does not flow automatically but has to be actively acquired. In the context of economic restructuring and metropolitan change, city officials seek alliances, it is argued, that will enhance their ability to achieve visible policy results.

These alliances between public officials and private actors constitute regimes through which governance

rests less on formal authority than on loosely structured arrangements and dealmaking. With an intensification of economic and social change in the 'postindustrial' city, new sociopolitical cleavages – green, yuppie, populist, neo-liberal – have been added to traditional class- and race-based cleavages, so that these regimes have become more complex and, potentially, more volatile. Meanwhile, the scale and extent of economic restructuring has meant that greater competition for economic development investments *between* municipalities has established a new dynamic whereby the intensity of political conflict *within* them is muted. Such considerations require us to take a broader view of urban politics.

Structuralist interpretations of the political economy of contemporary cities

Many scholars have turned to structuralist theories of political economy in response to the need to relate urban spatial structure to the institutions of urban society. At their most fundamental level, structuralist theories of political economy turn on the contention that all social phenomena are linked to the prevailing mode of production discussed in Chapter 2. This is the material economic base from which everything else – the social **superstructure** – derives.

In historical terms, the economic base is the product of a dialectical process in which the prevailing ideology, or 'thesis', of successive modes of production is overthrown by contradictory forces (the 'antithesis'), thus bringing about a *transformation* of society to a higher stage of development: from subsistence tribalism through feudalism to capitalism and eventually, as Marx believed, to socialism. The base in Western society is of course the capitalist mode of production and, like other bases, it is characterized by conflict between opposing social classes inherent in the economic order. The superstructure of capitalism encompasses everything that stems from and relates to this economic order, including tangible features such as the morphology of the city as well as more nebulous phenomena such as legal and political institutions, the ideology of capitalism and the counter-ideology of its antithesis.

111

As part of this superstructure, one of the major functions of the city is to fulfil the imperatives of capitalism, the most important of which is the *circulation* and **accumulation** of capital. Thus the spatial form of the city, by reducing indirect costs of production and costs of circulation and consumption, speeds up the rotation of capital, leading to its greater accumulation. Another important role of the city, according to structuralist theory, is to provide the conditions necessary for the perpetuation of the economic base. In short, this entails the **social reproduction** of the relationship between labour and capital and, therefore, the stabilization of the associated social structure. One aspect of this is the perpetuation of the economic class relationships through ecological processes, particularly the development of a variety of suburban settings with differential access to different kinds of services, amenities and resources.

The role of government is particularly important in this respect because of its control over the patterns and conditions of provision of schools, housing, shopping, leisure facilities, and the whole spectrum of collective consumption. Moreover, it can also be argued that urban neighbourhoods provide distinctive milieux from which individuals derive many of their consumption habits, moral codes, values and expectations. The resulting homogenization of life experiences within different neighbourhoods reinforces the tendency for relatively permanent social groupings to emerge within a relatively permanent structure of residential differentiation.

The division of the proletariat into distinctive, locality-based communities through the process of residential differentiation also serves to fragment class consciousness and solidarity while reinforcing the traditional authority of elite groups, something that is also strengthened by the symbolic power of the built environment. In short, the city is at once an expression of capitalism and a means of its perpetuation. It is here that we can see the notion of a sociospatial dialectic in its broadest terms.

Meanwhile, it is also recognized that the structure of the city reflects and incorporates many of the *contradictions* in capitalist society, thus leading to local friction and conflict. This is intensified as the city's economic landscape is continually altered in response to the 'creative destruction' of capital's drive towards the accumulation of profits. Residential neighbourhoods

are cleared to make way for new office developments; disinvestment in privately rented accommodation leads to the dissolution of inner-city communities; the switch of capital to more profitable investment in private housing leads to an expansion of the suburbs; and so on.

This continual tearing down, re-creation and transformation of spatial arrangements brings about locational conflict in several ways. Big capital comes into conflict with small capital in the form of retailers, property developers and small businesses. Meanwhile, conflict also arises locally between, on the one hand, capitalists (both large and small) and, on the other, those obtaining important use and exchange values from existing spatial arrangements. This includes conflict over the nature and location of new urban development, over urban renewal, road construction, conservation, land use zoning and so on: over the whole spectrum, in fact, of urban affairs.

Underlying most structuralist analyses of the political economy of cities is the additional hypothesis involving the role of the state as a **legitimating agent**, helping to fulfil the imperatives of capitalism in a number of ways. These include defusing discontent through the pursuit of welfare policies, the provision of a stable and predictable environment for business through the legal and judicial system, and the propagation of an ideology conducive to the operation and maintenance of the economic base through its control and penetration of socializing agencies such as the educational system, the armed forces and the civil service.

We should at this point remind ourselves again of the major criticism of structuralist theory – that it does not give sufficient recognition to the influence of human **agency**: the actions of individuals are seen in structuralist theory as a direct function of economic and sociopolitical structures. It has been argued, however, that there are elemental and universal human drives and behavioural responses that give life and structure to the city and that people are capable of generating, independently, important ideas and behaviours. Furthermore, it is argued that these products of the human spirit can sometimes contribute to the economic and sociopolitical apparatus that structuralist theory attributes exclusively to the material economic base. These ideas are the basis of the poststructuralist approaches described in Chapter 1; we will encounter them in subsequent chapters. Nevertheless, the overall contribution of structuralist theory is clearly significant. It provides a clear break with earlier, narrower conceptions of urban sociospatial relationships and a flexible theoretical framework for a wide range of phenomena. As we shall see, it has been deployed in the analysis of a variety of issues.

The local state and the sociospatial dialectic

Despite the clear importance of local government – even simply in terms of the magnitude of expenditure on public services – there is no properly developed and generally accepted theory of the behaviour and objectives of local government, or the 'local state'. Within the debate on this question, however, three principal positions have emerged:

➤ that the local state is an adjunct of the national state, with both acting in response to the prevailing balance of class forces within society (the 'structuralist' view);

➤ that the local state is controlled by officials, and that their goals and values are crucial in determining policy outcomes (the 'managerialist' view);

➤ that the local state is an instrument of the business elite (the 'instrumentalist' view).

The managerialist view (also termed **managerialism**) has generated widespread interest and support, and it is clear that a focus on the activities and ideologies of professional decision-makers can contribute a lot to the understanding of urban sociospatial processes (as we shall see in Chapter 6 in relation to the social production of the built environment). We shall also see, however, that the managerialist approach does not give adequate recognition either to the influence of local elites and pressure groups or to the economic and political constraints stemming from the national level.

Because of such shortcomings, attention has more recently been focused on the structuralist and instrumentalist positions, both of which share certain views on the role of the local state. These may be summarized in terms of three broad functions:

➤ Facilitating private production and capital accumulation (through, for example, the provision of the urban infrastructure; through planning processes that ease the spatial aspects of economic restructuring; through the provision of technical education; and through 'demand orchestration' which, through public works contracts, etc., brings stability and security to markets).

➤ Facilitating the reproduction of labour power through **collective consumption** (through, for example, subsidized housing).

➤ Facilitating the maintenance of social order and social cohesion (through, for example, the police, welfare programmes and social services, and 'agencies of legitimation' such as schools and public participation schemes).

Where the structuralist view differs from the instrumentalist view is not so much in the identification of the functions of the state as in the questions of for whom or for what the functions operate, whether they are class-biased, and the extent to which they reflect external political forces.

Regulation theory and urban governance

Some theorists have attempted to reconcile elements of all three models – managerialist, structuralist and instrumentalist. This, in turn, has led to the suggestion that the regulationist approach (introduced in Chapter 2) provides the most appropriate framework for understanding the local state (Dunford, 1990; Goodwin *et al.*, 1993). **Regulation theory** is based on the concept of successive **regimes of accumulation** that represent particular organizational forms of capitalism (e.g. Fordism), with distinctive patterns and structures of economic organization, income distribution and collective consumption. Each such regime develops an accompanying **mode of regulation** that is a collection of structural forms (political, economic, social, cultural) and institutional arrangements that define the 'rules of the game' for individual and collective behaviour. The mode of regulation thus gives expression to, and serves to reproduce, fundamental social relations. It also serves to guide and to accommodate change within the political

Changing governance in the city: redeveloped warehouses in London's Docklands. Photo Credit: Paul Knox

economy as a whole. In this context, the local state can be seen as both an object and an agent of regulation, a semi-autonomous institution that is itself regulated so that its strategies and structures can be used to help forge new social, political and economic relations within urban space.

One of the most important regulationist-inspired concepts of recent years is the notion of **governance**. The term is used to indicate the shift away from *direct* government control of the economy and society via hierarchical bureaucracies towards *indirect* government control via diverse non-governmental organizations (Jessop, 1995). Jessop argues that governance is part of the shift towards a new mode of regulation dictated by the demands of the new Fordist regime of accumulation. As introduced in Chapter 2, this new regime is characterized by continuous innovation to achieve global competitiveness via economies of scope in specialized products. This approach, it is argued, is necessary to compensate for the saturation of mass markets and the

inability of states to engage in Keynesian policies of demand management in a deregulated global financial system. This requires a new mode of regulation in which the interests of welfare are subordinated to those of business. This in turn has required circumventing many of the traditional democratic institutions such as local government that are perceived to be resistant to the new agenda of welfare reductions, flexibility and private–public partnerships. In addition, governance has been encouraged by the extensive critique of the public sector mounted from all sides of the political spectrum in recent years – but especially from the Right – which sees public bureaucracies as inherently inefficient and self-seeking (see also Box 5.4).

In the welfare sphere, governance has meant a shift towards increasing use of a diverse range of voluntary and charitable bodies. Wolch has argued that this had led to the creation of a **shadow state** '. . . a para state apparatus with collective service responsibilities previously shouldered by the public sector, administered

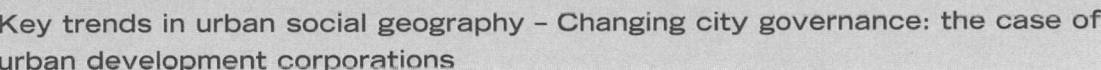

Box 5.4

Key trends in urban social geography – Changing city governance: the case of urban development corporations

One of the best examples of new forms of urban governance has been the activities of the British urban development corporations (UDCs) (Imrie and Thomas, 1999). These were quasi-public authorities constituted for between 5 and 15 years designated with the task of regenerating run-down inner-city areas. They were instigated under the Thatcher governments of the 1980s as part of a New Right project to infuse all aspects of British life with market forces. Their role was to provide infrastructure such as land and transport facilities to encourage private sector investment – hence this approach has been termed property-led development. UDCs have been established throughout Britain in cities including: Bristol, Leeds, Sheffield, Birmingham, Manchester and in London's Docklands.

These corporations were controversial for a number of reasons. First, there has been much dispute over the number of jobs generated by this public investment; many have argued that this publicly subsidized property speculation, far from generating new jobs, has actually retarded job growth. Certainly most commentators agree that the numbers of jobs created have been relatively small and expensive. Second, and perhaps more importantly, these corporations have bypassed traditional forms of local accountability. To facilitate a market-driven approach these corporations were dominated by representatives from the business community and local community groups were somewhat excluded. The UDCs are one of a series of quangos that have taken over many aspects of governance in the UK in recent years (Peck and Jones, 1995).

Broadly similar policies have been enacted in the United States. For example, empowerment zones are also intended to attract private sector funds into deprived areas but in this case there is a much explicit concern to secure local community support for any new initiatives. So far empowerment zones have been concentrated in the deindustrializing cities of the Northeast such as Detroit, where 80 local projects are aimed at improving training and social housing (Hall, 1998).

Key concepts associated with changing governance (see Glossary)

Civic boosterism, empowerment zones, 'hollowing-out', property-led development, quangos.

Further reading

Imrie, R. and Thomas, H. (1999) *British Urban Policy and the Urban Development Corporations* (second edition) Paul Chapman, London

---

outside traditional democratic politics, but yet controlled in formal and informal ways' (Wolch, 1989, p. 201; see also Wolch, 1990). In Los Angeles County Wolch found over 8500 voluntary bodies including both local neighbourhood organizations and large trusts and foundations. The scope of their activities ranged from social welfare to the arts, and their total spending amounted to a staggering 60 per cent of per capita municipal expenditures on public services. In the United Kingdom the shift towards contracting-out of services has meant that many welfare functions are now *funded* by the public sector but *delivered* by private sector organizations. This shift towards a diverse range of service providers is often termed **welfare pluralism** (see also Chapter 13).

Table 5.1 lists some new developments in British local governance following changes in the mode of regulation. As Goodwin and Painter (1996) observe, such tables can amount to little more than cataloguing a series of changes that are supposed to fit in with the Fordist/post-Fordist division without looking at the links between these new forms of governance and the new mode of regulation. What does seem clear, however, is that women have been somewhat marginalized in many of the new business-led **quango**s (quasi-autonomous non-governmental organizations) that dominate local policy (Tickell and Peck, 1996).

Redefining citizenship

One of the key elements of the reconfiguration of state activities around the new agendas of competitiveness, innovation and flexibility has been the reconstitution of notions of citizenship (Smith, 1989b; Kearns, 1995; Holston, 2001). This shift reflects the growing influence

Table 5.1 New developments in British local governance

Sites of regulation	British local governance in Fordism	New developments
Financial regime	Keynesian	Monetarist
Organizational structure of local governance	Centralized service delivery authorities pre-eminence of formal, elected local government	Wide variety of service providers Multiplicity of agencies of local governance
Management	Hierarchical Centralized Bureaucratic	Devolved 'Flat' hierarchies Performance driven
Local labour markets	Regulated Segmented by skill	Deregulated Dual labour market
Labour process	Technologically undeveloped Labour intensive Productivity increases difficult	Technologically dynamic (information-based) Capital intensive Productivity increases possible
Labour relations	Collectivized National bargaining Regulated	Individualized Local and individual bargaining 'Flexible'
Form of consumption	Universal Collective rights	Targeted Individualized 'contracts'
Nature of services provided	To meet local needs Expandable	To meet statutory obligations Constrained
Ideology	Social democratic	Neoliberal
Key discourse	Technocratic/managerialist	Entrepreneurial/enabling
Political forms	Corporatist	Neocorporatist (labour excluded)
Economic goals	Promotion of full employment Economic modernization based on technical advance and public investment	Promotion of private profit Economic modernization based on 'low-wage, low-skill, 'flexible' economy
Social goals	Progressive redistribution/social justice	Privatized consumption/active citizenry

Source: Goodwin and Painter (1996).

of continental European intellectual traditions in the sphere of welfare. Unlike the tradition established in the United States and United Kingdom in the nineteenth century, which was rooted in ideas of society as nothing more that a set of separate individuals, the European approach envisages society as a collectivity bound together in a mutual contract by obligations as well as rights. The notion of citizenship has therefore been expanded beyond the realm of entitlements to benefits to include reciprocal obligations to aim for self-reliance and paid work. This strategy, which is spreading throughout the Western world, is bound up with notions of 'workfare' that are discussed in greater detail in Chapter 13. Cynics would argue that this new version of citizenship is simply a way of saving welfare budgets by coercing unsuitable people into employment. Nevertheless, the new focus

on social exclusion recognizes that citizenship involves more than just material factors and also includes senses of belonging and democratic empowerment.

5.5 The question of social justice in the city

The issues of citizenship, patriarchy, racism, collective consumption, the law, the state and civil society reviewed in this chapter all raise, in one way or another, the question of social justice in the city. David Harvey, following Engels, recognizes that conceptions of justice vary not only with time and place, but also with the persons concerned (Harvey, 1992). It follows that it is essential to examine the material and moral bases for the production of the lifeworlds from which divergent concepts of social justice emerge. This we shall do in subsequent chapters. Meanwhile, however, it is useful to clarify some basic concepts and principles with respect to social justice: no student of social geography will be able to avoid confrontation with the moral (or theoretical) dilemmas of social justice for long.

There are, as Harvey points out, many competing concepts of social justice. Among them are: the positive law view (that justice is simply a matter of law); the utilitarian view (allowing us to discriminate between good and bad law on the basis of the greatest good of the greatest number); and the natural rights view (that no amount of greater good for a greater number can justify the violation of certain inalienable rights). Making clear sense of these concepts and understanding the 'moral geographies' of postindustrial cities and postmodern societies (which contain fragmented sociocultural groups and a variety of social movements, all eager to articulate their own definitions of social justice) is a task that even Harvey finds challenging. Having wrestled with the fundamental issues ever since writing the immensely influential volume on *Social Justice and the City*, published in 1973, Harvey avoids the fruitless task of reconciling competing claims to conceptions of social justice in different contexts. Rather, he follows Iris Young (1990) in focusing on sources of *oppression*, from which he draws six propositions in relation to just planning and policy practices:

➤ They must confront directly the problem of creating forms of social and political organization and systems of production and consumption that minimize the exploitation of labour power both in the workplace and the living place.

➤ They must confront the phenomenon of marginalization in a non-paternalistic mode and find ways to organize and militate within the politics of marginalization in such a way as to liberate captive groups from this distinctive form of oppression.

➤ They must empower rather than deprive the oppressed of access to political power and the ability to engage in self-expression.

➤ They must be sensitive to issues of cultural imperialism and seek, by a variety of means, to eliminate the imperialist attitude both in the design of urban projects and modes of popular consultation.

➤ They must seek out non-exclusionary and non-militarized forms of social control to contain the increasing levels of institutionalized violence without destroying capacities for empowerment and self-expression.

➤ They must recognize that the necessary ecological consequences of all social projects have impacts on future generations as well as upon distant peoples, and take steps to ensure a reasonable mitigation of negative impacts.

As Harvey recognizes, this still leaves a great deal for us to struggle with in interpreting the moral geographies of contemporary cities. Real-world geographies demand that we consider all six dimensions of social justice rather than applying them in isolation from one another; and this means developing some sense of consensus over priorities and acquiring some rationality in dealing with place- and context-specific oppression.

Box 5.5

Films related to urban social geography – Chapter 5

Chinatown (1974) A retro detective story of the *film noir* genre. Although a fictionalized account of the expansion of Los Angeles (see Chapter 14), many argue that this film captures the essence of local politics in the US.

Another on the 'must see' list for those interested in relationships between cinema and the city.

Devil in a Blue Dress (1995) Another detective story in a genre that has been termed the 'New Black Realism'. Deals with issues of race, identity and power but in a subtle way in an exciting story.

Chapter summary

5.1 The processes at work in cities have both shaped, and been shaped by, institutional frameworks. The law is a crucial link between the public and private spheres of city life. The changing economic base of urban areas has important implications for city life.

5.2 The fragmentation of cities into multiple political and administrative jurisdictions is justified in the interests of local democracy but often leads to considerable inequalities in resource allocation.

5.3 The functioning of local democracy in influenced by electoral frameworks, elected officials and city bureaucrats as well as the parapolitical structure of informal groups focused around residents, organized labour and various business interests.

5.4 Who has most power to effectively control cities is a question that has produced numerous theories principally: the pluralist, managerialist, instrumentalist and structuralist interpretations. Regulation theory is seen by some as a way of overcoming the limitations of these previous theories.

5.5 The functioning of institutional frameworks in mediating the activities of citizens and the state raises crucial issues of social and territorial justice.

Key concepts and terms

accumulation
agency
balkanization
citizenship
civic entrepreneurialism
civil society
coalition building
collective consumption
decision rules
de jure territories
empowerment zones
fiscal imbalance
fiscal mercantilism
fiscal stress
gerrymandering
governance
jurisdictional partitioning
laissez-faire

legitimating agent
local state
malapportionment
managerialism
manipulated city hypothesis
metropolitan fragmentation
mode of regulation
municipal socialism
NIMBY
parapolitical structure
paternalism
pluralistic model
pro-growth coalitions
property-led development
public sphere
quango
racism
regime of accumulation

regime theory
regulation theory
sexism
shadow state
social movements
social reproduction
state
structuralism
subsidiarity
suburban exploitation thesis
superstructure
sustainability
territorial justice
urban development corporations
voluntarism
welfare pluralism
zoning

Suggested reading

The author who has contributed most to the understanding of the spatial and institutional frameworks of the law and the state is Gordon Clark. His essay on 'Geography and law' in *New Models in Geography*, edited by R. Peet and N. Thrift (1990: Unwin Hyman, London) provides a good review of the interface between law and spatial change; while his co-authored book (with Michael Dear) on *State Apparatus: Structures and Language of Legitimacy* (1984: Allen and Unwin, London) provides both breadth and depth of coverage on the question of how to theorize the state from a geographical perspective. A more recent collation of work on the relationships between law and geography is Nicholas Blomley *et al. The Legal Geographies Reader* (2000: Blackwell, Oxford). Another authoritative author in this field is Ron Johnston: see his essay on 'The territoriality of law' (*Urban Geography*, 1990: **11**, 548–565), his review of the local state, local government, and local administration, in *The State in Action*, edited by J. Simmie and R. King (1990: Belhaven Press, London), and his book on *Political, Electoral, and Spatial Systems* (1979: Clarendon Press, Oxford), which also deals with the spatial organization of political boundaries in cities. An excellent review of the interrelationships between class and community in contemporary cities can be found in Kevin Cox's essay on 'The politics of turf and the question of class,' in *The Power of Geography*, edited by Jennifer Wolch and Michael Dear (1989: Unwin Hyman, London). An extensive review of the changes that occurred in British and American metropolitan areas in the 1980s with respect to socioeconomic change, political change, and metropolitan restructuring can be found in Brian Jacobs's book: *Fractured Cities: Capitalism, Community and Empowerment* in Britain and America (1992: Routledge, Chapman, and Hall, Andover); a more theoretical approach to the same issues can be found in Michael Peter Smith's book on *City, State and Market* (1988: Blackwell, Oxford). Also highly recommended is David Judge, Gerry Stoker and Harold Wolman (eds) *Theories of Urban Politics* (1995: Sage, London) and Mickey Lauria's *Reconstructing Urban Regime Theory: Regulating Urban Politics in a Global Economy* (1996: Sage, Beverly Hills). For a lively overview on recent developments in US cities see Joe R. Feagin's *The New Urban Paradigm: Critical Perspectives on the City* (1997: Rowman and Littlefield, Lanham) and Jon C. Teaford *Post-suburbia: Government and Politics in the Edge Cities* (1997: Johns Hopkins University Press, Baltimore) is worth seeking out. *The Social Control of Cities* by Sophie-Gendrot (1999: Blackwell, Oxford) provides a comparative analysis of the relationships between national elites and marginalized urban populations. For developments in urban policy in the United Kingdom see Rob Imrie and Huw Thomas (eds) *British Urban Policy and the Urban Development Corporations* (second edition, 1999: Paul Chapman, London); Rob Atkinson and Graham Moon *Urban Policy in Britain: The City, The State and the Market* (1999: Macmillan, London); Patsy Healey *et al.* (eds) *Managing Cities: the New Urban Context* (1995: John Wiley, London), Nick Oatley (ed.) *Cities, Economic Competition and Urban Policy* (1998: Paul Chapman, London) and Dilys M. Hill *Urban Policy and Politics in Britain* (2000: Palgrave, Basingstoke). Issues of governance in EU cities are considered in W. Salet *et al. Metropolitan Governance and Spatial Planning* (2002: Spon, London) and T. Herrschel and P. Newman *Governance of Europe's City Regions* (2002: Spon, London). For a review of changes in citizenship see the special issue of *Urban Studies*, in February 2003, Volume 24 Issue 2. The question of social justice is considered in David Harvey's essay on 'Social justice, postmodernism, and the city,' in the *International Journal of Urban and Regional Research* (1992: **16**, 588–601). A comprehensive guide to issues of social justice in a geographical setting is David Smith *Geography and Social Justice* (1994: Blackwell, Oxford) together with his more recent volume *Moral Geographies: Ethics in a World of Difference* (2000: Edinburgh University Press, Edinburgh) and his 'Social justice revisited' (2000: *Environment and Planning A*, **52**, 1149–1162). Finally, Mike Davis provides a very readable account of the political biography of one city – Los Angeles – in the second chapter ('Power Lines') of his *City of Quartz: Excavating the Future in Los Angeles* (1990: Verso, London). The third chapter of this book ('Homegrown Revolution') has an absorbing account of homeowners' associations; the rest of the book is in fact equally absorbing and provides an excellent stimulus and accompaniment to the study of urban social geography.

Chapter 6

Structures of building provision and the social production of the urban environment

Key questions addressed in this chapter

➤ What are the key processes that affect the allocation of housing in cities?

➤ In what ways does the operation of housing markets affect the residential structure of cities?

In this chapter, we explore another of the fundamental dimensions of the sociospatial dialectic: the intermediate-level structures and processes associated with the production of the built environment. 'Production' here is used in its widest sense – not just the construction of the built environment but also the exchange, distribution and use of the different elements and settings that provide the physical framework for the economic, social, cultural and political life of cities. At one level, all these

aspects of production can be seen in terms of the broad machinations of economics: supply and demand, working within (and interacting with) long-wave economic cycles and conditioned by the evolving institutional structures described in Chapter 5.

Yet the production of the built environment is not simply a function of supply and demand played out on a stage set by broad economic and institutional forces. It is also a function of time- and place-specific social relations that involve a variety of key actors (including landowners, investors, financiers, developers, builders, design professionals, construction workers, business and community leaders, and consumers). At the same time, the state – both local and national – must be recognized as an important agent in its own right and as a regulator of competition between various actors.

These sets of relations represent *structures of building provision* through which we can understand the social production of the built environment. These structures of building provision need to be understood in terms of their specific linkages (functional, historical, political, social and cultural) with the broader structural elements (economic, institutional) of the political economy (D'Arcy and Keogh, 1997; McAuslan, 2003). A comprehensive survey of the structures of building provision is beyond the scope of this book. We shall, instead, illustrate the social production of the built environment: first of all by showing how the dynamics of housing supply are socially constructed through the dynamics of the major housing submarkets and, second, by showing how some of the key actors in these submarkets are implicated in the structures of building provision.

6.1 Housing submarkets

Much of the importance of the structures of building provision to the sociospatial dialectic has to do with the special nature of housing as a commodity:

> It is fixed in geographic space, it changes hands infrequently, it is a commodity which we cannot do without, and it is a form of stored wealth which is subject to speculative activities in the market . . . In addition, the house has various forms of value to the user and above all it is the point from which the user relates to every other aspect of the urban scene.
>
> (Harvey, 1972, p. 16)

These qualities make for highly complex urban housing markets in which the needs and aspirations of different socioeconomic groups are matched to particular types of housing through a series of different market arrangements. In short, there exists in each city a series of distinctive **housing submarkets**. To the extent that these submarkets are localized, they have a direct expression in the residential structure of the city. At the same time, the spatial outcome of each submarket is significantly influenced by the actions of key decision-makers and mediators such as landowners, developers, estate agents and housing managers, whose motivation and behaviour effectively structures the supply of housing from which relocating households make their choices.

It is important to bear in mind that the housing available in any particular submarket is a complex package of goods and services that extends well beyond the shelter provided by the dwelling itself. Housing is also a primary determinant of personal security, autonomy, comfort, well-being and status, and the ownership of housing itself structures access to other scarce resources, such as educational, medical and leisure facilities. The net utility of these services is generally referred to as the **use value** of housing. This value is fixed not by the attributes of housing alone, for utility is very much in the eyes of the beholder and will vary a good deal according to life-course, lifestyle, social class and so on. The use value of housing will be a major determinant of its **exchange value** in the marketplace, although the special properties of housing as a commodity tend to distort the relationship. In particular, the role of housing as a form of stored wealth means that its exchange value will be influenced by its potential for reaping unearned income and for increasing capital.

In general it is useful to think in terms of housing markets as the focus for a variety of 'actors' operating within the various constraints of political and institutional contexts, the result of which are spatial outcomes that can be identified in terms of land use changes, occupancy patterns, social area changes, housing prices and housing quality (Fig. 6.1). Traditional definitions of housing submarkets have been couched in terms of the attributes of housing stock (type of dwelling, type of tenancy and price), household type (family status, economic status and ethnicity) or location (Fig. 6.2). But it is unusual for housing submarkets to form such neat, discrete compartments within any given city.

In the remainder of this chapter, detailed consideration is given to the housing groups and agencies of housing supply, beginning with a summary and explanation of the major trends in the transformation of urban housing: the increase in the construction of dwellings for homeownership, the decrease in the availability of cheaper privately rented dwellings, and the increase (in many countries) and subsequent decline in the construction and letting of dwellings by public authorities. In doing so, we shall see that *the dynamics of housing supply are socially constructed*: they are, ultimately, a product of the interdependence of political, economic and

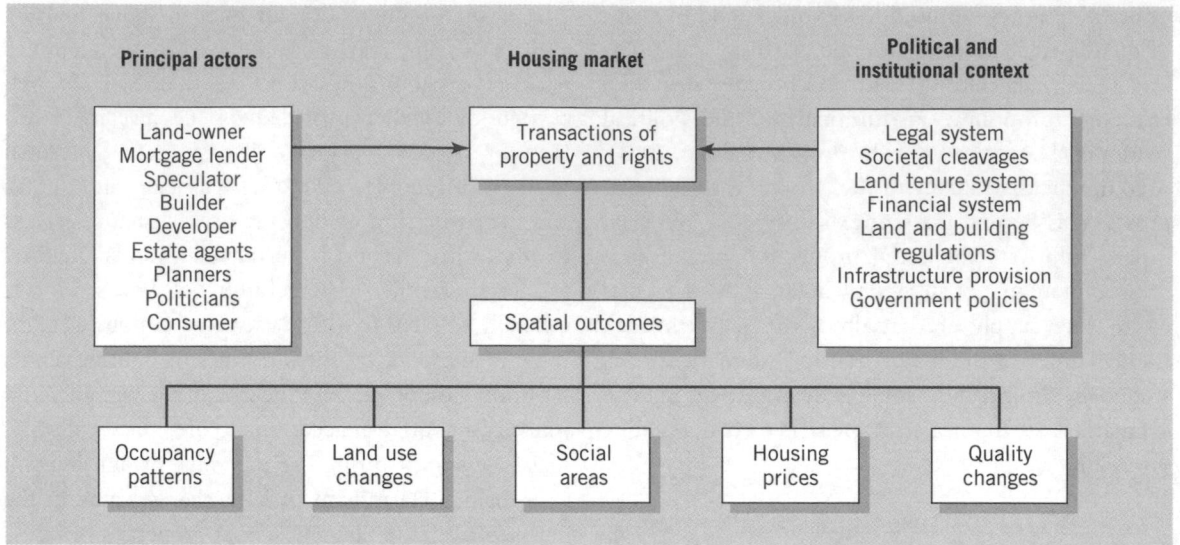

Figure 6.1 Actors and institutions in the housing market.
Source: Bourne (1981), Fig. 4.8, p. 85.

ideological factors. Most of these factors recur throughout the Western world; but their precise composition and their interrelationships make for significant variability between individual cities and nations.

The growth of homeownership

The growth of homeownership is characteristic of all Western countries and it has had a marked effect not only on residential differentiation but also on the whole space-economy of urbanized societies. In the United States, the overall proportion of owner-occupied dwellings rose from 20 per cent in 1920 to 44 per cent in 1940 and 65 per cent in 2000; in Britain, the proportion of owner-occupied dwellings rose steadily from 10.6 per cent in 1914 to 28 per cent in 1953, accelerated to reach 52 per cent by 1973, and was estimated to be just under 70 per cent in 2004. There is, of course, plenty of variability within countries: individual cities can develop their own **culture of property** as part of their structures of building provision. These local cultures of property are based on distinctive sets of social institutions and patterns of social behaviour that in turn derive from local income, class and ethnic composition.

In most countries, the supply of owner-occupier housing has been *deliberately stimulated* by government policies. Part of the motivation for these policies can be attributed to the notion of giving the electorate what it wants. But homeownership also became a key element in **Keynesianism:** macroeconomic policy involving government intervention to regulate business cycles. The labour-intensive nature of residential construction, together with the economic multiplier effects of house construction, transportation and domestic equipment, makes it an effective mechanism through which the economy can be regulated and stimulated. Political and social stabilization has also been a major motive for government intervention in promoting homeownership. As the US National Committee on Urban Problems observed more than 35 years ago, 'Homeownership encourages social stability and financial responsibility. It gives the homeowner a financial stake in society . . . It helps eliminate the "alienated tenant" psychology' (US National Committee on Urban Problems, 1968, p. 401).

Among the various policy instruments that have been used to encourage homeownership in Britain are:

➤ grants to building societies in order to keep interest rates below market rates and to encourage the purchase of pre-1919 dwellings that were formerly rented;

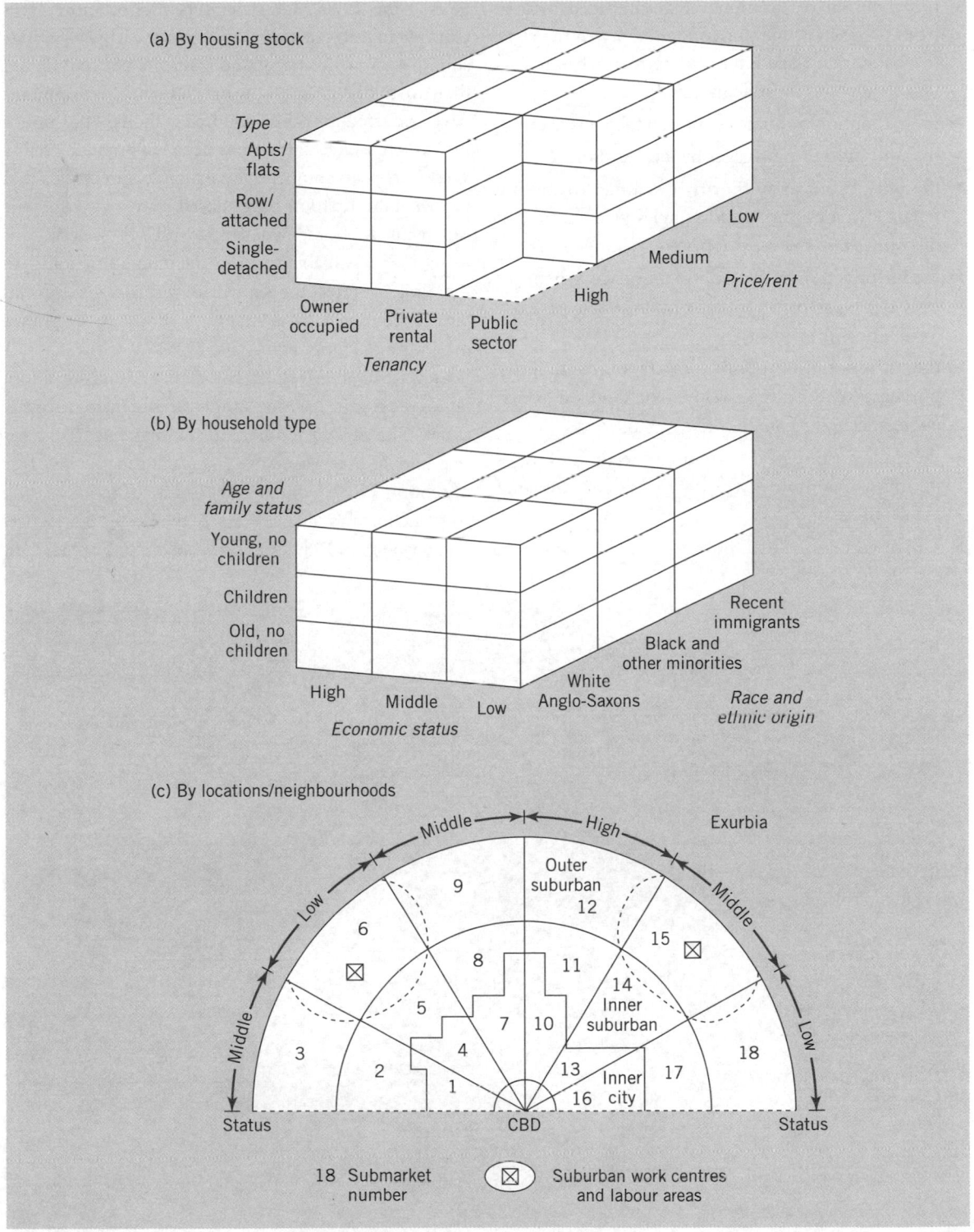

Figure 6.2 Intra-urban housing submarkets: traditional definitions.
Source: Bourne (1981), Fig. 4.9, p. 89.

➤ the abolition of taxation on the imputed income from property while preserving the taxpayer's right to deduct mortgage interest repayments from gross taxable income (now abandoned);

➤ the exemption of homes from capital gains taxes;

➤ the provision of mortgages by local authorities;

➤ the sale of local authority dwellings to 'sitting tenants' or newly married couples at a substantial discount from the market price;

➤ the introduction of the 'option mortgage' scheme to provide cheap loans for first-time housebuyers from lower-income groups;

➤ the utilization of public powers of compulsory purchase to acquire development land on which owner-occupier houses could be built;

➤ the underwriting of the 'voluntary sector' of house-building for homeownership through housing associations; and

➤ The discounted sale of the stock of public housing.

During the 1970s, tax relief subsidies to mortgagees grew fivefold in real terms, and by the early 2000s state subsidies to owner-occupation were 40 per cent higher than to public sector housing. Similar developments have occurred in other countries, though the policy instruments have sometimes been different. In many West European countries, for example, direct subsidies to owner-occupation have played a major role in the expansion of homeownership, with subsidized loans being made available to lower-income households. In the United States, the growth of homeownership has been encouraged by a variety of mortgage insurance programmes (Drier and Atlas, 1995).

In spite of all this state protection and subsidization, homeownership has not simply become progressively easier. The reason is inflation. During the 1980s, for example, house prices rose faster than the overall rate of economic inflation, largely because of rapidly rising construction and finance costs. In the United States, new homeowners in 1989 were on average using 50 per cent

Sharing the burden of financial risk: shared ownership housing in the UK. Photo Credit: Andrew Vowles.

more of their incomes to pay for their homes than new homeowners in 1981; and similar trends have continued throughout the Western world in the 1990s and the early part of the twenty-first century. Such figures, of course, obscure some very large variations within cities. The reason is that the dynamics of urban change are reflected in neighbourhood price levels. Detailed analyses of housing markets have shown that these variations are not simply a reflection of housing age, quality, or location but also a function of the complex sociospatial dialectic of urban restructuring.

Nevertheless, the overall effect of bouts of house-price inflation has *not* generally been to dampen long-term demand. On the contrary, inflation underscores for many households the urgency of participating in the potential benefits of homeownership. In general, the growth of homeownership is sustained in three ways: (1) at the expense of households' consumption in other fields, (2) by the increasing tendency for women to re-enter the labour market after having (or having raised) children, and (3) by moves among developers to build smaller houses at higher densities in order to reduce construction costs.

Homeownership and social polarization

There is at the same time an increasing number of households for whom house-price inflation *has* put homeownership clearly out of reach. Indeed, there is strong evidence that social *polarization* has been taking place between housing tenure categories. In Britain, for example, the rate of increase in homeownership among lower-status groups (unskilled manual) in the past 20 years or so has been less than half that of professionals/managers/employers. Conversely, unskilled and semi-skilled manual households have been increasingly localized in public housing.

This clearly has important implications for urban social geography. In addition to the implications for residential segregation, the polarization of socioeconomic groups by housing tenure raises some critical issues in relation to patterns of income and wealth, class structure, and social conflict. As more and more households become outright owners (having paid off mortgages), so an important new mechanism for the *transfer of wealth* is being created. And, as more and more of these owners are drawn from the higher socioeconomic groups, communities will become increasingly polarized economically (Hamnett, 1999).

The relationships between owner-occupation and *class structure and class-consciousness* are less clear. There is a case for regarding homeownership as an instrument through which sections of the working class have been *incorporated* within the conservatism of a property-owning democracy: 'Lawns for Pawns' (Edel *et al.*, 1984). But although incorporation theory provides a reasonable interpretation of the effects of homeownership, it tends to overemphasize the extent to which deliberate incorporation has been the cause of expanded homeownership. On the other hand, it has been argued that homeownership has intensified differences in terms of orientations toward household consumption, cutting across the traditional divisions of class.

Not that homeownership simply reflects and reinforces the positions that people occupy in the labour market. The uneven rates of house price inflation (and, therefore, of capital gains) across neighbourhoods tends to fragment rather than to unite social groups. Moreover, as different groups of owner-occupiers seek to protect and enhance the exchange value of their property, so they will inevitably come into *conflict* with one another and with other interest groups. The resolution of this conflict is part of the constant redefinition and recreation of urban structure, both physical and social.

The decline of private renting

The corollary of the growth in homeownership has been the decline of privately rented housing. In cities everywhere in the early to mid-twentieth century, between 80 and 90 per cent of all households lived in privately rented accommodation, whereas the equivalent figure now stands at between 25 and 35 per cent in North American cities and between 10 and 20 per cent in most European and Australian cities. Nowhere has this decline been more marked than in Britain, where just over 10 per cent of the housing stock is now rented from private landlords, compared with around 60 per cent in 1947 and 90 per cent in 1914.

In general terms, this decline reflects (1) the response by landlords to changes in the relative rates of return provided by investments in rental accommodation and (2) the response by households to the artificial financial advantages associated with homeownership and (in Europe) public tenure. It is, therefore, unrelated to any decline in the demand for privately rented accommodation as such: it is the product of wider economic and political changes.

The effects of rent controls

It is not difficult to understand landlords' desire to disinvest in rental accommodation. Before 1914, investment in rented property produced income that was almost double the return of gilt-edged securities, even allowing for maintenance and management costs; but after the Second World War landlords in Britain could obtain only around 6 per cent on their investment, compared with the 9 per cent obtainable from long-dated government securities. One of the major factors influencing the relatively low returns on investment in rental housing (thus impeding its supply) has been the existence of *rent controls*. These were introduced in many countries to curtail profiteering by landlords in the wake of housing shortages during the First World War. Once introduced, however, rent controls have tended to persist because of government fears of unpopularity with urban electorates.

The effect of such controls has been to restrict the ability of landlords to extract an adequate profit in the face of the costs of covering loan charges, maintenance and management. This situation has been worsened by taxation policies that do not allow landlords to deduct depreciation costs from taxes and by the introduction and enforcement of more rigorous building standards and housing codes. Meanwhile, the incomes of tenants in the privately rented sector have, in general, risen more slowly than the average. With inflation increasing landlords' costs sharply, many have responded by selling their property, either to sitting tenants or to developers interested in site redevelopment. These general trends have been arrested in the United Kingdom in recent years as increasing house prices have encouraged many to enter the buy-to-let market. However, the long-term significance of this development remains to be seen.

The spatial effects of disinvestment

In some inner-city neighbourhoods, the deterioration of the housing stock has reached the stage where landlords can find no buyers and so are forced to abandon their property altogether. In other areas, where there is a high level of demand for accommodation, specialist agencies have moved in to expedite disinvestment. This has been especially noticeable in London, where large numbers of purpose-built flats in interwar suburbs such as Ealing, Chiswick and Streatham and in some central areas – Kensington, Chelsea and Westminster – have been sold by specialist 'break-up' companies on behalf of large landlords such as property companies and insurance companies.

Overall, nearly 4 million dwellings have sold by landlords to owner-occupiers in Britain since the Second World War. Relatively little new property has since been built for private renting, so that what is left of the privately rented sector is old (about 70 per cent of the existing stock in Britain was built before 1950) and, because of a succession of rent controls, most of it has deteriorated badly. This deterioration has itself led to a further depletion of the privately rented stock in many inner-city areas, as urban renewal schemes have demolished large tracts of housing. It is also worth noting that many European cities suffered a considerable loss of privately rented housing through bomb damage during the Second World War.

This decline in the quantity and quality of privately rented accommodation has affected the social geography of the city in several ways:

➤ It has hastened the decay of inner-city areas while reinforcing the shift of a large proportion of the lower-middle and more prosperous working classes to owner-occupied housing in the suburbs.

➤ It has led to a re-sorting and realignment of inner-city neighbourhoods and populations as the various groups requiring cheap rented accommodation are squeezed into a smaller and smaller pool of housing. These groups encompass a variety of 'short stay' households, including young couples for whom private rental accommodation is a temporary but essential stepping-stone either to owner-occupied or to public housing. In addition there are the more permanent residents who have little chance of obtaining a mortgage, saving for a house or being allocated a house in the public sector. These include some indigenous low-income households, low-income migrants, transient individuals, single-parent families and elderly households on fixed incomes.

➤ Fierce competition for the diminishing supply of cheap rental housing between these economically similar but socially and racially very different groups inevitably results in an increase in social conflict that in turn leads to territorial segregation and the development of 'defended neighbourhoods'.

Finally, it should be noted that the shrinkage of the privately rented sector has been selective. In larger cities, the demand for centrally situated luxury flats has been sufficient to encourage investment in this type of property. Thus, in cities such as London, Paris, Brussels and Zürich, the more expensive element of the privately rented sector has been preserved intact, if not enhanced. It must also be recognized that in some of the larger and more affluent cities of Australia and North America the privately rented sector has maintained its overall share of the housing stock through the construction of new high-income apartments for rent, at about the same rate as low-income rental accommodation has been disappearing. Shortages in the supply of land and capital in the faster-growing cities of North America have further restored the position of rental housing.

The development of public housing

Like the other major changes in the long-term pattern of housing supply, the emergence of public housing is a product of wider economic and political factors rather than the result of secular changes in the underlying pattern of housing need or demand. Public housing is supplied in a variety of ways. In Britain, until recently the bulk of all public housing was purpose-built by local authorities, but the not-for-profit voluntary sector is now playing the dominant role through the work of housing associations; in the Netherlands, Denmark and Sweden much public housing is supplied by way of cooperatives; while in Germany the public housing

Public sector housing in Southampton, typical of the high-rise movement of the 1960s. Photo Credit: Andrew Vowles.

programme has been dominated by Neue Heimat, an adjunct of the trade union movement.

But, whatever the organizational framework, the quality and extent of public housing supply are ultimately dependent upon the resources and disposition of central and local governments and public institutions. For this reason, it is difficult to make sense of trends in the provision of public housing without recourse to specific examples. Here, attention is focused on the example of public housing in the United Kingdom.

Public housing in the United Kingdom

Although it has been much reduced in recent years by policies of privatization (see below and Box 6.1), public sector housing still accounts for a large proportion of the housing stock in UK cities: 20 per cent on average, rising to well over 50 per cent in Scottish cities. The provision of low-rent public housing dates from the late nineteenth century, when it emerged as part of the reformist public health and town planning movements. Nevertheless, public housing was slow to develop. The nineteenth-century legislation was permissive: local authorities could build housing for the poor but were under no obligation to do so, and there was no question of financial support from the central government. Not surprisingly, most local authorities did nothing, preferring to rely on the activities of philanthropic and charitable housing trusts.

The first major step towards large-scale public housing provision came in 1919, when an acute housing shortage (that had developed because of the virtual cessation of building during the war years) was made even more pressing by Lloyd George's highly publicized election promise of 'Homes fit for heroes'. It became politically necessary to make some effort to control and organize the supply of new houses, particularly of working-class houses to let. The response was to give local authorities the responsibility and funding to provide such housing.

After the Second World War there was again a backlog of housing, this time intensified by war damage. In addition, the incoming Labour government was heavily committed to the public sector and in 1949 passed a Housing Act that removed the caveat restricting local authorities to the provision of housing for the 'working classes'. From this date, local authorities were free to gear the supply of public housing to the more general needs of the community. The immediate result was a surge in housebuilding to make up the postwar backlog – the so-called 'pack 'em in' phase. Later, with the return of the Conservatives to power, the supply of public housing was more closely tied to slum clearance programmes and the needs of specific groups such as the elderly and the poor. Subsequently, although public housing became something of a political-football, the level of exchequer subsidies was steadily raised, and a succession of legislation gave local authorities increasing powers and responsibility to build public housing for a wider section of the community. By the early 1970s, the housing stock of every British city had been substantially altered by the addition of large amounts of public housing.

It was not until the Thatcher governments of the 1980s that this growth came to be significantly checked. Simultaneously, the stock of *existing* public housing began to be dissolved by the Conservatives' policy of encouraging the sale of local authority housing to sitting tenants at a discount of up to 60 per cent of the assessed value of the property. This retreat from public housing was part of a general '**recapitalization**' instigated by the emergence of the 'New Right' and its neoliberal policies (see Chapter 1) in UK politics. It was accompanied by cutbacks in public expenditure, reductions in taxation and the privatization of public services. In practice, the cuts in Britain were imposed disproportionately on local government expenditures, and on housing in particular. The reason for this lopsidedness is that it would have been much more difficult to implement similar cuts in social security, defence, education or health: they would have directly undermined the political constituency of the governing Conservative Party.

Sociospatial differentiation within the public sector

The legacy of these public housing policies has had a profound effect on the morphology and social geography of British cities. Tracts of public housing (or housing that was previously owned by local authorities and has

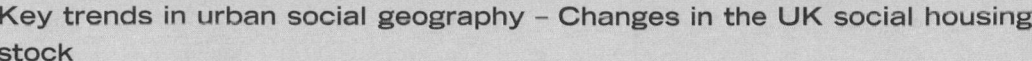

Box 6.1

Key trends in urban social geography – Changes in the UK social housing stock

The enormous changes made to the structure of welfare systems in recent years is emphasized by the actions taken in the sphere of social housing by the UK Labour administrations headed by Prime Minister Tony Blair. Housing that is owned and controlled by local authorities was at one time (together with the National Health Service) a 'jewel in the crown' of the UK welfare state. However, years of neglect by previous administrations (social housing was a prime target for public expenditure cuts by Conservative governments in the 1980s) has led to an enormous backlog of repairs. The Labour administrations have therefore attempted to introduce private sector capital and management expertise into transforming the social housing stock.

Local authority tenants are able to vote for one of three 'transfer' options:

➤ transfer the stock to a private sector registered social landlord (RSLs) – usually a housing association;

➤ adopt a private finance initiative (PFI);

➤ vote for an arm's length management organization (ALMOS).

This process of stock transfer is enormously controversial and some residents and councils have attempted to resist the process. Critics argue that a fourth option of enabling local authorities to continue managing and repairing stock would be much cheaper. They also argue that the information given to tenants before ballots on stock transfer amounts to propaganda. Rents have often risen when stock is transferred to an RSL and there have been greater eviction rates for tenants in arrears.

Advocates of stock transfer argue that it is defenders of traditional local authority housing who are most prone to propagate myths and disinformation. Supporters of change point out that most RSLs are in any case housing associations that are heavily regulated by the UK Housing Corporation. These are not-for-profit social businesses that plough surpluses back into homes, services and neighbourhoods. In addition, new tenancy contracts, it is argued, allow greater democratic participation in local decision-making and the development of strategies to deal with issues such as neighbourhood decline and antisocial behaviour.

Key concepts associated with changes in the UK housing stock (see Glossary)

Asset sales, demunicipalization, 'hollowing-out', housing associations, privatization, quasi-state, residualization, shadow state, voluntary sector.

Links with Other Chapters

Chapter 13: Services Sector Restructuring

---

been in some way transferred) are to be found throughout the urban fabric, with particular concentrations in suburban locations. In general terms, and certainly in comparison with the location of public housing in North American cities, the location of public housing in British cities is remarkable for its integration with owner-occupied housing and for its occupation by a wide band of the socioeconomic spectrum. The former is partly explained by the extensive planning powers of British local authorities; and the latter by standards of construction that compare favourably with those found at the lower end of the private market – a factor that is especially important when comparing the costs of renting public housing versus buying private housing.

This fits conveniently with the proposition that societies with a high degree of social stratification (such as Britain) require only a **symbolic distancing** of social groups, in contrast to the more overt territorial segregation of social groups required in more 'open' societies such as the United States. It is interesting to speculate on the role of architecture in this respect for, as many critics have pointed out, the aesthetic sterility of much local authority housing seems far in excess of any limitations on design imposed by financial constraints alone.

This image of spatial and social integration should not be taken too far, however. Public housing developments do not find a ready welcome near established owner-occupier neighbourhoods because the general perception of their morphological and social characteristics leads to fears among owner-occupiers of a fall in their existing quality of life and (perhaps more significantly) in the future exchange value of their houses. Vigorous opposition is therefore common; and, despite the legislative power of urban governments to override such opposition, it is usually deemed to be politically wiser to seek out the least contentious locations for public housing. In addition, financial pressures on local authorities tend to encourage the purchase of cheaper land whenever possible. For these reasons, many public housing developments tend to be located on extremely peripheral sites, thus effectively isolating their residents from the rest of the city, at least until residential infill and the general socioeconomic infrastructure catch up with the initial housing construction.

It must also be emphasized that a considerable amount of differentiation exists *within* the stock of public housing. Much of this differentiation can be explained in the context of the chronology of the supply of public housing. Six broad periods can be identified in the British case:

1. Early estates (built during the 1920s) consisted mostly of 'cottage-style' semi-detached dwellings.

2. The succeeding generation of council estates built in the 1930s was dominated by three- and four-storey walk-up flats built for slum-clearance families. These acquired a social character quite different from the earlier estates and have subsequently developed a poor reputation in the popular imagination, even if not always justified in practice.

3. A different character again was produced by the postwar boom in public housing construction. The accommodation provided at this time – in the face of waiting lists of tens of thousands in every city and in the context of strict financial constraints and a severe shortage of conventional building materials – created vast tracts of functional but austere housing on the outskirts of cities, much of it in the form of low-rise multifamily units. Because these peripheral estates were catering for those at the top of the waiting list (and who were therefore deemed to be most needy), *there developed a sequential segregation along socioeconomic lines.* The first estates to be completed

were thus dominated by the unskilled and semi-skilled who were the least able to compete in the private sector, and with large families whose previous accommodation was overcrowded. As these households were siphoned from the top of the waiting list, later estates were given over to relatively smaller and more prosperous households.

4. After the backlog had been cleared, architectural and planning experiments provided further differentiation of the public housing stock, leading specific housing schemes to acquire varying levels of popularity and, therefore, of status.

5. The public housing boom of the 1960s brought another set of distinctive housing environments, this time dominated by maisonettes and high-rise blocks of flats, most of which were located in inner-city areas on slum-clearance sites. The heavy emphasis on high-rise developments in the 1960s was the product of several factors. First was the infatuation of architects and planners with the 'Modern movement' in architectural design and the doctrine of high-rise solutions to urban sprawl. This was reinforced by the feeling on many city councils that large high-rise buildings were 'prestige' developments with which to display civic pride and achievement. The pattern of central government subsidies also favoured the construction of high-density, high-rise housing schemes. Meanwhile, large construction and civil engineering companies pushed to obtain contracts for high-rise buildings in order to recoup the considerable investment they had made in 'system' building.

6. After 1968 there was a rapid retreat from this kind of development, partly as a result of the publicity given to the damaging effects of high-rise living on family and social life, partly because of shortcomings in the design and construction of high-rise buildings (the partial collapse of the Ronan Point flats in east London were crucial in this respect), and partly because the big construction firms began to turn their attention to the 'office boom' that began in the late 1960s. Instead of low-cost, high-rise, high-density living, local authorities have opted for the development of low-rise, small-scale housing schemes with 'vernacular' architectural touches and the provision of at least some 'defensible space'.

There is, therefore, a considerable stratification of the public housing stock that is reflected in the morphology of the city as a whole. In Newcastle upon Tyne, for example, the distribution of flats and maisonettes reflects the large-scale building schemes undertaken from the late 1950s through the early 1970s in the city's outer fringes and in inner city redevelopment schemes (Fig. 6.3). Morphological patterns such as these are also reflected in social patterns. Within most British cities there are distinctive and significant patterns of social segregation within the public sector (though they are not as marked as in the private sector), with particular concentrations of deprived, unskilled and semi-skilled manual households in older estates. Such patterns form the basis for further segregation as a result of the actions of housing managers and other local authority officials (see below).

During the 1980s the sale of public housing under the right-to-buy legislation had a profound effect upon the social geography of British cities. To begin with, local authority housing declined as a proportion of the total housing stock, from about one-third to a fifth. Furthermore, working class areas of British cities no longer display housing of uniform appearance dictated by public officials, as ex-tenants have sought to express their new ownership through additions and modifications to their properties such as new windows, doors and stone-cladding of the outer walls. However, the uptake of owner occupation was highly uneven, both socially and geographically. Perhaps inevitably, it was the more affluent local authority tenants, often with multiple household incomes, who were more likely to purchase their own homes; in contrast, the unemployed, single-parent families and the elderly were less likely to purchase their council homes.

In geographical terms, tenants have been much more likely to purchase properties in suburban estates that have a 'good' reputation. In contrast, inner-city 'sink' estates with large proportions of flats, and socially deprived families saw far fewer sales. Social polarization of public housing has also been fostered by the policy of raising rents to market levels. The aim

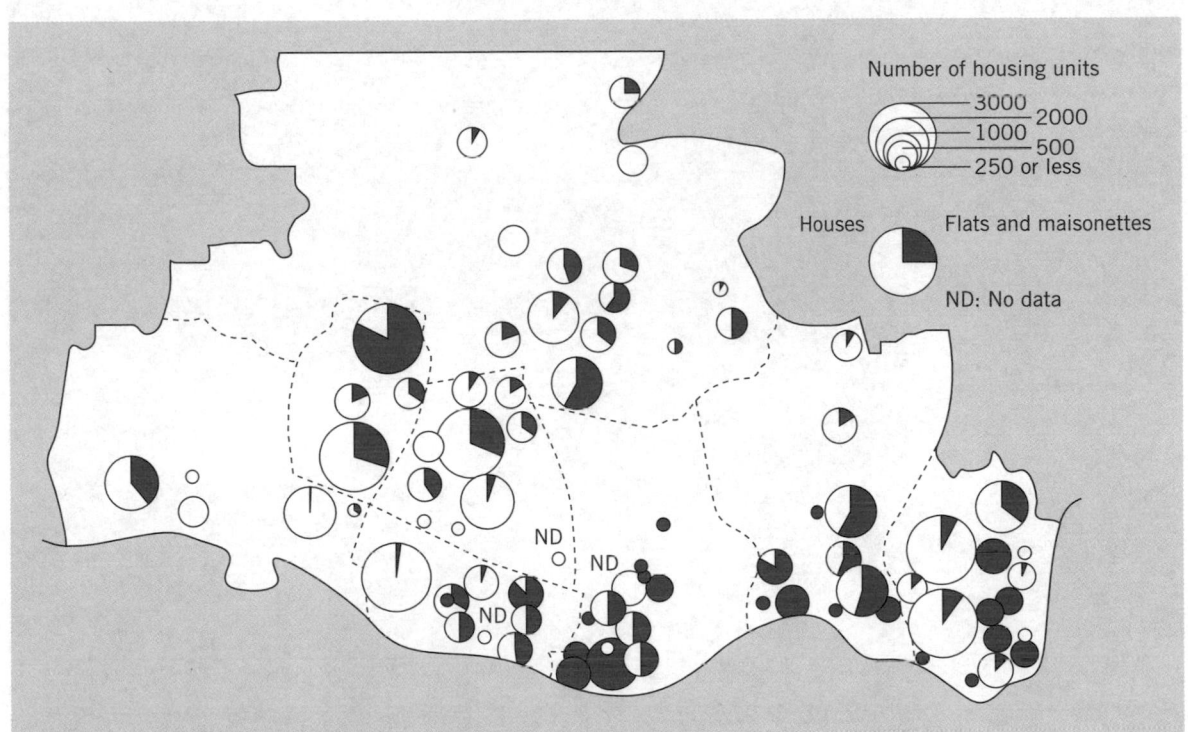

Figure 6.3 Distribution of council flats and maisonettes in Newcastle upon Tyne.
Source: Taylor and Hadfield (1982), Fig. 12.2, p. 248.

was to stimulate private investment in rented accommodation and withdraw the public sector from the provision of social housing. Thus, rather than capital subsidies for new construction, public money would be spent on housing benefit for those on low incomes. The net result is that about two-thirds of those in local authority housing in the United Kingdom now receive housing benefit.

Combined with general reductions in the construction of local authority housing, and wider economic forces leading to growing social inequality, the net effect has been **residualization** – the restriction of public sector housing to deprived minorities. However, the sale of the more desirable local authority dwellings has meant that there is no longer the stepladder of advancement available for many of these families out of the less desirable estates.

The voluntary sector: the 'third arm' of housing provision

The sale of public housing is just one of a whole raft of policies in the sphere of housing in British cities in recent years that have attempted to shift from collective public forms of provision towards individualized, privatized forms of provision. Figure 6.4 summarizes the most important of these initiatives and the ways in which they comprise new hybrid forms of provision that cross-cut the divide between the public and private sectors (Clarke and Bradford, 1998). One such policy is build-for-sale whereby private land is sold to developers at reduced prices. The 'third arm' of the housing system – the non-profit making voluntary sector, has played a crucial role in this context in

Social housing in Geneva. Photo Credit: Alan Latham.

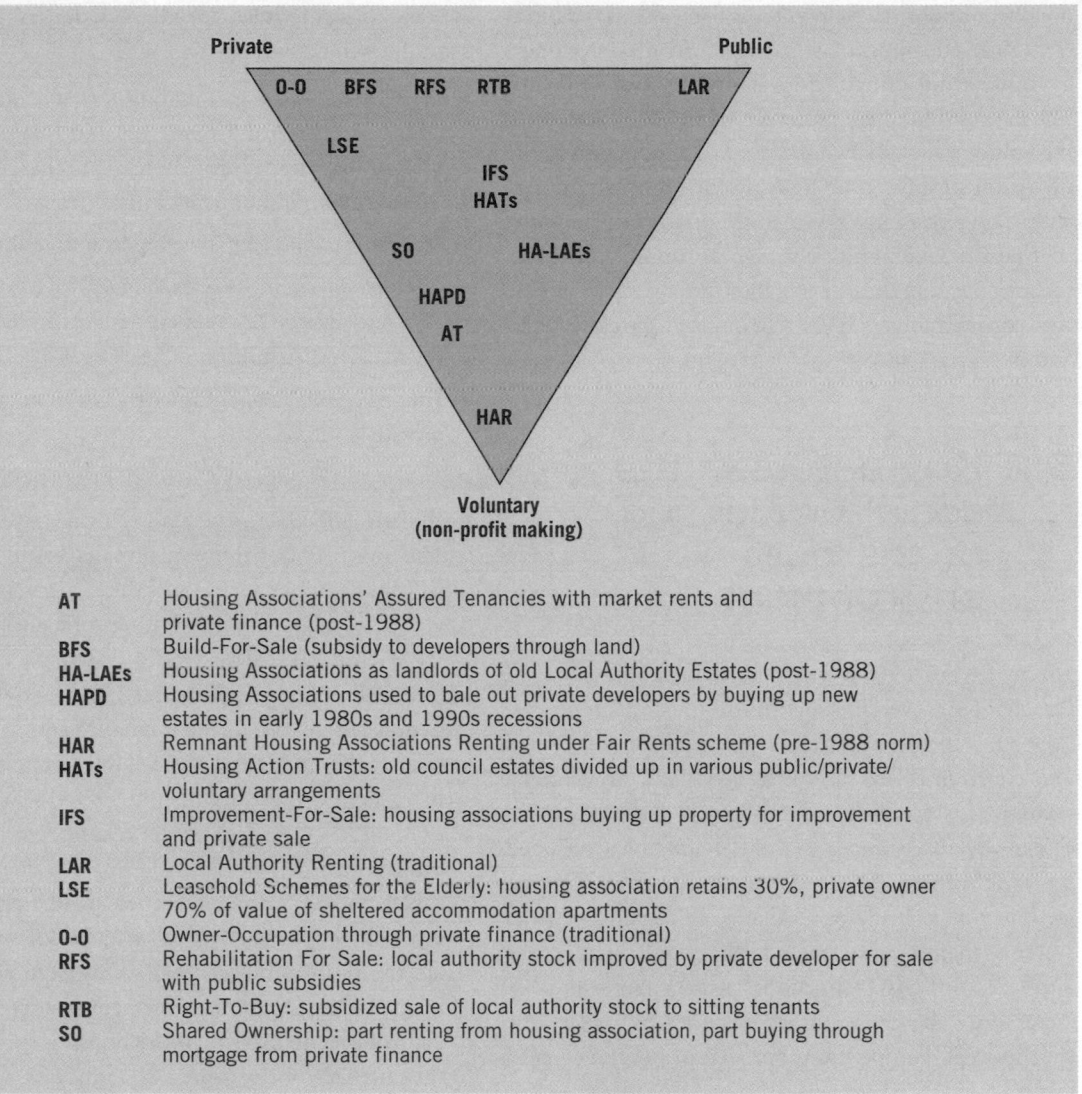

Figure 6.4 The blurring of housing provision in the UK in the 1980s and 1990s.
Source: Clarke and Bradford (1998), Fig. 1.

extending home ownership to groups that were previously excluded.

In the United Kingdom this sector comprises **housing associations** funded by the public sector through a quango known as the Housing Corporation. This sector has traditionally built properties for specific disadvantaged groups that were previously somewhat neglected by the local authorities. However, as part of their new role they were encouraged to take over local authority estates wholesale. They were also encouraged

to buy and improve properties for sale and to engage in shared ownership schemes, whereby the housing association retained a portion of the equity in the property.

However, under the 1988 Housing Act, introduced by a Conservative government, they were required to seek funding from the private sector and to charge market rents. While those on benefits could afford these higher rents, many of those on low incomes were excluded. This created a form of 'poverty trap' for those on benefits, since a low-paying job would mean that they could no longer afford the market rents. Housing associations were also encouraged to bale out private developers by purchasing difficult-to-sell private developments. The net result of these policies has been colonization of the voluntary sector by the state and a reduction in the provision of affordable housing for the most vulnerable in society. Many low-income groups were tempted (perhaps unwisely) into owner-occupation, and repossessions of properties among these groups subsequently rose considerably with the recessions of the early 1990s and the early 2000s.

6.2 Key actors in the social production of the built environment

Given the existence of these housing submarkets and the changing overall structure of housing supply, we now turn to the way these opportunities are shaped and constrained by various agencies and professional mediators. This focus stems largely from the view of class relations and social differentiation developed by Max Weber. Weberian analysis centres on an 'action frame of reference' that seeks to explain social systems mainly in terms of the people who make and sustain them. Institutional arrangements and key 'actors' are therefore studied in order to explain the outcome of competition between conflicting social groups.

The development of this approach in relation to the structures of building provision can be traced to the work of Ray Pahl. In a provocative and influential essay he argued that the proper focus or urban research should be the interplay of spatial and social constraints that determine opportunities of access to housing and urban resources. Furthermore, he suggested, the key to understanding the social constraints could be found in the activities, policies and ideologies of the managers or controllers of the urban system (Pahl, 1969). Very broadly, this is the basis of what has become known as the *managerialist thesis* and the study of **managerialism**. In the context of housing, the managers of scarce resources (or the 'middle dogs' or **'social gatekeepers'**

as they are sometimes called) include key personnel from the following spheres:

➤ *Finance capital*, e.g. building society and savings and loan association managers and others engaged in lending money for house purchase, housing development and housing improvements.

➤ *Industrial capital*, e.g. developers and builders.

➤ *Commercial capital*, e.g. exchange professionals such as estate agents, lawyers and surveyors engaged in the market distribution of housing.

➤ *Landed capital*, e.g. landowners and rentiers such as private landlords.

➤ *State agencies*, e.g. social security agencies.

➤ *Agencies of the local state* (local government). The most directly influential managers to be found within the public sector are housing managers *per se* and their related staff of lettings officers and housing visitors. However, under the policy of **welfare pluralism** being implemented by many Western nations there is a greater reliance upon a diverse range of voluntary and charitable organizations in the sphere of housing. In the United Kingdom for example, this involves greater reliance upon housing associations (Warrington, 1995). The managers of this sector can also be regarded as urban managers, especially since they have come to shadow and replicate the responsibilities that were previously borne by the public sector. To the extent that city planners control certain aspects of the housing environment they may also be regarded as 'managers' within the housing system.

What these groups have in common is a job at the interface between available resources and a client (or supplicant) population. It is in terms of their cumulative day-to-day decision-making that sociospatial differentiation takes place, but their influence can be shown to extend beyond day-to-day decision-making. Such is the power of the institutions of housing supply that they not only shape people's actual opportunities but also their *sense of possibilities*. The criteria through which these groups allocate resources are sometimes called **decision** (or **eligibility**) **rules**. These rules are necessary to simplify the frequent and repetitive but often complex and controversial decisions that the managers have to make. Sometimes the rules are explicit in the form of policy documents but often they are implicit in the hidden or tacit understandings that are employed within organizations.

A succession of empirical studies has left no doubt that there are, in every sphere of the structures of building provision, managers whose activities exert a considerable impact on the social production of the built environment – particularly in Europe, where the expansion of welfare capitalism has produced a powerful and easily identifiable bureaucratic influence on the housing scene.

Yet it is important to set the managerialist perspective against the wider sweep of the urban political economy. This question of the relative power of gate-keepers is important, and it is essential to recognize (1) that 'managerial' decisions are themselves subject to constraints determined by the wider economic, political and ideological structure of society and (2) that there are forces completely beyond the control of the managers that exert a significant influence on urban patterns. Urban managers, then, must be seen as actors of significant but limited importance in the context of a sociospatial dialectic in which economic, social and political processes set the limits for their activities while their professional *modus operandi* determines the detail of the resulting patterns. The following sections illustrate the influence of particular types of managers and social gatekeepers on the social production of the built environment.

Landowners and morphogenesis

Landowners stand at the beginning of a chain of key actors and decision-makers whose activities, like the households they ultimately supply, are not always 'rational' in economic terms. The main influence that landowners can exert is through the imposition of their wishes as to the *type of development* that takes place and, indeed, whether it takes place at all. Some owners hold on to their land for purely speculative reasons, releasing the land for urban development as soon as the chance of substantial profit is presented. This can have a considerable effect on the morphology of cities, not least in the way that plots tied up in speculative

schemes act as barriers to development, and in the sequence that land is released.

Because of the special properties of land as a commodity, many landowners are in fact reluctant to sell at all unless they need to raise capital. For many of the 'traditional' large landholders, land ownership is steeped in social and political significance that makes its disposal a matter of some concern. When landowners do sell, they sometimes limit the nature of subsequent development through restrictive covenants, either for idealistic reasons, or, more likely, to protect the exchange value of land they still hold.

It should also be recognized that there are different types of landowners, each with rather different priorities and time horizons. In Britain, three main types have been identified (Massey and Catalano, 1978):

- The biggest single group, 'former landed property', includes the Church, the Crown Estates, the landed aristocracy and the landed gentry. For this group, profitability is important but is strongly mediated by social and historical ties, and strategic decisions are made with an eye to the very long term.

- The second type, 'industrial land ownership', is dominated by owner-occupier farmers, a group that is of crucial importance to the land conversion process at the urban fringe. Their decision-making typically has to balance short-term financial considerations against longer-term social ones.

- The third type, 'financial land ownership', consists of financial institutions (such as insurance companies, pension funds and banks) and property companies. The importance of the former grew very rapidly in the postwar period as they channelled savings and profits into long-term investments, often involving city-centre land. The property companies, on the other hand, are less concerned with the long-term appreciation of assets and more with the exploitation of urban land markets for short- and medium-term profits.

Builders, developers and the search for profit

The profits to be made from property speculation give developers a strong incentive to insert themselves as key actors at the centre of structures of building provision

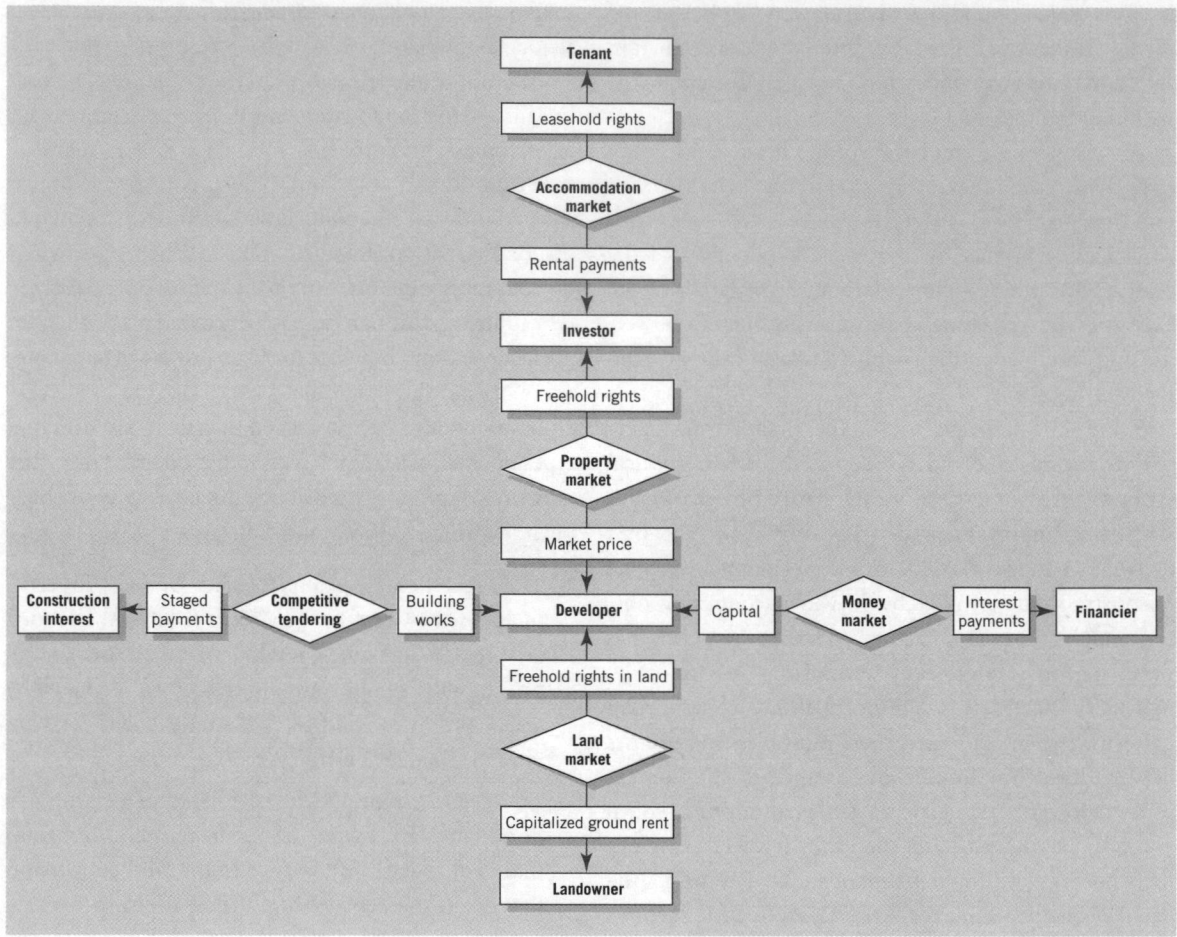

Figure 6.5 Major private-sector relations within the property development industry.
Source: MacLaran (1993), Fig. 7.1.

(Guy and Henneberry, 2002; MacLaran, 2003). This incentive is intensified by their interest in the speed of operation (because they have to finance land preparation and construction long before receiving income from the sale of completed projects). Figure 6.5 shows that the development function is pivotal:

It is the developers who initiate the development process – by recognising an opportunity to profit from a perceived demand for certain types of building in particular locations. They negotiate with landowners for the acquisition of development rights to sites, either purchasing a freehold or leasehold interest in the property or entering into joint development arrangements to share development profits with the site owner.

It is the developers who arrange the short-term financing for construction. They commission architects to devise a scheme, within certain cost constraints, that will be acceptable to the planning authorities. It is also they who engage the builders and use estate agents to seek suitable tenants or purchasers for the completed development. . . . developers might appropriately be regarded as the impresarios of the built environment.

(MacLaran, 2003)

The size of the larger companies enables them to exert strong economic and political pressure on local governments. Large builders and developers have come to represent a powerful lobby: so powerful, in fact, that they have been able to shape the nature of many of the government programmes affecting housing and urban development.

Developers exert their most direct effect on urban structure through their selection of *sites*. For developers of suburban residential subdivisions, the over-arching site criterion is that of price:

> Builders constructing lower-priced units are concerned primarily with minimizing costs. They seek to supply basic housing without frills; flat terrain with few trees is ideal. By contrast, builders of higher-priced houses look for sites with natural or social amenities. The costs of building on more rugged terrain and in denser vegetation are higher, but builders invariably find that the value of the completed house is increased by an even greater amount. When units will be sold for a higher price, builders and subdividers may create ponds out of marshes or add to the relief of an otherwise flat area by cutting streets lower and piling excess dirt up on the building sites. Characteristics that are limiting factors to the builder of lower-priced houses therefore are considered in a positive light in the trade-off calculations of the builders of higher-priced houses.
>
> (Baerwald, 1981, p. 351)

In contrast, the over-arching criterion for residential development in inner-city areas appears to be the availability of land.

Having secured the sites they need, developers are able to exert a considerable influence on urban structure through the physical and social character of the housing they build (Rabinowitz, 1996). They build what they think the market wants, and what is easiest and safest (in marketing terms) to produce. Given their need for plentiful supplies of cheap land and for speed of construction, together with the advantages to be derived from economies of scale and standardization, the result is a strong tendency to build large, uniform housing subdivisions on peripheral sites (Whitehand and Carr, 2001; Hayden, 2003). Moreover, the imperatives of profitability ensure that the housing is built at relatively high densities and to relatively low standards: the resultant stereotype being epitomized in Pete Seeger's classic folk song about 'little boxes made of ticky-tacky . . . (that) all look just the same'.

In Britain, houses at the lower end of the private market have for some time been of poorer quality and less generous space standards than the statutory minima governing public housing. Attempts by the National House Builders Registration Council to upgrade the minimum standards of private house construction have been resisted 'in the interests of the consumer', who would have to bear the extra cost. Housing opportunities for the broad spectrum of middle classes thus tend to be somewhat limited in scope.

In North America, the well-tried norm of detached, single-storey, single-family dwellings has been joined, because of the rising costs of suburban land, by suburban condominium apartment blocks. Variety is provided through deviations in façades and architectural detail: mock-Georgian doors and porticoes, mock-Tudor leaded windows and brick chimneys, and mock-Spanish Colonial stucco and wrought iron. Thrown together on a single housing development, these 'customized' features can sometimes create an unfortunate Disneyland effect. In contrast, British developments tend to be internally uniform down to the styling of doors, windows and garden furniture.

More important from the point of view of the socio-spatial dialectic, however, is the uniformity of size of house resulting from developers' interpretation of market forces. In Britain, the norm is a three-bedroom, two-storey, semi-detached or terrace house, while the lower end of the market is dominated by two-bedroom maisonettes and flats. Whole estates are built to this tried-and-trusted formula, with perhaps a sprinkling of strategically sited larger houses in order to increase sales through raising the apparent social status of the estate. Consequently, substantial numbers of 'atypical' middle-class families – single persons and large families – are effectively denied access to the new owner-occupier submarket and so must seek accommodation either in the older owner-occupier stock or in the privately rented sector. On the other hand, 'typical' families become increasingly localized – and isolated – in suburban estates.

Finally, it should be noted that the nature of developers' activities also depend on their size and corporate organization. The typical large developer, with a compelling need to keep an extensive and specialized organization gainfully employed, will assemble 'land banks', searching out and bidding for suitable land even before it has been put on the market: a strategy known as 'bird-dogging' in the United States. Large developers are also able to purchase materials inexpensively by the wagonload, but must maintain large inventories, develop efficient subcontracting relationships, use government financial aid and housing research, and deploy mass production methods on large tracts of land. In order to achieve this, they are inevitably concerned almost exclusively with mass-market suburban development.

Medium-sized companies cannot afford to take risks on land for which they might not secure permission to build, or to pay the interest on large parcels of land that take a long time to develop. When they do get sites of suitable size, they therefore seek to maximize profits either by building blocks of flats at high densities or by catering for the top end of the market, building low-density detached houses in 'exclusive' subdivisions.

This leaves small firms to use their more detailed local knowledge to scavenge for smaller infill sites, where they will assemble the necessary materials and personnel and seek to build as quickly as possible, usually aiming at the market for larger, higher-quality dwellings in neighbourhoods with an established social reputation. In this context it is significant for the future pattern of housing opportunities that although the building industry in most countries tends to be dominated by small firms, the tendency everywhere is for an increasing proportion of output to be accounted for by a few large companies with regional or national operations.

Discrimination by design: architects and planners

Members of the design professions have direct responsibility for the production of many aspects of the built environment, from individual buildings and detailed landscaping to land use regulations and strategic plans for urban development. In all of these tasks, they must work within the parameters set by clients, politicians, legal codes and so on; but to all these tasks they also bring a distinctive professional ideology and the opportunity to translate social and cultural values into material form:

> Their products, their social roles as cultural producers, and the organization of consumption in which they intervene create shifting landscapes in the most material sense. As both objects of desire and structural forms, their work bridges space and time. It also directly mediates economic power by both conforming to and structuring norms of market-driven investment, production, and consumption.
>
> (Zukin, 1991, p. 39)

The work of architects and planners can, therefore, be profitably interpreted in relation to their transcription of economic, social, cultural and political dynamics into the evolving physical settings of the city. Architects and planners are both products and carriers of the flux of ideas and power relationships inherent to particular stages of urbanization. Here, we illustrate the influential role of architects and planners in the sociospatial dialectic through one of the most important (if somewhat overlooked) dimensions: the patriarchal qualities of the built environment.

As a number of feminist theorists have established, the whole structure of contemporary cities and urban societies reflects and embodies fundamental gender divisions and conflicts. More specifically, urban structure reflects the construction of space into masculine centres of production and feminine suburbs of reproduction. Spaces outside the home have become the settings in which social relations are *produced*, while the space inside the home has become the setting in which social relations are reproduced. Suzanne Mackenzie, for example (1988), interpreted the evolution of urban structure in terms of a series of solutions to gender conflicts that are rooted in the separation of home and work that was necessary to large-scale industrialization in the nineteenth century. These are important aspects of the social construction of space and place that we shall explore in greater detail in Chapter 7. Here, we are concerned with the specific roles of architects and planners as agents of gender coding within the structures of building provision. Shared systems of belief about gender roles are created and sustained, in part, at least, through every aspect of urban design.

Women's spaces

One well-worn theme in architectural theory has been the manifestation of 'masculine' and 'feminine' elements of design. For the most part, this has involved a crude anatomical referencing: phallic towers and breast-like domes. Skyscrapers, for example, can be seen to embody the masculine character of capital. (Nevertheless, there are times when, as even Freud admitted, 'a cigar is just a cigar'.) As some feminist interpretations of architectural history have shown, however, the silences of architecture can be more revealing than crude anatomical metaphors. Thus, for example, Elizabeth Wilson has pointed to the way that modernist architecture, self-consciously progressive, had nothing to say about the relations between the sexes (Wilson, 1991, p. 94). It changed the shape of dwellings without challenging the functions of the domestic unit. Indeed, the Bauhaus School, vanguard of the Modern Movement, helped to reinforce the gender division of labour within households through Breuer's functional Modern kitchen.

The internal structure of buildings embodies the taken-for-granted rules that govern the relations of individuals to each other and to society just as much as their external appearance of buildings and the overall plan and morphological structure of cities. The floor-plans, decor and use of domestic architecture have in fact represented some of the most important encodings of patriarchal values. As architects themselves have so often emphasized, houses cannot be regarded simply as utilitarian structures but as 'designs for living'. The strong gender coding built into domestic architecture has been demonstrated in analyses ranging from Victorian country houses to bungalows and tenements (Spain, 1992; Weisman, 1992; Leslie and Reimer, 2003).

Today, the conventional interpretation of suburban domestic architecture recognizes the way that the ideals of domesticity and the wholesomeness of nuclear family living are embodied in the feminine coding given to the 'nurturing' environments afforded by single-family homes that centre on functional kitchens and a series of gendered domestic spaces: 'her' utility room, bathroom, bedroom, sitting room; 'his' garage, workshop, study. The importance of these codings rests in the way that they present gender differences as 'natural' and thereby universalize and legitimize a particular form of gender differentiation and domestic division of labour.

Women's places

City planning has a more overtly patriarchal and paternalistic ideology that has found expression in a number of ways. The key to the relationship between planning, society and urban structure can be found in the motivation, ideology and *modus operandi*, or praxis, of professional planners. The modern town planning movement grew from a coalition of sanitary reformers, garden city idealists and would-be conservers of the countryside and architectural heritage. For all its apparent progressivism, however, it was an essentially reactionary movement, in the sense that it aimed at containing the city and maintaining a (patriarchal) social and moral order.

Patrick Geddes, the visionary inspiration of the emerging planning movement in Britain, saw cities in the early twentieth century as 'sprawling man-reefs', expanding like 'ink-stains and grease-spots' over the 'natural' environment, creating nothing but 'slum, semi-slum and super-slum' with social environments that 'stunt the mind'. Cities, therefore, were to be thinned out, tidied up, penned in by green belts, fragmented into 'neighbourhood units' and generally made as like traditional villages as possible.

In the subsequent struggle to establish itself as a profession with intellectual standing as well as statutory powers, city planning developed a distinctive professional ideology that now constitutes the basic operating rationale by which planners feel able to justify their own activities and to judge the claims of others. This ideology contains strands of environmentalism, aesthetics, spatial determinism and futurism as well as a strong element of paternalism and an evangelical mantle that enables practitioners to turn a deaf ear to criticism. The cumulative result, as Ravetz put it (1980), has been to transform planning from an 'enabling' to a 'disabling' profession.

The patriarchal strand of planning ideology can be traced back to the formative years of the profession and the threat of new metropolitan environments to the established sociocultural order. Modern cities, in short, provided women with a potential escape from patri-archal relations. Part of the task set for themselves by liberal reformers and members of the early planning movement, therefore, was to create the physical con-ditions not only for economic efficiency and public health but also for social stability and moral order. As a result, town planning became 'an organized campaign to exclude women and children, along with other dis-ruptive elements – the working class, the poor, and minorities – from this infernal urban space altogether' (Wilson, 1991, p. 6).

The cumulative result has been the reinforcement and policing of the spatial separation of the 'natural', male, public domains of industry and commerce from the private, female domain of homemaking. Women were 'kept in their place' through comprehensive plans and zoning ordinances that were sometimes hostile, often merely insensitive to women's needs (Ritzdorf, 1989; Bondi, 1998a). Consequently, the contemporary city embodies serious gender inequalities and contrast-ing experiences of urban and suburban living.

Mortgage financiers: social and spatial bias as good business practice

The decisions of the senior managers of mortgage finance institutions – building societies, banks, savings and loan companies, etc. – represent one of the more striking examples of gatekeeping within the socio-spatial dialectic. It should be stressed at the outset that mortgage finance managers are not independent decision-makers. Much of their activity is closely circumscribed by head office policy, while many of their day-to-day decisions are dependent upon the activities of lawyers, real estate agents, surveyors, bank managers and the like. Nevertheless, mortgage finance managers enjoy a pivotal position in the 'magic-circle' of property exchange professionals, and although the self-image of the trade is that of a passive broker in the supply of housing, the mortgage allocation system 'exerts a decisive influence over who lives where, how much new housing get built, and whether neighbour-hoods survive' (Murphy, 1995).

In order to be properly understood, the activity of mortgage finance managers must be seen against the general background of their commercial objectives. The success of mortgage companies depends upon financial growth and security and the maintenance of large reserve funds. Their chief allegiance, therefore, is to the investor rather than the borrower. Not surprisingly, they operate a fairly rigid system of rules to protect their operations and encourage an ethos of conservative paternalism among their staff. Indeed, there is some evidence to support the idea of their managers as a rather narrowly defined breed: an 'ideal type' with a uniformity of atti-tudes resulting from the recruitment of a certain group (white, Anglo-Saxon, Protestant, moderately educated family men) and the absorption of company traditions and lending policies through a career structure with a high degree of internal promotion that rewards per-sonnel with a 'clean' record of lending decisions.

As a group, then, mortgage finance managers tend to have good reason to be cautious, investment-oriented and suspicious of unconventional behaviour in others. Likewise, the ground rules of lending policies are cauti-ous, devised to ensure financial security both in terms of the 'paying ability' of potential borrowers and the future exchange value of dwellings they are willing to finance.

Bias against people

In operating these ground rules, mortgage finance man-agers effectively act as social gatekeepers – wittingly or unwittingly – in a number of ways. It is normal prac-tice to lend only 80 to 90 per cent of the total cost or valuation of a house (whichever is the less), and for the maximum loan to be computed as a multiple of the household's main income (although some institutions also take into consideration a proportion of a second income, if there is one). It is in the evaluation of potential borrowers' ability to maintain the flow of repayments that the first major stratification by mortgage finance managers takes place.

Because of their desire for risk-minimization, loan officers tend to give a lot of weight to the general credit-worthiness of applicants. Credit register searches are used to reveal previous financial delinquency, evidence of which normally results in the refusal to advance a loan. If they pass this test, applicants are then judged principally in terms of the *stability* of their income and their future *expectations*. This, of course, tends to favour white-collar workers since their pay structure commonly has a built-in annual increment and is not subject to the ups and downs of over-time and short-time working. Conversely, several groups, including the self-employed, the low-paid and single women, will find that their chances of obtaining a mortgage are marginal.

There is also evidence that purely subjective factors influence mortgage managers' decisions. Managers appear to categorize applicants in terms of a set of operational stereotypes ranging from bad risks to good ones, although it has proved difficult to pin down these operational stereotypes in detail and to establish their generality within the professions. It is clearly difficult even for the managers themselves to articulate something that is an unconscious activity. Nevertheless, the criteria they employ in making subjective judgements about people seem to be closely related to their values of financial caution and social conventionality.

While deliberate sociospatial sorting by key actors attached to housing finance has been repeatedly documented, it is important not to oversimplify. The social production of the built environment does not play itself out with clearly defined edges, producing neat cleavages between black and white, rich and poor. The sociospatial dialectic is rich and complex, and a great variety of relationships are encompassed within the structures of building provision, many of them contributing towards the mosaic of spatial differentiation. To take just one example, the social production of certain gay neighbourhoods in New Orleans has been shown to have involved an intricate and illegal appraisal-fixing scheme whereby gay entrepreneurs secured financing for the purchase and renovation of inner-city homes and then marketed these opportunities to other gays (Knopp, 1992) (see also Chapter 11).

We should also note that in the United Kingdom restrictions on the allocation of mortgages eased somewhat in the latter part of the twentieth century as the increased competitive pressures associated with financial deregulation forced banks and building societies to extend their services to a wider range of social groups. Indeed, it is now widely agreed that many relatively poor groups were unwisely induced into home ownership in the 1980s – hence the high rates of repossessions of homes in the wake of the economic recessions of the early 1990s and early 2000s.

Bias against property

Sociospatial sorting also takes place through managers' evaluation of the *property* for which funds are sought. With any loan, the manager's first concern is with the liquidity of the asset, so that if the borrower defaults and the company is forced to foreclose, the sale of the property will at least cover the amount advanced. The assessment of this liquidity ultimately rests with professional surveyors, but mortgage managers tend to have clear ideas as to the 'safest' property in terms of price range, size and location, and surveyors tend to anticipate these criteria in formulating their survey reports.

Many managers evidently assume that market demand for properties that deviate from their ideal (a relatively new suburban house with three or four bedrooms) is very limited, and therefore regard them as greater risks and are more cautious about advancing loans for them. Managers tend to be particularly concerned with the size of dwellings because of the possibility of multiple occupation and the consequent problem of repossession if the borrower defaults. Their concern with *age* is related to the possibility that the property will deteriorate before the mortgage is fully redeemed; and their concern with *location* is related to the possibility of property values being undermined by changes in neighbourhood racial or social composition. Their concern with *price* reflects their anxiety that applicants should not overstretch themselves financially.

Mortgage finance managers thus effectively decide not only who gets loans but also what kinds of property they can aspire to. Households with more modest financial status, for example, will find it more difficult to buy older property even though the overall price may not be beyond their means, since loans for older property generally have to be repaid over a shorter period, thus increasing the monthly repayments.

The spatial outcome is often a dramatic contrast in lending levels for different neighbourhoods (Reibel, 2000). The most striking aspect of the gatekeeping activities of loan officers in this context has been the practice of refusing to advance funds on any property within neighbourhoods that they perceive to be bad risks – usually inner-city areas, as in Birmingham (Fig. 6.6). This practice is known as **'redlining'** and has been well documented in a number of studies, even

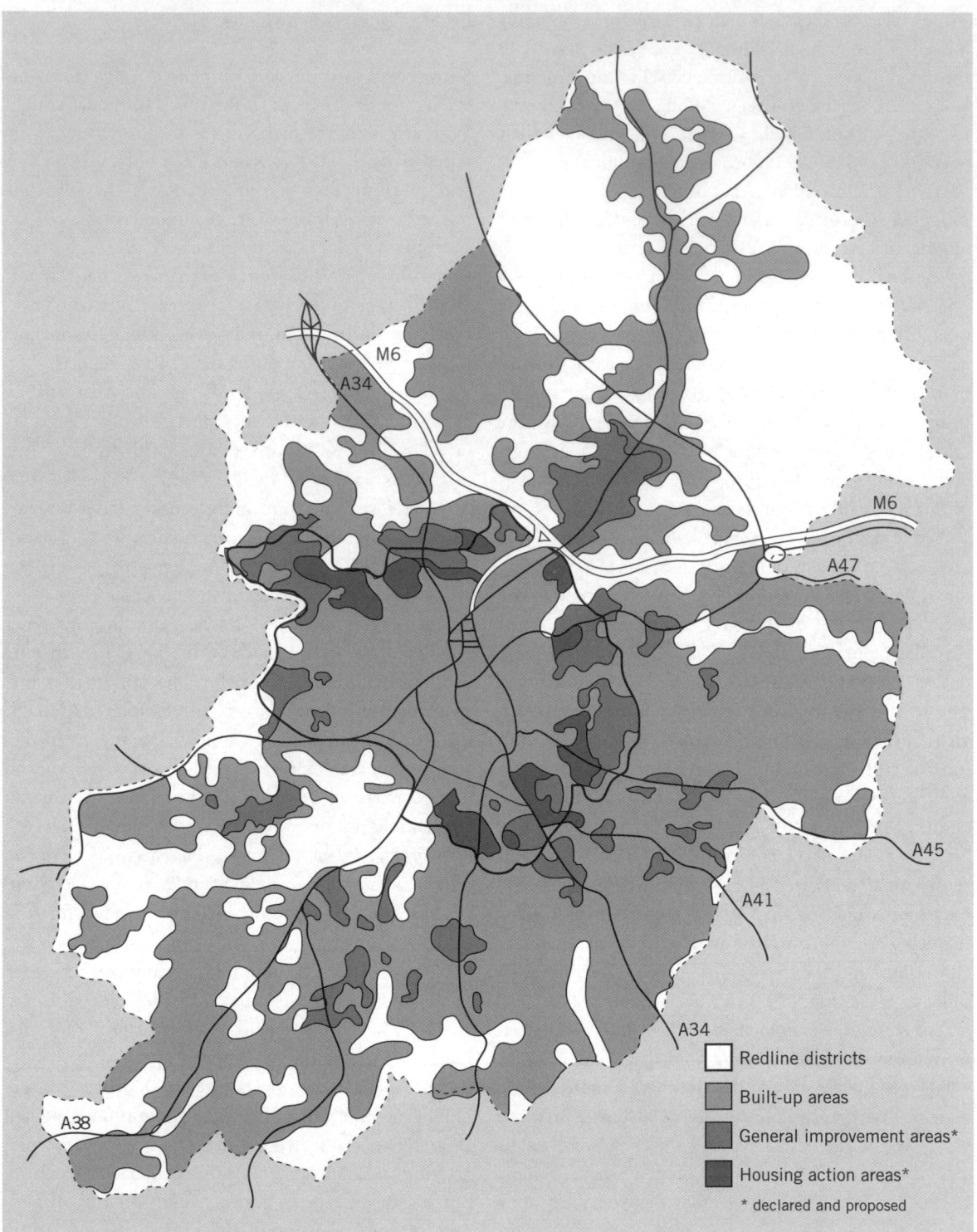

Legend:

- Redline districts
- Built-up areas
- General improvement areas*
- Housing action areas*
- * declared and proposed

Figure 6.6 The 'redline' district in Birmingham in relation to the city's inner area improvement programme.
Source: Weir (1976), p. 111.

though managers are usually reluctant to admit to redlining policies.

In the United States, some states have passed anti-redlining laws, while federal law requires lenders to disclose their policies and lending profiles in an attempt to discourage redlining (Perle *et al.*, 1994). Nevertheless, redlining continues to exist, largely through covert means: discouraging would-be borrowers with higher interest rates, higher down payments, lower loan-to-value rates and shorter loan maturity terms for property in redlined areas.

Although redlining may be an understandable and (in most countries) legitimate business practice, it has important consequences for the social geography of the city. The practice of redlining 'guarantees that property values will decline and generally leads to neighbourhood deterioration, destruction and abandonment. This process makes more credit available for the resale and financing of homes in other neighbourhoods, thus perpetuating differential neighbourhood quality, growth, decline and homeownership' (Darden, 1980, p. 98). This flow of capital to the suburbs, it should be noted, is closely tied to the wider operations of mortgage finance institutions; they are often heavily involved in financing and controlling the suburban activities of large construction companies. Such involvement is commonly reinforced by connections within the overall structures of building provision: overlapping directorships, for example, among building societies and housebuilding companies.

It should also be noted that a substantial proportion of the capital used to finance suburban house construction and purchase is derived from small investors in inner-city areas, so that the net effect of building society policies is to redistribute a scarce resource (investment capital) from a relatively deprived area to a relatively affluent one.

Real estate agents: manipulating and reinforcing neighbourhood patterns

Real estate agents are responsible for a wide range of activities connected with the exchange and management of residential property. They find houses and sometimes arrange finance for buyers; they attract purchasers and transact paperwork for sellers. In addition, they may also be involved in surveying, auctioneering, valuation, property management and insurance. They have close links with mortgage financiers, collecting mortgage repayments for companies and channelling investment funds to them. The mortgage financiers reciprocate by apportioning a quota of mortgage funds to be allocated by the estate agent and by paying a small commission on investment funds received through the agent. Estate agents then use their quota of mortgage funds to expedite the sale of properties on their books. Since real estate agents' profits are derived from percentage commissions on the purchase price of houses, one of their chief concerns is to maintain a high level of prices in the market while encouraging a high turnover of sales.

In many countries of Europe and in North America, estate agents account for between 50 and 70 per cent of all house sales; in Australia, the sale of houses has been almost entirely in the hands of estate agents. They are not simply passive brokers in these transactions, however; they influence the social production of the built environment in several ways. In addition to the bias introduced in their role as mediators of information, some estate agents introduce a *deliberate* bias by *steering* households into, or away from, a specific neighbourhood in order to maintain what they regard as optimal market conditions. Existing residents in a given neighbourhood represent potential clients for an agent, and if an agent is seen to be acting against their interests by introducing 'undesirable' purchasers to the area, the agent may suffer both by being denied any further listings and by any fall in prices that might result from panic selling. Thus the safest response for realtors is to keep like with like and to deter persons from moving to areas occupied by persons 'unlike' themselves. The most widespread discrimination undertaken by real estate agents is based on race and ethnicity, and the segregation resulting from this activity has been well documented (e.g. Feins and Bratt, 1983; James *et al.*, 1984; Downing and Gladstone, 1989).

Manipulating social geographies: blockbusting and gentrification

On the other hand, estate agents have been known to introduce black families to a white neighbourhood in the hope that whites will sell up quickly at deflated prices, allowing the agents to buy houses and then resell them to incoming black families at a much higher price: a practice known as '**blockbusting**'. Because the white residents of targeted neighbourhoods can and do distinguish between middle- and lower-class black families, blockbusters have sometimes resorted to a variety of tactics in order to give the impression that the incoming households represent a 'bad element': telephone calls, door-to-door solicitations and the posting of bogus 'For Sale' signs on front lawns; even, in extreme cases, hiring outsiders to commit petty acts of vandalism or to pose as indolent 'welfare cases'.

A similar process involves the purchase of older properties in prime development sites. These properties are promptly neglected and, as other residents see the neighbourhood beginning to deteriorate, more and more sell up to estate agents, who allow the properties to deteriorate along with the original 'seed' properties. As deterioration continues, the area becomes a fire risk, and as fire insurance companies refuse to renew insurance policies, more owners are persuaded to sell

out. When a sufficient number of dwellings have been acquired, the agents themselves are able to sell out at a considerable profit to developers seeking large plots of land for redevelopment schemes.

This kind of opportunism has also been shown to have been involved in the process of **gentrification**. It has been suggested, for example, that gentrification in parts of Islington, London, can be attributed as much to the activities of estate agents as to the incomers themselves. Estate agents were often the ones who persuaded mortgage financiers to give loans for the purchase and renovation of old working-class dwellings. In addition, some agents purchased and renovated property themselves before selling to incoming young professionals.

Evaluating the relative importance of the various factors associated with gentrification has led to a vigorous debate. On the one hand there is Neil Smith (1996) who is highly critical of explanations of gentrification that stress the importance of changing consumption patterns among new occupational groups or professional

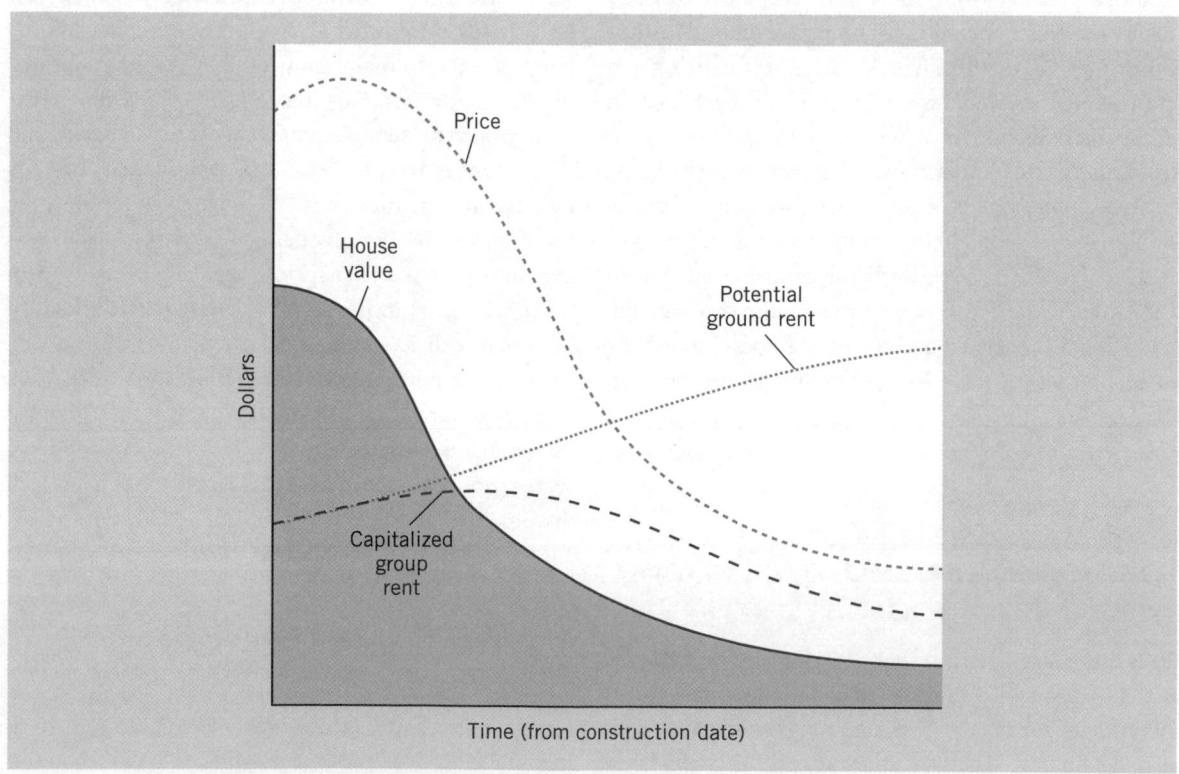

Figure 6.7 Neil Smith's model of the evolution of the rent-gap in US cities.
Source: Smith (1996), Fig. 3.2, p. 65.

144

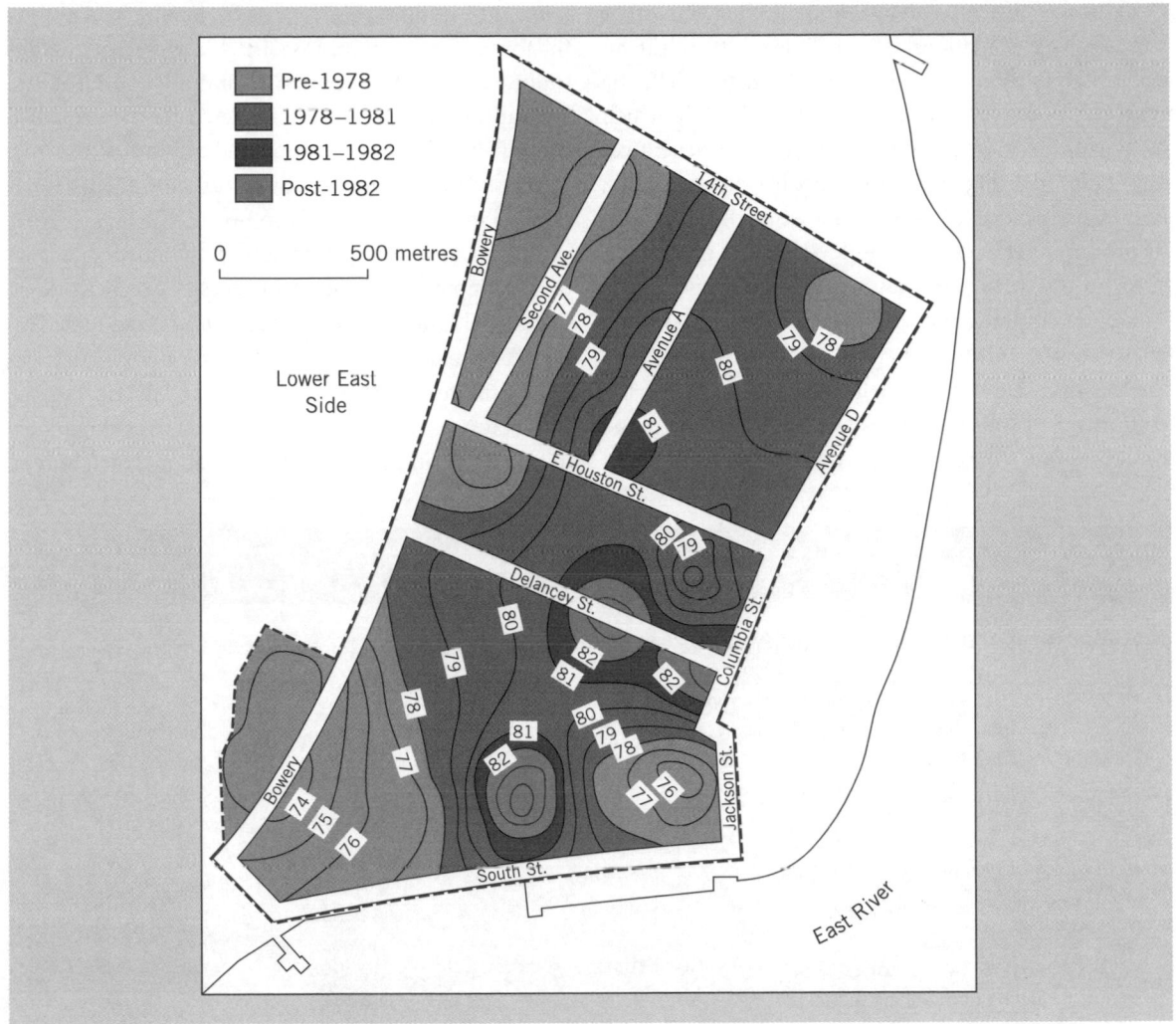

Figure 6.8 The gentrification frontier on the Lower East Side, 1974–1986.
Source: Smith (1996), Fig. 9.4, p. 205.

classes. Instead, Smith argues that gentrification must be seen as part of the process of capitalist economic development and, in particular, moves by capital to arrest the declining rate of profit. Underpinning his analysis is a structuralist perspective that sees socially necessary labour as the ultimate source of value (see Chapter 2). However, Smith acknowledges that in city economies the exchange values of properties is usually far removed from labour costs by powerful market forces of supply and demand.

Inner-city decline and suburban expansion has therefore led to a **rent gap** – a disparity between the potential rents that could be commanded by inner-city properties and the actual rents they are commanding

(see Fig. 6.7). This means that it becomes profitable for developers to buy up run-down properties cheaply, pay builders' costs together with interest charges on mortgage and construction loans, and sell the renovated property at a significant profit. Gentrification is therefore a back-to-the-city move by capital. Smith sees this move – together with deregulation, privatization and other neo-liberal reforms – as a form of revenge by the powerful in society for the moral and economic decline of city life following the social reforms of the 1960s (hence the term *revanchist* **city** – the French word *revanche* meaning revenge). Figure 6.8 shows some of the manifestations of this gentrification in New York.

145

Critics of this interpretation argue that, like other structuralist explanations, it leaves little room for human agency or consumer preferences. Thus, by itself, the rent-gap theory cannot explain which cities, and which areas within cities, are most likely to be regenerated. Ley (1996) focusing upon experience in Canadian cities, links gentrification of inner-city areas with the growth of producer services and the development of middle-class groups with new values and aspirations. These values are complex and multifaceted. Ley relates them in part to the cultural rebellion inspired by hippies in the 1960s, since those seeking lifestyles 'alternative' to the conformity of suburban areas were among some of the first to re-enter inner-city areas. However, subsequent influxes of middle-class groups have unleashed powerful forces for consumption in chic wine bars, coffee shops, restaurants, bookstores, clothing boutiques and the various cultural facilities offered by the gentrified central city. Such developments emphasize the role of space in the formation of culture and identities, as stressed in Chapter 3 (see also Box 6.2).

All writers on gentrification acknowledge that both economic and cultural processes are at work, so the crucial issue is *which factor is most important*. This might seem like some arcane academic debate but it has important implications for planning and political action. If the forces of capital are seen as overwhelmingly dominant, as in structuralist explanations, then human

Box 6.2

Key thinkers in urban social geography – David Ley

David Ley's influence upon urban social geography has not resulted from the formulation of any one particular view or theory of the city, but instead from a series of attitudes and principles that have been manifest in a series of influential empirical studies of inner-city developments in Canadian cities.

Ley has been highly critical of both positivist spatial science and structuralist Marxist interpretations of the city. Instead, he has championed humanistic and behavioural approaches that focus upon local cultures and the everyday subjective experiences of city dwellers. These approaches were manifest in his highly influential textbook *A Social Geography of the City* (1983). Rather than adopt an over-arching view, Ley has displayed a desire to draw upon a wide range of theories in interpreting developments such as community change.

Ley has also been notable for his combination of both qualitative methods, such as participant observation and in-depth interviews, with quantitative methods such as statistical surveys.

Hamnett (1998) sees Ley's approach as in the tradition of 'grounded theory', an approach that Locke (2001, p. 34) defines as 'research and 'discovery' through direct contact with the social world studied, coupled with a rejection of *a priori* theorising' (see also Strauss and Corbin, 1998).

Key concepts associated with David Ley (see Glossary)

Gentrification, grounded theory, phenomenology.

Further reading

Hamnett, C. (1998) *The New Urban Frontier* (book review essay) *Transactions of the Institute of British Geographers*, **23**, 412–416

Ley, D. (1977) Social geography and the taken for granted world *Transactions of the Institute of British Geographers*, **2**, 498–512

Ley, D. (1996) *The New Middle Class and the Remaking of the Central City* Oxford University Press, Oxford

Ley, D. (1999) Myths and meanings of immigration and the metropolis *The Canadian Geographer*, **43**, 2–19.

Ley, D. and Duncan, J. (1982) Structural Marxism and human geography: a critical assessment *Annals of the Association of American Geographers*, **72**, 30–59

Locke, K. (2001) *Grounded Theory in Management Research* Sage, London

Rodaway, P. (2004) David Ley, in P. Hubbard, R. Kitchin and G. Valentine (eds) *Key Thinkers on Space and Place* Sage, London

Strauss, A. and Corbin, J. (1998) *Basics of Qualitative Research, Techniques and Theories for Developing Grounded Theory* Sage, London

Links with Other Chapters

Chapter 1: Different Approaches Within Human Geography

Chapter 11: Gay Spaces

agency can achieve relatively little without wholesale reforms of the operation of capital markets. If, however, one allows more scope for the autonomous role of cultural movements, then these can influence the nature of capitalist development itself (see Badcock, 1989, 1995; Bondi, 1991; Hamnett, 1991, 1992; Clark, 1992; Smith, 1992; Lees, 1994, 2000; Boyle, 1995; Butler, 1997; Butler and Robson, 2001; and Slater 2002 for elaboration of this debate). What is clear from research is that the relative importance of economic and cultural factors varies in different cities; for example, the rent gap seems to have been much more important in New York than in Canadian cities.

Public housing managers: sorting and grading

Within the public sector the principal gatekeepers are the housing managers and their staff who operate the housing authority's admissions and allocation policies. In Britain, the discretion given to local authorities in formulating and operating such policies is very broad and is encumbered by a minimum of legal regulation. There is a requirement to re-house families displaced by clearance or other public action as well as those officially classed as overcrowded, but otherwise it is only necessary to give 'reasonable preference' to households in 'unsatisfactory' housing conditions. Since demand for public housing often exceeds supply, housing managers in most cities are in a position of considerable power and importance in relation to the spatial outcome of public housing programmes.

The rationing of available housing is carried out through a wide variety of eligibility rules and priority systems. Most local authorities operate waiting lists, although these vary in practice from a simple first-come, first-served basis to sophisticated queuing systems using 'points schemes' to evaluate need for a specific type of dwelling: points may be awarded, for example, for overcrowding, ill-health or disability, substandard accommodation, marital status, length of time on the waiting list and so on, together (in some authorities) with discretionary points awarded by housing managers to enable priority to be given to 'special cases'. A general representation of the allocation process in public sector housing is shown in Fig. 6.9.

Not surprisingly, different schemes have different outcomes, and families in identical circumstances may find themselves with quite different degrees of access to council housing, depending on the local authority within whose jurisdiction they live. In general, those households with least access to public housing in British cities include young single people without dependants, newcomers to the area and former owner-occupiers. Conversely, the letting policies of most authorities tend to favour households from slum clearance and redevelopment areas, households living in overcrowded conditions, small elderly households, new households who lack their own accommodations and are living with parents or in-laws, and households with young children.

Problem families and dump estates

In addition to the question of whether or not a household is offered accommodation there is the question of what sort of accommodation is offered, and in *what neighbourhood*. For housing managers it makes sense not only to allocate households to dwellings according to size characteristics but also to match 'good' tenants to their best housing in order to minimize maintenance costs, to ensure that the aged and 'problem families' are easily supervised, and (some would argue) to punish unsatisfactory tenants (those with records of rent arrears and unsociable behaviour in their previous accommodation) by sending them to 'dump' estates.

In this situation, problem families are often doubly disadvantaged by living in low-grade property while having to pay rent at comparable levels to those paid by families in more attractive housing schemes. The localization of problem families in this way can be traced to the policy of housing 'socially weak' families in specially designed austere and durable public housing schemes in France and the Netherlands in the 1930s. After 1945, many local authorities in Britain pursued similar, if less well-publicized, policies using obsolescent housing stock rather than purpose-built developments. By the 1960s, the segregation and localization of 'problem families' as well as grading of other tenants according to their worthiness for particular housing vacancies was common-place, exciting little or no attention (Huttman, 1991).

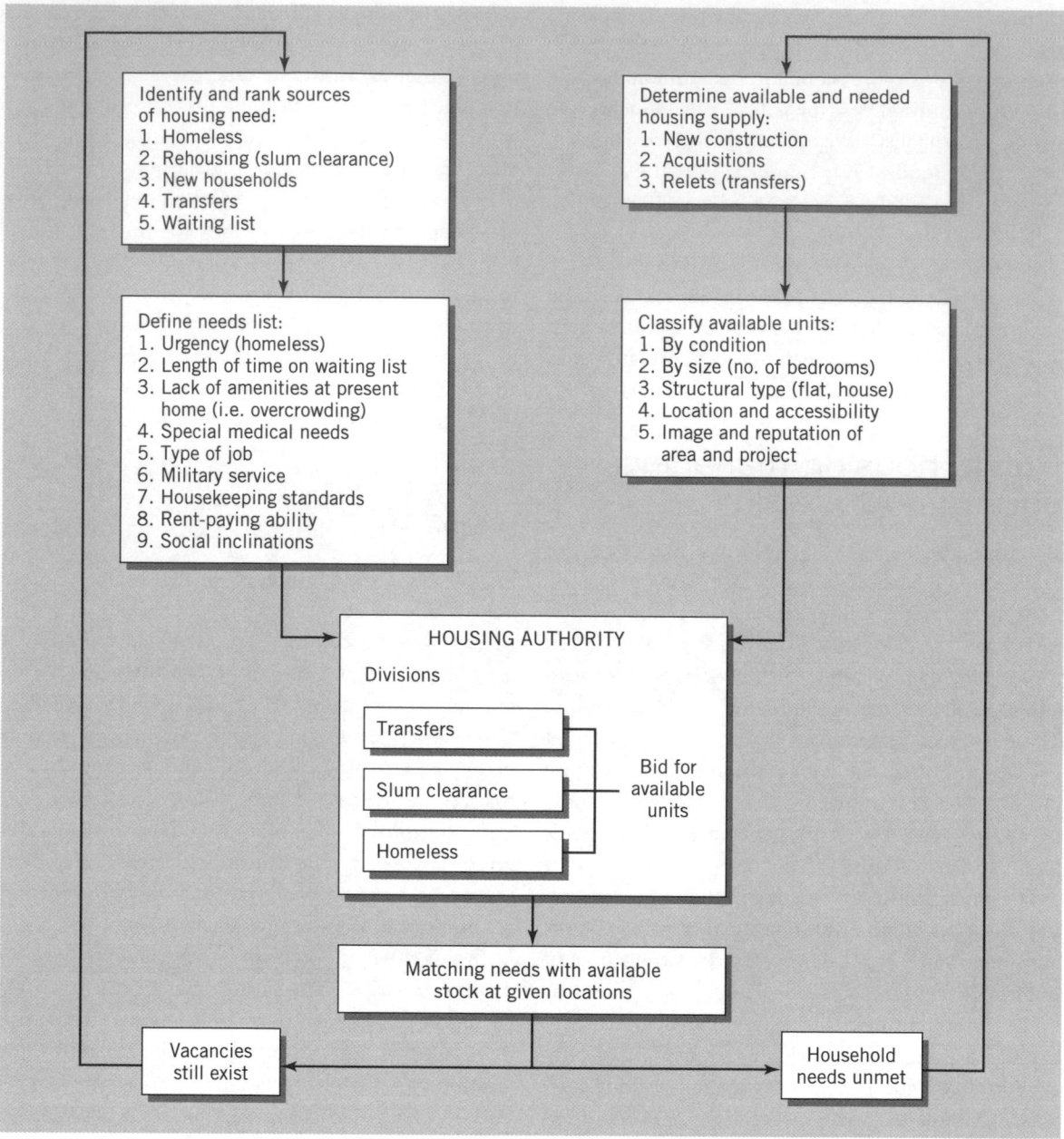

Figure 6.9 The allocation process for public-sector housing.
Source: Bourne (1981), Fig. 10.2, p. 222.

The 'moralistic' attitudes of local authorities were condemned (in suitably diplomatic language), however, by the Central Housing Advisory Committee:

the underlying philosophy seemed to be that council tenancies were to be given only to those who
• 'deserved' them and that the 'most deserving' should

get the best houses. Thus, unmarried mothers, cohabitees, 'dirty' families and 'transients' tended to be grouped together as 'undesirable'. Moral rectitude, social conformity, clean living and a 'clean' rent book . . . seemed to be essential qualifications for eligibility – at least for new housing.

(Central Housing Advisory Committee, 1969)

It is now recognized by housing managers that the localization of families in dump estates sets in motion a labelling process that results in the stigmatization of both the estate and its residents. Because of this stigmatization, accommodation in such areas becomes difficult to let. The problem is further exacerbated by societal reaction to dump estates, with media coverage helping to dramatize the situation and to reinforce 'moral panic' through the creation of sensational and sometimes distorted stereotypes. This, in turn, polarizes attitudes and behaviour both inside and outside dump estates, leading to an increase in antisocial behaviour on the part of the inhabitants,

and therefore to a confirmation of the stereotypes and a further reinforcement of the area's undesirable character.

Finally, it is important to bear in mind that housing managers do not have the power to determine the structure, form and quantity of the resources they distribute, even though they can control allocation procedures. Thus, while the basic operating principle in the social sector is 'need', families with the greatest need tend to end up in the least desirable accommodation. The end result is a hierarchy of council house estates in a manner not unlike the ranking of private estates by market mechanisms.

Chapter summary

6.1 There are many distinctive housing submarkets in cities that are manifest in the residential structure. While owner-occupation is a form of housing tenure that has been increasing in most Western societies in recent years, private renting has been in decline. The fortunes of both of these sectors has been affected by government policies. Social housing exists in many different forms but has had a profound effect upon the structure of many cities, especially those in Europe.

6.2 The built environment is not just a reflection of the economics of supply and demand but is affected by institutional factors and the interactions of numerous actors: governments (both local and national), landowners, investors, developers, builders, planners, architects, community activists and consumers. Social relations of class, gender and ethnicity affect the ways in which these agencies 'sort' different types of people into different residential areas.

Key concepts and terms

'blockbusting'
culture of property
decision rules
eligibility rules
exchange value
gentrification
housing associations

housing submarkets
Keynesianism
managerialism
recapitalization
'redlining'
rent gap
residualization

revanchist city
social gatekeepers
symbolic distancing
use value
welfare pluralism

Suggested reading

The concept of the structures of building provision is Michael Ball's, and his original formulations can be consulted in his essay on 'The built environment

and the urban question' (*Environment & Planning D: Society and Space*, 1986: **4**, 447–464). Another useful essay on this sort of approach is by Patsy Healey and Susan Barrett: 'Structure and agency in land and property development processes' (*Urban Studies*, 1990:

27, 89–104). A good introduction to housing sub-markets, along with coverage of many of the issues dealt with in this chapter, is provided by Larry Bourne's *The Geography of Housing* (1981: Edward Arnold, London), while for British coverage see Paul Balchin and Maureen Rhodes' *Housing Policy* (fourth edition 2002: Routledge, London), Andrew Golland and Ron Blake (eds) *Housing Development* (2003: Routledge, London), Glen Bramley *et al. Key Issues in Housing* (2004: Palgrave, Basingstoke) and Ray Forrest and James Lee *Housing and Social Change* (2003: Spon Press, London). The activities of property developers are considered in Simon Guy and John Henneberry's *Development and Developers: Perspectives on Property* (2002: Blackwell, London). A detailed view of the structure and processes characterizing an American metropolitan area is given by John Adams in *Our Changing Cities* (J.F. Hart, ed., 1991: Johns Hopkins University Press, Baltimore). The sociospatial issues surrounding homeownership are best approached through Geraldine Pratt's work: see 'Incorporation theory and the reproduction of community fabric,' in *The Power of Geography. How Territory Shapes Social Life* (J. Wolch and M. Dear, eds, 1989: Unwin Hyman, London). For home ownership also see Chris Hamnett's *Winners and Losers: Home Ownership in Modern Britain* (1999: UCL Press, London). Good sources on the role of key actors in the urban development process in America are Marc Weiss, *The Rise of the Community Builders* (1987: Columbia University Press, New York), and Joe Feagin and R. Parker, *Building American Cities* (1990: Prentice Hall, Englewood Cliffs); on British cities, see Jeremy Whitehand, *The Making of the Urban Landscape* (1992: Blackwell, Oxford). The roles of developers, planners, financiers, architects and others are also covered in *The Restless Urban Landscape* (P.L. Knox, ed., 1993: Prentice Hall, Englewood Cliffs). Liz Bondi provides a good review of the relationships between gender and urban design in her essay in *Progress in Human Geography* (1992: **16**, 157–170). Linda McDowell provides an excellent review of the broader issues of space, place and gender relations in the same journal (1993:

17, 157–179). More detailed sources include Leslie Weisman's *Discrimination By Design* (1992: University of Illinois Press, Urbana) and Daphne Spain's *Gendered Spaces* (1992: University of North Carolina Press, Chapel Hill). The privatization of public housing in the UK is examined in *Selling the Welfare State: The Privatization of Public Housing* (R. Forrest and A. Murie, 1991: Routledge, London) which can be supplemented by Alison Ravetz's *Council Housing and Culture: The History of a Social Experiment* (2001: Spon Press, London). For an overview of dilemmas in the sphere of housing policy see Peter Williams (ed.) *Directions in Housing Policy: Towards Sustainable Housing Policies for the UK* (1997: Paul Chapman Publishing, London), Jo Dean and Annette Hastings *Challenging Images: Housing Estates, Stigma and Regeneration* (2000: Policy Press, London), R. Burrows *Poverty and Home Ownership in Contemporary Britain* (2003: Policy Press, Bristol), Craig Gurney (ed.) *Placing Changes: Perspectives on Place in Housing and Urban Studies* (2005: Ashgate, London) and David Cowan and Alex Marsh (eds) *Two Steps Forward: Housing Policy into the New Millennium* (2002: Policy Press, Bristol). For the history of housing associations in the UK see Peter Malpass *Housing Associations and Housing Policy* (2000: Palgrave, Basingstoke). See also Susan J. Smith *et al.* 'Housing for health; does the market work?' (*Environment and Planning A*, 2004: **36**, 579–600). Neil Smith's extensive writings on gentrification are brought together in *The New Urban Frontier: Gentrification and the Revanchist City* (1996: Routledge, London) while for evidence from Canada see David Ley's *The New Middle Class and the Remaking of the Central City* (1996: Oxford University Press, Oxford) and for UK perspective see Tim Butler's *Gentrification and the Middle Classes* (1997: Ashgate, Aldershot). See also the theme issue on gentrification in *Environment and Planning A*, 2004, Volume 36 and the web site http://members.lycos.co.uk/gentrification together with Rowland Atkinson and Gary Bridge *Gentrification in a Global Perspective* (2004: Routledge, London).

Chapter 7

The social dimensions of modern urbanism

Key questions addressed in this chapter

➤ What are the principal attitudes towards cities in Western societies?

➤ How does urban living affect people's sense of identity?

➤ How are social networks structured?

➤ What are the main tenets of the Chicago School of human ecology?

➤ Why has the work of the Chicago School been so influential despite much criticism?

➤ How have the ideas of Georg Simmel influenced the study of social relationships in cities?

An important part of the geographer's task of illuminating the nature and causes of spatial differentiation is to lay bare the interrelationships between people and their environments. The spatial order of the city can only be properly understood against the background of the underlying dimensions of social organization and human behaviour in the city: the urban 'ultrastructure'. The suggestion is not, of course, that the spatial order of the city is subordinate to social factors; the influence of space and distance on individual behaviour and social organization will be a recurring theme in this chapter as we explore the sociospatial dialectic in close-up.

7.1 Urban life in Western culture

The traditional view of the relationship between people and their urban environments has, for the most part, been negative. It is a view that persists. Elizabeth Wilson, for example, claims that 'today in many cities we have the worst of all worlds: danger without pleasure, safety without stimulation, consumerism without choice, monumentality without diversity' (Wilson, 1991, p. 9). Public opinion and social theories about city life, together with the interpretations of many artists, writers, filmmakers and musicians, tend to err towards negative impressions. They tend to be highly deterministic, emphasizing the ills of city life and blaming them on the inherent attributes of urban environments.

Evidence from attitudinal surveys, for example, suggests that most people believe city environments to be unsatisfactory. Only one in five Americans thinks that cities represent the best kind of environment in which to live; 30 per cent nominate suburban environments, and 44 per cent nominate small town or rural environments. Data from European surveys show a similar anti-urban leaning. Moreover, these attitudes show up not only in hypothetical residential preferences but also in evaluations of actual communities. Satisfaction with the overall 'quality of life' and with several major components of well-being tends to decline steadily with the transition from rural to metropolitan environments. Such data, however, are notoriously difficult to interpret. A case in point is the apparent ambiguity of results that show people professing to prefer rural or small-town living but whose behaviour has brought them to the city, presumably in pursuit of a higher material level of living. The city thus emerges as neither good nor bad, but as a 'necessary evil'.

This lopsided ambivalence towards the city – a grudging functional attraction accompanied by an intellectual dislike – has long been reflected in the literature and art of Western society (including the idiom of popular songs). Raymond Williams, for example, showed that while British literature has occasionally celebrated the positive attributes of city life, there has been a much greater emphasis on its defects, paralleled by the persistence of a romanticist portrayal of country life. For every urban thrill and sophistication there are several urban laments and rural yearnings (Williams, 1973).

Some writers have been able to demonstrate how the city has in addition been regarded as a catalyst, a challenge and a 'stage' for the enactment of human drama and personal lifestyle by certain schools of thought. Charles Baudelaire, writing in the mid-nineteenth century, was among the first to see in modern cities the possibility for transcending traditional values and cultural norms. He saw that the city can turn people outward, providing them with *experiences of otherness*. The power of the city to reorient people in this way lies in its diversity. In the presence of difference, people at least have the possibility to step outside themselves, even if it is just for a short while. It is this quality that makes cities so stimulating to many of us. Yet the significance of the diversity of city life goes much further. As

we are exposed to otherness, so our impressions of city life and urban society and the meanings we draw from them are modified and renegotiated. In this way, our cultures (i.e. systems of shared meanings introduced in Chapter 3) take on a fluidity and a dynamism that is central to the sociospatial dialectic.

The cumulative image of the city has thus come to be 'a montage of mixed and clashing elements: . . . of senseless, brutal crime; of personal freedom and boundless hopes; of variety, choice, excitement; of callous and uncaring people; of social groups diverse enough to satisfy each individual's unique needs; of crass and crushing materialism; of experiment, innovation and creativity; of anxious days and frightful nights' (Fischer, 1976, p. 17).

Fischer suggested that it is possible to recognize four basic themes within this ambivalent imagery of cities, each expressed as polarities:

Rural		Urban
Nature	versus	Art
Familiarity	versus	Strangeness
Community	versus	Individualism
Tradition	versus	Change

Fischer pointed out that there is a tension in these pairings that derives from the fact that neither half is universally 'better' or 'worse' than the other. Instead, they pose dilemmas of personal choice. Depending on which horn of the dilemma they have grasped, philosophers and poets have become either pro-urbanists or (as is usually the case) anti-urbanists.

7.2 Urbanism and social theory

These polarities are also present in the stock of social theories concerning city life. The strangeness, artificiality, individualism and diversity of urban environments have been seen by many social scientists as fundamental influences on human behaviour and social organization. This deterministic and environmentalist perspective has had a profound effect on the study of urban social geography as well as on sociology and all the cognate disciplines. It stems from the writings of European social philosophers such as Durkheim,

The Galleria Vittorio, Emanuele II, Milan, a covered walkway typical of the nineteenth-century modernist metropolis. Photo Credit: Paul Knox.

Weber, Simmel and Tönnies, who were seeking to understand the social and psychological implications of the urbanism and urbanization associated with the Industrial Revolution of the nineteenth century.

The kernel of this classic sociological analysis is the association between the scale of society and its 'moral order'. Basically, the argument runs as follows. In pre-industrial society small, fairly homogeneous populations contain people who know each other, perform the same kind of work and have the same kind of interests: they thus tend to look, think and behave alike, reflecting a consensus of values and norms of behaviour. In contrast, the inhabitants of large cities constitute part of what Durkheim called a 'dynamic density' of population subject to new forms of economic and social organization as a result of economic specialization and innovations in transport and communications technology.

In this urbanized, industrial society there is contact with more people but close **'primary' relationships** with family and friends are less easily sustained. At the same time, social differentiation brings about a divergence of lifestyles, values and aspirations, thus weakening social consensus and cohesion and threatening to disrupt social order. This, in turn, leads to attempts to adopt 'rational' approaches to social organization, a proliferation of formal controls and, where these are unsuccessful, to an increase in social disorganization and deviant behaviour (see Box 7.1).

The Chicago School

The impact of these ideas on urban geography came chiefly by way of their adoption and modification in the 1920s and 1930s by researchers in the Department of Sociology in the University of Chicago (the so-called **Chicago School**) under the leadership of Robert Park, a former student of Georg Simmel. Like earlier theorists, Park believed that urbanization produced new environments, new types of people and new ways of life. The net result, he suggested, was 'a mosaic of little worlds which touch but do not interpenetrate' (Park, 1916, p. 608). He encouraged the 'exploration' and empirical documentation of these social worlds by his colleagues and, as a result, there developed an influential series of 'natural histories' of the distinctive groups and areas of Chicago in the 1920s: juvenile

gangs, hobos, the rooming house area, prostitutes, taxi-dancers, the Jewish ghetto and so on. These studies represented part of an approach to urban sociology which became known as **human ecology**, the principles of which are discussed below.

Urbanism as a way of life

A closely related and equally influential approach to urban sociology also sprang from Chicago a few years later: the so-called **Wirthian theory** of urbanism as a way of life. Wirth's ideas, although they contained much of the thinking inherent to human ecology, synthesized a wide range of deterministic principles relevant to individual as well as group behaviour. Wirth, like Park, had studied under Georg Simmel and was heavily influenced by Simmel's (1969) work on 'The metropolis and mental life'. Putting Simmel's ideas together with subsequent work from the human ecologists, Wirth produced his classic essay, 'Urbanism as a way of life' (Wirth, 1969), which became one of the most often quoted and reprinted articles in the literature of the city. Wirth attributed the social and psychological consequences of city life (i.e. 'urbanism') to the combined effects of three factors that he saw as the products of increasing urbanization:

➤ the increased size of populations;

➤ the increased density of populations; and

➤ the increased heterogeneity, or differentiation, of populations.

At the *personal* level the effect of these factors, Wirth suggested, is as follows: faced with the abundant and varied physical and social stimuli experienced in the large, dense and highly diverse city environment, the individual has to adapt 'normal' behaviour in order to cope. City dwellers thus become, for example, aloof, brusque and impersonal in their dealings with others: emotionally buffered in their relationships. Nevertheless, the intense stimuli of city environments will sometimes generate what has subsequently been dubbed a **'psychic overload'**, leading to anxiety and nervous strain. Furthermore, the loosening of personal bonds through this adaptive behaviour tends to leave people both *unsupported* in times of crisis and *unrestrained* in pursuing ego-centred behaviour. The

Box 7.1

Key trends in urban social geography – The growth of Bohemian enclaves

A major change in Western cities in recent decades has been the growth of so-called 'Bohemian enclaves'. Before considering these areas it is worth exploring the origin of this terminology. Bohemia is a region in what is currently the Czech Republic. Over many centuries peoples in this area developed a reputation for unconventional behaviour. Many of them were gypsies who had moved west to escape persecution by the Russian Empire but they also included liberals who could not cope with the authoritarianism to be found to the East. The term Bohemian thus became associated with migrant peoples and especially with gypsies. The famous *Encyclopedie ou Dictionnaire Raisonne Des Sciences des Arts et des Métiers* edited by Diderot and d'Alembert, published in Paris in 1775, referred to Bohemians as '. . . the vagabonds who are professional soothsayers and palm readers. Their talent is to sing and dance and steal . . .' (Blom, 2004).

In nineteenth-century Paris the term Bohemian became associated with various artistic groups (e.g. writers, poets, intellectuals, painters, sculptors, musicians, students) who were considered to be transient, unconventional and 'gypsy-like' in their behaviour. These artists (together with their associates – dealers, publishers, models, students, friends, mistresses,

etc.) congregated in the Montmartre area of Paris. This was an older, hilly, residential district, characterized by narrow streets of cheap rented housing that had escaped the rebuilding undertaken elsewhere in the city. The lower slopes of Montmartre closer to central Paris were full of facilities that fostered the Bohemian lifestyle: street cafés, theatres, cabarets, night-clubs and brothels. This nightlife of *fin-de-siecle* Paris was made famous in the paintings by the artist Toulouse-Lautrec. Although sometimes romanticized, this was an area plagued by alcoholism, disease and sexual abuse. Montmartre also became an area of transgression by 'respectable' upper-middle class members of society and as it developed into something of a tourist attraction it was overtaken by another Bohemian enclave, Montparnasse, on the left bank of the river Seine. Although the term Bohemian is mostly associated with nineteenth-century Paris, all major European cities had their Bohemian enclaves; thus Dr Johnson referred to the poor writers of 'Grub Street' in London (now the Barbican area).

In essence, Bohemian lifestyles were seen as being in opposition to 'bourgeois' values of hard work, thrift and nuclear family life. However, David Brooks has argued that in recent years the distinction between bourgeois and

Bohemian lifestyles has diminished. The growth of the 'cultural industries' in the late twentieth century has meant that various non-conformists, eccentrics and mavericks have become incorporated into mainstream capitalist enterprises. Thus, it seems that every major city is now trying to foster its own artistic quarter, and corporations now tolerate and even encourage talented, creative, unconventional employees.

Key concepts associated with Bohemian enclaves (see Glossary)

Bohemia, creative cities, creative class, cultural industries.

Further reading

Blom, P. (2004) *Encyclopedie: The Triumph of Reason in an Unreasonable Age* Fourth Estate, London

Brooks, D. (2000) *Bobos in Paradise: The New Upper Class and How They Got There* Simon and Schuster, New York

Hall, P. (1998) *Cities in Civilization* Phoenix, London, Chapter 6

Links with Other Chapters

Chapter 10: Box 10.2 The Development of 'Urban Nightscapes'

Chapter 11: Box 11.2 Richard Florida

---

net result, Wirth argues, is an increase in the incidence, on the one hand, of social incompetence, loneliness and mental illness and, on the other, of deviant behaviour of all kinds: from the charmingly eccentric to the dangerously criminal.

Wirth draws a parallel picture of *social* change associated with the increased size, density and heterogeneity of urban populations. The specialized neighbourhoods

and social groupings resulting from economic competition and the division of labour result in a fragmentation of social life between home, school, workplace, friends and relatives; and so people's time and attention are divided among unconnected people and places. This weakens the social support and control of primary social groups such as family, friends and neighbours, leading to a lack of social order and an increase in 'social disorganization'. Moreover, these trends are reinforced by the weakening of social norms (the rules and conventions of proper and permissible behaviour) resulting from the divergent interests and lifestyles of the various specialized groups in the city.

The overall societal response is to replace the support and controls formerly provided by primary social groups with 'rational' and impersonal procedures and institutions (welfare agencies, criminal codes supported by police forces, etc.). According to Wirth, however, such an order can never replace a communal order based on consensus and the moral strength of small primary groups. As a result, situations develop in which social norms are so muddled and weak that a social condition known as *anomie* develops: individuals, unclear or unhappy about norms, tend to challenge or ignore them, thus generating a further source of deviant behaviour.

Box 7.2

Key thinkers in urban social geography – Walter Benjamin

A growing influence upon urban studies in recent years is the social critic Walter Benjamin. Only a small proportion of his extensive writings were published during his lifetime (he committed suicide in 1940) and even today some of these have not been translated into English. Furthermore, his writings, through extremely stimulating, have an enigmatic, almost poetic, quality. This ambiguity has led to a lively 'Benjamin industry' reinterpreting what he *really* meant. Nevertheless, his insights have inspired many contemporary scholars.

The main reason for this attention is Benjamin's famous 'Arcades Project', a description of the throbbing crowds in the covered shopping passages in the cities of the late nineteenth and early twentieth centuries – but most notably in Paris. Benjamin captured the enormous impact of the modernist metropolis upon the individual with his concept of the urban *flaneur* – the casual stroller through the city. According to Benjamin, the *flaneur* is enthralled and excited by the dynamism and diversity of the city, but he is never overwhelmed by it. He is engrossed in the spectacle of the crowd but is incapable of participating in it fully.

The *flaneur* was essentially a phenomenon of the metropolis, since smaller urban areas did not display sufficient diversity to permit such voyeurism. Hence, the famous boulevards of Paris designed by Baron Haussmann gave plenty of scope for *flanerie* (see also Box 7.1 on Bohemia). The *flaneur* is also a product of the industrial capitalist city and the enormous display of wealth that it engendered. It is also generally accepted that the *boulevadier* is essentially a male character. In late nineteenth-century cities middle-class women, unless chaperoned, were very much confined to the domestic sphere – unaccompanied women on the streets were more likely to be street sellers or prostitutes. Benjamin felt that a supreme example of the *flaneur* was the poet Charles Baudelaire, who in his famous poem 'To a woman passing by' noted his flirtatious glances at a passing female stranger.

Benjamin saw much beauty in the city and thought many were blinded to this by their everyday urban existence. He was also a deeply troubled person and used his observations to uncover childhood memories, perhaps in the hope of understanding his younger experiences. Benjamin's reputation as an urban commentator continues to grow and some have even argued that the modern surfer of the Internet is a form of electronic *flaneur*.

Key concepts associated with Walter Benjamin (see Glossary)

Commodification, *flanerie*, *flaneur*, gaze, modernism.

Further reading

Benjamin, W. (1973) *Charles Baudelaire: a Lyric Poet in the Era of High Capitalism* Verso, London

Benjamin, W. (1999) *The Arcades Project* Belknap, Cambridge MA. See also, Benjamin, W. (2002) From the Arcades Project, in G. Bridge and S. Watson (eds) *The Blackwell City Reader* Blackwell, Oxford

Latham, A. (1999) The power of distraction: distraction, tactility and habit in the work of Walter Benjamin *Environment and Planning D*, **17**, 451–474

Tester, K. (ed.) (1994) *The Flaneur* Routledge, London

Wilson, E. (1995) The invisible flaneur, in K. Gibson and S. Watson (eds) *Postmodern Cities and Spaces* Blackwell, Oxford

Links with Other Chapters

Chapter 9: Place, Consumption and Cultural Politics

One of the persistent problems associated with Wirthian theory has been that the results of empirical research have been ambivalent. Most of the available evidence comes from four kinds of research: studies of helpfulness, conflict, social ties and psychological states. In general, the first two tend to support Wirthian theory, while the second two undermine it.

Studies of *helpfulness* have typically involved field experiments designed to gauge people's reactions to 'strangers' who, for example, reach 'wrong' telephone numbers with their 'last' coin, who have 'lost' addressed letters, who ask for directions and so on. The general drift of the results has been that city dwellers tend to be significantly less helpful than small-town residents. Studies of *conflict* show that both group conflict – racial, social and economic – and interpersonal conflict – certain categories of crime – are disproportionately likely to occur in large communities.

On the other hand, studies that have attempted to compare the number of quality of friendships or *personal relations* have generally shown no difference between different-sized communities, or have shown greater social integration among urbanites. Similarly, studies of *psychological states* such as stress and alienation show that the incidence of such phenomena is just as great, if not greater, in smaller communities.

The public and private worlds of city life

Accepting the validity of such findings, how might they be reconciled? One way is to re-examine the idea of urban environments, recognizing the distinction between the public and the private spheres of urban life. The former consists of settings where people are strangers, in which it requires a special etiquette: reserved, careful, non-intrusive. In the public sphere, people must be, or at least appear to be, indifferent to other people. Richard Sennett contends that modern cultures suffer from having deliberately divided off subjective experience from worldly experience, that we have – literally – constructed our urban spaces in order to maintain this divide:

> The spaces full of people in the modern city are either spaces limited to and carefully orchestrating consumption, like the shopping mall, or spaces limited to and carefully orchestrating the experience of tourism. . . . The way cities look reflects a great, unreckoned fear of exposure. 'Exposure' more connotes the likelihood of being hurt than of being stimulated. . . . What is characteristic of our city-building is to wall off the differences between people, assuming that these differences are more likely to be mutually threatening than mutually stimulating. What we make in the urban realm are therefore bland, neutralizing spaces, spaces that remove the threat of social contact: street walls faced in sheets of plate glass, highways that cut off poor neighborhoods from the rest of the city, dormitory housing developments.

> (Sennett, 1990, p. xii)

But avoiding 'exposure', whether through individual comportment or through urban design, is situational behaviour, not a psychological state, and says nothing about people's attitudes and actions in the private sphere. The city dweller 'did not lose the capacity for the deep, long-lasting, multifaceted relationship. But he *gained* the capacity for the surface, fleeting, restricted relationship' (Lofland, 1973, p. 178). Fischer drew on this distinction, suggesting that urbanism is not characterized by distrust, estrangement and alienation among neighbours although it is associated with estrangement and alienation from 'other people' in the wider community. In other words, 'urbanism produces fear and distrust of "foreign" groups in the public sphere, but does not affect private social worlds' (Fischer, 1981). In Wirthian terminology, this means that urbanism accommodates both 'moral order' and 'social disorganization'.

The self: identity and experience in private and public worlds

Questions about how individuals and social groups come to identify themselves and 'others' require us to reconsider how human subjects are constructed: how we come to think of ourselves within our worlds, both public and private. As indicated in Chapter 3, this is a fundamentally geographical issue: as 'knowledgeable' subjects, our intentionality and **subjectivity** are grounded in social relations and direct experiences that are geographically bounded. They are bounded, moreover, by spaces occupied by other 'knowledgeable subjects', which means that our 'selves' are, to a certain extent, constructed by others. As Andrew Sayer put it, 'what you are depends not just on what you have, together with how you conceive yourself, but on how others relate to you, on what they understand you to be and themselves to be' (Sayer, 1989, p. 211). Put another way, we have to accommodate to meanings, roles and identities imposed through the expectations of others.

In order to come to grips with this subjectivity we ought to begin with the 'unknowing subject' through psychoanalytic theory. For some social scientists, this has carried the attraction of allowing us to admit human emotions such as love, desire, narcissism, anxiety, hate and suffering to our models:

> These . . . feelings are the core of our being, the stuff of our everyday lives. They are the foundations of all society. They come before symbolic meaning and value, lead us continually to reinterpret, hide from, evade and recreate thoughts and values. They inspire our practical uses of rules and they are the reasons behind our reasoned accounts. . . . Without feelings, there would be no uses for rules, ideas, or social structures; and there would be none.
>
> (Douglas, 1977, p. 51)

The case for deploying psychoanalytic theories such as those of Freud and Lacan in the context of social geography has been made by several authors (e.g. Pile, 1993), although little empirical research has been carried out. Proponents of the relevance of psychoanalytic models to social geography point to the work of Alice Miller, who has shown how children repress and restructure their feelings and emotions in order to achieve a sense of safety and security amid the demands and expectations of others (Miller, 1987).

Miller calls this process of repression and restructuring 'the construction of the false self'. This can be seen as the first step towards the creation of knowledgeable subjects whose personal and social identities are conditioned by various dimensions of lived experience such as family life, school, community, work and class consciousness. It follows that personal and social identities have to be seen as malleable and flexible, continually subject to negotiation. They are 'stories told by ourselves about ourselves' in order to cope with our experiences and to operate successfully in the urban settings in which we find ourselves. They are also part of what de Certeau (1985) called the 'constant murmuring of secret creativity' resulting from individuals' attributing certain meanings to their relations with the world(s) around them.

The significance of this 'murmuring' is that it becomes embodied, in time, within cultural practices that constitute an 'upward cultural dynamics' that, in turn, engages dialectically with the cultural norms and meanings that are passed 'downward' by taste-makers, educators and all sorts of 'experts' in science, morality and art. In contemporary cities, the significance of upward cultural dynamics is widely accepted as being of increasing importance. In part, this can be attributed to the swing towards a more open social texture and people's ability to choose from a multiplicity of lifestyle options and sociocultural contexts (Giddens, 1991).

7.3 Social interaction and social networks in urban settings

Most people are involved in several different social relationships, some of which are interconnected to some extent. We not only have friends, but know friends-of-friends; and kinfolk do not exist in isolation: we may get to know a complete stranger because he or she is a member of the same club or organization as an uncle, an aunt or a cousin. The way in which these social linkages are structured is often very complex and, overall, they represent the foundations of social organization. Not surprisingly, therefore, the analysis of these linkages – known as social network analysis – has attracted a good deal of attention.

Social network analysis

Basically, social network analysis attempts to illustrate the structure of social interaction by treating persons as points and relationships as connecting lines. The analysis of social networks thus allows the researcher to map out the complex reality of the interpersonal worlds surrounding specific individuals, and has the advantage

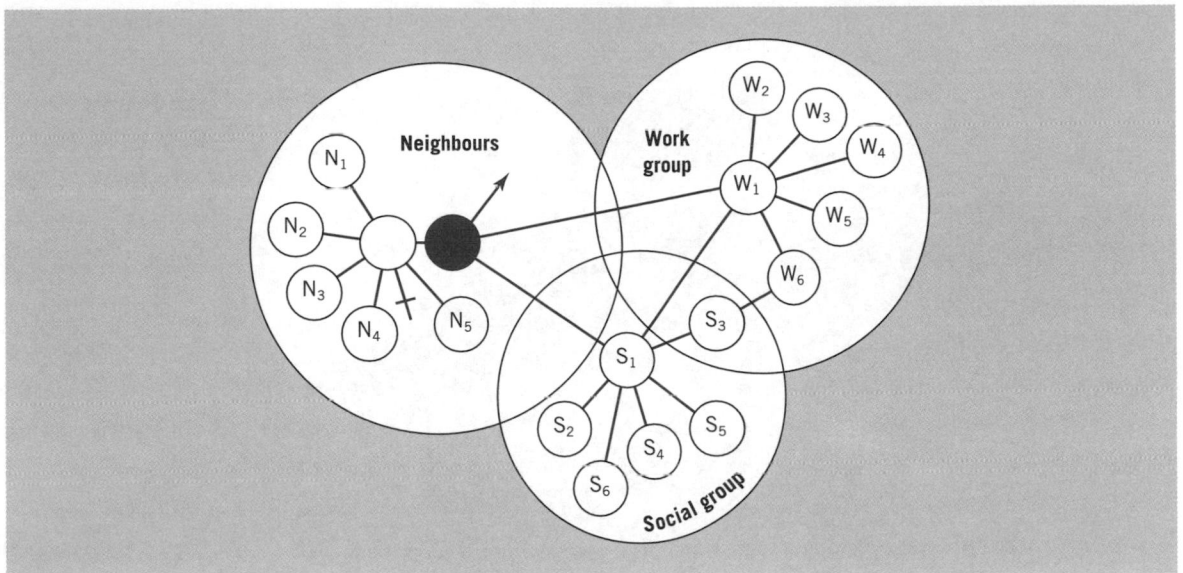

Figure 7.1 The morphology of a husband's social network.
Source: Smith and Smith (1978), p. 106.

of not being confined, a priori, to any specific level of analysis such as the family or the neighbourhood. As with the analysis of other kinds of network – in transport geography and physical geography, for example – this approach facilitates not only the 'mapping' of the morphology' of networks (Fig. 7.1), but also the quantification of certain key characteristics such as their 'connectedness', 'centrality', 'proximity' and 'range'.

Empirical research on social networks has suggested that the number of *potential* contacts for interaction in the social networks of 'typical' urbanites (defined as white, male, married, about 40 years old, with a child in elementary school) in North American cities is about 1500, with *actual* networks averaging about 400 contacts (see, for example, Killworth *et al.*, 1990). For the most part, these networks are loosely or moderately knit, with less than half of any one person's network knowing one another independently of that person. Furthermore, very few of these social ties provide significant levels of support and companionship. Wellman estimated that about 20 of the 400 in the typical network are 'active', about 10 are 'interactive', about 5 are 'intimate' and just one or two are 'confidants' (Wellman, 1987).

Any one person may belong to several different and non-overlapping social networks at the same time, and each of these networks may well have different properties: some may be spatially bounded while others are not; some may have dentritic structures while others are web-like, with interlocking ties, clusters, knots or subgraphs. Early formulations of types of social network were based on the notion of a continuum of networks ranging from *looseknit* (where few members of the network know each other independently) to *closeknit* (where most members of the network know each other); but this presents practical difficulties in operationalizing a definition of linkage. Should it extend beyond kinship and friendship to acquaintance or 'knowledge of' another person, or what? And how is friendship, for instance, to be measured?

In an attempt to minimize such confusion, a typology of social situations has been proposed that incorporates the notion of the complexity as well as the structure of social networks (Fig. 7.2). The typology can be illustrated by way of the extreme and limiting cases: 'A' and 'B' in the diagram. A is the traditional community as normally understood: social relationships are multiplex in that, for example, neighbours are workmates are kinsmen are leisure-time companions, and the social network has a dense structure in that everyone knows everyone else. B is the situation of idealized urban anonymous anomie: social

The positive image of the city: dining out in the Piazza Erbe, Mantua. Photo Credit: Paul Knox.

relationships are uniplex (the taxi driver and his fare), fleeting, impersonal and anonymous, and the social network structure is single-stranded in that only one person knows the others.

Structure			
Plexity	Dense	Looseknit	Single
Multiplex	**A**		
Simplex			
Uniplex			**B**

Figure 7.2 A typology of social situations.
Source: Bell and Newby (1976), p. 109.

Some researchers have suggested – in contrast to the assertions of Wirthian theory – that self-help networks emerge in cities in order to provide help in many different contexts, and that their existence prevents formal welfare agencies from being swamped.

The focus of these self-help networks is often the 'natural neighbour': a person with a propensity to become involved or make himself available in resolving the problems of other people, whether for self-aggrandizement, altruism or some other motive. They are usually untrained amateurs who may not consciously recognize their own role in helping others. Indeed, they may not actually provide any direct help themselves but act as 'brokers', putting people in touch with someone who can help.

But the study of social networks does not provide the urban social geographer with a sufficiently holistic approach: there remain the fundamental questions regarding the extent to which social networks of various kinds are spatially defined, and at what scales: questions that have as yet received little attention. This brings us to a consideration of the ideas of urban social ecology. Here, we must return once again to the foundational ideas of the Chicago School.

Urban ecology as shaper and outcome of social interaction

Because the deterministic ideas of Robert Park and his colleagues in the Chicago School of urban sociology have been so influential, they merit careful consideration. The most distinctive feature of the approach adopted by the human ecologists is the conception of the city as a kind of social organism, with individual behaviour and social organization governed by a 'struggle for existence'. The biological analogy provided Park and his colleagues with an attractive general framework in which to place their studies of the 'natural histories' and 'social worlds' of different groups in Chicago. Just as in plant and animal communities, Park concluded, order in human communities must emerge through the operation of 'natural' processes such as dominance, segregation, impersonal competition and succession. If the analogy now seems somewhat naive, it should be remembered that it was conceived at a time when the appeal of **social Darwinism** and classical economic theory was strong. Moreover, ecological studies of plants and animals provided a rich source of concepts and a graphic terminology with which to portray the sociology of the city.

One of the central concepts was that of **impersonal competition** between individuals for favourable locations within the city. This struggle was acted out primarily through market mechanisms, resulting in a characteristic pattern of land rents and the consequent **segregation** of different types of people according to their ability to meet the rents associated with different sites and situations. Economic differentiation was thus seen as the basic mechanism of residential segregation, and the local **dominance** of a particular group was ascribed to its relative competitive power.

Functional relationships between different individuals and social groups were seen as *symbiotic* and, where such relationships could be identified as being focused within a particular geographical area, the human ecologists identified *communities*, or **natural areas**: 'territorial units whose distinctive characteristics – physical, economic and cultural – are the result of the unplanned operation of ecological and social processes' (Burgess, 1964). As the competitive power of different groups altered and the relative attractiveness of different locations changed in the course of time, these territories were seen to shift. Once more, ecological concepts were invoked to describe the process, this time using the ideas of **invasion, dominance** and **succession** derived from the study of plant communities.

The spatial model

These concepts were all brought together by Burgess in his model of residential differentiation and neighbourhood change in Chicago. Observations on the location and extent of specific communities formed the basis for the identification of an urban spatial structure consisting of a series of concentric zones (Fig. 7.3). These zones were seen by Burgess as reflections of the differential economic competitive power of broad groups within society, whereas the further segregation of smaller areas within each zone – such as the ghetto, Chinatown and Little Sicily within the **zone in transition** – were seen as reflections of symbiotic relationships forged on the basis of language, culture and race.

The **concentric zone model** was set out in terms of dynamic change as well as the spatial disposition of different groups. Zones I to V represent, in Burgess's words, 'both the successive zones of urban extension and the types of areas differentiated in the process of expansion' (Burgess, 1924, p. 88). As the city grew, the changing occupancy of each zone was related to the process of invasion and succession, and Burgess was able to point to many examples of this in Chicago in the early 1900s as successive waves of immigrants worked their way from their initial quarters in the zone of transition (Zone II) to more salubrious neighbourhoods elsewhere. In his diagrammatic model, some of the early immigrant groups – the Germans are explicitly noted – have already 'made it' to the area of superior accommodation in

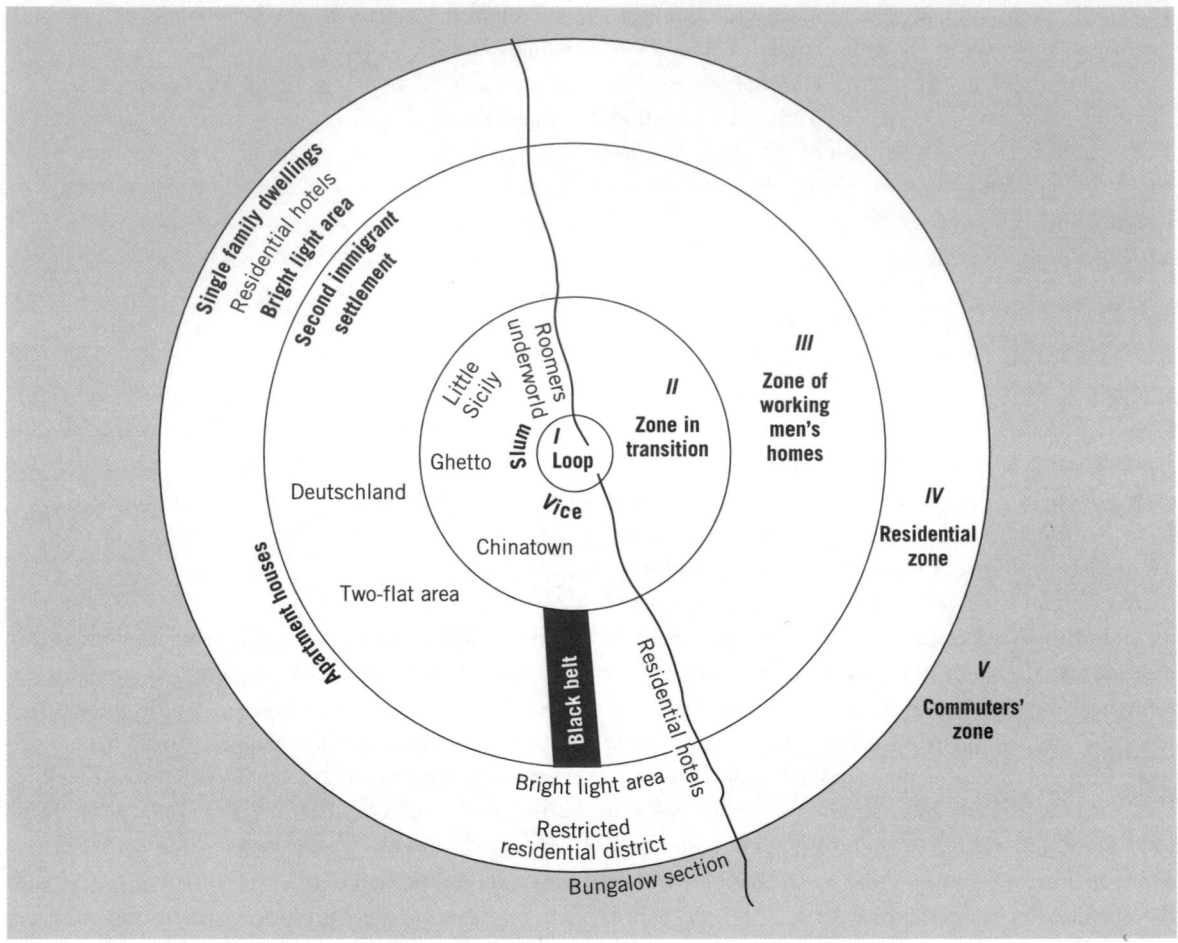

Figure 7.3 Burgess's zonal model applied to Chicago.
Source: Park *et al.* (1925), p. 53.

Zone III and become the dominant group, replacing second-generation American families who had moved out to colonize the outer residential zone (Zone IV).

Within this broad framework three types of study were produced by the school of human ecologists (Berry and Kasarda, 1977):

1. Studies focusing on the process of competition, dominance and succession and their consequences for the spatial distribution of populations and land use. Such work is best represented by the early writings of Park, Burgess and McKenzie described above.

2. Detailed descriptions of the physical features of 'natural' areas along with the social, economic and demographic characteristics of their inhabitants.

Well-known examples of this type of work include Wirth's study of *The Ghetto* (1928) and Zorbaugh's portrayal of Chicago's 'near' North Side in *The Gold Coast and the Slum* (1929). Zorbaugh's work provides a good example of the intimate portrayal of individual social worlds set in the framework of broader ecological theory. The near North Side area was part of the zone in transition and contained four distinctive natural areas: the Gold Coast, a wealthy neighbourhood adjacent to the lakeshore; a rooming house area with a top-heavy demographic structure and a high population turnover; a bright-lights district, Towertown, with brothels, dance-halls and a 'bohemian' population; and a slum area containing clusters of immigrant groups. Zorbaugh showed how the personality of these different quarters related to their physical attributes – the 'habitat' they offered – as well as to the attributes and ways of life of their inhabitants. Moreover, he was also able to illustrate the dynamism of the area, charting the territorial shifts of different groups resulting from the process of invasion and succession.

3. Studies of the ecological context of specific social phenomena such as delinquency, prostitution and mental disorders. A central concern was the investigation of ecologies that seemed to generate high levels of deviant behaviour, and typical examples include the work by Shaw *et al.* on *Delinquency Areas* (1929) and Faris and Dunham's work on *Mental Disorders in Urban Areas* (1939). Much of this work had a clear 'geographical' flavour since it often involved mapping exercises. It also provided the stimulus for a number of the more recent studies discussed in Chapter 10.

Criticisms of the ecological approach

Ecological research was neglected during the 1940s and 1950s following a series of theoretical and empirical critiques. The most general criticism was directed towards the biological analogies, which had been brought into great disrepute by the parallel concept of *Lebensraum*, part of the theory of geopolitics used to justify some of the territorial claims of Hitler's Third Reich. Other criticisms were more specific, centring on the excessive reliance on competition as the basis of social organization, the failure of its general structural concepts (such as the natural area and concentric zonation) to hold up under comparative examination, and its almost complete exclusion of cultural and motivational factors in explaining residential behaviour.

This last criticism was perhaps the most damaging of all. The first (and therefore best-known) critic of the Chicago School, on the grounds that they overlooked the

---

Box 7.3

Key thinkers in urban social geography – Robert Park (1864–1944)

When most geographers think of the Chicago School of human ecology, working in the University of Chicago in the 1920s and 1930s, they think of Ernest W. Burgess and his famous concentric ring model of city structure. However, there was one person who arguably had a much bigger influence on the School during his 30 years in Chicago – Robert Park. The scholars who worked in Chicago during the 1920s and 1930s were, in fact, a highly diverse bunch who it seems at the time were not so convinced as subsequent generations of interpreters that they formed a distinctive school! Nevertheless, Robert Park was the key scholar who gave the school an over-arching sense of purpose.

Park studied in Berlin under the pioneering German sociologist Georg Simmel and was greatly influenced by Simmel's ideas on the relationships between territory and behaviour. He was also influenced by another founding father of sociology – Ferdinand Tönnies – but unlike European scholars Park was hostile to 'armchair theorizing'. Instead, he urged his students to walk the streets and mix with all the diverse forms of human life to be found in cities. It is clear from Park's writings that, although he recognized that cities had their dark, dangerous and disordered sides, he found them exciting, stimulating and liberating places.

Key concepts associated with Robert Park (see Glossary)

Biological analogy, biotic forces, Chicago School, human ecology,

natural areas, social Darwinism, zone in transition.

Further reading

Bulmer, M. (1984) *The Chicago School of Sociology: Institutionalisation, Diversity and the Rise of Sociological Research* University of Chicago Press, Chicago, IL

Entrikin, J. (1980) Robert Park's human ecology and human geography *Annals of the Association of America Geographers*, **70**, 43–58

Links with Other Chapters

Chapter 10: The Geography of Urban Crime

Chapter 12: Residential Mobility and Neighbourhood Change

role of 'sentiment' and 'symbolism' in people's behaviour, was Walter Firey, who pointed to the evidence of social patterns in Boston where, although there were 'vague concentric patterns', it was clear that the persistence of the status and social characteristics of distinctive neighbourhoods such as Beacon Hill, The Common and the Italian North End could be attributed in large part to the 'irrational' and 'sentimental' values attached to them by different sections of the population (Firey, 1945). In short, social values could, and often did, override impersonal, economic competition as the basis for sociospatial organization. Firey's work is significant in that it directed the attention of geographers and sociologists to the importance of the subjective world in the understanding of social patterns in cities.

In fairness to the Chicago School, it should be acknowledged that they themselves did not regard their ideas on human ecology as either comprehensive or universally applicable. Park, for instance, clearly distinguished two levels of social organization: the *biotic* and the *cultural*. The former, he argued, was governed by **impersonal competition** whereas the latter was shaped by the consensus of social values. These cultural aspects of social organization clearly encompass Firey's notions of sentiment and symbolism, and Park and his colleagues were well aware of their influence. Park believed, however, that it was possible to study the biotic level of social organization separately, treating social values and communications as a kind of superstructure of the more basic level of the community. It is thus not so much the denial of non-biotic factors as the inadequacy of their treatment that led to the unpopularity of traditional human ecology.

Social interaction in urban environments

A quite different approach to the study of social organization in urban environments has developed from the pursuit of another of Georg Simmel's suggestions: that the essentials of social organization are to be found in the *forms of interaction* among individuals. At the most fundamental level, interactionist research seeks to establish the nature of non-random interaction patterns at the 'dyadic' or 'triadic' level: that is, between two or three individuals. A good deal of this research has involved the temporal and sequential characteristics of personal relationships, focusing on considerations of initiative, role and status, but it is the qualities of the *nature* and *intensity* of interaction that hold most interest for geographers.

It is common for the nature of interaction to be classified according to whether it takes place in the context of primary or secondary settings. **Primary relationships** include those between kinfolk – based on ties of blood and duty – and those between friends – based on ties of attraction and mutual interest. Beyond this distinction, the nature of primary relationships may be further qualified. For example, family relationships may be differentiated according to whether the setting is a 'nuclear' unit – husband, wife and offspring – or an 'extended' unit that includes members of more than two generations. Interaction between friends may be differentiated according to whether the friendship is based on age, culture, locality and so on.

Secondary relationships are more purposive, involving individuals who group together to achieve particular ends. Such relationships are conveniently subdivided into those in which there is some intrinsic satisfaction in the interaction involved – known as **'expressive' interaction** – and those in which the interaction is merely a means of achieving some common goal – **'instrumental' interaction**. Both kinds are normally set within a broad group framework. Expressive interaction, for example, is typically facilitated by voluntary associations of various kinds: sports, hobby and social clubs, and 'do-gooding' associations. Instrumental interaction, on the other hand, normally takes place within the framework of business associations, political parties, trade unions and pressure groups.

Social distance and physical distance

The difficulties of conceptual and empirical classifications of different types of interaction are compounded by the fact that the propensity for, and intensity of, interaction of all kinds is strongly conditioned by the effects of distance: both *social distance* and *physical distance*. There is, however, some overlap in practice between these two concepts of distance; and a further level of complexity is introduced by the fact that patterns of interaction are not only affected by the physical and social structure of cities but that they themselves also have an effect on city structure. Unravelling the processes involved in this apparently indivisible chain of events is a central concern of urban social geography. Before proceeding to a consideration of more complex situations, however, some initial clarification of the role of social and physical distance is in order.

The idea of **social distance** has a long history, and is sharply illustrated by Bogardus's (1962) attempt to measure the perceived social distance between native-born white Americans and other racial, ethnic and linguistic groups. He suggested that social distance could be reflected by a ranked scale of social relationships that people would be willing to sanction: the further up the scale, the closer the perceived distance between people:

1. to admit to close kinship by marriage;

2. to have as a friend;

3. to have as a neighbour on the same street;

4. to admit as a member of one's occupation within one's country;

5. to admit as a citizen of one's country;

6. to admit only as a visitor to one's country;

7. to exclude entirely from one's country.

It is now generally accepted that the less social distance there is between individuals, the greater the probability of interaction of some kind. Similarly, the greater the physical proximity between people – their 'residential propinquity' – the more likelihood of interaction of some kind. The exact influence of social and physical distance depends to some extent on the nature of the interaction concerned, however. Instrumental interaction related to trade unions or political parties, for instance, will clearly be less dependent on physical distance than instrumental interaction that is focused on a local action group concerned with the closure of a school, the construction of a power station, or the organization of a block party.

In most cases, of course, the influences of social and physical distance are closely interwoven and difficult to isolate. Membership of voluntary associations, for example, tends to reflect class and lifestyle, with participation depending largely on social distance. Middle-class groups, in particular, have a propensity to use voluntary associations as a means of establishing and sustaining social relationships. But, because of the close correspondence between social and residential segregation, membership of such associations is also strongly correlated with locational factors.

Geographers, of course, have a special interest in the role of distance, space and location. There is, however, no real consensus on the role of propinquity in stimulating or retarding social interaction. Melvin Webber (1964), while acknowledging the effects of propinquity, suggested that the constraints of distance are rapidly diminishing in the 'shrinking world' of modern technology and mass communications. He argued that improvements in personal mobility, combined with the spatial separation of home, workplace and recreational opportunities, have released people from neighbourhood ties. More recently, it has been suggested that 'the 800 number and the piece of plastic have made time and space obsolete' (Sorkin, 1992, p. xi). But not everyone, of course, benefits from mobility to the same extent: some people are 'localites', with restricted urban realms; others are 'cosmopolites', for whom distance is elastic and who inhabit a social world without finite geographical borders.

This tendency towards an aspatial basis for social interaction has been seen by others as a result not so much of increased personal mobility as a product of modern city planning and social values. Colin Ward, for example, argued that modern housing estates have 'annihilated' community spirit and replaced it with a parental authoritarianism that restricts the outdoor activities of children and so retards the development of locality-based friendships from the earliest years of a person's life (Ward, 1978). Similarly, Susan Keller has claimed that there has been a widespread decline of both organized and spontaneous neighbouring in America because of the combined effects of changes in economic organization and social values. She attributes the decline in neighbouring to four factors:

➤ The presence of multiple sources of information and opinion via mass media, travel, voluntary organizations and employment away from the local area.

➤ Better transport beyond local boundaries.

➤ Increased differentiation in people's interests and desires, and greater differentiation in rhythms of work, resulting in less inclination and ability, respectively, to interact with neighbours.

➤ Better social services and greater economic security.

Against such arguments we must set the observation that the residential neighbourhood continues to provide much raw material for social life, especially for relatively immobile groups such as the poor, the aged and mothers with young children. Even the more mobile must be susceptible to chance local encounters and the subsequent interaction that may follow; and most householders will establish some contact with neighbours from the purely functional point of view of mutual security.

The most telling argument in support of the role of propinquity is the way that residential patterns – whether defined in terms of class, race, ethnicity, lifestyle, kinship, family status or age – have persistently exhibited a strong tendency towards spatial differentiation. In a pioneering study, Duncan and Duncan showed that the residential segregation of occupational groups in Chicago closely paralleled their social distance and that the most segregated categories were those possessing the clearest rank, i.e. those at the top and the bottom of the socioeconomic scale (Duncan and Duncan, 1955). Subsequent studies of socioeconomic groups elsewhere and of racial and ethnic groups in a wide variety of cities have all reported a significant degree of residential segregation. The persistence of such patterns requires us to look more closely at the sociocultural bases of residential segregation.

Box 7. 4

Key films related to urban social geography – Chapter 7

Grand Canyon (1991) A little contrived, but a thought-provoking and ultimately life-affirming consideration of issues of race, identity and power in Los Angeles.

The Godfather (1971) Something of a soap opera, but nevertheless, a compelling depiction of turf wars among immigrant communities in early twentieth-century urban America. *The Godfather 2* (1974) is even better in this respect.

Falling Down (1992) A film that defies easy categorization, depicting the stresses of the contemporary urban jungle as they affect a middle-aged male.

It's a Wonderful Life (1946) Sickly and sentimental, but also with very dark undertones, this is a classic portrayal of US suspicion of big city life and a celebration of small town living.

Mean Streets (1973) A film dealing with life in the New York underclass, plus a path-breaking rock sound track.

Metropolis (1926) A mixture of science fiction and gothic horror, this early film by Fritz Lang is a nightmarish vision of modernism taken to extremes in a futuristic city in which a populace is enslaved. Visually impressive.

Short Cuts (1993) A brilliant interweaving of eight stories of suburban middle-class life.

Taxi Driver (1976) A now classic portrayal of the US urban dystopia. But beware if you do not like nasty violent endings.

Trainspotting (1995) Both comic and at times very disturbing, a portrayal of the cynicism surrounding a group of drug takers. Interesting for a portrayal of Edinburgh that is far from the usual cultural and tourist image.

Tokyo Story (1953) According to many film critics this is one of the greatest movies ever made (it is in many film fans' top ten). Moving, sad and poignant, but never sentimental, the cult director Ozu portrays the differences between the values of older generations in rural Japan and their modern urban-dwelling offspring. Essential viewing for all true film buffs.

Chapter summary

7.1 Western attitudes towards urban living are characterized by hostility, yet cities are also seen as centres of diversity and opportunity.

7.2 The Chicago School of urban sociology interpreted cities through analogy with ideas from plant and animal ecology. Wirthian ideas stressed the destabilizing effect of increasing size, density and heterogeneity of population on existing social norms. Subsequent research has stressed that personal identities are critically influenced by factors such as class, community, work and family.

7.3 People belong to different non-overlapping social networks that vary in structure and intensity. Although highly influential, the ideas of the Chicago School failed to adequately reconcile the relationships between 'economic' and 'cultural' factors in cities. Physical and social distance are intimately connected in urban areas.

Key concepts and terms

Chicago School	instrumental interaction	social Darwinism
concentric zone model	invasion	social distance
dominance	natural areas	subjectivity
expressive interaction	primary relationships	succession
human ecology	'psychic overload'	Wirthian theory
identity	secondary relationships	zone in transition
impersonal competition	segregation	

Suggested reading

The suggested reading for Chapter 3 provides a general introduction to the issues considered in this chapter. A good introduction to the specifically urban context of this chapter is provided by Claude Fischer's book on *The Urban Experience* (1976: Harcourt, Brace, Jovanovich, New York), while a useful reference is the collection of classic essays edited by P. Sennett: *Classic Essays on the Culture of Cities* (1969: Appleton-Century-Crofts, New York). Contemporary work on the social dimensions of urbanism is represented by de Certeau's *The Practice of Everyday Life* (1985: University of Columbia Press, Berkeley), Richard Sennett's *The Fall of Public Man* (1977; Faber and Faber, New York) and by Anthony Giddens's *Modernity and Self-identity: Self and Society in the Late Modern Age* (1991: Polity Press, Cambridge). See also

Jan Lin and Christopher Mele *The Urban Sociology Reader* (2004: Routledge, London). An excellent review of the spatial dimensions of interpersonal relations is provided by Henry Irving in a series edited by David Herbert and Ron Johnston (*Geography and the Urban Environment*, 1978: 1, 249–284). For work on urban networks see also C. Fischer *To Dwell Amongst Friends: Personal Networks in Town and City* (1982: Chicago University Press, Chicago), Hannerz's *Transnational Connections: Culture, People, Places* (1996: Routledge, London) and B. Wellman and S. Berkowitz's edited volume *Social Structures: A Network Approach* (1988: Cambridge University Press, Cambridge). For a discussion of how discourses surrounding the mythical nuclear family continue to shape urban space see Stuart C. Aitken's *Family Fantasies and Community Space* (1998: Rutgers University Press, New Brunswick).

Chapter 8

Segregation and congregation

Key questions addressed in this chapter

➤ Which 'minority groups' are most residentially segregated in Western cities?

➤ What are the processes responsible for these patterns?

There are several rational reasons for segregation within urban society. The spatial segregation of different 'communities' helps to minimize conflict between social groups while facilitating a greater degree of social control and endowing specific social groups with a more cohesive political voice. Another important reason for the residential clustering of social groups is the desire of its members to preserve their own group identity or lifestyle. One of the basic mechanisms by which this segregation can be achieved is through group norms that support marriage within the group and oppose marriage between members of different social, religious, ethnic or racial groups. The organization of groups into different territories facilitates the operation of this mechanism by restricting the number of 'outside' contacts. Thus people tend to marry their equals in social status; neighbours tend to be social equals; and so they marry their neighbours. There are also, of course, several negative reasons for the persistence of residential segregation. Beginning with fear of exposure to 'otherness', these extend to personal and institutionalized discrimination on the basis of class, culture, gender, sexual orientation, ethnicity and race.

8.1 Social closure, racism and discrimination

'What we are talking about here', observes Chris Philo, is 'the contest of cultures bound up in processes of socio-spatial differentiation . . .' and '. . . the clash of moralities (of differing assumptions and arguments about worth and non-worth) which are both constituted through and constitutive of a society's socio-spatial hierarchy of "winners" and "losers" ' (Philo, 1991, p. 19).

One concept that is useful here is Frank Parkin's (1979) notion of **social closure**, whereby 'winners' are characterized by their ability to exercise power in a downward direction, excluding less powerful groups from desirable spaces and resources. Parkin also calls this **exclusionary closure**: an example would be the explicitly exclusionary practices of housing classes defined through membership of homeowners' associations (see Chapter 6).

Another means of differentiating 'winners' from 'losers' is through the social construction of **racism**. Peter Jackson has defined racism as 'the assumption, consciously or unconsciously held, that people can be divided into a distinct number of discrete "races" according to physical, biological criteria and that systematic social differences automatically and inevitably follow the same lines of physical differentiation' (Jackson, 1989, pp. 132–133), a definition that can be extended to include cultural differences. Racism produces pejorative associations aimed both at individuals (e.g. sexuality, criminality) and of social groups (e.g. family structures, cultural pathologies).

The critical point here is that racism is not a uniform or invariable condition of human nature but, rather, consists of sets of attitudes that are rooted in the changing material conditions of society. We can, therefore, identify a multiplicity of racisms within contemporary cities, depending on the particular circumstances of different places. Susan Smith (1988) adopted this perspective in suggesting that the interaction of political culture with economic contingency produced three distinctive phases of racism in postwar Britain:

➤ *1945–1960*: a period during which blacks and Asians, although frequently regarded as culturally backward or morally inferior, were regarded as intrinsically British, sharing equally with whites the status and privilege of Commonwealth citizenship. 'It was widely assumed that immigrant status, like the problems accompanying it, would be a temporary prelude to assimilation and absorption' (Smith, 1988, p. 425).

➤ *1961–1975*: civil unrest in the Notting Hill area of London (in 1958) marked the turning point at which blacks and Asians ceased to be regarded as fellow citizens and began to be depicted as alien, with alien cultures, different temperaments, backgrounds and ways of life. 'Immutable differences, indexed by colour, were overlaid on the malleable cultural boundaries previously assumed to distinguish immigrant from "host" ' (Smith, 1988, p. 428).

➤ *1976–*: a period of social authoritarianism, in which neoliberal economic philosophies have defined issues of race as being insignificant to the concerns of politics and the economy, while at the same time a resurgence of moral conservatism, in appealing to a revival of national pride, has reinforced racism, albeit in the disguised language of 'culture' or 'ethnicity'.

Smith's emphasis on the interaction of political culture and economic circumstances is particularly important to our understanding of racial segregation in societies (such as Britain and America) where institutional discrimination carries racism into the entire housing delivery system. Such discrimination permeates the legal framework, government policies (those relating, among others, to urban renewal, public housing and suburban development), municipal land use ordinances, and, as we saw in Chapter 6, the practices of builders, landlords, bankers, insurance companies, appraisers and real estate agents. The impersonal web of exclusionary practices that results from this institutional discrimination has reinforced the racism and discrimination of individuals to the point where segregated housing has led to de facto segregated schools, shopping areas and recreational facilities. All this spatial segregation, in turn, serves to reproduce racism and to sustain material inequalities between racial categories.

8.2 The spatial segregation of minority groups

Given these caveats about racism and discrimination, we can interpret minority group residential congregation and segregation as being inversely related to the process of **assimilation** with the host society, a process that is itself governed by various forms of group behaviour designed to minimize real or perceived threats to the group from outsiders. But before going on to examine this behaviour and its spatial consequences in detail, it is first necessary to clarify the meaning of terms such as 'minority group', 'host society', 'segregation' and 'assimilation'.

Issues of definition and measurement

The term 'minority group' is widely used to mean any group that is defined or characterized by race, religion, nationality or culture. Implicit in its use is the idea that their presence in the city stems from a past or continuing stream of in-migration. Minority groups in this sense therefore include African-Americans, Puerto Ricans, Italians, Jews, Mexicans, Vietnamese and (Asian) Indians in American cities; Afro-Caribbeans, Asians and Irish in British cities; Algerians and Spaniards in French cities; Turks and Croats in German cities; and so on. While the host society may not be homogeneous, it always contains a charter group that represents the dominant matrix into which new minority groups are inserted. In North America, Australia and Britain the charter group in most cities is white, with an 'Anglo-Saxon' culture. The degree to which minority groups are spatially segregated from the charter group varies a good deal from city to city according to the group involved.

Segregation is taken here to refer to situations where members of a minority group are not distributed absolutely uniformly across residential space in relation to the rest of the population. This clearly covers a wide range of circumstances, and it is useful to be able to quantify the overall degree of segregation in some way. Several indexes of segregation are available, although the sensitivity of all of them depends on the scale of the areal units employed. One of the most widely used methods of quantifying the degree to which a minority group is residentially segregated is the index of dissimilarity, which is analogous to the Gini index of inequality and which produces a theoretical range of values from 0 (no segregation) to 100 (complete segregation).

Index values calculated from census tract data in US cities show that African-Americans are generally the most segregated of the minorities in America, with index values at the tract level commonly exceeding 80. Puerto Ricans and Cubans have also been found to be very highly segregated in American cities, with index values at the tract level commonly exceeding 60; as have the new immigrant groups of the 1980s, 1990s and 2000s – Mexicans and Asians (Garcia, 1985; see also Roseman et al., 1998; and Box 8.1).

By comparison, minority-group residential segregation in European cities is relatively low (Madanipour et al., 1998; Hudson and Williams, 1999). In Britain, for example, index values calculated for immigrant minority groups – Afro-Caribbeans, Pakistanis, Indians and Africans – at the enumeration district level range between 40 and 70 (Phillips and Karn, 1991; Peach, 1998). The major exception is the Bangladeshi group, which in London, for example, is highly concentrated in a single borough (Tower Hamlets) and has indexes of dissimilarity similar to African-Americans in the United States. The *Gastarbeiter* (guest worker) population of continental European cities is much less segregated: index values for Turks, Greeks, Spaniards and Portuguese in German, Dutch and Swiss cities, for example, range between 35 and 50. At more fine-grained levels of analysis the degree of segregation can be much higher, with index values of between 80 and 90 for Asians, Afro-Caribbeans, Turks and North Africans at the scale of individual streets in northwest European cities. This emphasizes the vulnerability of statistical indexes, and makes inter-city comparisons difficult.

Another practical difficulty in making precise statements about the degree of residential segregation is that minority groups may subsume important internal differences. Statements about the segregation of Asians in British cities, for instance, often overlook the tendency for Indians, Pakistanis and Bangladeshis to exist in quite separate communities, even though these communities may appear to outsiders to be part and parcel of the same community. Muslims are separated from Hindus, Gujerati speakers from Punjabi speakers, and East African Asians from all other Asians; and these segregations are preserved even within public sector housing. Similarly, the distinctive island communities of the West Indies have been reflected in the map of London for much of the period since the Second World War:

There is an archipelago of Windward and Leeward islanders north of the Thames; Dominicans and St Lucians have their core areas in Notting Hill; Grenadians are found in the west in Hammersmith and Ealing; Montserratians are concentrated around

Box 8.1

Key trends in urban social geography – The Latinization of US cities

The most important development in the evolving social geography of US cities is their growing Latinization – the evolution of neighbourhoods dominated by peoples with Latin American cultural roots (variously referred to as Latin-Americans, Latinos or Hispanics). The terminology can be controversial (see Davis, 2000) but the category includes those whose ancestry relates back to many central and South American nations, predominantly Mexico, but also Puerto Rico, Cuba, El Salvador, the Dominican Republic, Chile, Colombia, Ecuador and Guatemala.

Whether by preference or constraint, Latinos are heavily concentrated in urban areas, and in many US cities – especially those in the South-West, such as Los Angeles, Houston, San Diego and Phoenix – Latinos have displaced African-Americans as the dominant ethnic minority group. Furthermore, as a result of recent in-migration (both legal and illegal) and a much higher rate of fertility than the Anglo population, Latinos are increasing at a much faster rate than most other sections of US society.

The mix of Latino groups varies from city to city; in Los Angeles the Latino population is over three-quarters Mexican in origin, in Miami it is two-thirds Cuban and in New York it is almost half Puerto Rican. Mike Davis (2000) argues that Latino neighbourhoods (traditionally known as *barrios*) broadly conform to one of four different types:

➤ *Primate barrio with small satellites*: a simple ethnic wedge among broad concentric rings, a pattern typical of Latinos in Los Angeles before 1970 and in present day Washington, DC, Houston, Atlanta, Philadelphia and Phoenix.

➤ *Polycentric barrios*: multiple enclaves as in contemporary Chicago which has four roughly equal sized Latino districts.

➤ *Multicultural mosaic*: a diverse range of neighbourhoods with Hispanics of differing backgrounds, the category refers exclusively to New York which has at least 21 major Latino districts.

➤ *City within a city*: Davis reserves this term exclusively for contemporary Los Angeles with its scores of Latino neighbourhoods sprawling in both rings and sectors around the older blue-collar industrial neighbourhoods of the metropolis. This pattern reflects the occupations taken up by Latinos, predominantly in relatively low-paying labour-intensive services and light industry.

The complex implications of this growing Latinization of US cities have yet to be fully appreciated by urban analysts. There are many different manifestations of Latino culture in US cities. For example, while overwhelmingly Catholic and Spanish-speaking, some are adherents of Protestant evangelical religious denominations. Similarly, Cuban exiles in Miami may have differing attitudes to village peoples uprooted by poverty from Mexico to Los Angeles. It seems that diverse new hybrid identities are emerging, the nature of which depends upon the blend of differing Latino groups (e.g. the nature of the national and sub-regional identities) and their relationship with main ethnic groups.

Ever hostile to suburban conservatism, Mike Davis sees the growth of Latino neighbourhoods as a welcome splash of cultural diversity regenerating run-down inner cities. He also envisages Latinos as heading class-based alliances of workers fighting for better wages and employment rights. Although widely disenfranchised or reluctant to participate in electoral politics, Latinos are overwhelmingly Democratic in allegiance. Nevertheless, a significant proportion of Latinos align themselves with Republican values of self-reliance, the importance of religion and opposition to abortion.

Key concepts associated with Latinization (see Glossary)

Borderlands, hybridity, heterotopia, liminal space, Mestizo, pariah city, third space.

Further reading

Davis, M. (2000) *Magical Urbanism: Latinos Reinvent the US City* Verso, London

Stoke Newington, Hackney and Finsbury Park; Antiguans spill over to the east in Hackney, Waltham Forest and Newham; south of the river is Jamaica.

(Peach, 1984)

What is clear enough from the available evidence, however, is that most minorities tend to be highly segregated from the charter group. This segregation has been shown to be greater than might be anticipated from the socioeconomic status of the groups concerned.

In other words, the low socioeconomic status of minority groups can only partially explain their high levels of residential segregation. The maintenance of the minority in-migrant group as a distinctive social and spatial entity will depend on the degree to which assimilation occurs.

This process can take place at different speeds for different groups, depending on the perceived social distance between them and the charter group. Moreover, **behavioural assimilation** – the acquisition by the minority group of a cultural life in common with the charter group – may take place faster than **structural assimilation** – the diffusion of members of the minority group through the social and occupational strata of the charter group society (Hiebert and Ley, 2003). We should recall at this point that assimilation is not simply the process of one culture being absorbed into another. Both mainstream and minority cultures are changed by assimilation through the creation of new hybrid forms of identity (Chapter 3). Nevertheless, in general, the rate and degree of assimilation of a minority group will depend on two sets of factors: (1) external factors, including charter group attitudes, institutional discrimination and structural effects, and (2) internal group cohesiveness. Between them, these factors determine not only the degree and nature of conflict between minority groups and the charter group, but also the spatial patterns of residential congregation and segregation.

External factors: discrimination and structural effects

Minority groups that are perceived by members of the charter group to be socially undesirable will find themselves spatially isolated through a variety of mechanisms. One of the most obvious and straightforward of these is the '*blocking*' strategy by existing occupants of city neighbourhoods as they seek to resist the 'invasion' of minority groups. Meanwhile, established tightly knit minority group clusters tend to be the most resistant to invasion by others, actively defending their own territory in a variety of ways (ranging from social hostility and the refusal to sell or rent homes to petty violence and deliberate vandalism) against intruding members of minority groups.

Where this strategy of '**voicing**' opposition is unsuccessful, or where the territory in question is occupied by socially and geographically more mobile households, the charter group strategy commonly becomes one of '**exit**'. The invasion of charter group territory generally precipitates an outflow of charter group residents that continues steadily until the critical point is reached where the proportion of households from the invading minority group is large enough to precipitate a much faster exodus. This is known as the '**tipping point**'. The precise level of the tipping point is difficult to establish, although it has been suggested that for whites facing 'invasion' by African-Americans, the tipping point may be expected to occur when African-American occupancy reaches a level of about 30 per cent (Rose, 1970). The subsequent withdrawal of charter group residents to other neighbourhoods effectively resolves the territorial conflict between the two groups, leaving the minority group spatially isolated until its next phase of territorial expansion.

The spatial isolation of minority groups is also contrived through discrimination in the housing market, thus limiting minority groups to small niches within the urban fabric. Although formal discriminatory barriers are illegal, minorities are systematically excluded from charter group neighbourhoods in a variety of ways. As we have seen in Chapter 6, the role of real estate agents and mortgage financiers in the owner-occupied sector is particularly important, while the general gate-keeping role of private landlords also tends to perpetuate racially segregated local submarkets. There is also a considerable weight of evidence to suggest that immigrants and minorities are discriminated against in the public sector.

In Britain, racial minorities have found themselves disadvantaged within the public sector in three respects. First, they initially had more difficulty in gaining access to any public housing because of their limited period of residence in a particular local government area; second, they have often been allocated to poor-quality property, particularly older flats; and, third, they have been disproportionately allocated to unpopular inner-city housing estates, thus intensifying the localization of the non-white population in the inner city. These disadvantages are partly the result of unintentional discrimination (such as the residential requirements associated with **eligibility rules**), and partly the result of more deliberate discrimination through the personal prejudices, for example, of housing visitors, who may have little or no understanding of the cultural background and family life of immigrant households (Phillips and Karn, 1991). The effects of this type of discrimination are intensified by the discriminatory policies of city planners. Again, some of this discrimination is unintentional, as in the omission of minority neighbourhoods from urban renewal and rehabilitation schemes; but much is deliberate, as in the manipulation of land use plans and zoning regulations in order to exclude non-whites from suburban residential areas of US cities.

The net effect of this discrimination is to render much of the housing stock unavailable to members of minority groups, thus trapping them in privately rented accommodation and allowing landlords to charge inflated rents while providing little security of tenure. In an attempt to escape from this situation, some householders become landlords, buying large, deteriorating houses and subletting part of the house in order to maintain mortgage repayments and/or repair costs. Others manage to purchase smaller dwellings that are shared with another family or a lodger, but many can only finance the purchase through burdensome and unorthodox means. Asians, in particular, have been found to exhibit a strong propensity towards home ownership in preference to tenancy, notwithstanding the extra financial costs.

The localized nature of cheaper accommodation (whether for sale or rent) is an important aspect of urban structure (sometimes referred to as a 'fabric effect') that tends to segregate minority groups from the rest of the population by channelling them into a limited niche. Moreover, since many minority groups have an atypical demographic structure, with a predominance of young adult males and/or large, extended families, their housing needs – single-room accommodation and large dwellings respectively – can be met only in very specific locations. In many cities, therefore, the distribution of clusters of minorities is closely related to the geography of the housing stock.

Recent evidence from the UK suggest that ethnic minorities have made limited progress in the housing market since the 1950s and 1960s, although the progress is highly variable between different groups. Nevertheless, although less explicit than in the past, institutional discrimination still plays a key role. For example, a study in 1990 by the Commission for Racial Equality using actors found that one in five accommodation agencies were employing discriminatory practices that limited the options for members of ethnic groups (Phillips, 1998). A review by Ratcliffe (1997) concluded that although levels of segregation of ethnic minority groups were declining, there were still variations in housing quality between ethnic minorities and the wider population that could be explained by economic factors alone. This suggests that discriminatory factors may have reduced but not entirely disappeared. For example, in recent years competition among private sector financial institutions has led them to envisage ethnic minorities as a new source of profit and actively seek their custom but there is evidence that perceptions of risk vary with differing groups. Research in Bedford suggested that Asians were valued for their reliability and thrift but Caribbeans were regarded with suspicion (Sarre *et al.*, 1989).

Underlying both charter group discrimination and the localization of minority groups in particular pockets of low-cost housing is their position in the overall social and economic structure of society. In this context, discrimination by working-class members of the charter group is related to the role attributed to minority groups in job and housing markets as competitors whose presence serves to depress wages and erode the quality of life. In short, minority groups are treated as the scapegoats for the shortcomings of the economic system.

But it is the concentration of minority groups at the lower end of the occupational structure that is the more fundamental factor in their localization in poor housing. Because of their lack of skills and educational qualifications, members of minority groups often tend to be concentrated in occupations that are unattractive to members of the charter group, that are often unpleasant or degrading in one way or another, and that are usually associated with low wages. The majority of such occupations are associated with the CBD and its immediate surrounds, and the dependence of minority groups on city centre and inner-city job opportunities is widely cited as a prime determinant in the location of minority residential clusters. This factor, in turn, is reinforced by the location of inexpensive accommodation in inner-city neighbourhoods surrounding the CBD. The isolation of minority groups in this sector of housing and labour markets was intensified by the massive deindustrialization of many large cities in the 1970s and 1980s and the suburbanization of new job opportunities (see Chapter 2). This has effectively trapped many of the poor in inner-city locations because of their inability to meet the necessary transportation costs.

Recent evidence from Britain shows limited but highly uneven progress in terms of the socioeconomic advancement of ethnic minority groups (Phillips, 1998). Indian men, for example, display an occupational structure similar to that for white men and are just as likely to hold professional or managerial jobs; but Afro-Caribbean men are significantly underrepresented in the higher occupational groups. Similarly, Pakistani and Bangladeshi men are less likely than whites to be in skilled manual or professional occupations. In the case of women, however, there are fewer differences between white and black employment patterns, largely the result of fewer women being in professional or managerial groups. Interestingly, young adult members of ethnic minority groups reveal somewhat different patterns of occupational development than their elders. Thus, young black Caribbean adults have a higher propensity to work in white-collar occupations than their parents. However, young adults of Indian descent are less likely than their parents to be in professional occupations, although they are well represented in junior white-collar occupations. Overall, unemployment among ethnic minorities in the UK is higher than for the white population and this represents a significant impediment to socioeconomic advance and residential relocation.

Box 8.2

Key trends in urban social geography – The growth of the eruv

Zones called 'eruvs' now exist in over 200 cities around the world. There is one in every major city in the United States (the Washington eruv includes the White House and Supreme Court) and one in every town in Israel (Morris, 2002). The latter case provides a clue as to their character, for they are formally designated zones in which those of the Orthodox Jewish faith can undertake tasks that would normally be forbidden on the Jewish Sabbath (e.g. pushing wheelchairs or prams, carrying keys or prayerbooks). Such prohibitions stem from Old Testament teachings since, under Orthodox Jewish biblical law, Jews are not permitted to carry any item in a public domain on the Sabbath. Many Jewish homes are, in fact, established as eruvs to permit Orthodox families to perform tasks that would otherwise be forbidden on the Sabbath.

The creation of such zones has not been without controversy, especially where Jews are a minority of the total population. Especially contentious have been those cases where the zone is delineated by poles, typically of 10 metres (approximately 30 feet) in height and linked by wire lines. However, many eruvs do not have such poles; others have boundaries that include railway tracks, roads, and even cliffs and beaches. Non-Jewish inhabitants sometimes fear that eruvs will lead to an 'invasion' of the designated area by those of the Jewish faith. As with many NIMBY protests, objections to eruvs are often couched in other terms, such as fears for nature and wildlife (e.g. birds getting trapped in wires or damage to trees). Others are bemused by a faith that includes such restrictions but also enables followers to evade such prohibitions, while even some Jewish people feel that such zones make their religion open to parody. Nevertheless, there are many examples of such zones where Jewish and non-Jewish groups live in peaceful coexistence.

Key concepts associated with the eruv (see Glossary)

Clustering, community action, 'ethnic village', exclusionary zoning, invasion, NIMBY, residential differentiation, structural assimilation, 'turf' politics.

Further reading

Morris, S. (2002) Testing the boundaries *Guardian*, 10 August, p. 8.

Links with Other Chapters

Chapter 3: What is Culture?

Chapter 9: Urban Villages: Community Saved?, Box 9.2 The Development of New 'Sacred Spaces'

Congregation: internal group cohesiveness

While charter group attitudes and structural effects go a long way towards explaining residential segregation, they do not satisfactorily explain the clustering (**congregation**) of minority groups into discrete, homogeneous territories. Such clusters must also be seen as defensive and conservative in function, partly in response to the external pressures outlined above. In short, when people – especially immigrants and in-migrants – feel threatened they often have a strong urge to internal cohesion, so that the social and cultural coherence of the group may be retained. Four principal functions have been identified for the clustering of minority groups: defence, support, preservation and attack.

Clustering for defence

The defensive role of minority clusters is most prominent when charter group discrimination is widespread and intense, so that the existence of a territorial heartland enables members of the minority group to withdraw from the hostility of the wider society. The term **ghetto** was first used in Venice in the Renaissance era to describe the district in which Jewish people were forced to reside. The Jewish community made a crucial contribution to the status of Venice as a centre of trade and commerce through their money-lending activities. However, this success eventually bred hostility and Jews were forced into a small district of the city previously dominated by foundry activities. Large doors at either end of the neighbourhood were used to enforce a curfew. (Today, with a much reduced Jewish population, this district is crumbling and neglected but tours are available and plans are afoot for preservation and renewal.) Jewish ghettos in medieval European cities therefore functioned as defensive clusters.

Anderson (1988) has shown how 'Chinatown' developed in Vancouver in the late nineteenth century in response to threats of mob rule and riot towards the Chinese population on the part of the wider white populace. Similarly, working-class Catholic and Protestant communities in Belfast have become increasingly segregated from one another in response to their need for physical safety. Nowhere has this phenomenon

been more marked than on the Shankhill-Falls 'Divide' between the Protestant neighbourhood of Shankhill and the Catholic neighbourhood of Clonard-Springfield. Transitional between the two, and marking the Divide between the two groups, is the Cupar Street area, which had acquired a mixed residential pattern in the years up to 1968. When the 'troubles' broke out in 1969, however, the territorial boundary between the two groups took on a much sharper definition. It is estimated that within the following seven years between 35 000 and 60 000 people from the Belfast area relocated to the relative safety of their own religious heartland, thus reinforcing the segregation of Protestants and Catholics into what became known in army circles as 'tribal areas'. In a very different vein, many of the most recent examples of clustering for defence involve affluent, high-status households living in gated communities in 'Fortress America' (Low, 2003).

Clustering for mutual support

Closely related to the defensive functions of minority clusters is their role as a haven, providing support for members of the group in a variety of ways. Such support ranges from formal minority-oriented institutions and business to informal friendship and kinship ties. Clustered together in a mutually supportive haven, members of the group are able to avoid the hostility and rejection of the charter group, exchanging insecurity and anxiety for familiarity and strength. This 'buffer' function of minority clusters has been documented in a number of studies. The existence of ethnic institutions within the territorial cluster is one of the most important factors in protecting group members from unwanted contact with the host community. Sikh temples and Moslem mosques in British cities, for example, have become the focus of Sikh and Pakistani local welfare systems, offering a source of food, shelter, recreation and education as well as being a cultural and religious focus.

More generally, most minority groups develop informal self-help networks and welfare organizations in order to provide both material and social support for group members. In addition, the desire to avoid outside contact and the existence of a local concentration of a minority population with distinctive, culturally based needs serve to provide protected niches for ethnic enterprise, both legitimate and illegitimate. Minority enterprise is an important component of community cohesion in minority neighbourhoods everywhere, providing an expression of group solidarity as well as a means of economic and social advancement for successful entrepreneurs and an alternative route by which minority workers can bypass the white-controlled labour market.

In their classic study of the African-American community in Chicago, Drake and Cayton described the doctrine of the 'double-duty dollar', according to which members of the community should use their money not only to satisfy their personal needs but also to 'advance the race' by making their purchases in African-American-owned businesses (Drake and Cayton, 1962). In Britain, the most distinctive manifestations of minority enterprise are the clusters of banks, butchers, grocery stores, travel agencies, cinemas and clothing shops that have developed in response to the food taboos, specialized clothing styles and general cultural aloofness of Asian communities combined with the economic repression of British society.

Clustering for cultural preservation

This brings us to a third major function of minority residential clustering: that of preserving and promoting a distinctive cultural heritage. Minority group consciousness sometimes results from external pressure, as in the use of Jamaican Creole by young Afro-Caribbeans in London as a private language to shut out the oppressive elements of the white world. But for many groups there exists an inherent desire to maintain (or develop) a distinctive cultural identity rather than to become completely assimilated within the charter group (see also Box 8.3). Residential clustering helps to achieve this not only through the operation of ethnic institutions and businesses but also through the effects of residential propinquity on marriage patterns.

Many commentators have emphasized the self-segregating tendencies of Asian communities in British cities in this context, while the persistence of Jewish residential clusters is often interpreted as being closely related to the knowledge among Jewish parents that residence in a Jewish neighbourhood confers a very

high probability of their children marrying a Jewish person. The residential clustering of some minority groups is also directly related to the demands of their religious precepts relating to dietary laws, the preparation of food and attendance for prayer and religious ceremony. Where such mores form an important part of the group's culture, they are followed more easily where the group is territorially clustered. On the other hand, where group consciousness is weak and the group culture is not especially distinctive, ties between group members tend to be superficial – sentimental rather than functional – with the result that residential clustering as well as group solidarity is steadily eroded.

A classic example of cultural expression within a residential area is the Harlem district of New York (Zukin, 1995; Smith, 1996). Previously inhabited by German-Jewish immigrants, from the 1920s onwards the area became occupied by African-Americans. Musicians, entertainers and artists of many kinds flocked to the area and there emerged a world-famous expression of African-American culture that, over the years, Harlem has continued to reinterpret and express. However, one must be careful not to over-glamorize such areas: together with cultural expression there is another side to Harlem – one of poverty, racial discrimination and urban decay.

Spaces of resistance: clustering to facilitate 'attacks'

The fourth major function of minority spatial concentration is the provision of a 'base' for action in the struggle of its members with society in general in what are often termed **spaces of resistance**. This 'attack' function is usually both peaceful and legitimate. Spatial concentrations of group members represent considerable electoral power and often enable minority groups to gain official representation within the institutional framework of urban politics. This has been an important factor in the political power base of African-Americans in the United States, where the Black Power movement was able to exploit the electoral power of the ghetto with considerable success – to the extent that African-American politicians now constitute an important (and sometimes dominant) voice

Box 8.3

Key trends in urban social geography – Changing patterns of segregation in the United States

Although there is a great deal of work on the residential segregation of ethnic groups within cities, very few studies have compared, in any detailed fashion, the changing degree of segregation over time across an *entire* urban system. One recent exception is the work of Johnston, Poulsen and Forrest on the United States (Johnston *et al.*, 2004). They examined the degrees of segregation of the three largest ethnic groups self-identified in the 1980 and 2000 US Censuses – Blacks (African-Americans), Hispanics (Latinos) and Asians (see also Johnston *et al.*, 2002, 2003).

The results are complex but may be summarized as follows:

➤ *The larger the ethnic group the greater the degree of segregation.* Hence blacks, the largest ethnic group, show the greatest degree of segregation. This is explained by the assumption that larger groups will appear more intimidating and therefore be subject to greater degrees of discrimination and disadvantage.

➤ *The larger the urban area the greater the degree of segregation.* This applies to all the three ethnic groups and is related to

the previous observation; larger cities are more likely to have larger minorities with segmented labour and housing markets.

➤ *There is an east–west divide in segregation.* Blacks and Hispanics were less segregated in California than elsewhere, while levels of segregation were higher in the northeastern cities. Asians show higher levels of segregation in California in 2000, reflecting their recent entry into the United States.

➤ *For Asians and Hispanics, the more diverse the city, the greater the degree of segregation.* It would seem that the greater the ethnic diversity, the greater are the social tensions and the retreat by ethnic groups into enclaves. Cities that provide 'ports of entry' for rapidly increasing new Asian and Hispanic immigrants have the highest levels of segregation (see also Box 8.1).

➤ *Although they tend to show the highest levels of segregation, over the period 1980 to 2000, Blacks have shown declining levels of segregation, while the segregation*

levels of Asians and Hispanics have increased. This decline in segregation has been less in the large cities of the Northeast.

Key concepts associated with changing patterns of segregation (see Glossary)

Congregation, enclave, ethnic group, ghetto, racism, segregation.

Further reading

Johnston, R., Poulsen, M. and Forrest, J. (2002) From modern to post-modern? Contemporary ethnic segregation in four US metropolitan areas *Cities*, **19**, 161–172

Johnston, R., Poulsen, M. and Forrest, J. (2003) And did the walls come tumbling down? Ethnic segregation in four US metropolitan areas, 1980–2000 *Urban Geography*, **24**, 560–581

Johnston, R., Poulsen, M. and Forrest, J. (2004) The comparative study of ethnic residential segregation in the USA, 1980–2000 *Tijdschrift voor Economische en Sociale Geografie*, **95**, 550–569

Links with Other Chapters

Chapter 12: Box 12.2 Ron Johnston

---

in the urban political arena. Gay neighbourhoods also represent a potentially effective electoral base. Perhaps the best-known illustration of this is West Hollywood, where voters elected in 1984 to create a self-governing municipality and subsequently elected a city council dominated by gays (Moos, 1989).

Minority clusters also provide a convenient base for illegitimate attacks on the charter group. Insurrectionary groups and urban guerrillas with minority affiliations are able to 'disappear' in their own group's

territory, camouflaged by a relative anonymity within their own cultural milieu and protected by a silence resulting from a mixture of sympathy and intimidation. An obvious example of this is the way in which the IRA and Loyalist para-military organizations took advantage of their respective territorial heartlands in Belfast in the 1980s and 1990s; and, indeed, the way in which the IRA used Irish communities in Birmingham, Liverpool, London and Southampton as bases for terrorist attacks.

Colonies, enclaves and ghettos

The spatial expression of segregation and congregation, then, is determined by the interplay of discrimination, fabric effects and the strength of internal group cohesion. Where the perceived social distance between the minority group and the charter group is relatively small, the effects of both charter group discrimination and internal cohesion are likely to be minimized and so minority residential clusters are likely to be only a *temporary* stage in the assimilation of the group into the wider urban sociospatial fabric. Such clusters may be termed **colonies**. They essentially serve as a port-of-entry for members of the group concerned, providing a base from which group members are culturally assimilated and spatially dispersed. Their persistence over

time is therefore dependent on the continuing input of new minority group members. Examples of this type of pattern include the distribution of European minority groups in North American cities during the 1920s and 1930s, of similar groups in Australasian cities during the 1950s and 1960s (Poulsen and Johnston, 2000), and of the Maltese in London during the 1950s.

Minority clusters that persist over the longer term are usually a product of the interaction between discrimination and internal cohesion. Where the latter is the more dominant of the forces, the resultant residential clusters may be termed **enclaves**; and where external factors are more dominant, the residential clusters are generally referred to as **ghettos**. In reality, of course, it is often difficult to ascertain the degree to which segregation is voluntary or involuntary, and it is more realistic to think in terms of a continuum rather than a

The ghetto in Venice. Photo Credit; Paul Knox.

twofold classification. This enclave/ghetto continuum exhibits several distinctive spatial patterns:

➤ The first of these is exemplified by Jewish residential areas in many cities, where an initial residential clustering in inner-city areas has formed the base for the subsequent formation of new suburban residential clusters (Fig. 8.1). The fact that this suburbanization represents a general upward shift in socioeconomic status and that it is usually accompanied by the transferral of Jewish cultural and religious institutions to the suburbs suggests that this type of pattern is largely the result of voluntary segregation. It has been suggested, in fact, that congregation rather

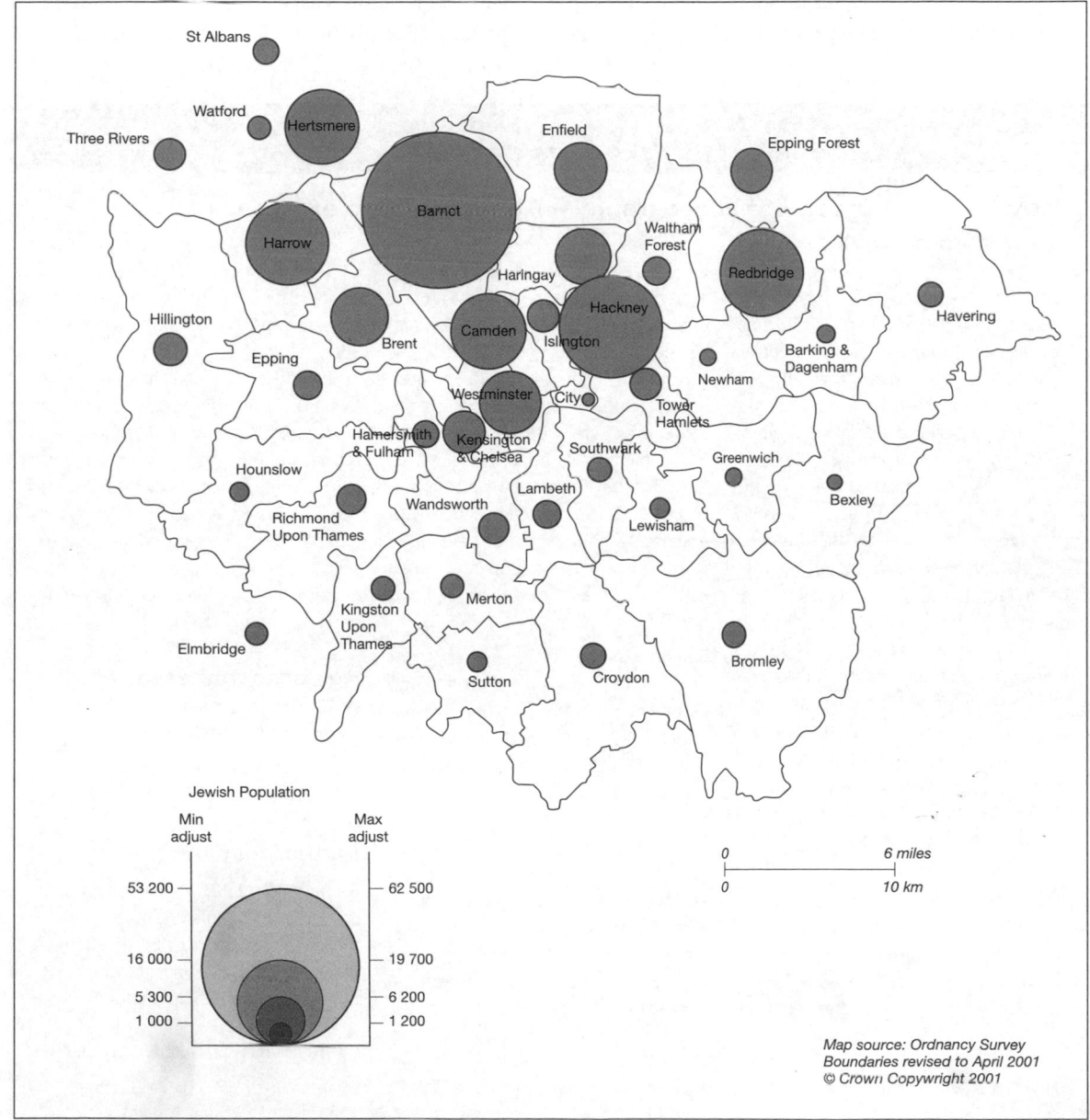

Figure 8.1 Number of Jewish people in Greater London, by boroughs, 2001.
Source: JPR Report No. 5, Long Term Planning for British Jewry, Final Report and Recommendations, 2003, p. 54.

than segregation is the most appropriate term for the Jewish residential patterns (Waterman and Kosmin, 1988).

➤ The second distinctive expression of the enclave/ghetto takes the form of a concentric zone of minority neighbourhoods that has spread from an initial cluster to encircle the CBD. Such zones are often patchy, the discontinuities reflecting variations in the urban fabric in terms of house types and resistant social groups. The growth of African-American

areas in many US cities tends to conform to this pattern, as does the distribution of Asians, Irish and Afro-Caribbeans in British cities and the distribution of Mediterraneans, Surinamers and Antilleans in Rotterdam.

➤ Where a minority group continues to grow numerically, and provided that a sufficient number of its population are able to afford better housing, residential segregation is likely to result in a sectoral spatial pattern. The distribution of African-Americans in

---

Box 8.4

Key trends in urban social geography – Rap as cultural expression (and commercialization)

One of the famous – not to say notorious – examples of cultural expression to emerge from the black ghettos of US cities (originally the South Bronx, New York), which has become a worldwide commercial phenomenon, is rap music (Berman, 1995). Devised by male black youths who were too poor to afford lessons or expensive musical instruments, it initially involved a single kid with a microphone and a speaker that played a drum synthesizer track. This evolved into a dual form with an MC (master of ceremonies) in the foreground who provided the vocals while behind him a DJ (disc jockey) created a diverse background of rhythms and sounds. Technology was important here; the introduction of the digital sampler in the mid-1980s enabled DJs to create complex mixes of different styles.

What caught the interest of youth culture, both black and otherwise (as well as some intellectuals), was the complex torrent of lyrics that emerged from this music. These lyrics gave expression to the mixture of alienation, marginalization and frustration experienced by black youths, combined also with

aggression and assertiveness. The first international hit was 'The Message' by *Grandmaster Flash and the Furious Five* which listed the plight of those in the South Bronx. Throughout the 1980s rap grew in popularity and was copied by white groups such as *The Beastie Boys*. However, in the early 1990s there emerged from Los Angeles 'gansta rap' that celebrated violence and was frequently laced with misogynistic images of rape and brutality.

Commercial interests, always alert to opportunities for profit-making, promoted rap music on an international scale and the style passed over into the 'gangsta rap' movie genre exemplified by *New Jack City*. As some rappers became fabulously wealthy they moved out of black areas into the fortified mansions of the rich and surrounded by image consultants. This led to charges that their audacious messages of black separatism and hatred towards whites lacked the 'authenticity' of the early cries of the South Bronx. Some rappers have been involved in violence, shootings and firearms offences and others have preached against violence. However,

it would appear that violent imagery sells records. Thus it was after New York rappers preached anti-violence in the late 1980s that they were overtaken in sales by the gun-carrying rappers of Los Angeles. While much rap music grabs attention through its stark, crude, hostile images, it has become a diverse and complex set of cultural forms. For example, in true postmodern fashion, a strong element of irony and pastiche has emerged in much of the genre.

Key concepts associated with rap as cultural expression (see Glossary)

Alienation, cultural industries, hybridity.

Further reading

Berman, M. (1995) 'Justice/just us': rap and social injustice in America, in A. Merrifield and E. Swyngedouw (eds) *The Urbanization of Injustice* Lawrence and Wishart, London

Links with Other Chapters

Chapter 3: Box 3.1 Hybridization: The Case of Popular Music

many of the more prosperous and rapidly expanding cities of the United States tends towards this model, although sectoral development is often truncated because of economic constraints operating at the suburban margin. The distribution of the black population in the Baltimore metropolitan area pro- vides a good example of this type of pattern. As Fig. 8.2 shows, the city's African-American popula- tion has been concentrated in the central city, with two sectors of census tracts – one to the northwest and one to the northeast of the CBD – were more than 80 per cent African-American in 2000.

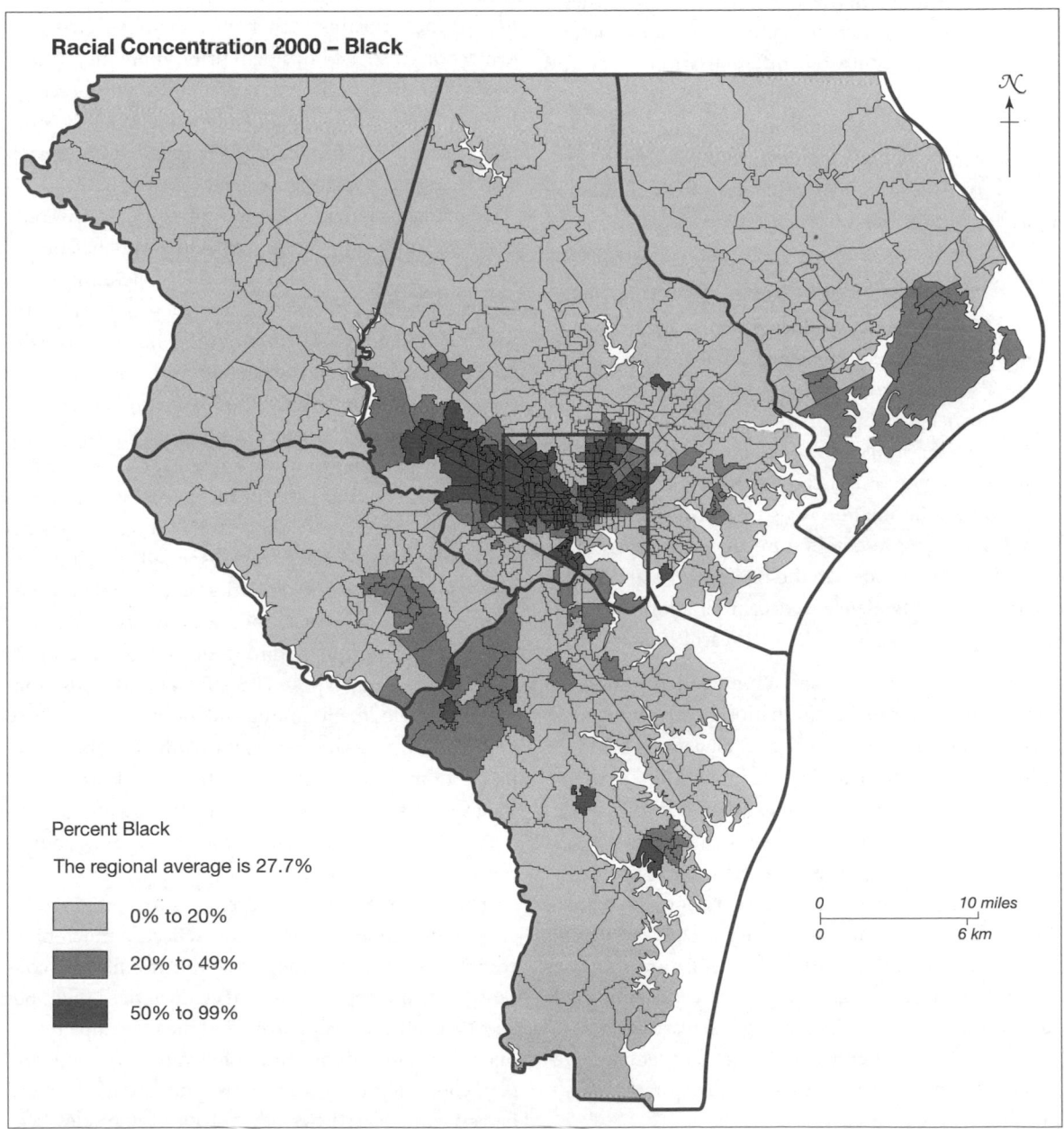

Figure 8.2 Blacks as percentage of total population in Baltimore, 2000.
Source: US Census Bureau, 2000 Census.

Illustrative example 1: structural constraints and cultural preservation in the United Kingdom

It will be clear by now that most large cities contain a variety of minorities, each responding to a different mix of internal and external factors, each exhibiting rather different spatial outcomes, and each changing in different rates and in different ways. Peach (1998) provides a summary of the relative importance of both structural constraints and cultural differentiation strategies among South Asian and Caribbean ethnic minorities in Britain. He notes that minority groups that arrived in Britain at more or less the same time after the Second World War faced similar problems of discrimination but have become significantly differentiated from one another in terms of housing tenure and residential location:

> In brief, the Indian profile appears as white-collar, suburbanized, semi-detached and owner occupying; the Pakistani profile as blue-collar inner city and owner-occupying in terraced housing; the Bangladeshi profile is blue-collar and council-housed in inner-city terraced and flatted properties; the Caribbean population is also blue-collared with substantial representation in council housing, but far less segregated than the Bangladeshis and with a pronounced tendency to decentralisation.
>
> (Peach, 1998, p. 1657)

Peach argues that one should not underestimate the continuing power of discrimination in explaining these patterns. He also cautions that explanations based on cultural traditions can be misinterpreted as being based on **essentialism** – the ideas that these differences are intrinsic elements of peoples rather than social constructs (see Chapter 3). Nevertheless, whereas the situation in the 1960s was often represented in blanket terms of white discrimination, current explanations also place stress upon the cultural differentiation strategies of the different ethnic groups. In particular, ethnic cultural values strongly influence age of marriage, family size, household structures and degree of female independence, and these factors in turn impact upon housing tenure and location.

For example, Afro-Caribbean societies are characterized by matrifocal (i.e. female-dominated) households, cohabitation and visiting relationships. Marriage in this context is often a middle-class institution adopted *after* the rearing of a family. This tradition of female independence continues in the Afro-Caribbean population in Britain, which has the highest rate of female participation in the formal economy of any ethnic group (70 per cent in 2001). It is the high proportion of single female-headed households among Caribbean ethnic groups that explains their higher than expected concentration in social housing. Furthermore, the relative poverty of these households, combined with processes leading to residualization of the social housing stock (see Chapter 12), has led to their frequent concentration in less desirable high-rise blocks of flats.

In marked contrast, the ethnic minorities with a South Asian background – India, Pakistan and Bangladesh – are characterized by patrifocal (i.e. male-dominated) nuclear families with a strong tendency for large multi-family households. Furthermore, lone parents with dependent children and cohabitation are unusual among these groups. The Islamic groups from Pakistan and Bangladesh have a strong tradition of *purdah,* which involves sheltering women from outside society and so female participation rates in the formal economy are extremely low. The tendency for arranged marriages among the South Asian groups, especially Muslims and Sikhs, means that these societies tend to be ethnically homogeneous. However, the differing socioeconomic status of the various communities tends to result in differing housing types. The tendency towards non-manual employment among Indian men is associated with suburban owner-occupation whereas the higher proportion of manual work among Pakistanis and Bangladeshis is associated with poorer quality owner-occupied housing in the former case and council housing in the latter case.

The outcome of these processes is that ethnic residential segregation in Britain now reflects a much more complex pattern than the stereotype of inner city concentration and deprivation that characterized the 1960s and 1970s. It is certainly true that the overall picture is one of economic disadvantage for minority groups and they continue to experience discrimination and racial harassment. Nevertheless, there have been modest economic gains for some groups and moderate amounts of decentralization from inner-cities into suburban

The Medina Mosque in Southampton: a sacred space illustrating a distinctive subculture. Photo Credit: Andrew Vowles.

areas (although not into rural areas of Britain). Afro-Caribbeans, although showing signs of continuing economic marginalization, have become more culturally integrated (and have in turn influenced) British working-class society. In addition, largely through their presence in social housing schemes, they show decreased signs of residential segregation (Phillips, 1998).

While Bangladeshis have a similar socioeconomic and housing profile to Caribbean groups, they display much higher and continuing degrees of residential segregation. In general, ethnic minorities from Muslim societies – such as Pakistan and Bangladesh – show a continuing tendency for segregation whether they are predominantly in owner-occupation or the social housing sector. The somewhat higher socioeconomic status of Indian groups is associated with moderate deconcentration especially in London into Redbridge and Harrow (Rees and Phillips, 1996). However, as Phillips (1998) notes, socioeconomic advancement is no guarantee of suburbanization or dispersal since numerous cultural factors (such as the aversion of Muslims on religious grounds to taking out loans for purchasing more expensive housing) may serve to anchor ethnic group members in a particular location.

Illustrative example 2: migrant workers in continental European cities

During the past 40 years an important new dimension has been added to the social geography of continental European cities with the arrival of tens of thousands of migrant workers, most of them from the poorer regions of the Mediterranean. Although estimates vary a good deal, there were in the mid-2000s some 22 million aliens living in European countries. About half of these were young adult males who had emigrated in order to seek work, originally, at least, on a temporary basis. The rest were wives and families who had joined them.

At the heart of this influx of foreign-born workers were the labour needs of the more developed regions, coupled with the demographic 'echo effect' of low birth rates in the 1930s and 1940s: a sluggish rate of growth in the indigenous labour force. As the demand for labour in more developed countries expanded in the 1960s, so indigenous workers found themselves able to shun low-wage, unpleasant and menial occupations; immigrant labour filled the vacuum. At the same time, the more prosperous countries perceived that foreign workers might provide a buffer for the indigenous labour force against the effects of economic cycles – the so-called *konjunkturpuffer* philosophy.

The peak of these streams of immigrants occurred in the mid-1960s and early 1970s. The onset of deep economic recession in 1973, however, brought a dramatic check to the flows. Restrictions on the admission of non-European Community immigrants began in West Germany in November 1973, and within 12 months France, Belgium and the Netherlands had followed with new restrictions. By this time there were nearly 2 million foreign workers in West Germany, over 1.5 million in France, half a million in Switzerland, and around a quarter of a million in Belgium and Sweden.

Within each of these countries the impact of migrant labour has been localized in larger urban areas, reflecting the immigrants' role as replacement labour for the low-paid, assembly-line and service sector jobs vacated in inner-city areas by the upward socioeconomic mobility and outward geographical mobility of the indigenous

population. Thus, for example, more than 2 million of the 6 million aliens living in France in 2000 lived in the Paris region, representing over 15 per cent of its population. Other French cities where aliens account for more than 10 per cent of the populations include Lyons, Marseilles, Nice and St Etienne. In Germany, 25 per cent of Frankfurt's population is foreign born, as is more than 15 per cent of the population of Köln, Munich, Düsseldorf and Stuttgart. In Switzerland, Basel, Lausanne and Zürich all have between 15 and 20 per cent foreign born, while Geneva (a special case) has 35 per cent.

Since nearly all of the immigrants were initially recruited to low-skill, low-wage occupations, they have inevitably been channelled towards the cheapest housing and the most run-down neighbourhoods. The position of immigrants in the labour market is of course partly self-inflicted: for many the objective has been to earn as much as possible, as quickly as possible, the easiest strategy being to take on employment with an hourly wage where overtime and even a second job can be pursued. Most immigrants, however, tend to be kept at the foot of the economic ladder by a combination of institutional and social discrimination.

Similarly, immigrants' position in housing markets is partly self-inflicted: inexpensiveness is of the essence. But the localization of immigrants in camps, factory hostels, *hôtels meublés* (immigrant hostels), *bidonvilles* (suburban shanty towns also known as *banlieue*) and inner-city tenements is also reinforced by bureaucratic restrictions and discrimination. Concentrated in such housing, immigrants find themselves in an environment that creates problems both for themselves and for the indigenous population. Trapped in limited niches of the housing stock, they are vulnerable to exploitation. One response to excessively high rents has been the notorious 'hotbed' arrangement, whereby two or even three workers on different shifts take turns at sleeping in the same bed. In France a more common response has been to retreat to the cardboard and corrugated iron *bidonvilles*, where although there may be no sanitary facilities immigrants can at least live inexpensively among their own compatriots.

In terms of spatial outcomes, a fairly consistent pattern has emerged, despite the very different minority populations involved in different cities – Serbs, Croats and Turks in Duisburg, Frankfurt, Köln and Vienna; Algerians, Italians and Tunisians in Paris; and Surinamese and Turks in Rotterdam, for example. In short, just as the migrants are replacing the lower echelons of the indigenous population in the labour market, so they are acting as a partial replacement for the rapidly declining indigenous population in the older neighbourhoods of privately rented housing near to sources of service employment and factory jobs. Quite simply, the overriding priority for migrant workers is to live close to their jobs in cheap accommodation. This applies to all national groups, so that similar spatial patterns have persisted even as (in Hamburg's case) the culturally more alien Turkish, Greek and Portuguese populations replaced older and more familiar groups such as Italians and Spaniards (Friedrichs and Alpheis, 1991). In some cities, a secondary pattern has developed as foreign-born workers have been allocated space in public housing projects. The allocation procedures of housing officials therefore tend to result in concentrations of foreigners in dilapidated, peripheral public housing estates.

A new twist to the story of ethnic minority groups in Europe has been provided by the massive economic changes sweeping the continent following the end of the 'long boom' and the advent of globalization, deindustrialization, high levels of long-term unemployment and increasing social polarization (see Chapter 2). At the same time, the social welfare policies of EU states have been curtailed as they attempt to cut budget deficits. As a result, social polarization has begun to affect the charter group as well as ethnic minorities, leading to heightened social tensions in many cases. Indeed, it has been estimated that by the mid-1990s no fewer than one in six of *all* individuals in the European Union was living in poverty – not just ethnic minorities but also young adults, women and the elderly (Madanipour *et al.*, 1998).

The promotion of greater social and economic cohesion was one of the key objectives of the 1991 Maastricht Treaty that paved the way for greater integration of the EU states. However, this treaty created two classes of immigrants within Europe: first, there are the 5 million European residents living in the boundaries of other European nations with the right to movement within the European Union; and second, there are the 10 million non-European immigrants from outside the European Union who, although long-settled, do not have full citizenship or the right to move to other EU member states. The aim of this division is to create a 'Fortress Europe' policy and inhibit immigration, including that from refugees and asylum-seekers outside the European Union.

These formally sanctioned processes of exclusion interact with race and ethnicity to form complex and different patterns in different nations. At one extreme, non-nationals in the United Kingdom enjoy virtually all the formal rights of citizenship of nationals and so (as discussed above) processes of social exclusion operate more clearly on the basis of race. At the other extreme, non-nationals in Germany are excluded from a wide range of formal social institutions such as social housing or democratic channels, and so race and ethnicity get compounded with other issues of 'Europeanness' (Allen, 1998).

To illustrate, Kurpick and Weck (1998) note that until the early 1980s most middle- and lower-income workers with German citizenship felt relatively secure, but then increasing levels of long-term unemployment meant that they were increasingly cut off from the mainstream. Social polarization processes have also left second and third generation immigrants in danger of becoming permanently excluded from economic and social life in Germany. So far, these social polarization processes have not resulted in the severe manifestations of decay as in distressed inner-city neighbourhoods of the United States or the sort of violent actions seen on the periphery of some French cities. The Marxloh neighbourhood of Duisburg in the distressed Ruhr region of Germany has a 25 per cent unemployment rate and has 36 per cent foreign nationals, mostly of Turkish origin. In response, policies have been instigated by the city of Duisburg to encourage local businesses. Interestingly, it has been found that the German population in this neighbourhood is more socially isolated and lacking in behavioural attitudes conducive to entrepreneurship than the Turkish population, with its lively and diverse ethnically based social and economic networks.

Box 8.5

Key films related to urban social geography – Chapter 8

Bend It Like Beckham (2002) Extremely popular portrayal of inter-generational cultural tensions among Sikhs in Britain.

Boyz n the Hood (1991) A brutal experience set in the gangland of South Central Los Angeles, this film depicts the problems of growing up in an environment characterized by poverty, violence and crack cocaine dealing. A controversial film after some thought it would lead to 'copy-cat' violence.

Do the Right Thing (1989) Lively, tragi-comic film about conflicts between Italian-Americans, African-Americans and Koreans in Brooklyn.

East is East (1999) Set in the northern English cities of Salford and Bradford in the 1970s, this films portrays issues of cultural adaptation in a way that is both comic and moving.

La Haine (1995) A much-praised film about the tensions and stresses facing youths from an ethnic enclave in France.

Menace II Society (1993) Another brutal violent film set in the Watts ghetto.

Mississippi Massala (1991) Another film dealing with issues of migration, identity and cultural hybridity, in this case in the context of Ugandan Asians in the southern United States.

New Jack City (1991) An 'over the top' portrayal of crime and violence but worth seeing for the way it represents ghetto life in mainstream movies.

Chapter summary

8.1 The residential segregation of minority groups in Western cities is the product of various processes of exclusionary closure and institutional racism.

8.2 Minority groups reveal differing degree of residential segregation in cities. These patterns reflect hostility among the wider population, discrimination in employment and housing markets, and clustering for defence, mutual support and cultural preservation.

Key concepts and terms

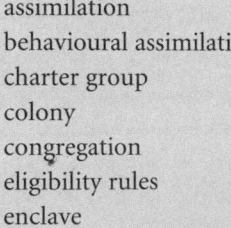

assimilation	essentialism	racism
behavioural assimilation	exclusionary closure	segregation
charter group	'exit' option	social closure
colony	'fabric effect'	spaces of resistance
congregation	ghetto	structural assimilation
eligibility rules	index of dissimilarity	'tipping point'
enclave	minority group	'voice' option

Suggested reading

There are several books that provide good resources on the topics of race, racism and spatial segregation. These include Peter Jackson's edited volume, *Race and Racism: Essays in Social Geography* (1987: Allen and Unwin, London) and Susan Smith's *The Politics of Race and Residence* (1989: Polity Press, Cambridge). A useful introductory review essay is provided by Fred Boal in *Progress in Social Geography*, edited by Michael Pacione (1987: Croom Helm, London, pp. 90–128). Examples of detailed case studies can be found in a volume edited by Peter Jackson and Susan Smith: *Social Interaction and Ethnic Segregation* (1981: Academic Press, London), S. Musterd and W. Ostendorf *Urban Segregation and the Welfare State: Inequality and Social Exclusion in Western Cities* (1998: Routledge, London) and the special edition of the journal *Urban Studies* on ethnic segregation in cities (1998: 35, No. 10). For issues of racism see C. Nash, 'Cultural geography; anti-racist geographies' (*Progress in Human Geography*, 2000: 27, 637–648). Broader theoretical issues, along with recent trends, are discussed in *Racism, the City and the State*, edited by Malcolm Cross and Michael Keith (1993: Routledge, London). A set of case studies on issues of ethnicity around the globe may be found in *EthniCity: Geographic Perspectives on Ethnic Change in Modern Cities* (1995: Rowman and Littlefield, Lanham MD)

edited by Curtis C. Roseman, ⌐ Gunter Thieme. See also J.C. Bog (eds) (1996) *Race, Poverty, and A* University of North Carolina NC) and T. Boswell and A.D. Cr segregation by socioeconomic cla Miami: 1990' (*Urban Geography*, 1997: **18**(6), 474–496). For recent developments in the sphere of segregation in Europe see the essays in *Social Exclusion in European Cities* edited by Ali Madanipour, Goran Cars and Judith Allen (1998: Jessica Kingsley, London) and *Divided Europe* edited by Ray Hudson and Alan Williams (1999: Sage, London). In addition to the above, reference should also be made to processes leading to exclusion and 'othering' as identified in Chapter 3. The issue of asylum seekers and refugees is discussed in Vaughan Robinson *et al. Spreading the Burden? A Review of Policies to Disperse Asylum Seekers and Refugees* (2003: Policy Press, Bristol). For a critique of recent work on ethnic segregation see Ceri Peach 'Social geography: new religions and ethnoburbs – contrasts with cultural geography (*Progress in Human Geography*, 2002: **26**, 252–260). For a response see Loretta Lees, 'Urban geography: 'New' urban geography and the ethnographic void' (*Progress in Human Geography*, 2003: **27**(1), 107–113). For an overview of multiculturalism in cities see Michael Keith's *After the Cosmopolitan?* (2004: Routledge, London).

Chapter 9

Neighbourhood, community and the social construction of place

Key questions addressed in this chapter

➤ What has been the effect of urbanization upon community life?

➤ How do people construct images of urban environments and how do these affect the way they live their lives?

➤ What are the social meanings incorporated within the built environment?

A key theme running throughout this book is the fact that cities involve the interchange between many different cultures in relatively confined spaces, often leading to new cultural forms but also to social segregation (see Chapters 3 and 7). These cultural exchanges involve peoples with many differing and complex social networks – some overlapping, some separate. For urban social geographers, key questions include: whether some of these networks constitute a 'community'; whether this concept is synonymous with 'neighbourhood' or 'locality'; and, if so, in what circumstances.

According to classic sociological theory, communities should not exist at all in cities; or, at best, only in a weakened form. This idea first entered sociological theory in the nineteenth century by way of the writings of Ferdinand Tönnies, who argued that two basic forms of human association could be recognized in all cultural systems (Tönnies, 1963). The first of these, *Gemeinschaft*, he related to an earlier period in which the basic unit of organization was the family or kin-group, with social relationships characterized by depth, continuity, cohesion and fulfilment. The second, *Gesellschaft*, was seen as the product of urbanization and industrialization that resulted in social and economic relationships based on rationality, efficiency and contractual obligations among individuals whose roles had become specialized. This perspective was subsequently reinforced by the writings of sociologists such as Durkheim, Simmel, Sumner and, as we have seen,

Wirth, and has become part of the conventional wisdom about city life: it is not conducive to 'community', however it might be defined. This view has been characterized as the '**community lost**' argument.

9.1 Neighbourhood and community

There is, however, a good deal of evidence to support the idea of socially cohesive communities in cities. Writers have long portrayed the city as an inherently human place, where sociability and friendliness are a natural consequence of social organization at the neighbourhood level (see, for example, Jacobs, 1961). Moreover, this view has been sustained by empirical research in sociology and anthropology, which has demonstrated the existence of distinctive social worlds that are territorially bounded and that have a vitality that is focused on local 'institutions' such as taverns, pool halls and laundromats (e.g. Liebow, 1967; Suttles, 1968). 'The landscape of modernity, then, is much more than the simple product of industrial relocation, the real estate market, the architect's office, the planner's dreams, the government's regulators, and the engineer's system. It is also the product of diverse people shaping neighborhoods' (Ward and Zunz, 1992).

Urban villages: community saved?

Herbert Gans, following his classic study of the West End of Boston, suggested that we need not mourn the passing of the cohesive social networks and sense of self-identity associated with village life because, he argued, these properties existed within the inner city in a series of 'urban villages' (Gans, 1962). This perspective has become known as '**community saved**'. The focus of Gans's study was an **ethnic village** (the Italian quarter), but studies in other cities have described urban villages based on class rather than ethnicity. The stereotypical example of an urban village is Bethnal Green, London, the residents of which became something of a classic sociological stereotype. They exhibited 'a *sense* of community . . . a feeling of solidarity between

people who occupy the common territory' based on a strong local network of kinship, reinforced by the localized patterns of employment, shopping and leisure activities (Young and Wilmott, 1957, p. 89, emphasis added, see also Box 9.1).

Similar situations have been described in a series of subsequent studies of inner-city life on both sides of the Atlantic, most recently by Thomas Jablonsky in his study of Chicago's 'Back of the Yards' neighbourhood, where 'community spirit . . . was dependent upon – indeed, was generated by – spatial forces. The *culture* of the community evolved in part from spatial habits and territorial loyalties' (Jablonsky, 1993, p. 152, emphasis added). Although the utility of such studies is limited by their rather different objectives and by the diversity of the neighbourhoods themselves, the localized social networks they describe do tend to have common origins. In short, urban villages are most likely to develop in long-established working-class areas with a relatively stable population and a narrow range of occupations.

The importance of permanence and immobility in fostering the development of local social systems has been stressed by many writers. The relative immobility of the working classes (in every sense: personal mobility, occupational mobility and residential mobility) is a particularly important factor. Immobility results in a strengthening of *vertical* bonds of kinship and *horizontal* bonds of friendship. The high degree of residential propinquity between family members in working-class areas not only makes for a greater intensity of interaction between kinfolk but also facilitates the important role of the matriarch in reinforcing kinship bonds. The matriarch has traditionally played a key role by providing practical support (e.g. looking after grandchildren, thus enabling a daughter or daughter-in-law to take a job) and by passing on attitudes, information, beliefs and norms of behaviour. Primary social interaction between friends is also reinforced by the residential propinquity that results from immobility. Relationships formed among a cohort of children at school are carried over into street life, courtship and, later on, the pursuit of social activities in pubs, clubs and bingo halls.

Another important factor in fostering the development of close-knit and overlapping social networks in

Box 9.1

Key debates in urban social geography – What were working-class communities really like?: The case of the East End of London

A major source of controversy over many years in urban studies concerns the true nature of the older working-class communities of industrial cities. A key work here is Young and Wilmott's now classic book *Family and Kinship in East London*, which examined working-class life in the 1950s and became one of the most popular and influential works of social science ever to be published in Britain. Many have argued that Young and Wilmott's portrayal of Bethnal Green as full of close-knit family ties, mutual self-help and local social solidarity presented a somewhat romantic and sentimental view. Indeed, many years later Michael Young admitted that he missed a great deal during his researches (he initially had problems in understanding the Cockney accents) and that he may have exaggerated the presence in this community of certain qualities, such as warm family ties (characteristics that he found missing in his own more austere middle-class upbringing).

Critics argue that in addition to exaggeration, Young and Wilmott missed the double-edged character of close-knit communities; intense scrutiny of the behaviour of friends and relatives could be oppressive as well as supportive. Furthermore, women bore a large burden of responsibility in these communities. It is also argued that the problems associated with the re-location of Bethnal Green families to new council estates in outer suburbs of Greenwich were exaggerated. Over time, community ties were recreated in the new estates.

Another charge is that Young and Wilmott ignored the long and complex history of London's East End and therefore downplayed the social diversity stemming from previous waves of immigrants. This meant that they also ignored the conflicts between these groups. For example, in previous centuries there were riotous clashes between Irish immigrants and Huguenot silk weavers and later conflicts between various Jewish sects (especially tensions between Jewish immigrants who had previously been urban-based compared with those from agricultural communities). In the 1930s white working-class East Enders had rioted against Jewish immigrants and more recently there have been tensions between Bangladeshi settlers and the white working class. Ironically, many of the Bangladeshi immigrants appear to have the attributes that Young and Wilmott perceived in the traditional working class communities of Bethnal Green – extended families and mutual support systems.

Young and Wilmott therefore missed the parochialism, zenophobia and racism that were often to be found in London's East End. But perhaps the most enduring feature of this area is its capacity, despite many tensions, to absorb in a relatively successful way over many years diverse immigrants from many areas (the most recent group has been asylum seekers from Somalia).

Key concepts related to London's East End (see Glossary)

Community, 'ethnic village', neighbourhood, othering, primary relationships.

Further reading

Young, M. and Wilmott, P. (1957) *Family and Kinship in East London* Routledge and Kegan Paul, London

Links with Other Chapters

Chapter 4: Box 4.3 Are Western Cities Becoming Socially Polarized?

Chapter 13: Box 13.1 The Emergence of Clusters of Asylum Seekers and Refugees

---

working-class areas is the economic division of society that leaves many people vulnerable to the cycle of poverty. The shared and repeated experience of hard times, together with the cohesion and functional interdependence resulting from the tight criss-crossing of kinship and friendship networks, generates a mutuality of feeling and purpose in working-class areas: a mutuality that is the mainspring of the social institutions, ways of life and 'community spirit' associated with the urban village.

The fragility of communality

The cohesiveness and communality arising from immobility and economic deprivation is a fragile phenomenon, however. The mutuality of the urban village is underlain by stresses and tensions that follow from social intimacy and economic insecurity, and several studies of working-class neighbourhoods have described as much conflict and disorder as cohesion and communality. The one factor that has received most attention in this respect is the stress resulting from the simple shortage of space in working-class areas. High densities lead to noise problems, inadequate play space and inadequate clothes-drying facilities and are associated with personal stress and fatigue. Children, in particular, are likely to suffer from the psychological effects of the lack of privacy.

The fragility of working-class communality stems from several sources, including the conflict of values that can arise from the juxtaposition of people from a variety of ethnic and cultural backgrounds, notwithstanding their common economic experiences. Another threat to communality is the disruption of social relationships that occurs as one cohort of inhabitants ages, dies and is replaced by younger families, who, even though they may be essentially of the same class and lifestyle, represent an unwitting intrusion on the quieter lives of older folk. A third factor is the disruption associated with the presence of undesirable elements – 'problem families', transients and prostitutes – in the midst of an area of otherwise 'respectable' families. The relative strength of these stressors may be the crucial factor in tipping the balance between an inner-city neighbourhood of the 'urban village' type and one characterized by the anomie and social disorganization postulated by Wirthian theory.

Suburban neighbourhoods: community transformed?

In contrast to the close-knit social networks of the urban village, suburban life is seen by many observers as the antithesis of 'community'. Lewis Mumford, for example, wrote that the suburbs represent 'a collective attempt to lead a private life' (Mumford, 1940, p. 215). This view was generally endorsed by a number of early studies of suburban life, including the Lynds' (1956) study of Muncie, Indiana, and Warner and Lunt's (1941) study of 'Yankee City' (New Haven). Further sociological work such as Whyte's (1956) *The Organization Man* and Stein's (1960) *The Eclipse of Community* reinforced the image of the suburbs as an area of loose-knit, secondary ties where lifestyles were focused squarely on the nuclear family's pursuit of money, status and consumer durables and the privacy in which to enjoy them.

Subsequent investigation, however, has shown the need to revise the myth of suburban 'non-community'.

Although there is little evidence for the existence of suburban villages comparable to the urban villages of inner-city areas, it is evident that many suburban neighbourhoods do contain localized social networks with a considerable degree of cohesion: as Gans (1967) showed, for example, in his classic study of Levittown. Suburban neighbourhoods can be thought of as 'communities of limited liability' – one of a series of social constituencies in which urbanites may choose to participate. This view has been translated as '**community transformed**' or 'community liberated'. Instead of urban communities *breaking up*, they can be thought of as *breaking down* into an ever-increasing number of independent sub-groups, only some of which are locality-based.

It has been suggested by some that the social networks of suburban residents are in fact more localized and cohesive than those of inner-city residents, even if they lack something in *feelings* of mutuality. This perspective emphasizes the high levels of 'neighbouring' in suburbs and suggests that this may be due to one or more of a number of factors:

➤ that the detached house is conducive to local social life;

➤ that suburbs tend to be more homogeneous, socially and demographically, than other areas;

➤ that there is a 'pioneer eagerness' to make friends in new suburban developments;

➤ that suburban residents are a self-selected group having the same preferences for social and leisure activities;

➤ that physical distance from other social contacts forces people to settle for local contacts.

The cohesiveness of suburban communities is further reinforced by social networks related to voluntary associations of various kinds: parent–teacher associations (PTAs), gardening clubs, country clubs, rotary clubs and so on. Furthermore, it appears from the evidence at hand that suburban relationships are neither more nor less superficial than those found in central city areas.

Nevertheless, there are some groups for whom suburban living does result in an attenuation of social contact. Members of minority groups of all kinds and people with slightly atypical values or lifestyles will not easily be able to find friends or to pursue their own interests in the suburbs. This often results in such people having to travel long distances to maintain social relationships. Those who cannot or will not travel must suffer a degree of social isolation as part of the price of suburban residence.

Splintering urbanism and the diversity of suburbia

It should also be acknowledged that the nature and intensity of social interaction in suburban neighbourhoods tend to vary according to the *type* of suburb concerned. One outcome of the 'splintering urbanism' described in Chapter 1 is that suburbia has become increasingly differentiated. This has involved a reorganization of cultural space around different lifestyles related variously to careerist orientations, family orientations, 'ecological' orientations, etc., and constrained by income and life-cycle characteristics. It has also involved an increasing tendency for people to want to withdraw into a 'territorially defended enclave' inhabited by like-minded people, in an attempt to find refuge from potentially antagonistic rival groups.

The net result is the emergence of distinctive 'voluntary regions' in the suburbs, a process that has been reinforced by the proliferation of suburban housing types that now extend from traditional tracts of detached single-family dwelling to include private master-planned communities, specialist condominium apartments, townhouse developments and exclusive retirement communities. As a result, an 'archipelago' of similar suburban communities, with outliers in every metropolitan ring, now extends from coast to coast.

In detail, suburbia is a mosaic of sociodemographic settings, each with a rather different pattern of social interaction. Nevertheless, we can identify four principal types (after Muller, 1981):

1. *Exclusive upper income suburbs*: these neighbourhoods are typically situated in the outermost parts of the city and consist of large detached houses built in extensive grounds, screened by trees and shrubbery. This makes casual interaction with neighbours rather difficult, and so local social networks tend to be based more on voluntary associations such as churches and country clubs.

2. *Middle-class family suburbs*: the dominant form of the middle-class suburb is the detached single-family dwelling, and the dominant pattern of social interaction is based on the nuclear family. As in the more exclusive suburbs, socializing with relatives is infrequent, and emphasis on family privacy tends to inhibit social interaction. Since the care of children is a central concern, much social contact occurs through family-oriented organizations such as the PTA, the Scouts/Guides and organized sports; and the social cohesion of the neighbourhood derives to a large extent from the overlap of the social networks resulting from these organizations. However, with the tendency for young people to defer marriage, for young couples to defer child-rearing, and for land and building costs to escalate, there has emerged a quite different type of middle-class suburb based on apartment living. Social interaction in these neighbourhoods tends to be less influenced by local ties, conforming more to the idea of an aspatial community of interest.

3. *Suburban cosmopolitan centres*: this apparently contradictory label is given to the small but rapidly increasing number of suburban neighbourhoods that serve as voluntary residential enclaves for professionals, intellectuals, students, artists, writers and the creative classes (Florida, 2002): people with broad rather than local horizons, but whose special interests and lifestyle nevertheless generate a cohesive community through a series of overlapping social networks based on cultural activities, professional interests and voluntary organizations. These neighbourhoods are very much a contemporary phenomenon, and are chiefly associated with suburbs adjacent to large universities and colleges.

4. *Working-class suburbs*: blue-collar suburban neighbourhoods have multiplied greatly in number since the Second World War to the point where they are almost as common as middle-class family suburbs in many cities. But although these neighbourhoods are also dominated by single-family dwellings, patterns of social interaction are quite different. An intensive use of communal outdoor space makes for a high level of primary local social interaction, and community cohesion is reinforced by a 'person-oriented'

rather than a material- or status-oriented lifestyle. Moreover, the tendency for blue-collar workers to be geographically less mobile means that people's homes are more often seen as a place of permanent settlement, with the result that people are more willing to establish local ties.

Suburban neighbourhoods in many European cities do not conform particularly well to this typology, largely because of greater economic constraints on the elaboration of different lifestyles and the regulatory constraints that are central to European land use planning. Exclusive upper income suburbs are sustained by fewer cities; and there are few examples to be found of the 'suburban cosmopolitan centre'. Moreover, the magnitude of the public housing sector in many European countries means that a large proportion of the suburbs are publicly owned working-class neighbourhoods: altogether different from the North American suburban working-class neighbourhood.

Status panic and crisis communality

One thing that suburban neighbourhoods everywhere seem to have in common is a lack of the mutuality, the permanent but intangible 'community spirit' that is characteristic of the urban village. An obvious explanation for this is the newness of many suburban communities: they have not had time to fully develop a locality-based social system. An equally likely explanation, however, is that the residents of suburban neighbourhoods are simply not likely to develop a sense of mutuality in the same way as urban villagers because they are not exposed to the same levels of deprivation or stress.

This reasoning is borne out to a certain extent by the 'crisis communality' exhibited in suburban neighbourhoods at times when there is an unusually strong threat to territorial exclusivity, amenities or property values. Examples of the communality generated in the wake of status panic are well documented, and the classic case is that of the Cutteslowe Walls. In 1932 Oxford City Council set up a housing estate on a suburban site to the north of the city and directly adjacent to a private middle-class estate. The home-owners, united by their fear of a drop both in the status of their neighbourhood

and in the value of their property and drawn together by their mutual desire to maintain the social distance between themselves and their new proletarian neighbours, went to the length of building an unscalable wall as a barrier between the two estates (Collinson, 1963). Other documented examples have mostly been related to NIMBY (Not In My Backyard) threats posed by urban motorways, airports or the zoning of land for business use.

Communities and neighbourhoods: definitions and classifications

As we have seen, the nature and cohesiveness of social networks vary a lot from one set of sociospatial circumstances to another, and it is not easy to say which situations, if any, reflect the existence of 'community', let alone which of these are also congruent with a discrete geographical territory. Nevertheless, it is possible to think in terms of a loose hierarchical relationship between neighbourhood, community and communality. Thus **neighbourhoods** are territories containing people of broadly similar demographic, economic and social characteristics, but are not necessarily significant as a basis for social interaction. **Communities** exist where a degree of social coherence develops on the basis of interdependence, which in turn produces a uniformity of custom, taste and modes of thought and speech. Communities are 'taken-for-granted' worlds defined by reference groups which may be locality-based, school-based, work-based or media-based. *Communality*, or 'communion', exists as a form of human association based on affective bonds. It is community experience at the level of consciousness, but it requires an intense mutual involvement that is difficult to sustain and so only appears under conditions of stress.

In the final analysis, each neighbourhood is what its inhabitants think it is. This means that definitions and classifications of neighbourhoods and communities must depend on the geographical scales of reference used by people. In this context, it may be helpful to think of *immediate* neighbourhoods (which are small, which may overlap one another, and which are characterized by personal association rather than interaction through formal groups, institutions or organizations), *traditional* neighbourhoods (which are characterized by social interaction that is consolidated by the sharing of local facilities and the use of local organizations) and *emergent* neighbourhoods (which are large, diverse and characterized by relatively low levels of social interaction).

A rather different way of approaching neighbourhoods and communities is to focus on their *functions*. It is possible, for example, to think in terms of neighbourhoods' *existential* functions (related to people's affective bonds and sense of belonging), *economic* functions (geared to consumption), *administrative* functions (geared to the organization and use of public services), *locational* functions (relating to the social and material benefits of relative location), *structural* functions (related to the social outcomes of urban design), *political* functions (geared to the articulation of local issues) and *social reproduction* functions (related to the broader political economy of urbanization).

9.2 The social construction of urban places

'Place', observes David Harvey, 'has to be one of the most multi-layered and multi-purpose words in our language' (Harvey, 1993, p. 4). This layering of meanings reflects the way that places are socially constructed – given different meanings by different groups for different purposes. It also reflects the difficulty of developing theoretical concepts of place:

> There are all sorts of words such as milieu, locality, location, locale, neighbourhood, region, territory and the like, which refer to the generic qualities of place. There are other terms such as city, village, town, megalopolis and state, which designate particular kinds of places. There are still others, such as home, hearth, 'turf,' community, nation and landscape, which have such strong connotations of place that it would be hard to talk about one without the other.
>
> (Harvey, 1993, p. 4)

In this context it is helpful to recognize the **'betweenness' of place**: that is, the dependence of place on perspective. Places exist, and are constructed, from a subjective point of view; while simultaneously they are constructed and seen as an external 'other' by outsiders. As Nicholas Entrikin put it, 'Our neighborhood is both an area centered on ourselves and our home, as well as an area containing houses, streets and people that we may view from a decentered or an outsider's perspective. Thus place is both a center of meaning and the external context of our actions' (Entrikin, 1991, p. 7). In addition, views from 'outside' can vary in abstraction from being in a specific place to being virtually 'nowhere' (i.e. an abstract, perspectiveless view) (Sack, 1992) (see also Box 9.2).

These distinctions are useful in pointing to the importance of understanding urban spaces and places in terms of the insider, the person who normally lives in and uses a particular place or setting. Yet insideness and outsideness must be seen as ends of a continuum along which various modes of place-experience can be identified. The key argument here is that places have meaning in direct proportion to the degree that people feel 'inside' that place: 'here rather than there, enclosed rather than exposed, secure rather than threatened' (Relph, 1976). One important element in the construction of place is to define the other in an exclusionary and stereotypical way. This is part of the human strategy of **territoriality**: 'a spatial strategy to make places instruments of power' (Sack, 1992). Self-definition comes in relation to the other, the people and places outside the boundaries (real and perceived) that we establish.

Another key element in the construction of place is the existential imperative for people to define themselves in relation to the material world. The roots of this idea are to be found in the philosophy of Martin Heidegger, who contended that men and women originate in an alienated condition and define themselves, among other ways, spatially. Their 'creation' of space provides them with roots, their homes and localities becoming biographies of that creation (Heidegger, 1971). Central to Heidegger's philosophy is the notion of 'dwelling': the basic capacity to achieve a form of spiritual unity between humans and the material world. Through repeated experience and complex associations,

Box 9.2

Key trends in urban social geography – The development of new 'sacred spaces'

Geographers have until recently paid relatively little attention to religious issues. One possible reason for this neglect may be the (mistaken?) assumption that Western societies are becoming increasingly secular. Chris Park argues that religious issues are relatively marginal in a great deal of academic analysis because of 'the assumed rationality of post-Enlightenment science, which dismisses as irrational (and thus undeserving of academic study) such fundamental issues as mystery, spirituality and awe' (Park, 1994, p. 1). We should also note here that the key social thinkers of the nineteenth century who laid the foundations for much contemporary social theory – Marx, Weber and Durkheim – all stressed the ways in which religion has been used to bolster the existing social order in society, justifying inequalities and placating the poor with the hope of a better after-life.

Recently, however, there has been an increase in geographical work on religion. The reasons for this should have become clear by now from previous chapters on culture, identity, space and ethnicity. For many people in Western societies religious values continue to play a key role in the formation of their sense of identity.

Indeed, the very idea that 'religion' can be defined as a distinct separate sphere of life is a particular Western notion; for many religions including Sikhism and Islam (which are now extensively practised in Western cities) the very idea of separating religious and non-religious spheres is anathema. Although there has been a decline in the numbers attending established forms of Christian worship, there has been a substantial increase in alternative forms of worship such as evangelicalism.

Furthermore, places of religious worship such as churches, chapels, cathedrals, temples and mosques can have a powerful symbolic value. These are places where members of a religious community come together to reinforce their beliefs through various rituals. Some religions such as Sikhism regard the whole of the world as a sacred space full of the presence of God and yet like most religions, Sikhism also has buildings and spaces that are of special spiritual significance.

Most religions proclaim moral values that are universally relevant and that have been communicated to humankind by an omniscient being(s) through various prophets and gurus. This stance stands in sharp contrast with secular humanist approaches that stress that values are specific to particular times and places. In practice, many religious belief systems have been modified over the years to accommodate changing societal attitudes towards issues such as the role of women and science. However, in response to postmodern moral relativism, and state secularism, we have recently seen the growth of various fundamentalist and evangelical religious movements asserting moral absolutes over issues ranging from homosexuality to abortion, dress and diet.

Key concepts associated with sacred spaces (see Glossary)

Essentialism, identities, signification.

Further reading

Jackson, R.H. and Henrie, R. (1993) Perception of sacred space *Journal of Cultural Geography*, **3**, 94–107

Park, C. (1994) *Sacred Worlds* Routledge, London

Links with Other Chapters

Chapter 8: Box 8.2 The Growth of the Eruv

our capacity for dwelling allows us to construct places, to give them meanings that are deepened and qualified over time with multiple nuances.

Here, though, Heidegger introduced an additional argument: that this deepening and multiple layering of meaning is subverted in the modern world by the spread of telecommunications technology, rationalism, mass production and mass values. The result, he suggested, is that the 'authenticity' of place is subverted. City spaces become inauthentic and 'placeless', a process that is, ironically, reinforced as people seek authenticity through professionally designed and commercially constructed spaces and places whose invented traditions, sanitized and simplified symbolism, and commercialized heritage all make for convergence rather than spatial identity.

Yet the construction of place by 'insiders' cannot take place independently of societal norms and representations of the world: what Larissa Lomnitz called the 'cultural grammar' that codifies the social construction of spaces and places (Lomnitz and Diaz, 1992). Both our territoriality and our sense of dwelling are informed by broadly shared notions of social distance, rules of comportment, forms of social organization, conceptions of worth and value, and so on. We see here, then, another important dialectical relationship: between social structures and the everyday practices of the 'insiders' of subjectively constructed spaces and places. We live, as noted before in this book, both in and through places. Place, then, is much more than a container or a mental construct. It is both text and context, a setting for social interaction that, among other things (Thrift, 1985):

➤ structures the daily routines of economic and social life;

➤ structures people's life paths, providing them with both opportunities and constraints;

➤ provides an arena in which everyday, 'common-sense' knowledge and experience is gathered;

➤ provides a site for processes of socialization and social reproduction; and

➤ provides an arena for contesting social norms.

Urban lifeworlds, time–space routinization and intersubjectivity

This dialectical relationship lends both dynamism and structure to the social geography of the city:

> The social reality of the city is not simply given.
> It is also constructed and maintained
> intersubjectively in a semiclosed world of
> communication and shared symbolization.
> The routines of daily life create a particular view
> of the world and a mandate for action. It is the
> unself-conscious, taken-for-granted character
> of the life-world that makes it so binding on
> its members, that ensures that its realities will
> remain secure.
>
> (Ley, 1983, p. 203)

The crucial idea here is that of the **lifeworld**, the taken-for-granted pattern and context for everyday living through which people conduct their day-to-day lives without having to make it an object of conscious attention (Madsen and Plunz, 2001). Sometimes, this pattern and context extend to conscious attitudes and feelings: a self-conscious sense of place with an inter-locking set of cognitive elements attached to the built environment and to people's dress codes, speech patterns, public comportment and material possessions. This is what Raymond Williams termed a *structure of feeling* (Williams, 1973). The basis of both individual lifeworlds and the collective structure of feeling is **intersubjectivity**: shared meanings that are derived from the lived experience of everyday practice. Part of the basis for intersubjectivity is the *routinization* of individual and social practice in time and space. As suggested by Figure 9.1, the temporality of social life can be broken out into three levels, each of which is interrelated to the others (Simonsen, 1991). The *longue durée* of social life is bound up with the historical development of institutions (the law, the family, etc.). Within the *dasein*, or lifespan, social life is influenced by the life cycle of individuals and families and (interacting with the *longue durée*) by the social conditions characteristic of their particular generation. And within the *durée* of daily life, individual routines interact with both the structure of institutional frameworks and with the rhythm of their life cycle.

The *spatiality* of social life can also be broken down into three dimensions. At the broadest scale there is institutional spatial practice, which refers to the collective level of the social construction of space. 'Place' can then be thought of as related to the human consciousness and social meanings attached to urban spaces. Finally, individual spatial practice refers to the physical presence and spatial interaction of individuals and groups. These three levels of spatiality, in turn, can be related to the three levels of temporality of social life, as depicted in Fig. 9.2. We are, thus, presented with a multidimensional framework within which time–space routinization is able to foster the intersubjectivity upon which people's lifeworlds depend.

The best-known element of this framework to geographers is the time–geography of daily life that has been elaborated by Torsten Hägerstrand (Carlstein

	Longue durée	*Dasein* (lifespan)	*Durée* of daily life
Longue durée	Institutional time History	Coupling of history and life history Generation	Dialectics between life institutions and daily life
Dasein (lifespan)		Life history, the 'I'	Relation between life strategies and daily life
Durée of daily life			Day-to-day routines (time use)

Figure 9.1 Interrelations among the dimensions of temporality.
Source: Simonsen (1991), Fig. 1, p. 427.

Time Space	*Longue durée*	*Dasein*	*Durée* of daily life
Institutional spatial practice	Sociospatial development (historical geography)	Life strategies in spatial context	Geographical conditioning of daily routines
Place	Local history, culture and tradition	Biography in time and space Identity	Spatially based 'natural attitudes'
Individual spatial practice	Historical conditioning of spatial practices	Relation between life strategies and spatial practices	Daily time–space routines (time–geography)

Figure 9.2 Temporality, spatiality and social life.
Source: Simonsen (1991), Fig. 3.

et al., 1978). His basic model (Fig. 9.3) captures the constraints of space and time on daily, individual spatial practices. It illustrates the way that people trace out 'paths' in time and space, moving from one place (or 'station') to another in order to fulfil particular pur-poses (or 'projects'). This movement is conceptualized as being circumscribed by three kinds of constraint: (1) **capability constraints** – principally, the time available for travelling and the speed of the available mode of transportation, (2) **authority constraints** – laws and

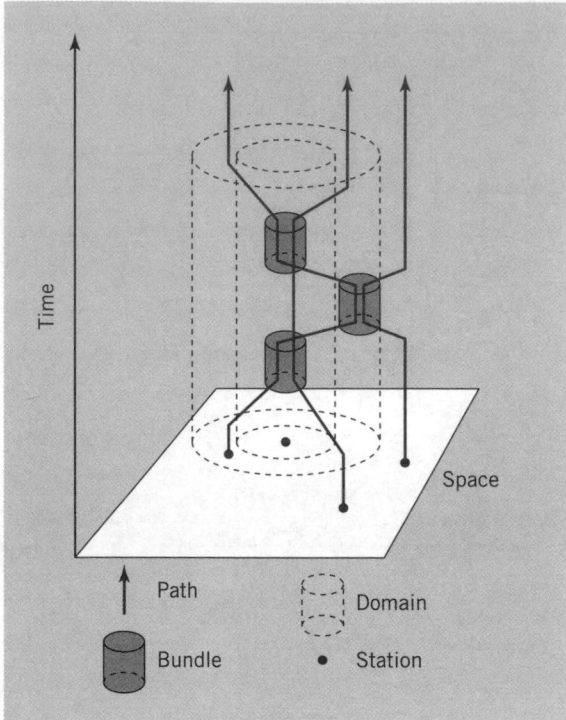

Figure 9.3 Concepts and notation of time–geography (after Hägerstrand).
Source: Gregory (1989), Fig. 1.4.4, p. 82.

customs affecting travel and accessibility, and (3) **coupling constraints** – resulting from the limited periods during which specific projects are available for access. The particular significance of time–geographies in the present context is that groups of people with similar constraints are thrown together in 'bundles' of time–space activity: routine patterns that are an important precondition for the development of intersubjectivity.

Structuration and the 'becoming' of place

These issues are central to **structuration theory**, which addresses the way in which everyday social practices are structured across space and time. Developed by Anthony Giddens (1979, 1981, 1984, 1985: see also Bryant and Jary, 1991), structurationist theory accepts and elaborates Karl Marx's famous dictum that human beings 'make history, but not in circumstances of their own choosing'. Reduced to its essentials, and seen from

a geographical perspective, structurationist theory holds that human landscapes:

> are created by knowledgeable actors (or *agents*) operating within a specific social context (or *structure*). The structure–agency relationship is mediated by a series of institutional arrangements that both enable and constrain action. Hence three 'levels of analysis' can be identified: structures, institutions, and agents. Structures include the long-term, deep-seated social practices that govern daily life, such as law and the family. Institutions represent the phenomenal forms of structures, including, for example, the state apparatus. And agents are those influential human actors who determine the precise, observable outcomes of any social interaction.
>
> (Dear and Wolch, 1989, p. 6)

We are all actors, then (whether ordinary citizens or powerful business leaders, members of interest groups, bureaucrats or elected officials), and all part of a dualism in which structures (the communicative structures of language and signification as well as formal and informal economic, political and legal structures) enable our behaviour while our behaviour itself reconstitutes, and sometimes changes, these structures. Structures may act as constraints on individual action but they are also, at the same time, the medium and outcome of the behaviour they recursively organize. Furthermore, structurationist theory recognizes that we are all members of *systems* of social actors: networks, organizations, social classes and so on.

Human action is seen as being based on 'practical consciousness', meaning that the way in which we make sense of our own actions and the actions of others, and the way we generate meaning in the world is rooted in routinized day-to-day practices that occupy a place in our minds somewhere between the conscious and the unconscious. Recursivity, the continual reproduction of individual and social practices through routine actions (*time–space routinization*), contributes to *social integration*, the development of social systems and structures among agents in particular locales.

In addition, structures and social systems can be seen to develop across broader spans of space and time through *system integration*, which takes place through

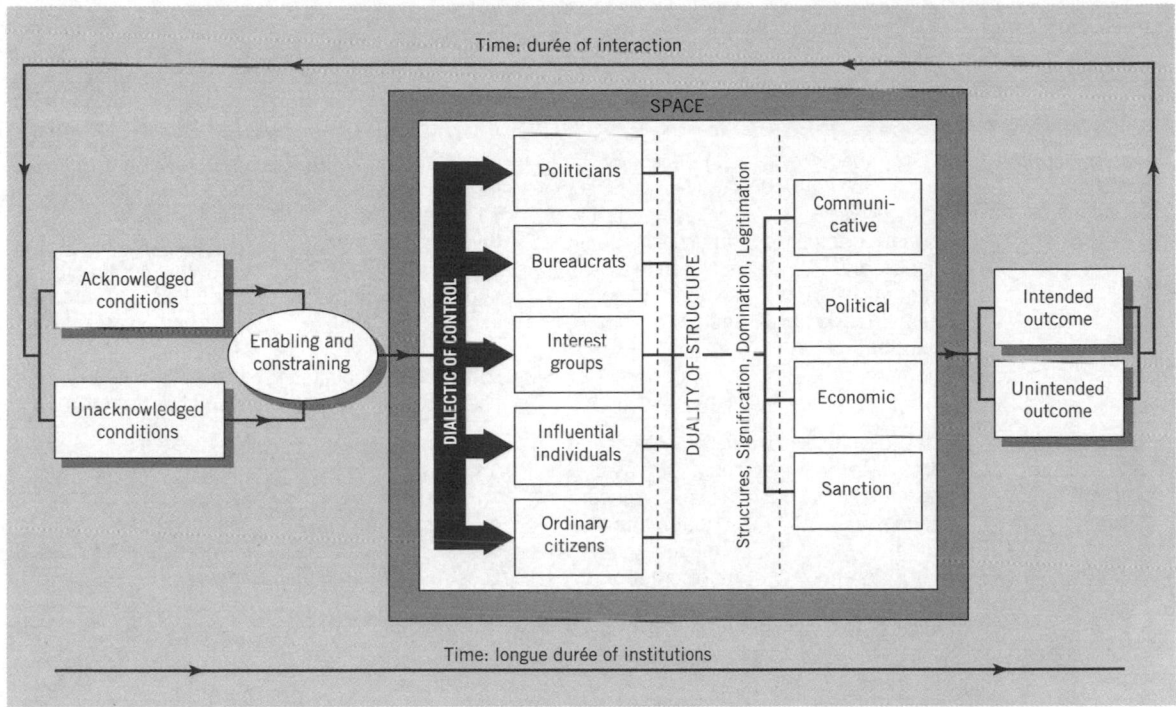

Figure 9.4 A model of the structuration of urban space.
Source: Moos and Dear (1986), p. 245.

time–space **distanciation**: the 'stretching' of social relations over time and space as ideas, attitudes and norms are spread through print and electronic media, for example. All this recursivity and integration does not make for stasis, however, since the structurationist approach sees all human action as involving unanticipated or unacknowledged conditions and as having unintended consequences that modify or change the nature of recurrent practices (Fig. 9.4).

This kind of perspective leads us to see urban spaces and places as constantly changing, or '*becoming*'. Place, in other words, is an historically contingent *process* in which practice and structure become one another through the intertwining of recursive individual and social practices and structured relations of power. At the same time, place involves processes (socialization, language acquisition, personality development, social and spatial division of labour, etc.) through which individual biographies and collective ways of life also become one another. The structurationist approach has become an important influence in contemporary human geography, particularly in urban social geo

graphy because of its central concern with the sociospatial dialectic. It has, nevertheless, proven difficult to incorporate into substantive accounts of city and/or neighbourhood formation. It has also been criticized for its emphasis on recursivity (to the relative neglect of the unforeseen and the unintended), for its inattention to the role of the unconscious, and for its neglect of issues of culture, gender and ethnicity.

Constructing place through spatial practices

David Harvey's 'grid' of spatial practices (Table 9.1) provides one way of accommodating a broader, richer array of issues in addressing the ways in which places are constructed and experienced, how they are represented, and how they become used as symbolic spaces. The matrix is useful in focusing our attention on the dialectical interplay between experience, perception and imagination; and in clarifying the relationships between distanciation and the appropriation, domination and

Table 9.1 A 'grid' of spatial practices

	Accessibility and distanciation	Appropriation and use of space	Domination and control of space	Production of space
Material spatial practices (experience)	Flows of goods, money, people, labour, power, information, etc.; transport and communications systems; market and urban hierarchies; agglomeration	Land uses and built environments; social spaces and other 'turf' designations; social networks of communication and mutual aid	Private property in land; state and administrative divisions of space; exclusive communities and neighbourhoods; exclusionary zoning and other forms of social control (policing and surveillance)	Production of physical infrastructures (transport and communications; built environments; land clearance, etc.); territorial organization of social infrastructures (formal and informal)
Representations of space (perception)	Social, psychological and physical measures of distance; map-making; theories of the 'friction of distance' (principle of least effort, social physics, range of good, central place and other forms of location theory)	Personal space; mental maps of occupied space; spatial hierarchies; symbolic representation of spaces; spatial 'discourses'	Forbidden spaces; 'territorial imperatives'; community; regional cultures; nationalism; geopolitics; hierarchies	New systems of mapping, visual representation, communication, etc.; new artistic and architectural 'discourses'; semiotics
Spaces of representation (imagination)	Attraction/repulsion; distance/desire; access/denial; transcendence 'medium is the message'	Familiarity; hearth and home; open places; places of popular spectacle (streets, squares, markets); iconography and graffiti, advertising	Unfamiliarity; spaces of fear; property and possession; monumentality and constructed spaces of ritual; symbolic barriers and symbolic capital; construction of 'tradition'; spaces of repression	Utopian plans; imaginary landscapes; science fiction ontologies and space; artists' sketches; mythologies of space and place; poetics of space; spaces of desire

Source: Harvey (1989b) pp. 220–221.

production of places. It does not, though, summarize a theory: it is merely a framework across which we can interpret social relations of class, gender, community and race.

The three dimensions on the vertical axis of the grid are drawn from Lefebvre's (1991) distinction between the experienced, the perceived and the imagined:

➤ *Material spatial practices* refer to the interactions and physical flows that occur in and across space as part of fundamental processes of economic production and social reproduction.

➤ *Representations of space* include all of the signs, symbols, codifications and knowledge that allow material spatial practices to be talked about and understood.

➤ *Spaces of representation* are mental constructs such as utopian plans, imaginary landscapes, paintings and symbolic structures that imagine new meanings or possibilities for spatial practices.

The four dimensions across the horizontal axis of the grid have to be seen as mutually interdependent. *Accessibility and distanciation* are two sides of the same coin: the role of the friction of distance in human affairs. Distance, as we saw in Chapter 8, is both a barrier to and a defence against social interaction. Distanciation 'is simply a measure of the degree to which the friction of space has been overcome to accommodate social interaction' (Harvey, 1989b, p. 222). The *appropriation of space* refers to the way in which space is occupied by individuals, social groups, activities (e.g. land uses) and objects (houses, factories, streets). The *domination of space* refers to the way in which the organization and production of spaces and places can be controlled by powerful individuals or groups: through private property laws, zoning ordinances, restrictive covenants, gates (and implied gates), etc. The *production of space* refers to the way in which new systems of territorial organization, land use, transport and communications, etc. (actual or imagined) arise, along with new ways of representing them (see also Box 9.3).

We shall draw on this grid throughout the remainder of this chapter as we examine the ways in which material and social worlds are given meaning through cultural politics, in which political and economic power is

---

Box 9.3

Key thinkers in urban social geography – Henri Lefebvre

The publication in 1991 of an English translation of Henri Lefebvre's *The Production of Space* (first published in French in 1974) was treated in some geographical circles like the discovery of the Holy Grail! In the two previous decades only a few geographers 'in the know' had spoken in awe of a French scholar who had developed a radical reformulation of Marxian theory that put space rather than time at the heart of his analysis. In the event, Lefebvre's book proved to be a densely written and, at times, impenetrable text that inevitably led to differing interpretations over the *real* meaning of his work. Nevertheless, Lefebvre's influence upon urban social geography should not be underestimated. In particular, he had a big influence upon the work of David Harvey, arguably the most influential human geographer in the late twentieth century (see also Box 1.1). A key insight has been Lefebvre's distinction between differing conceptions of space:

On the one hand Lefebvre referred to *representations of space* – the dominant ways in which cities are portrayed, such as in planning documents (sometimes termed 'conceived space'). These are bound up with existing capitalist norms involving issues such as property rights and legalization surrounding labour laws. On the other hand he noted that there are *spaces of representation*, the personal feelings that people have towards the spaces they inhabit through everyday interactions (sometimes termed 'perceived space'). The crucial point is that these two notions of space are interlinked and may be in conflict. For example, the way in which a city centre is portrayed in official governmental literature, extolling its virtues for shoppers and inward investment, may differ from the ways in which local inhabitants think of such a space. Relatively low-paid service workers may have a different idea of a re-gentrified city centre than well-paid high-tech workers or visiting tourists.

Lefebvre was especially interested in the processes that led to the creation of these dominant representations of space and the exclusion of other visions of space. As a social visionary, Lefebvre hoped that people could envisage new notions of space. Since he put space at the heart of his analysis, radical social change must involve more than changing the means of production, but also new concepts of space in which people live their everyday lives.

Lefebvre has been criticized for ignoring the new *cultural politics* (Blum and Nast, 1996) but he has inspired a number of Marxian geographers who have examined recent changes in cities (e.g. Smith, 1984; Merrifield, 1993).

Key concepts associated with Henri Lefebvre (see Glossary)

Material practices, representations of space, spaces of representation.

Further reading

Blum, V. and Nast, H. (1996) Where's the difference? The heterosexualization of alterity in Henri Lefebvre and Jacques Lacan *Environment and Planning D: Society and Space*, **14**, 559–580

Merrifield, A. (1993) Space and place: a Lefebvrian reconciliation *Transactions of the Institute of British Geographers*, **18**, 516–531

Shields, R. (1999) *Henri Lefebvre: Love and Struggle: Spatial Dialectics* Routledge, London

Shields, R. (2004) Henri Lefebvre, in P. Hubbard, R. Kitchin and G. Valentine (eds) *Key Thinkers on Space and Place* Sage, London

Smith, N. (1984) *Uneven Development: Nature, Capital and the Production of Space* Blackwell, Oxford

projected through urban form, and in which space and place are appropriated through symbolism and coded meanings.

Place, consumption and cultural politics

An important lesson is implicit in the grid of spatial practices outlined by Harvey: it is that we should not treat 'society' as separate from 'economy', 'politics', 'culture' or 'place'. We are thus pointed to the domain of '**cultural politics**', defined by Peter Jackson as:

> . . . the domain in which meanings are constructed and negotiated, where relations of dominance and subordination are defined and contested. . . .
> In opposition to the unitary view of culture as the artistic and intellectual product of an elite, 'cultural politics' insists on a plurality of cultures, each defined as a whole 'way of life', where ideologies are interpreted in relation to the material interests they serve. From this perspective, the cultural is always, simultaneously, political.
>
> (Jackson, 1991a, p. 219)

Our experiences of material and social worlds are always mediated by power relationships and culture. 'Social' issues of distinction and 'cultural' issues of aesthetics, taste and style cannot be separated from 'political' issues of power and inequality or from 'gender' issues of dominance and oppression. As noted in Chapter 3, the construction of place is therefore bound up with the construction of class, gender, sexuality, power and culture. To quote Peter Jackson again, 'class relations have a cultural as well as an economic dimension and . . . patriarchy cannot be confined to questions of sexuality, marriage, or domesticity. "Home" and "work" cannot readily be separated, as relations of dominance and subordination established in one domain carry over into the other' (Jackson, 1989, p. 115).

Habitus

An important contribution to this perspective has been made by French sociologist Pierre Bourdieu. His concept of *habitus*, like Raymond Williams's concept of a 'structure of feeling', noted above, deals with the construction of meaning in everyday lifeworlds. *Habitus* evolves in response to specific objective circumstances of class, race, gender relations and place. Yet it is more than the sum of these parts. It consists of a distinctive set of values, cognitive structures and orienting practices: a collective perceptual and evaluative schema that derives from its members' everyday experience and operates at a subconscious level, through commonplace daily practices, dress codes, use of language, comportment and patterns of material consumption. The result is a distinctive cultural politics of 'regulated improvisations' in which 'each dimension of lifestyle symbolizes with the others' (Bourdieu, 1984, p. 173).

According to Bourdieu, each group will seek to sustain and extend its *habitus* (and new sociospatial groups will seek to establish a *habitus*) through the appropriation of **symbolic capital**: consumer goods and services that reflect the taste and distinction of the owner. In this process, not every group necessarily accepts the definitions of taste and distinction set out by the elite groups and tastemakers with the 'cultural capital' to exercise power over the canons of 'good' taste and 'high' culture. In any case, such definitions are constantly subject to devaluation by the popularization of goods and practices that were formerly exclusive.

The fact that symbolic capital is vulnerable to devaluation and to shifts in avant-garde taste makes it even more potent, of course, as a measure of distinction. As a result, though, dominant groups must continually pursue refinement and originality in their lifestyles and ensembles of material possessions. Less dominant groups, meanwhile, must find and legitimize alternative lifestyles, symbols and practices in order to achieve distinction. Subordinate groups are not necessarily left to construct a *habitus* that is a poor copy of others', however: they can – and often do – develop a *habitus* that embodies different values and 'rituals of resistance' (Hall and Jefferson, 1976; Jackson, 1991a) in which the meaning of things is appropriated and transformed.

All this points once again to the importance of *consumption* and of the **aestheticization** of everyday life introduced in Chapter 3. Consumption purports to dispel the dread of being in a world of strangers. Advertisements tell us what to expect, what is acceptable and unacceptable, and what we need to do in order to belong. They are primary vehicles for producing and transmitting cultural symbols. [Consumption] not only produces and circulates meaning, it . . . interweaves and alters forces and perspectives, and it empowers us in our daily lives to change our culture, to transform nature, and *to create place*.

(Sack, 1992)

Consumption is inherently spatial. The propinquity of object and place allows the former to take on the cultural authority of the latter; objects displayed beside each other exchange symbolic attributes; places become transformed into commodities. The consumer's world consists not only of settings where things are purchased or consumed (shops, malls, amusement parks, resorts – see Wrigley and Lowe, 1996) but also of settings and contexts that are created with and through purchased products (homes, neighbourhoods – see Duncan and Duncan, 2004). All of these settings are infused with signs and symbols that collectively constitute 'maps of meaning'.

9.3 The social meanings of the built environment

At the most general level, the landscape of cities can be seen as a reflection of the prevailing ideology (in the sense of a political climate, *Zeitgeist*, or 'spirit') of a particular society. The idea of urban fabric being seen – in part, at least – as the outcome of broad political, socioeconomic and cultural forces has been explicit in much writing on urbanization. We can illustrate this with reference to the simple example of the symbolization of wealth and achievement by groups of prosperous merchants and industrialists. Early instances of this include the industrial capitalists of Victorian times, who felt the compulsion to express their achievements in buildings (Domosh, 1996). The Cross Street area of central Manchester is still dominated by the imposing gothic architecture commissioned by the city's Victorian elite who, preoccupied with the accumulation and display of wealth but with a rather philistine attitude towards aesthetics, left a clear impression of their values on the central area of the city.

As the petite bourgeoisie of small-scale merchant and industrial capital lost ground to corporate and international capital, so the symbolization of achievement and prosperity became dominated by corporate structures. Huge office blocks such as the Prudential Building in Boston and the Pirelli Building in Milan were clearly intended as statements of corporate power and achievement, notwithstanding any administrative or speculative functions. At a more general level, of course, the whole complex of offices and stores in entire downtown areas can be interpreted as symbolic of the power of the 'central district elite' in relation to the rest of the city.

Meanwhile, other institutions have added their particular statements to the palimpsest of the urban fabric. The sponsors of universities, trade union headquarters, cultural centres and the like, unable (or unwilling) to make use of the rude message of high-rise building, have generally fallen back on the combination of neo-classicism and modernism that has become the reigning international style for any building aspiring to carry authority through an image of high-mindedness rather than raw power.

The appropriation of space and place: symbolism and coded meanings

While the built environment is heavily endowed with social meaning, this meaning is rarely simple, straightforward or unidimensional. To begin with there is an important distinction between the *intended* meaning of architecture and the *perceived* meaning of the built environment as seen by others. This distinction is essential to a proper understanding of the social meaning of the built environment. David Harvey's study of the Sacre-Coeur in Paris, for example, demonstrates how the intended symbolism of the building – a reaffirmation of Monarchism in the wake of the Paris Commune – 'was for many years seen as a provocation to civil war', and is still interpreted by the predominantly republican population of Paris as a provocative rather than a unifying symbol (Harvey, 2003).

Another critical point is that the social meaning of the built environment is not static. The meanings associated with particular symbols and symbolic environments tend to be modified as social values change in response to changing lifestyles and changing patterns of socioeconomic organization. At the same time, powerful symbols and motifs from earlier periods are often borrowed in order to legitimize a new social order, as in Mussolini's co-opting of the symbols of Augustan Rome in an attempt to legitimize Fascist urban reorganization; and (ironically) in the adoption of a selection of motifs from the classical revival in Europe by Jefferson and the founding fathers responsible for commissioning public and ceremonial architecture in Washington, DC.

How can all such observations be accommodated within a coherent framework of analysis that addresses the fundamental questions of communication by whom, to what audience, to what purpose and to what effect? These are the questions that have prompted a number of writers to build on structuralist social theory in such a way as to accommodate the social meaning of the built environment. According to this perspective, the built environment, as part of the socioeconomic superstructure stemming from the dominant mode of production (feudalism, merchant capitalism, industrial capitalism, etc.), reflects the Zeitgeist of the prevailing system; it also serves, like other components of the superstructure, as one of the means through which the necessary conditions for the continuation of the system are reproduced. One of the first people to sketch out these relationships between social process and urban form was David Harvey, who emphasized the danger of thinking in terms of simple causal relationships, stressing the need for a flexible approach that allows urbanism to exhibit a variety of forms within any dominant mode of production, while similar forms may exist as products of different modes of production (Harvey, 1978).

Within the broad framework sketched by Harvey, others have contributed detailed studies of the form, interior layout and exterior design of various components of the built environment as a response to the reorganization of society under industrial capitalism (Ford, 1994; Schein, 1997). To take just a few examples: Rob Shields (1989) has drawn upon Lefebvre's concepts of spaces of representation and representations of space (see Harvey's matrix of spatial practices in Table 9.1) in analyzing the 'social spatialization' of the built environment, using the West Edmonton Mall as a case study; while Jon Goss (1993) and Margaret Crawford (1992) have each explored the structuration of place through the retail spaces of shopping malls. Christine Boyer (1992) has described the 'commodification of history' in waterfront redevelopments; and Sharon Zukin (1991) has described the social construction of the built environment in relation to several different kinds of 'landscapes of power', including industrial neighbourhoods, malls, gentrified neighbourhoods and Disney World.

Barbara Rubin, in her documentation of the emergence of 'signature' structures and franchise architecture, noted that 'good taste' in urban design generally has become part of an ideology that has been used to control and exploit urban space. 'In the ideology of American aesthetics, it is understood that those who make taste make money, and those who make money make taste' (Rubin, 1979, p. 360). This brings us back to the key role of certain actors – design professionals, in this case – within the social production of the built environment.

Architecture, aesthetics and the sociospatial dialectic

The architect's role as an arbiter, creator and manipulator of style can be interpreted as part of the process whereby changing relationships within society at large become expressed in the 'superstructure' of ideas, institutions and objects. This allows us to see major shifts in architectural style as a dialectical response to the evolving Zeitgeist of urban-industrial society – as part of a series of broad intellectual and artistic reactions rather than the product of isolated innovations wrought by inspired architects.

Thus, for example, the Art Nouveau and Jugendstil architecture of the late nineteenth century can be seen as the architectural expression of the romantic reaction to what Lewis Mumford called the 'palaeotechnic' era of the industrial revolution: a reaction that was first expressed in the Arts and Crafts movement and in Impressionist painting. By 1900 the Art Nouveau style was firmly established as the snobbish style, consciously elitist, for all 'high' architecture.

Postmodern 'high-tech' architecture: The Lloyds building and the Swiss Re Insurance building a.k.a. The London 'Gherkin'. Photo Credit: Paul Knox.

The dialectic response was a series of artistic and intellectual movements, beginning with Cubism, that went out of their way to dramatize modern techno-logy, seeking an anonymous and collective method of design in an attempt to divorce themselves from 'capitalist' canons of reputability and power. Thus emerged the Constructivist and Futurist movements, the Bauhaus school and, later, Les Congrès Inter-nationaux d'Architecture Moderne (CIAM) and the Modern Architecture Research Group (MARS), who believed that their new architecture and their new con-cepts of urban planning were expressing not just a new aesthetic image but the very substance of new social conditions that they were helping to create.

The subsequent fusion and transformation of these movements into the glib 'Esperanto' of the Interna-tional Style and the simultaneous adoption of the style as the preferred image of corporate and bureaucratic conservatism, solidity and respectability provides an important example of the way in which the dominant social order is able to protect itself from opposing ideo-logical forces. In this particular example, the energy of opposing ideological forces – idealist radicalism – has been neatly diverted into the defence of the *status quo*. The question is: how?

Commodification

One answer is that the professional ideology and career structure within which most practising architects (as opposed to the avant-garde) operate is itself heavily oriented towards Establishment values and sensitively tuned to the existing institutional setting and economic order. Consequently, the meaning and symbolism of new architectural styles emanating from radical quar-ters tend to be modified as they are institutionalized and converted into commercialism; while the core movement itself, having forfeited its raw power in the process of commodification, passes quietly into the mythology of architectural education and the coffee-table books of the cognoscenti. There is a direct paral-lel here in the way that the liberal ideology of the town planning movement was transformed into a defensive arm of urbanized capital, systematically working to the advantage of the middle-class community in general and the business community in particular. According

to this perspective, architects, like planners, can be seen as unwitting functionaries, part of a series of 'internal survival mechanisms' that have evolved to meet the imperatives of urbanized capital.

Architecture and the circulation of capital

Another way in which architects serve these imperatives is in helping to stimulate consumption and extract sur-plus value. The architect, by virtue of the prestige and mystique socially accorded to creativity, adds exchange value to a building through his or her decisions about design, so that the label 'architect designed' confers a presumption of quality even though this quality may not be apparent to every observer. Moreover, as one of the key arbiters of style in modern society, the architect is in a powerful position to stimulate consumption merely by generating and/or endorsing changes in the nuances of building design.

The professional ideology and career structure that reward innovation and the ability to feel the pulse of fashion also serve to promote the circulation of capital. Without a steady supply of new fashions in domestic architecture (reinforced by innovations in kitchen technology, heating systems, etc.), the filtering mechanisms on which the whole owner-occupier housing market is based would slow down to a level unacceptable not only to builders and developers but also to the exchange professionals (surveyors, real estate agents, etc.) and the whole range of financial institu-tions involved in the housing market. The rich and the upper-middle classes, in short, must be encouraged to move from their comfortable homes to new dwellings with even more 'design' and 'convenience' features in order to help maintain a sufficient turnover in the housing market.

One way in which they are enticed to move is through the cachet of fashionable design and state-of-the-art technology. Hence the rapid diffusion of innovations such as energy-conserving homes; and the desperate search for successful design themes to be revived and 're-released', just like the contrived revivals of haute couture and pop music. In parts of the United States, the process has advanced to the stage where many upper-middle class suburban developments resemble

Box 9.4

Key thinkers in urban social geography – Nigel Thrift

It is impossible to venture very far into human geography before coming across the work of Nigel Thrift. Indeed, if Richard Florida is the geographical equivalent of the populist business studies expert Michael Porter (see Box 11.2), then Nigel Thrift must be the equivalent of the highly influential social scientist Anthony Giddens. Like Giddens, Thrift has drawn upon a wide range of ideas from many sources. One of the most prolific writers in the field, with scores of books and articles to his name, the scope of Thrift's work is enormous, ranging from the nature of capitalism (1997), to globalization and regional development (Amin and Thrift, 1992), the character of money (Leyshon and Thrift, 1997) and the role of time in social life (Thrift, 1977) (to mention but a few!). It is impossible to summarize this vast corpus of work in any detail but, nevertheless, underpinning this extraordinary variety are a number of key themes.

Above all, Thrift has resisted the 'top down' structuralist theorizing of Marxian approaches. While acknowledging the existence of complex networks of power, knowledge and authority, these are seen as the outcome of actions by people who make choices based on various types of knowledge that they employ in the course of their everyday lives. Issues such as performance, subjectivity, discourse, representation and identity therefore figure highly in Thrift's work.

Taken as a whole, Thrift's work represents a remarkable reconstitution, in a geographical setting, of diverse but interrelated ideas from a wide range of highly influential thinkers in the social sciences including: Torsten Hagerstrand's work on time-space budgets; Anthony Giddens' work on structuration theory; Michel Foucault's work on knowledge and power; Manuel Castells' work on 'spaces of flows'; Bruno Latour's work on actor network theory; and Gilles Deleuze's work on performance.

Key concepts associated with Nigel Thrift (see Glossary)

Embeddedness, performativity, subjectivity.

Further reading

Amin, A. and Thrift, N. (1992) Neo-Marshallian nodes in global networks *International Journal of Urban and Regional Research*, **116**, 571–587

Leyshon, A. and Thrift, N. (eds) (1997) *Money/Space: Geographies of Monetary Transformation* Routledge, London

Thrift, N. (1977) Time and theory in human geography, part one *Progress in Human Geography*, **1**, 65–101

Thrift, N. (1997) The rise of soft capitalism, in A. Herod, S. Roberts and G. Toal (eds) *An Unruly World? Globalization and Space* Routledge, London

Links with Other Chapters

Chapter 4: Box 4.1 The Role of Non-representational Theory

---

small chunks of Disneyland, with mock-Tudor, Spanish Colonial, neo-Georgian, Victorian gothic and log cabin de luxe standing together: style for style's sake, the *zeit* for sore eyes. And, in some cities, new housing for upper-income groups is now promoted through annual exhibitions of 'this year's' designs, much like the Fordist automobile industry's carefully planned obsolescence in design.

But it is by no means only 'high' architecture and expensive housing that help to sustain urbanized capital. One of the more straightforward functions of architecture in relation to the structuration of class relations through residential settings is the symbolic distancing of social groups. The aesthetic sterility of most British public housing, for example, serves to distance its inhabitants from other, neighbouring, social groups. At a further level, it can be argued that the scarcity of symbolic stimuli typical of many planned, postwar working-class environments may act as a kind of intellectual and emotional straitjacket, minimizing people's self-esteem and sense of potential while fostering attitudes of deference and defeatism. Although the process is at present very poorly understood, the role of the architect is clearly central to the eventual outcome not only in terms of the social order of the city but also in terms of the existential meaning of urban settings.

This brings us back to a final but crucial consideration: the role of the self in the interaction between society and environment. One framework that accommodates

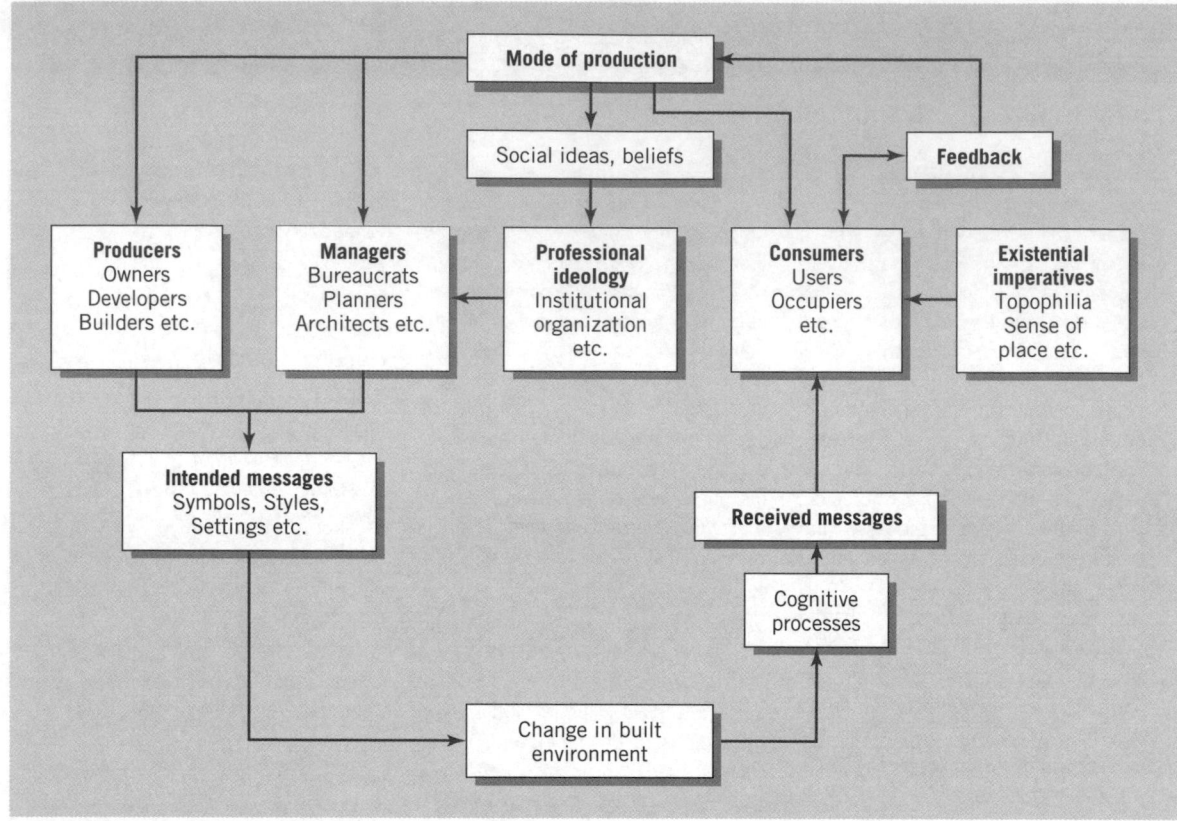

Figure 9.5 Signs, symbolism and settings: a framework for analysis.
Source: Knox (1984).

this is shown in Fig. 9.5. Accepting architectural design as part of the superstructure of culture and ideas stemming from the basic socioeconomic organization of society (whether as part of the prevailing ideology or as part of the counter-ideology), this framework focuses attention on (i) the intended messages emanating from particular owners/producers and mediated by professional 'managers' (architects, planners, etc.), and (ii) the received messages of environmental 'consumers' as seen through the prisms of cognitive processes and existential imperatives and the filter of the dominant ideology.

Chapter summary

9.1 Despite many views to the contrary, cities have not resulted in the destruction of community networks but these have been radically transformed through decentralization, suburbanization and social polarization.

9.2 Places develop complex multilayered meanings depending upon the views of those who live within them, as well as those who live outside. These meanings have an important influence upon the ways people go about their everyday lives.

9.3 The landscape of cities tends to reflect the prevailing ideology of the times – a complex mixture of political, economic and cultural forces.

Key concepts and terms

aestheticization	cultural politics	material practices
authority constraint	*daesin*	neighbourhood
'betweenness' (of place)	distanciation	place
capability constraint	ethnic village	representations of space
community	*Gemeinschaft*	spaces of representation
'community lost' argument	*Gesellschaft*	structuration theory
'community saved' argument	*habitus*	symbolic capital
'community transformed'	intersubjectivity	territoriality
argument	lifeworld	Zeitgeist
coupling constraint	*longue durée*	

Suggested reading

Good reviews of concepts of community and neighbourhood from a sociospatial perspective can be found in Barry Wellman's *The Community Question Re-evaluated* (1987: Centre for Urban and Regional Studies, University of Toronto) and in the introductory text by Wayne Davies and David Herbert: *Communities Within Cities* (1993: Pinter, London). See also Gerard Delanty's *Community* (2003: Routledge, London). Further treatment of the ideas of intersubjectivity and lifeworlds can be found in Chapter 2 of Peter Jackson and Susan Smith's *Exploring Social Geography* (1984: George Allen & Unwin, London). Concepts of space and place are treated exhaustively in G. Benko and U. Strohmayer *Space and Social Theory* (1997: Blackwell, Oxford); in Nick Entrikin's *Betweenness of Place* (1991: Johns Hopkins University Press, Baltimore); and in Bob Sack's *Place, Modernity, and the Consumer's World* (1992: Johns Hopkins University Press, Baltimore). See also D. Brooks 'Patio man and the sprawl people' (*Weekly Standard*, 2002: 19–29) and D.D. Martin 'Enacting neighborhood' (*Urban Geography*, 2003: 25(5), 361–385). David Harvey's essay on 'From space to place and back again' (in J. Bird, ed., *Mapping the Futures: Local Cultures, Global Change*, 1998: Routledge, London, pp. 3–29) provides a good introduction to the idea of the social construction of place, while a detailed review of structurationist concepts is given in Chapter 4 of Paul Cloke, Chris Philo and David Sadler's *Approaching Human Geography* (1991: Guilford Press, London). For an introduction to the issues of place, consumption and cultural politics, see Peter Jackson's essay, 'Towards a cultural politics of consumption' in *Mapping the Futures: Local Cultures, Global Change* (cited above; pp. 207–228). Sharon Zukin's *Landscapes of Power* (1991: University of California Press, Berkeley) and Michael Sorkin's edited volume *Variations on a Theme Park* (1992: Noonday, New York) provide a lot of interesting examples of the social meanings of the built environment, while some of the theoretical aspects of symbolism and meaning are addressed in an essay by A.P. Lagopoulos (*Environment & Planning D: Society and Space*, 1993: 11, 255–278). *Landscapes of Privilege* (2003: Routledge, London) by James and Nancy Duncan shows how suburban landscapes are bound with processes of exclusion.

Chapter 10

Environment and behaviour in urban settings

Key questions addressed in this chapter

➤ What are the geographies of crime and other deviant behaviour in cities?

➤ How can we best explain these patterns?

➤ How do people form images of urban areas, and how do these images relate to their behaviour?

Geographers have for a long time been interested in the relationships between urban settings and certain aspects of people's behaviour. As we know well enough by now, these relationships are *reciprocal*: 'a neighbourhood takes its character from the values and life-styles of its residents; however, reciprocally, its personality is also a context that acts to reinforce and narrow a range of human responses' (Ley, 1983). The emphasis of most research in this area, though, has been on the way in which the 'personality' of urban settings influences individual and group behaviour and, in particular, the way in which 'deviant' behaviour is related to urban settings.

It should be very clear at the outset that this sort of approach can easily fall into a deterministic frame of thinking, where 'space' is a cause. Behavioural geography has its roots in the classic observed-stimulus → observed-response **behaviourism** of psychologist J.B. Watson (1913). The sophistication of contemporary social geography is that behaviour is no longer described solely in terms of stimulus and reaction. Rather, stimuli are thought of in terms of information (of any kind) that are filtered through the elements of cognition, reflection and consciousness before provoking behavioural responses (**behaviouralism**) (Fig. 10.1).

Central to this whole perspective is the idea of **environmental conditioning**. A classic example is provided by Newson and Newson's (1965) work on patterns of infant care. Impoverished neighbourhoods, they argued, are characterized by a poverty of sensory stimulation, by crowded environments that inhibit play, and by unenlightened attitudes towards child rearing. The net result, they concluded, is that the environmental conditioning experienced by children growing up in such settings produces 'under-socialized' individuals with a competitive nature attuned only to the immediate social

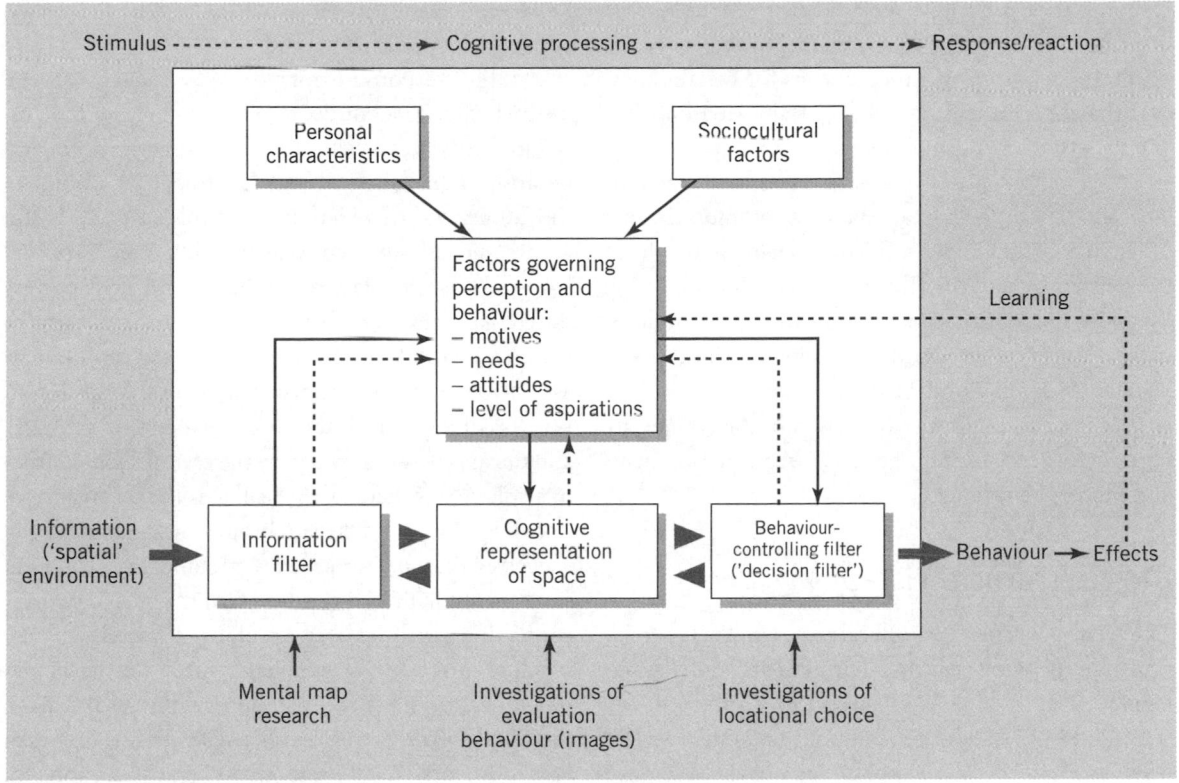

Figure 10.1 A model of behaviour in human geography.
Source: Werlen (1993).

group, leaving them ill-equipped to cope with the more subtle forms of competition that prevail in the world beyond. By extension, it was argued that this kind of environmental conditioning tends to curtail creativity, adaptability and flexibility. The result is doubly-disadvantaged individuals who, on the one hand, seek short-term gratification and are weakly attuned to the established norms and rules of society at large while, on the other, are unable to articulate a coherent alternative or opposition to these norms and rules.

But this kind of argument begs all sorts of questions about the mechanisms and processes involved. How, for example, are distinctive local values sustained in local settings; and to what extent do localized values and attitudes affect the incidence of particular patterns of behaviour? How important is the built environment? And what is the role of broader class-based factors? As we shall see from an examination of ideas about deviant behaviour, there is a broad spectrum of theories.

10.1 Theories about deviant behaviour

A **deviant subgroup** may be defined as a group within society that has norms substantially different from the majority of the population. The notion of deviance covers a multitude of social sins, but geographers have been most interested in behaviour with a distinctive pattern of intra-urban variation, such as prostitution, suicide, truancy, delinquency and drug addiction. In fact, most aspects of deviant behaviour seem to exhibit a definite spatial pattern of some sort, rather than being randomly distributed across the city. But, whereas there is little disagreement about the nature of the patterns themselves, theory and research in geography, sociology and environmental psychology are less conclusive about explanations of the patterns. Some writers, for instance, see deviant behaviour as a pathological response to a particular social and/or physical environment. Others argue that certain physical or social attributes act as environmental cues for certain kinds of behaviour; others still that certain environments simply attract certain kinds of people. Until quite recently, almost all the theorizing about spatial variations in deviant behaviour shared a common element of environmental determinism, usually traceable to the determinists of the Chicago School. In this section the more influential aspects of this theory are outlined before going on to examine briefly the intra-urban geography of one kind of deviant behaviour – crime and delinquency – as an illustration of the complexity of the actual relationships between urban environments and human behaviour.

Determinist theory

There is no need to reiterate at length the relationships between urban environments and deviant behaviour postulated by adherents to Wirthian theory (see pp. 154–57). The general position is that deviant behaviour is a product either of adaptive behaviour or maladjustment to city life, or to life in certain parts of the city. Thus the aloofness and impersonality that are developed in response to the competing stimuli and conflicting demands of different social situations are thought to lead to a breakdown of interpersonal relationships and social order and to an increase in social isolation, which in turn facilitates the emergence of ego-centred, unconventional behaviour and precipitates various kinds of deviant behaviour.

Evidence to support these tenets of **determinist theory** has been assembled on several fronts. The idea of **psychological overload** resulting from complex or unfamiliar environments has been investigated by environmental psychologists and popularized by Alvin Toffler (1970), who suggested that the need to 'scoop up and process' additional information in such situations can lead to 'future-shock': the human response to overstimulation. The nature of this response has been shown by psychologists to take several forms: 'Dernier's strategy', for example, involves the elimination from perception of unwelcome reality, and in an extreme form can result in the construction of a mythological world that becomes a substitute for the real world and

in which deviant behaviour may be seen by the person concerned as 'normal'.

Another response is for people to 'manage' several distinct roles or identities at once. According to determinist theory, this is characteristic of urban environments because of the physical and functional separation of the 'audiences' to which different roles are addressed: family, neighbours, co-workers, club members and so on. Thus, people tend to be able to present very different 'selves' in different social contexts. Again, the extreme form of this behaviour may lead to deviancy. The city, with its wide choice of different roles and identities becomes a 'magic theatre', an 'emporium of styles', and the anonymity afforded by the ease of slipping from one role to another clearly facilitates the emergence of unconventional and deviant behaviour. It has also been suggested that *further* deviancy or pathology may result from the strain of having to sustain different and perhaps conflicting identities over a prolonged period.

Most interest, however, has centred on the *impersonality* and *aloofness* that apparently results from the *psychological overload* associated with certain urban environments. There are many manifestations of this impersonality, the most striking of which is the collective paralysis of social responsibility that seems to occur in central city areas in crisis situations. The classic and oft-quoted example is the murder of Catherine Genovese, who was stabbed to death in a respectable district of Queens in New York, the event evidently witnessed by nearly 40 people, none of whom attempted even to call the police. Other evidence of the lack of 'bystander intervention' and of an unwillingness to assist strangers comes from experiments contrived by psychologists. Such behaviour is itself deviant to some extent, of course; but its significance to determinist theory is in the way in which it fosters the spread of more serious forms of deviancy by eroding social responsibility and social control.

The overall result of these various forms of adaptive behaviour is thus a weakening of personal supports and social constraints and a confusion of behavioural norms. This, in turn, gives a further general impetus to deviant behaviour. Feelings of isolation among the 'lonely crowd' are associated with neurosis, alcoholism and suicide; and the anomic state induced by the weakening of behavioural norms and intensifying levels of incivility is associated with various forms of crime and delinquency.

It is, however, difficult to establish either proof or refutation of the connections between stress, adaptive behaviour, social isolation, social disorganization, anomie and deviancy because of the difficulty of controlling for the many intervening variables such as age, class, education and personality. Nevertheless, many investigations of intra-urban variations in deviant behaviour

have found it useful to invoke determinist theory in at least partial explanation of the patterns encountered. The geography of looting during electricity blackouts in New York City (Fig. 10.2), for example, was closely correlated with patterns of poverty. 'Apparently, people excluded from effective participation in the affluent society did not, and cannot be expected to, act with restraint when the enforcement of law and order is immobilized' (Wohlenberg, 1982). To the extent that

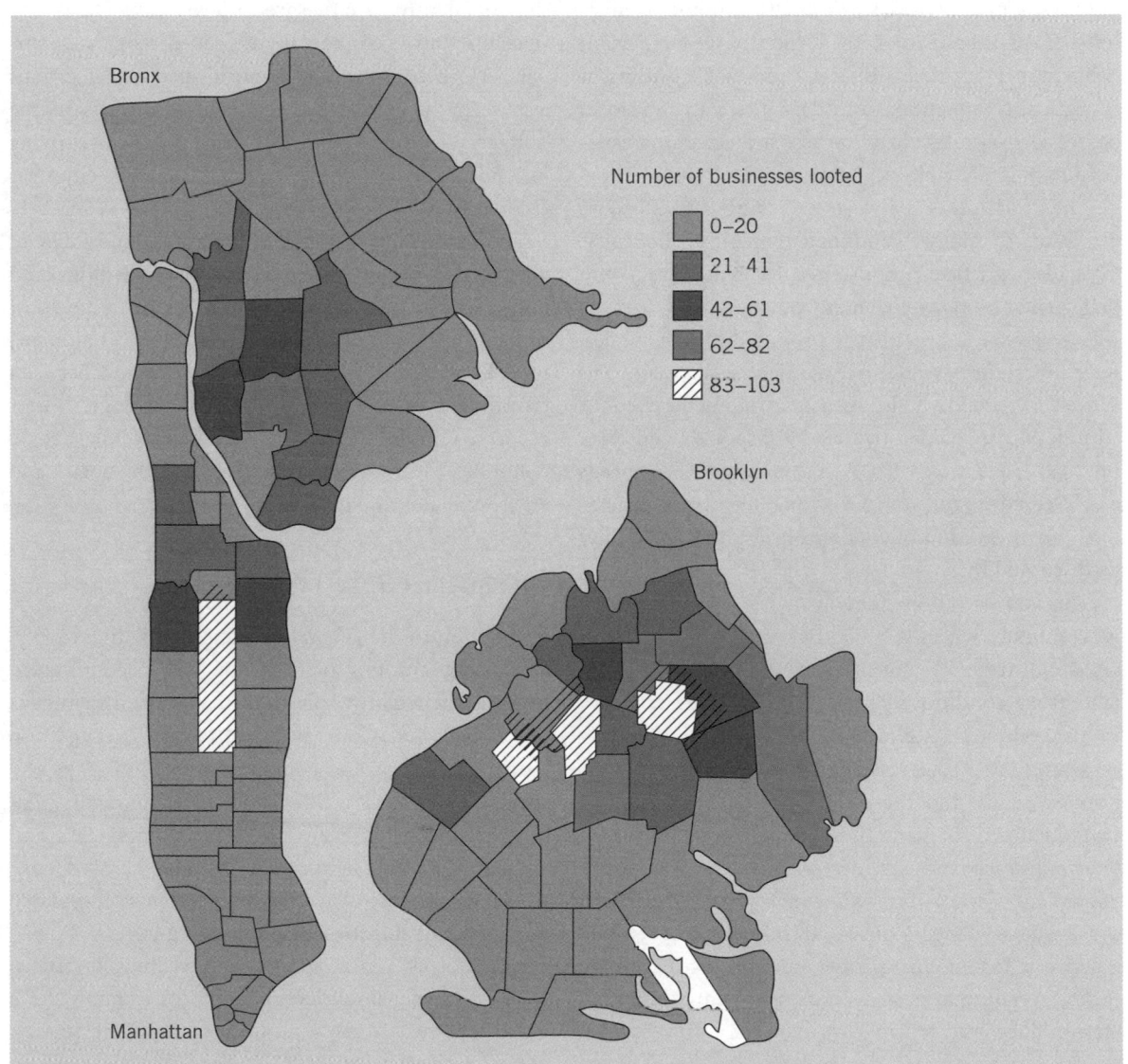

Figure 10.2 Looting in New York: businesses approved for Reestablishment Grants by the mayor's Emergency Aid Commission.
Source: Wohlenberg (1982), Fig. 2, p. 36.

determinist theory is founded on the effects of urbanism on human behaviour, the inference must be that some parts of the city are more 'urban' (in the Wirthian sense) than others, with more social disorganization, a greater incidence of anomie and, consequently, a higher incidence of deviant behaviour.

Crowding theory

There is now a considerable body of literature linking high residential densities, irrespective of other characteristics of urbanism, with a wide range of deviant behaviour. High densities and a sense of crowding, it is argued, create strains and tensions that can lead to aggression, withdrawal or, if these strategies are unsuccessful, mental or physical illness.

The initial link between crowding and stress is attributed by many to an innate sense of **territoriality**. This idea has been popularized by ethnologists who believe that humans, like many other animals, are subject to a genetic trait that is produced by the species' need for territory as a source of safety, security and privacy. Territoriality is also seen as satisfying the need for stimulation (provided by 'border disputes') and for a physical expression of personal identity. These needs are believed to add up to a strong 'territorial imperative': a natural component of behaviour that will clearly be disrupted by crowding.

This approach draws heavily on behavioural research with animals, where the links between crowding, stress and abnormal behaviour can be clearly established under laboratory conditions. Calhoun (1962), for example, in his celebrated studies of rat behaviour, showed that crowding led to aggression, listlessness, promiscuity, homosexuality and the rodent equivalent of juvenile delinquency. Projecting these ideas directly to human behaviours leads to the idea of crowded urbanites as 'killer apes'. Critics of crowding theory have emphasized the obvious dangers involved in extending animal behaviour to humans: people are not rats; it is by no means certain that humans possess any innate sense of territoriality; and in any case even the most crowded slums do not approach the levels of crowding to which experimental animals have been subjected.

It is difficult, however, to establish conclusively whether or not there is any connection between territoriality, crowding and deviant behaviour in human populations. Territoriality may exist in humans through cultural acquisition even if it is not an innate instinct, since territoriality in the form of property rights does provide society with a means of distinguishing social rank and of regulating social interaction. There is a considerable body of evidence to support the idea of territorial behaviour in urban men and women, whatever the source of this behaviour may be. Individuals' home territory represents a *haven*, and an expression of identity. At the group level, gang 'turfs' are rigorously and ceremonially defended by gang members. More complex and sophisticated social groups also seem to exhibit territorial behaviour, as in the 'foreign relations' of different social groups occupying 'defended neighbourhoods' in American inner-city areas (Marcuse, 1997).

Accepting that humans do acquire some form of territoriality, it does seem plausible that crowding could induce stress and so precipitate a certain amount of deviant behaviour. The evidence, however, is ambiguous. Some studies report a clear association between crowding and social and physical pathology, others report contradictory findings, and the whole debate continues to attract controversy in all the social and environmental disciplines.

Design determinism

In addition to the general debate on crowding theory, a growing amount of attention has been directed towards the negative effects of the built environment on people's behaviour. In broad terms the suggestion is that the design and configuration of buildings and spaces sometimes creates micro-environments that discourage 'normal' patterns of social interaction and encourage deviant behaviour of various kinds. A considerable amount of evidence has been accumulated in support of this idea. The inhibiting effects of high-rise and deck-access apartment dwellings on social interaction and child development, for example, have been documented in a number of different studies; and from these it is a short step to studies that point to the correlation between certain aspects of urban design and the incidence of particular aspects of deviancy such as mental illness and suicide.

The nature of these relationships is not entirely clear. One interesting proposition was put forward by Peter Smith (1977), who suggested that the configuration of buildings and spaces creates a 'syntax' of images and symbolism to which people respond through a synthesis of 'gut reactions' and intellectual reactions. Environments that are dominated by an unfamiliar or illogical visual language are thus likely to appear threatening or confusing: qualities that may well precipitate certain aspects of malaise or deviant behaviour.

A better-known and more thoroughly examined link between urban design and deviant behaviour is Oscar Newman's (1972) concept of '**defensible space**'. Newman suggested that much of the petty crime, vandalism, mugging and burglary in modern housing developments is related to an attenuation of community life and a withdrawal of local social controls caused by the inability of residents to identify with, or exert any control over, the space beyond their own front door. This, he argued, was a result of the 'designing out' of territorial definition and delineation in new housing developments, in accordance with popular taste among architects.

Once the space immediately outside the dwelling becomes public, Newman suggested, nobody will feel obliged to 'supervise' it or 'defend' it against intruders. Newman's ideas have been supported by some empirical work and enthusiastically received in the professions concerned with urban design, where they have created a new conventional wisdom of their own: defensible space is now an essential component in the praxis of urban design. On the other hand, Newman's work has been heavily criticized for the quality of his statistical analysis and for his neglect of the interplay of physical and social variables.

Alienation

The concept of **alienation** is a central construct of Marxian theory, where it is associated with the loss of control that workers have over their labour power under a capitalist mode of production. Alienation is also conceptualized in Marxian theory as a mechanism of social change contributing towards the antithesis of the dominant mode of production. However, the concept of alienation also has wider sociopolitical connotations with some relevance to the explanation of deviant behaviour. In its wider sense, alienation is characterized by feelings of powerlessness, dissatisfaction and distrust, and a rejection of the prevailing distribution of wealth and power. These feelings usually stem from people's experience of some aspect of social, political or economic system. Some people may be alienated because they feel that the structure of these systems prevents their effective participation; others may be alienated because they disagree with the very nature of the systems – perhaps because of their ineffectiveness in satisfying human needs.

Whatever the source, such feelings are clearly experience-based and therefore spatially focused, to a certain extent, on people's area of residence. This makes alienation an attractive explanatory factor when considering spatial variations in people's behaviour, as the early deterministic theorists were quick to note. The major interest in this respect has been the relationship between alienation and political behaviour, but it has also been suggested that certain aspects of deviant behaviour may be related to feelings of alienation. Such behaviour may be manifested in apathy: mildly unconventional in itself but more significant if it is prevalent enough to erode social order. Alternatively, alienation may precipitate deviance directly through some form of activism, which can range from eccentric forms of protest to violence and terrorism.

Compositional theory

Compositional theory is the product of another school of thought that has developed out of the writings of the Chicago determinists. Compositionalists emphasize the cohesion and intimacy of distinctive social worlds based on ethnicity, kinship, neighbourhood, occupation or lifestyle, rejecting the idea that these social networks are in any way diminished by urban life (e.g. Gans, 1962). They also minimize the psychological effects of city life on people's behaviour, suggesting, instead, that behaviour is determined largely by economic status, cultural characteristics, family status and so on: the same attributes that determine which social worlds they live in.

Compositional theory is not framed explicitly to analyze deviant behaviour, but it does offer a distinctive perspective on the question. Deviancy, like other forms

Beware the Dog!: fear of crime leading to defensive measures in the city (Milan).
Photo Credit: Paul Knox.

of behaviour, is seen as a product of the composition of local populations, with the social mores, political attitudes and cultural traits of certain groups being more productive of unconventional or deviant behaviour than others.

The pattern of sexually transmitted disease in London serves to illustrate this compositionalist perspective. The incidence of this particular manifestation of deviant behaviour had for many years a very marked peak in the bed-sitter land of West-Central London, especially around Earls Court. The explanation, in compositionalist terms, is the high proportion of young transients in the area – mostly young, single people living in furnished rooms – whose sexual mores are different from those of the rest of the population and whose vulnerability to venereal and other sexually transmitted diseases is increased by the presence of a significant proportion of young males who have themselves been infected before arriving in London. According to London's urban folklore, much of the blame in this respect is attached to Australians who arrive in London having visited Bangkok.

Subcultural theory

Subcultural theory is closely related to compositional theory. Like the latter, subcultural theory subscribes to the idea of social worlds with distinctive socio-demographic characteristics and distinctive lifestyles that propagate certain forms of behaviour. In addition, however, subcultural theory holds that these social worlds, or subcultures, will be *intensified* by the conflict and competition of urban life; and that new subcultures will be spawned as specialized groups generated by the arrival of immigrants and by the structural differentiation resulting from industrialization and urbanization reach the 'critical mass' required to sustain cohesive social networks. Thus:

> Among the subcultures spawned or intensified by urbanism are those that are considered to be either downright deviant by the larger society – such as delinquents, professional criminals, and homosexuals; or to be at least 'odd' – such as artists, missionaries of new religious sects, and intellectuals; or to be breakers of tradition – such as life-style experimenters, radicals and scientists.
>
> (Fischer, 1976)

What is seen as deviancy by the larger society, however, is seen by the members of these subcultural groups as a normal form of activity and part of the group's internal social system.

Subcultural theory does not in itself carry any explicitly spatial connotations but the continued existence of subcultural groups depends to a large extent on avoiding conflict with other groups. Conflict may be avoided by implicit *behavioural* boundaries beyond which groups 'promise' not to trespass: a kind of social contract; but the most effective means of maintaining inter-group tolerance is through *spatial* segregation. This idea makes subcultural theory attractive in explaining spatial variations in deviant behaviour. It has proved useful, for example, in studies of delinquent behaviour.

Subcultural theory also fits in conveniently with the idea of **cultural transmission**, whereby deviant norms are passed from one generation to another within a local environment. This process was identified over 140 years ago by Mayhew (1862) in the 'rookeries' of London, where children were 'born and bred' to the business of crime; and it was given prominence by Shaw and McKay (1942) in their classic study of delinquency in Chicago.

Another concept relevant to the understanding of deviant behaviour within a localized subculture is the so-called **neighbourhood effect**, whereby people tend to conform to what they perceive as local norms in order to gain the respect of their local peer group. Empirical evidence for this phenomenon has been presented in a number of studies. Many aspects of people's behaviour seem to be directly susceptible to a neighbourhood effect. The paradoxical syndrome of 'suburban poverty' in new owner-occupier subdivisions, for example, can be seen as a product of neighbourhood effects that serve to impose middle-class consumption patterns on incoming families, many of whom have incomes that are really insufficient to 'keep up with the Jones's's' but who nevertheless feel obliged to conform with their neighbours' habits.

Structuralist theory

This perspective views the rules of social behaviour and the definitions of deviant behaviour as part of society's **superstructure**, the framework of social and philosophical organization that stems from the economic relationships on which society is based (see Chapter 2). Definitions of deviance, it is argued, protect the interests of the dominant class, thereby helping that class to continue its domination. In modern society, deviant behaviour can be seen as a direct result of stresses associated with the contradictions that are inherent to the operation of the economic system.

One apparent contradiction in this context involves the necessary existence of a **'reserve army'** of surplus labour that is both vulnerable, in the sense of being powerless, but at the same time dangerous, because its members represent a potentially volatile group. The need to maintain this reserve army and to defuse unrest among its members explains the substantial social expenditure of modern welfare states; while the need to control the behaviour of its members explains the rules and definitions attached to many aspects of 'deviant' behaviour associated with the stress of unemployment and the repression and degradation of being supported at a marginal level by the welfare state.

Another important contradiction, it is argued, is that while capital accumulation requires fit and healthy workers, it also tends to debilitate them through the effects of the stresses that result from the various controls that are exerted on the labour force. Examples of these controls include the patterns of socialization that are part of the superstructure of society, in which individuals are rewarded for being competitive but not too individualistic, and in which they are encouraged to spend their rewards on the acquisition of material possessions:

These sources of stress are endemic in the capitalist system, but they are unequally allocated between the classes; workers experience more than their share of the costs or stresses, and less than their share of the benefits. It is no surprise, therefore, that the working classes are disproportionately represented in the prevalence data for mental illness, drug and alcohol abuse, and crime.

(Smith, 1984)

Multifactor explanations: the example of crime and delinquency

The difficulty of reconciling the apparently conflicting evidence relating to these different theories has, inevitably, led to a more flexible approach in which multifactor explanations of deviant behaviour are admitted without being attached to a specific theoretical perspective. This is common to all branches of social deviance research, although it is probably best illustrated in relation to crime and delinquency. Empirical studies of spatial variations in crime and delinquency have lent support, variously, to theories of crowding, social disorganization, anomie, design determinism and deviant subcultures; but it is difficult to assemble evidence in support of any one theory in preference to the rest. In the absence of any alternative all-embracing theoretical perspective, an eclectic multifactor approach thus becomes an attractive framework of explanation.

Data problems

The evidence that can be drawn from studies of spatial variations in crime and delinquency is, like much social geographical research, subject to important qualifications relating to the nature of the data and methods of research that have been employed. It is, therefore, worth noting some of the difficulties and pitfalls involved in such research before going on to illustrate the complexity of interrelationships between environment and behaviour suggested by the results of empirical research.

One of the most fundamental problems concerns the *quality of data*. Most research has to rely on official data derived from law enforcement agencies, and these data are usually far from comprehensive in their coverage. Many offences do not enter official records because they are not notified to the police; and data on offenders are further diluted by the relatively low detection rate for most offences. More disconcerting is the possibility that the data that are recorded do not provide a representative sample. Many researchers have argued that official data are biased against working-class offenders, suggesting that the police are more likely to allow parental sanctions to replace legal sanctions in middle-class areas, that working-class areas are more intensively policed, and that crime reporting by adults is similarly biased.

Conversely, 'white-collar' crimes – fraud, tax evasion, expense account 'fiddles' and so on – tend to be under-reported and are more difficult to detect, even where large amounts of money are involved. Some critics have suggested that this bias has been compounded by the predilection in empirical research for data relating to blue-collar crimes. This may be attributable in part to the differential availability of data on different kinds of offence, but it also seems likely that data on white-collar crimes have been neglected because they are, simply, less amenable to deterministic hypotheses.

Because data for many important crime-related variables are available only for groups of people rather than individuals, many studies have pursued an ecological approach, examining variations in crime between territorial groups. Such an approach is inherently attractive to geographers but it does involve certain limitations and pitfalls. The chief limitation of ecological studies is that they cannot provide conclusive evidence of causal links. Thus, although certain categories of offenders may be found in crowded and/or socially disorganized areas, their criminal behaviour may actually be related to other causes – alienation or personality factors, for example – and the ecological correlation may simply result from their gravitation to a certain kind of neighbourhood.

The chief pitfall associated with ecological studies is the so-called **ecological fallacy**: the mistake of drawing inferences about *individuals* on the basis of correlations calculated for areas. One pertinent example is the frequently encountered association within British cities between crime rates and neighbourhoods containing large numbers of immigrants. The inference drawn by many is that immigrants and their subcultures are

Box 10.1

Key debates in urban social geography – How to understand geographies of childhood and youth cultures

Despite the considerable attention given to juvenile delinquency in studies of crime and disorder in cities, urban geographers have, until recently, paid relatively little to issues of childhood and youth culture. These categories are in any case 'social constructs' that have varied over time. For example, although the existence of 'childhood' might seem to be obvious, remarkably, in Western societies the concept only came to the fore in the fifteenth century. Previously, it appears that all young people, once they were beyond the stage of infantile dependency, were considered to be miniature adults with no special considerations (Skelton and Valentine, 1998). Furthermore, to begin with, it was only the upper classes who could give their children special considerations in the form of education and upbringing. Thus, it was only in the late nineteenth and early twentieth centuries that the concept of childhood really came to the fore, through factors such as the growth on universal education and developmental psychology.

The concept of 'adolescence' is even more recent than childhood, being essentially a twentieth-century notion. It has been argued that with the development of industrial capitalism in the eighteenth century the middle classes began to extend the period of children's schooling to better equip them for the needs of the new economy. This period was further extended during the nineteenth century and it was during this period that there emerged a series of themes that have dominated the discourses surrounding adolescence ever since – rebellion, violence, crime, juvenile delinquency, unruly antisocial behaviour and the need for control. However, the adolescent and 'teenager' really came to the forefront of popular consciousness during the so-called long boom of Fordism in the 1950s and 1960s. The emphasis upon consumption in this period led to the development of a series of goods such as records,

clothing, portable radios, record players and cars – together with services such as clubs and dance halls – that were aimed specifically at young adults in the 16 to 25 age bracket. It was during this period that there emerged a series of 'moral panics' about the lifestyles of various youth subcultures: 'rockers', 'mods', 'hippies', 'punks', 'skinheads', 'ravers', 'yuppies' and so on. The result, which continues into the twenty-first century, is 'the popular imagining of youth as consumption-oriented, into sub-cultural styles based on music and drugs, and free to embark on adventurous travel' (Skelton and Valentine, 1998, p. 1).

As James (1990) notes, adolescence is therefore an 'in-between' or liminal age between childhood and adulthood. As such, adolescence highlights the ways in which notions of ageing are socially constructed. Rather than being just a biological essence, being a teenager is also bound up with particular ways of acting – or performativity as – introduced in Chapter 3. This obviously includes particular styles of dress, types of music, ways of walking and modes of speech. This disjuncture between biological and social definitions of adulthood and social responsibility can lead to conflicts and tensions. Thus we find that teenagers can act in ways that are more or less adult, depending upon the context. This ambiguity is also reflected in the fact that legal definitions of adulthood reflected in stipulations concerning when young people can legally drive a car, drink alcohol, have sexual intercourse, earn money or join the armed forces, vary considerably between different societies.

In keeping with these prevailing discourses, most geographical work on youths has focused upon the problems of juvenile delinquency as discussed in this chapter. A more recent focus has been upon the surveillance strategies used to govern the behaviour of potentially unruly

adolescents in 'public' spaces such as shopping malls. Nevertheless, Western societies tend to place great store on youthful attributes of appearance, physical fitness and athleticism. This contrasts with the attitudes to be found in many Asian societies which tend to venerate old age to a greater extent. This privileging of youth in the West is manifest in advertising. It is therefore difficult to escape the conclusion that current attitudes towards ageing in Western societies are, at least in part, bound up with issues of consumption.

Key concepts associated with childhood and youth cultures (see Glossary)

Alienation, consumption, exclusion, family status, identities, Panopticon, positional good, 'scanscape', social constructionism, spaces of resistance, surveillance.

Further reading

James, S. (1990) Is there a 'place' for children in geography? *Area*, **22**, 278–283

Matthews, H. and Lamb, M. (1999) Defining an agenda for the geography of children *Progress in Human Geography*, **23**, 61–90

Skelton, T. and Valentine, G. (1998) *Cool Places: Geographies of Youth Cultures* Routledge, London

Valentine, G. (1996) Angels and devils: moral landscapes of childhood *Environment and Planning D*, **14**, 581–599

Links with Other Chapters

Chapter 3: The Cultures of Cities; Box 3.1 Hybridization: The Case of Popular Music

Chapter 4: Box 4.2 The Growth of Student Enclaves

Chapter 8: Box 8.4 Rap as Cultural Expression (and Commercialization)

Chapter 10: Box 10.1 The Development of 'Urban Nightscapes'

particularly disposed towards crime and delinquency; but empirical research at the level of the individual has in fact shown that immigrants are very much less involved in the crime and disorder that surround them in the areas where they live than their white neighbours.

The geography of urban crime

Bearing these limitations in mind, what conclusions can be drawn from empirical studies about the factors that precipitate crime and delinquency? First, it is useful to distinguish between factors influencing the pattern of *occurrence* of crime and delinquency and those influencing the pattern of *residence* of offenders.

Most cities exhibit very distinctive areas where the occurrence of crime and delinquency is well above average. In many cities, the pattern conforms to the archetypal distribution identified in Chicago in the 1920s, with low rates in the suburbs increasing steadily to a peak in the inner city and CBD. The most notable exceptions are in European cities, where substantial numbers of low-income, 'problem' households have become localized in suburban public housing estates.

In detail, however, patterns of occurrence vary considerably by the *type of offence.* In his pioneering study of crime in Seattle, Schmid demonstrated the concentration of shoplifting and cheque fraud offences in the CBD, of larceny and burglary in suburban areas, and of robbery and female drunkenness in the 'skid row' area of the city (Schmid, 1960). In a later study of the same city, the dominant pattern of crime occurrence was found to be associated with inner-city areas of low social cohesion, where there was a concentration of burglary, car theft and handbag snatching (Schmid and Schmid, 1972). Studies of other cities have demonstrated a similar general association between the occurrence of crime and poverty, and detailed ecological analyses have revealed a distinct association between low-income neighbourhoods and crimes of violence, including murder, rape and assault. There is also evidence to suggest that transitional areas – with a high proportion of land devoted to manufacturing and wholesaling, a decaying physical environment and an ageing population – are associated with a separate and equally distinctive concentration of offences that include larceny, robbery and car theft as well as assault and murder.

Other important relationships to emerge from empirical studies are the correlation between property crimes – burglary, larceny and car theft – and stable, mid- and upper-income suburban neighbourhoods, and between violent crimes and black neighbourhoods. A *compositional* perspective is useful in interpreting these various findings: the idea here being that communities move through 'life cycles' or 'careers' in their experience of criminality as the demographic composition of their population changes in response to neighbourhood deterioration and family life-cycle changes. Because the peak years for offence rates are the teens and early 20s, neighbourhoods with high proportions of youths of this age can be expected to exhibit high levels of criminality, especially if the neighbourhood is caught in a spiral of economic decline and physical decay that heightens youths' feelings of relative deprivation. The compositional perspective has been developed into what has become known as the *routine activities* theory of crime, in which demographic or social class characteristics lead to certain activity routines that bring together the three prerequisites for crime: the presence of a motivated offender, a suitable target and the absence of a capable guardian.

Spatial variations in *opportunities* for crime have been shown to be critical in studies of occurrence patterns of several different kinds of offences, with a marked relationship between property values and house-breaking offences. Accessibility, visibility, control of property, residential density and state of physical repair are the most significant aspects of the micro-environment of violent crime. Newman's concept of defensible space is clearly relevant here.

In summary, there are qualities attached to the offence location that relate to the *built* environment – its design, detailed land use – and to the social environment – status, local activity patterns, local control systems. Figure 10.3 represents an attempt to capture this, emphasizing sociodemographic composition, routine activities and opportunities.

Patterns of *residence* of offenders are subject to a much wider range of explanatory factors although, like patterns of occurrence, they display a consistent social order and clustering that make them suitable for ecological analysis. Although there are variations by type of offence and age of offender, the classic pattern is the

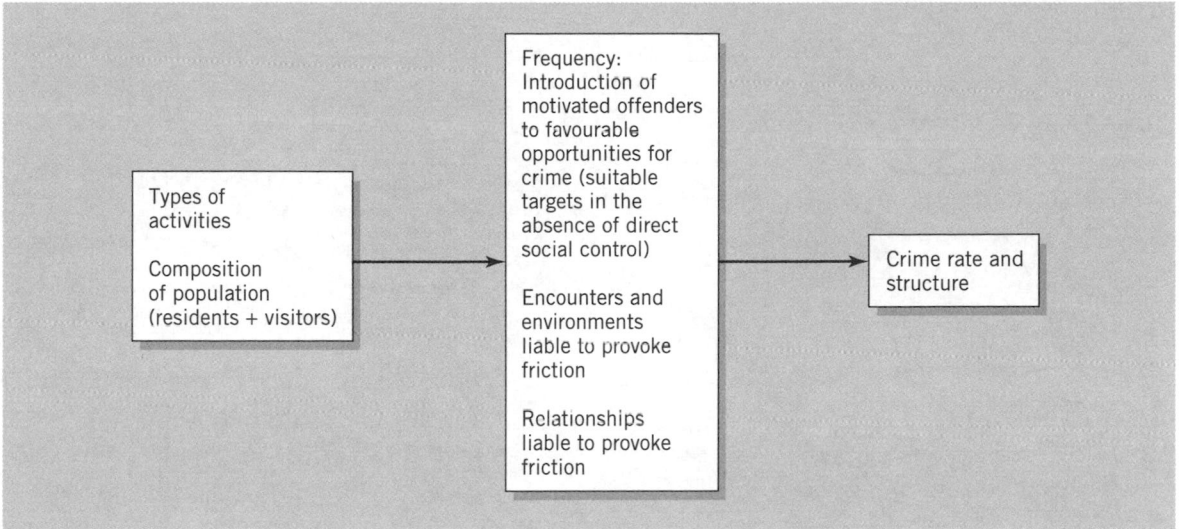

Figure 10.3 Variation in crime (offence rate) and structure in the urban environment.
Source: Bottoms and Wiles (1992), Fig. 1.1.

one described by Shaw and McKay (1942) for Chicago and other American cities: a regular gradient, with low rates in the suburbs and a peak in the inner city. Such gradients have typified not only North American cities, but virtually all Western cities for which evidence is available. Recently, however, departures from this pattern have become more apparent as the spatial structure of the Western city has changed.

Many cities have experienced an outward shift of offenders' residences with changes in residential mobility and housing policies. British studies, in particular, have identified localized clusters of offenders in peripheral local authority housing estates, which suggests that the social environment is at least as important as the physical environment in explaining offenders' patterns.

Most geographical research has set aside the possible influence of personal factors (such as physical and mental make-up) and factors associated with the family, school and workplace in order to concentrate on the social and physical context provided by the neighbourhood. From these studies there is an overwhelming weight of evidence connecting known offenders with inner-city neighbourhoods characterized by crowded and substandard housing, poverty, unemployment and demographic imbalance. In cities where peripheral clusters of offenders are found, there appears to be an additional syndrome linking offenders with public

housing developments containing high proportions of families of particularly low social and economic status, many of whom have been dumped in problem estates through the housing allocation mechanisms of public authorities.

A few studies have followed up this general ecological approach with an examination of the less tangible local factors that may be related to crime and delinquency: the dominant values and attitudes associated with different areas. Susan Smith (1986), for example, argued that the distribution of crime reflects the lifestyle and activity patterns of a community and that the effects of crime, in turn, help to shape these *routine urban behaviours*.

While evidence can be cited in support of particular theories and concepts, empirical research has also demonstrated that it is possible to find support for quite different theories within the same pool of evidence. In this situation it seems sensible to accept a multifactor explanation. David Herbert provided a useful framework within which to subsume the various factors that appear to be involved (Fig. 10.4). Areas of crime and delinquency are linked to several local environmental contexts and generally related to a nexus of social problems. Poverty is the central focus of the model, and is seen as the product of structural factors which, through differential access to educational

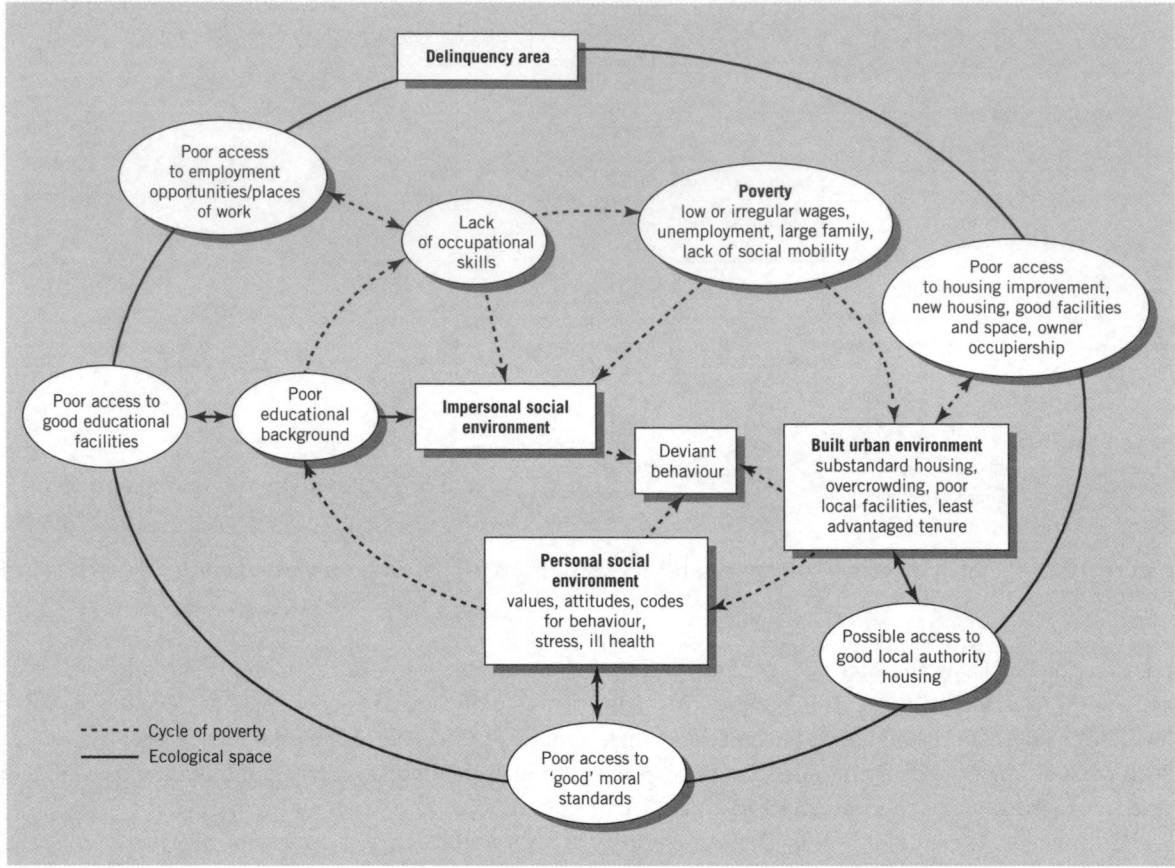

Figure 10.4 Delinquency residence: the cycle of disadvantage and its spatial connotations.
Source: Herbert (1977), p. 277.

facilities and employment opportunities, produce an 'impersonal social environment' (i.e. local population) consisting of 'losers' – the aged, the unemployed, misfits and members of minority groups.

One aspect of crime-related research in which geographers have made significant contributions in recent years is women's fear of violent crime (Pain, 1991; Valentine, 1992; Pawson and Banks, 1993). Such studies have mapped the areas with high rates of reported crime and correlated these with areas in which women claim they are most worried about being victimized, the objective being to evaluate the relationships between fear and risk. We should note that there is a serious methodological problem with such studies in that violent crime, and especially that of a sexual nature, is thought to be seriously underreported, both to the police and to social scientists.

Nevertheless, these studies present something of a paradox. On the one hand women mostly fear strangers in public places. However, studies have shown that women are much more likely to be raped or experience violence at the hands of men whom they know; hence they are much more at risk in their own homes or in semi-private places than in public spaces. However, this does not mean that women's fear of strangers in public spaces is 'irrational'. As Pain (1997) notes, women are flooded with media reporting of violent sexual crime, which tends to exaggerate the risks of certain types of behaviour while ignoring others. And, of course, fear is of major concern even if it is exaggerated. Valentine (1992) argues that women's fear of violent crime is linked to the social construction of space within a patriarchal society. 'Traditional' ideologies about the role of women place them in the home while the

Box 10.2

Key trends in urban social geography – The development of 'urban nightscapes'

The redevelopment of city centres has been associated with what Paul Chatterton and Robert Hollands (2003) describe as 'urban nightscapes' – districts full of activities centred around nocturnal leisure-based consumption and hedonism among young adults (i.e. clubs, pubs and bars). Many have attributed such developments to the changing character of young adulthood. Delayed marriage and child-rearing, prolonged participation in higher education and increasing dependency upon parental accommodation have all led many young adult groups to have increased disposable incomes for drinking and clubbing. In addition, it is argued that these young adults have a desire for new experiences through experiments with new forms of identity in a carnivalesque atmosphere.

However, Chatterton and Hollands locate such developments firmly within a political economy perspective, noting the collusion between local governments, property developers and major corporations. Thus, major brewers and leisure-based corporations have used various branding strategies to entice cash-rich young people into 'cool' venues. As with many aspects of consumption in contemporary Western societies, what was once seen as 'alternative' and transgressive (e.g. self-organized parties and raves) has been incorporated into mainstream culture (see also Chapter 11 on 'gay spaces').

However, 'urban nightscapes' reflect a complex mixture of both deregulation and reregulation. For example, in the UK recent policies aimed at relaxing licensing laws and prohibitions on planning permissions for casinos have led to a 'moral panic'. Many have objected to the violence, petty crime and ill-health that are seen to be associated with excessive alcohol consumption (so-called 'binge drinking') in city centres. Although many lament the difficulty of instigating a European-style 'café culture' on the streets of Britain, the general trend in the United States, Australia and indeed Europe, is for increased surveillance, policing and control of urban nightscapes. It is also important to recall that many young people are excluded from expensive leisure spaces through poverty and unemployment. The redevelopment of 'urban nightscapes' has therefore sometimes led to the demise of older notions of public space, diversity and universal access (see also Box 5.2 on Barcelona and Box 11.3 on Dublin).

Key concepts associated with urban nightscapes (see Glossary)

Commodification, externalities, liminal space, postindustrial cities, 'scanscape', surveillance.

Further reading

Chatterton, P. and Hollands, R. (2003) *Urban Nightscapes: Youth Cultures, Pleasure Spaces and Power* Routledge, London

Clarke, D. (1997) Consumption and the city, modern and postmodern *International Journal of Urban and Regional Research*, **21**, 18–37

Skelton, T. and Valentine, G. (eds) (1998) *Cool Places: Geographies of Youth Cultures* Routledge, London

public sphere is dominated by men. Crime or the fear of crime may be seen as yet another way in which a particular section of society is able to dominate space (as outlined in Chapter 3).

10.2 Cognition and perception

Cognition and perception are associated with images, inner representations, mental maps and schemata that are the result of processes in which personal experiences and values are used to filter the barrage of environmental stimuli to which the brain is subjected, allowing the mind to work with a partial, simplified (and often distorted) version of reality. The same environmental stimuli may evoke different responses from different individuals, with each person effectively living in his or her 'own world'. Nevertheless, it is logical to assume that certain aspects of imagery will be held in common over quite large groups of people because of similarities in their socialization, past experience and present urban environment.

What are these images like? What urban geographies exist within the minds of urbanites, and how do they relate to the objective world? It is possible to give only tentative answers to these questions. It is clear, though, that people do not have a single image or **mental map** that can be consulted or recalled at will. Rather, we appear to possess a series of latent images that are unconsciously operationalized in response to specific behavioural tasks. In this context, a useful distinction can be made between:

▶ the *designative* aspects of people's imagery that relate to the mental or cognitive organization of space necessary to their orientation within the urban environment; and

▶ the *appraisive* aspects of imagery that reflect people's feelings about the environment and that are related to decision-making within the urban environment.

Designative aspects of urban imagery

The seminal work in this field was Kevin Lynch's book *The Image of the City*, published in 1960 and based on the results of lengthy interviews with (very) small samples of middle- and upper-class residents in three cities: Boston, Jersey City and Los Angeles (Lynch, 1960, 1984). In the course of these interviews, respondents were asked to describe the city, to indicate the location of features that were important to them, and to make outline sketches, the intention being to gently tease out a mental map from the subject's consciousness. From an examination of the resultant data, Lynch found that people apparently structure their mental image of the city in terms of five different kinds of elements: *paths* (e.g. streets, transit lines, canals), *edges* (e.g. lakeshores, walls, steep embankments, cliffs) *districts* (e.g. named neighbourhoods or shopping districts), *nodes* (e.g. plazas, squares, busy intersections) and *landmarks* (e.g. prominent buildings, signs, monuments). As Lynch pointed out, none of these elements exists in isolation in people's minds. Districts are structured with nodes, defined by edges, penetrated by paths and sprinkled with landmarks. Elements thus overlap and pierce one another, and some may be psychologically more dominant than others.

Lynch also found that the residents of a given city tend to structure their mental map of the city with the same elements as one another, and he produced ingenious maps with which to demonstrate the collective image of Boston (Fig. 10.5), using symbols of different boldness to indicate the proportion of respondents who had mentioned each element. Another important finding was that, whereas the collective image of Boston was structured by a fairly dense combination of elements, those of Los Angeles and Jersey City were much less complex. Lynch suggested that this reflected a difference in the *legibility* or imageability of the cities resulting from differences in the 'form qualities' of the built environment. These, he argued, include the clarity and simplicity of visible form, the continuity and 'rhythm' of edges and surfaces, the dominance (whether in terms of size, intensity of interest) of one morphological unit over others, and the presence or absence of directional differentiation in terms of asymmetries, gradients and radial features.

Although Lynch's work has been criticized for its intuitive approach to the identification of image elements, and the validity of attempting to aggregate the imagery of people with quite different backgrounds and experience has been questioned, these techniques have found wide application. One consistent finding derives from the differences that exist between the social classes in their images of the city. Basically, middle-class residents tend to hold a more comprehensive image than lower-class residents, covering a much wider territory and including a larger number and greater variety of elements. This is certainly true for Los Angeles, where ethnicity is closely associated with socioeconomic status. The high-status, white residents of Westwood (a 'foothills' neighbourhood situated between Beverly Hills and Santa Monica) have a well-formed, detailed and generalized image of the entire Los Angeles Basin (Fig. 10.6a), whereas the middle-class residents of Northridge (a suburb in the San Fernando Valley) have a less comprehensive image that is oriented away from the city proper (Fig. 10.6b). At the other end of the socioeconomic ladder, residents of the black ghetto neighbourhood of Avalon, near Watts, have a vaguer image of the city which, in contrast to the white images that are structured around the major east–west boulevards and freeways, is dominated by the grid-iron layout of streets between Watts and the city centre (Fig. 10.6c). Reasons for these differences are not hard to find. The greater wealth and extended education of higher-status whites confers a greater mobility, a

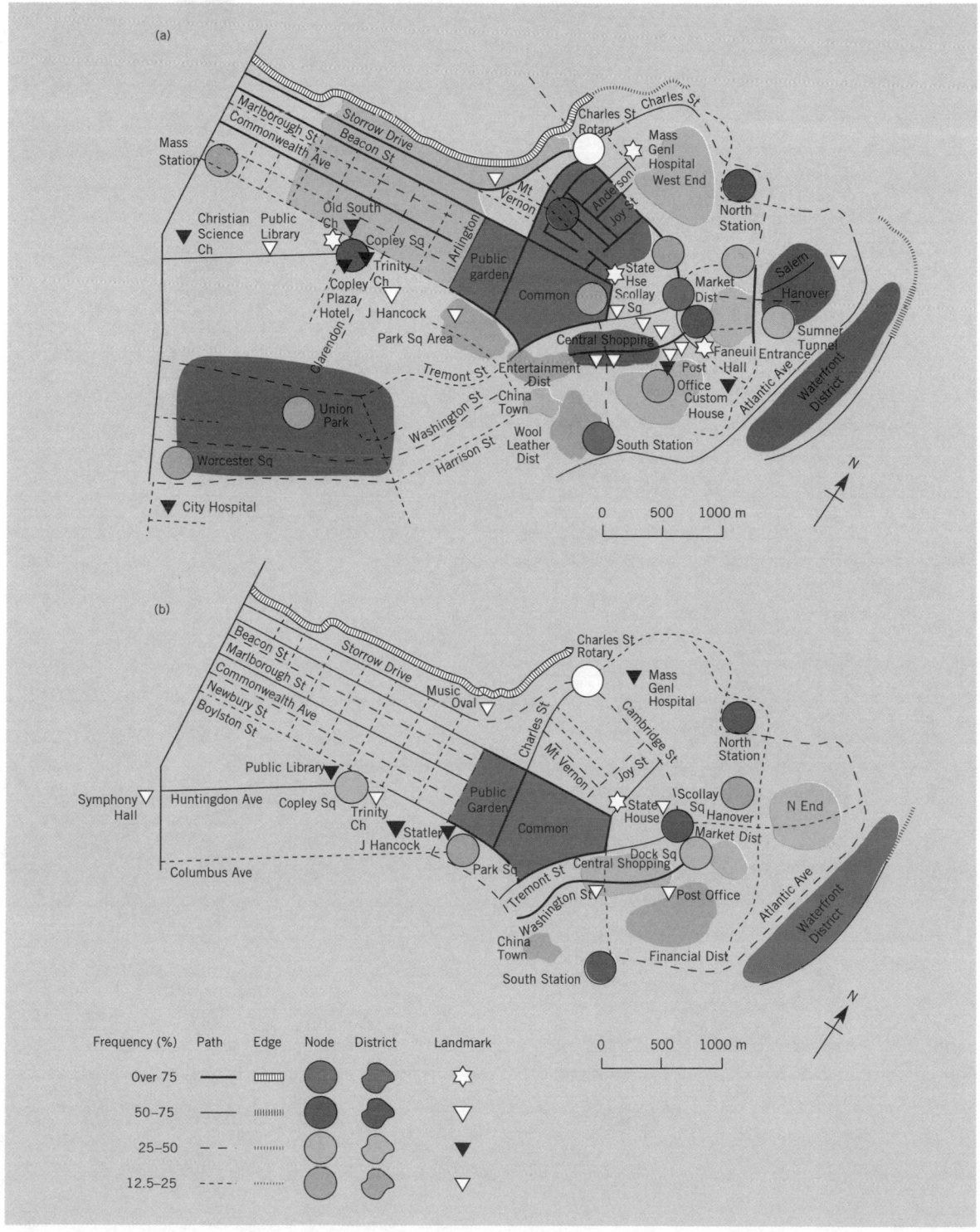

Figure 10.5 Designative images of Boston as derived from: (a) verbal interviewing; (b) sketch maps.
Source: Lynch (1960), p. 146.

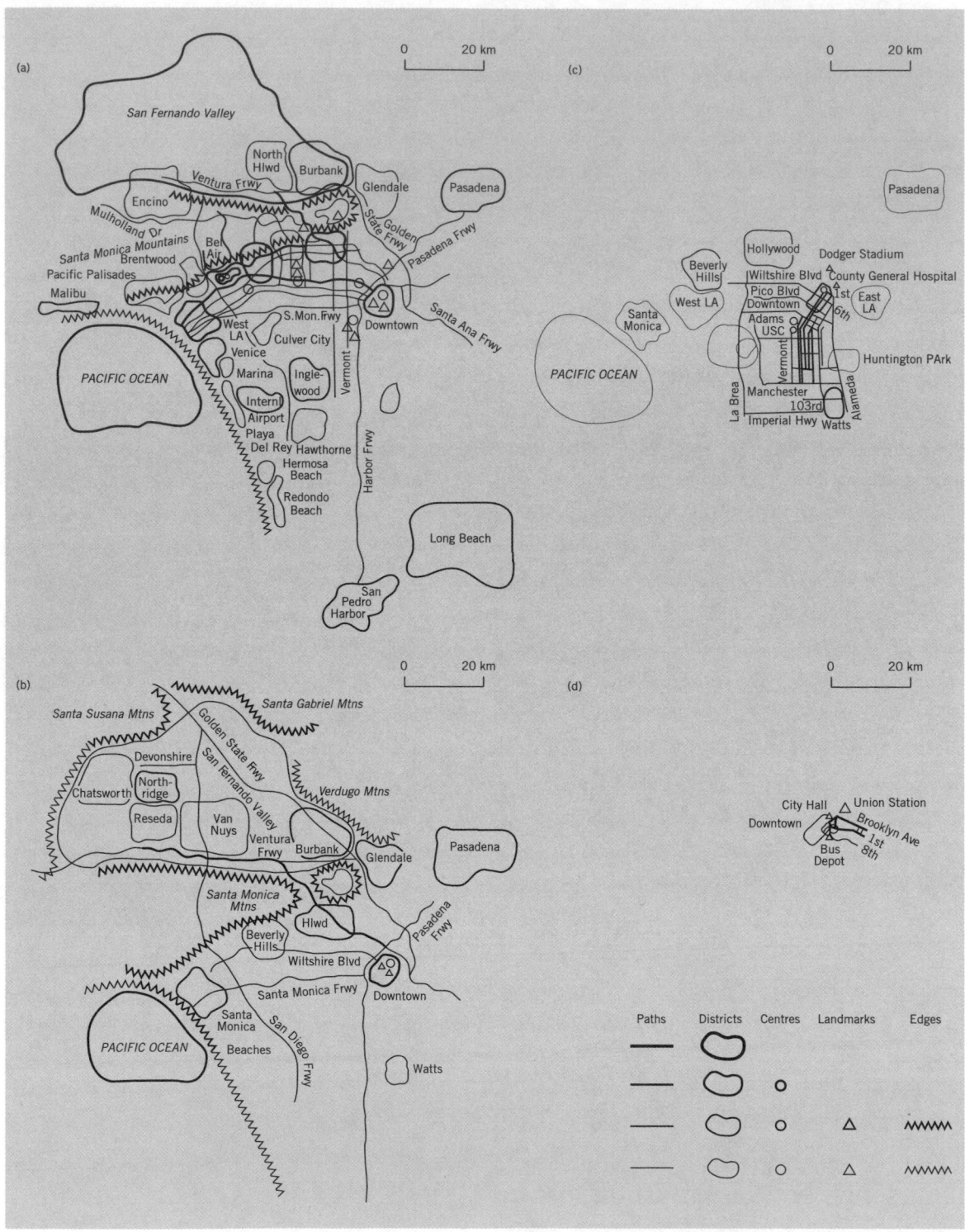

Figure 10.6 Designative images of Los Angeles as seen by residents of (a) Westwood; (b) Northridge; (c) Avalon; (d) Boyle Heights.

Source: Orleans (1973), pp. 120–123.

greater propensity to visit other parts of the city, and a tendency to utilize a wider range of information sources. In contrast, the less mobile poor, with a shorter journey to work, and with less exposure to other sources of environmental information, will naturally tend to have a local rather than a metropolitan orientation: something that will be buttressed by racial or ethnic segregation. Where language barriers further reinforce this introversion, the likely outcome is an extremely restricted image of the city, as in the Spanish-speaking neighbourhood of Boyle Heights (Fig. 10.6d).

Cognitive distance

Underlying the organization of people's mental maps is the **cognitive distance** between image elements, and this is another aspect of imagery that has been shown to exhibit interesting and important regularities. Cognitive distance is the basis for the spatial information stored in cognitive representations of the environment. It is generated from a variety of mechanisms that include the brain's perception of the distance between visible objects, the use-patterns and structure of the visible environment, and the impact of symbolic representations of the environment such as maps and road signs. For the majority of people, intra-urban cognitive distance is generally greater than objective distance, regardless of city size and their usual means of transport, although there is evidence to suggest that this overestimation declines with increasing physical distance.

It has been suggested that people's images and cognitive distance estimates are a function of the number and type of environmental stimuli, or *cues*, they encounter along the paths, or *supports*, that they normally use, and that the actual form of the city is of greater importance in determining the cue selection process than any personal characteristics, including length of residence. It is also suggested that different types of urban structure will result in the selection of different cues, thus generating a different metric of cognitive distance and producing different kinds of mental maps. Residents of concentrically zoned cities might be expected to respond more to changes in land use, for example, than residents of sectorally structured cities, who might be expected to respond more to traffic-related cues along the typical path from suburb to city centre and back.

Given the distorting effect of the values attached to different 'origins' and 'destinations', it seems likely that people possess a basic image of the city consisting of the branching network of their 'action space' that undergoes topological deformation, perhaps hourly, as they move about the city from one major node – home, workplace, city centre – to another. Who, for instance, has not experienced a homeward trip to be shorter than the identical outward journey? The relationship of such a cognitive structure to the more general Lynch-type image of the city has not yet been properly explored, but it seems logical to expect that most people will possess an interlocking hierarchy of images that relates directly to the different geographical scales at which they act out different aspects of their lives.

Appraisive aspects of urban imagery

In many circumstances it is not so much the structural aspects of people's imagery that are important so much as the meaning attached to, or evoked by, the different components of the urban environment in their mental map. Behaviour of all kinds obviously depends not only on *what* people perceive as being *where*, but also on how they *feel* about these different elements. A specific node or district, for example, may be regarded as attractive or repellent, exciting or relaxing, fearsome or reassuring or, more likely, it may evoke a combination of such feelings. These reactions reflect the *appraisive* aspects of urban imagery.

In overall terms, the appraisive imagery of the city is reflected by the desirability or attractiveness of different neighbourhoods as residential locations. This is something that can be measured and aggregated to produce a map of the collective image of the city that can be regarded as a synthesis of all the feelings, positive and negative, that people have about different neighbourhoods.

The cognitive dimensions of the urban environment

Given that people are able to make these overall evaluations of residential desirability, the question arises as to their derivation. In other words, what are the components of people's overall evaluation of a given place or neighbourhood, and how do they feel about these particular aspects of the environment?

There remains a good deal of investigation to be undertaken before the composition of appraisive imagery in cities can be fully understood. There are many facets to the dialectic between places and people's perceptions of them. In addition to this layering of imagery, we must recognize that both people and neighbourhoods are continually changing.

One attempt to come to grips with residents' perceptual responses to change found that *neighbourhood stability* was the dominant cognitive concern (Aitken, 1990) – other, more specific aspects of appraisive imagery have been elicited by researchers pursuing particular themes. David Ley (1974), for instance, illustrated the local geography of perceived danger in an inner-city neighbourhood in Philadelphia, showing how most people recognized – and avoided – the danger points near gang hang-outs, abandoned buildings and places where drugs were peddled. The imagery of fear is often time-dependent: public parks, for example, may be felt to be tranquil and safe places by day but might induce quite different feelings at night. It is also gender-dependent, women being subject to fear of crime and harassment in a much greater range of settings and to a much greater degree than men. This is an important (but under-researched) topic, since the spatial patterns of women's perceptions of risks, of the actual risks they are exposed to, and of their behavioural responses, have implications for their equal participation in society (Smith, 1989a; Pain, 1991).

Another important aspect of appraisive imagery is the way in which some areas of larger cities become *stigmatized*, their inhabitants being labelled as 'work-shy', 'unreliable' or 'troublesome', thus making it difficult for them to compete in local housing and job-markets. Another concerns the role of clothes and personal objects (rather than buildings and social characteristics) in contributing towards our feelings about different parts of the city. Many of our material objects are used, consciously or not, to communicate what we like or believe in: the pair of shoes, the book, the wall poster and the cut of a pair of jeans become briefly exhibited signs and badges that not only help their owners to say something about themselves but that also help others to attach meaning and significance to their owners and to *their owners' environment.*

Images of the home area

Just as individual personality is reflected in home and possessions, so collective personality and values are translated into the wider environment of cultural landscapes. The existence of such relationships between places and people leads to the idea of a 'sense of place', which incorporates aspects of imageability, the symbolic meaning of places, and 'topophilia' – the affective bond between people and place (Tuan, 1974).

In the specific context of urban social geography, the most important aspect of this sense of place is probably the attachment people feel to their *home area*. There is no doubt that the immediate physical and social environment is crucially important in the early psychological and social development of the individual, and it seems that this generates a strong bond – often amounting almost to reverence – for the territorial homeland: a phenomenon that Yi-Fu Tuan (1976) called 'geopiety'. Such feelings are clearly related to the idea of territoriality, and there is plenty of evidence to suggest that they exist as a kind of latent 'neighbourhood attachment' in most people who have lived in a particular area for any length of time. The most striking evidence of such feelings emerges after people have been forced to leave their home neighbourhood in the cause of redevelopment or renewal schemes, when many report feelings of grief at the loss of their old neighbourhood. People's home area seems to be closely related to their 'activity space' around the home. Here is one person's description of his own 'home area':

> The Greater London Council [was] responsible for a sprawl shaped like a rugby ball about twenty-five miles long and twenty miles wide [40 × 32 km]; my city is a concise kidney-shaped patch within that space, in which no point is more than about seven miles from any other. On the south, it is bounded by the river, on the north by the fat tongue of Hampstead Heath and Highgate Village, on the west by Brompton cemetery and on the east by Liverpool Street station.

I hardly ever trespass beyond those limits and when I do I feel I'm in foreign territory, a landscape of hazard and rumour. Kilburn, on the far side of my northern and western boundaries, I imagine to be inhabited by vicious drunken Irishmen; Hackney and Dalston by crooked car dealers with pencil moustaches and goldfilled teeth; London south of the Thames still seems impossibly illogical and contingent, a territory of meaningless circles, incomprehensible one-way systems, warehouses and cage-bird shops. Like any tribesman hedging himself in a stockade of taboos, I mark my boundaries with graveyards, terminal transportation points and wildernesses. Beyond them, nothing is to be trusted and anything might happen.

The constrictedness of this private city-within-a-city has the character of a self-fulfilling prophecy. Its boundaries, originally arrived at by chance and usage, grow more not less real the longer I live in London. I have friends who live in Clapham, only three miles away, but to visit them is a definite journey, for it involves crossing the river. I can, though, drop in on friends in Islington, twice as far away as Clapham, since it is within what I feel to be my own territory.

(Raban, 1975, pp. 166–167)

Box 10.3

Key debates in urban social geography – The first and second nature of cities

This book is primarily concerned with the internal structure of cities, what the eminent political geographer Peter Taylor (2004) has called their 'first nature'. However, as Taylor goes on to point out, cities also have a 'second nature', the interconnections between them. In the past 50 years geographers have somewhat neglected this second aspect. One of the likely reasons for this neglect is the fact that this second nature is rather complex and difficult to analyze, consisting of complex exchanges of trade, money, peoples and ideas. Another probable reason is that our thinking about society is heavily influenced by notions of the nation state. This has led to a focus upon connections between a supposed national hierarchy of urban centres to the neglect of connections between cities in different states throughout the globe.

Taylor points out that for many centuries, before nation states became such dominant entities, there were extensive links between cities around the world. Most notably from the twelfth century onwards, Venice was linked via complex trade routes with Constantinople, Samarkand and on to Beijing and other Asian centres. Another important example is the extensive trade linkages between the scores of cities in northern Europe that flourished from the twelfth to the seventeenth centuries in what became known as the 'Hanseatic League'.

According to Taylor, one consequence of this neglect of city interconnections is a serious question mark over the 'world city' hypothesis (see also Short et al., 1996). Thus, Sassen's (2001) highly influential ideas concerning the strategic role of what she calls 'global cities' as strategic command centres for business and financial services, though intuitively plausible, are not substantiated by a great deal of hard evidence. Taylor's own quantitative analysis of linkages between business services in many cities reveals complex networks that form what he calls a 'hinterworld'.

Key concepts related to the first and second nature of cities (see Glossary)

Global cities, spaces of flows, world cities.

Further reading

Sassen, S. (2001) *The Global City* (second edition) Princeton University Press, Princeton, NJ

Short, J.R., Kim, Y., Kuss, M. and Wells, H. (1996) The dirty little secret of world city research *International Journal of Urban and Regional Research*, **20**, 697–717

Taylor, P. (2004) *World City Network; A Global Urban Analysis* Routledge, London

Links with Other Chapters

Chapter 1: Box 1.3 Growing Cosmopolitanism in Western Cities: The Example of London

Chapter 2: Box 2.2 Manuel Castells; Globalization: Knowledge Economies and the Informational City

Chapter 12: Box 12.3 The Rise of 'Transnational Urbanism'

Chapter summary

10.1 The geographical distribution of the sites of reported crime and the location of apprehended offenders display distinct patterns in cities. Many theories have been suggested to explain these patterns but the need for a multifactor explanation is paramount.

10.2 People form mental images or representations of cities based upon key legible features such as key landmarks and transport networks. These images influence the people's views about the desirability of different areas within cities.

Key concepts and terms

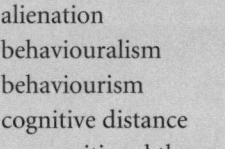

alienation	defensible space	neighbourhood effect
behaviouralism	determinist theory	'psychological overload'
behaviourism	deviant subgroup	reserve army
cognitive distance	ecological fallacy	subcultural theory
compositional theory	environmental conditioning	superstructure
cultural transmission	mental map	territoriality

Suggested reading

A good place to get started with an introduction to behavioural geography is D.J. Walmsley's *Urban Living: The Individual in the City* (1988: Longman, Harlow). Also useful as introductory reviews are John Gold's *An Introduction to Behavioural Geography* (1980: Oxford University Press, Oxford) and Reg Golledge and R. Stimson's *Analytical Behavioural Geography* (1987: Croom Helm, London). A review of geographical approaches to social deviance in the city is provided by David Herbert in his essay in Mike Pacione's edited volume, *Social Geography: Progress and Prospect* (1987: Croom Helm, London); while Pacione himself reviews a wide range of behavioural literature in his essay on urban liveability (*Urban Geography*, 1990: **11**, 1–30). The relationships between urban settings and criminality are dealt with in several accessible sources: David Evans and David Herbert (eds), *The Geography of Crime* (1989: Routledge,

London), Susan Smith, *Crime, Space, and Society* (1986: Cambridge University Press, Cambridge), H. Jones (ed.) *Crime and the Urban Environment* (1993: Avebury, Aldershot), Ralph Taylor's chapter on urban communities and crime in *Urban Life in Transition* (1991: M. Gottdiener and C. Pickvance (eds), Sage, Newbury Park) and Meagan E. Cahill and Gordon F. Mulligan 'The determinants of crime in Tucson, Arizona' (*Urban Geography*, 2003: **24**, 582–610). Anthony Bottoms and Paul Wiles provide a structurationist perspective on crime and place that is useful in bridging the behavioural emphasis of this chapter with the material covered in the previous two chapters: their essay on 'Explanations of crime and place' is in *Crime, Policing and Place*, edited by David Evans, Nicholas Fyfe and David Herbert (1992: Routledge, London, pp. 11–35). On mental maps, perception and the cognitive dimensions of urban settings, see Douglas Pocock and Ray Hudson's *Images of the Urban Environment* (1978: Macmillan, London).

Chapter 11

Bodies, sexuality and the city

Key questions addressed in this chapter

➤ Why have scholars become interested in relationships between bodies and the city?

➤ What is meant by sexuality?

➤ In what ways have cities influenced, and in turn have been influenced by, sexuality?

➤ In what ways do cities oppress disabled people?

The 'cultural turn' (introduced in Chapter 3) has brought about considerable interest in the role of people's bodies in contemporary life, especially in the city. One reason for this is that (as noted in Chapter 1) the body has frequently been used as a metaphor to describe cities. Thus, notions of 'circulation' and references to 'arteries' and 'nerve centres' frequent descriptions of urban transport systems. In addition, allusions to disease and social pathology underpin many writings about urban problems. More importantly, however, the body is also of interest because it is an important signifier of cultures in city spaces. Hence, people's bodily appearance and dress provide important signals about culture and social values.

Bodies come in many shapes and sizes but there are strong social pressures for people to conform to certain standards of appearance and the associated social values that accompany these forms. These pressures are reinforced by powerful images in films, television, advertising and magazines. Typically in contemporary Western societies these images stress relatively slim young women and relatively muscular young men (aided of course by computer-enhanced photos that can eliminate blemishes!). Not only are extremes of weight and stature avoided, but these images also stress heterosexuality and the absence of disability. These dominant images are highly specific in time and place; for example, Hollywood films from the early part of the twentieth century show a preference for a rather fuller female figure, and in the 1960s men wore long hair in a manner that would have been considered effeminate only a few years before.

The term **corporeality** is used to indicate the ways in which these body images are not just the result of biological differences between people but are socially constructed through various signs and systems of meaning. Elizabeth Grosz has argued that 'the city is one the crucial factors in the social production of (sexed) corporeality' (Grosz, 1992, p. 242). An important task for geographers therefore is to examine '. . . the ways in which bodies are physically, sexually and discursively or representationally produced, and the ways in turn, that bodies reinscribe and project themselves onto their sociocultural environment so that the environment both produces and reflects the form and interests of the body' (Grosz, 1992, p. 242).

Cultural theorists stress the unstable and malleable character of bodily identities. Gender, sexuality, cultural and other physical differences between people are not natural entities but 'cultural performances' related to particular spaces. This does not mean that bodily appearances are just a question of superficial external image; they also reflect people's interior sense of themselves which (as we saw in Chapter 3) is constituted by a whole host of historical and geographical factors. Once again, then, space is a crucial factor in social processes. The ways in which identities are socially constructed through particular ways of acting (and not as the result of some biological essence) is sometimes termed **performativity**. The processes through which the body is socially constructed in spaces by wider systems of meaning is also known as **embodiment**.

This work on embodiment represents another attempt to avoid the mind/body split that is evident in Cartesian reasoning about the world. As we saw in Chapter 1, this detached view is based on the mistaken assumption that the writer '. . . has no particular history, is member of no community, *has no body*' (Young, 1978, cited in Duncan, 1996, emphasis added). In contrast, the concept of **embodied knowledge** (also termed local or situated knowledge) explicitly acknowledges that ideas, concepts and analyses emerge from people in specific places at particular times.

11.1 Gender, heteropatriarchy and the city

As we have indicated at various stages in this book, cities are social, as well as physical, constructs. Social control within spaces is exercised through expected patterns of behaviour and the exclusion of groups who transgress (or who are *expected* to transgress) these codes of behaviour. One such code of behaviour relates to public displays of physical affection; while some may be embarrassed by people of *different* sex kissing in public, this tends not to arouse the same degree of discomfort – or even antipathy or outrage – as when people of the *same* sex show mutual attraction in public (Valentine, 1996).

Bodily appearances through factors such as dress, age, ethnicity and the like provide important signifiers about whether people meet these behavioural expectations. One of the most important components of these bodily identities is gender (even if on some occasion this can be difficult to determine!). In social research the term **gender** is typically used to refer to social, psychological or cultural differences between men and women, rather than biological differences of sex. The assumption behind this distinction between gender and sex is that the way we act is primarily the result of socially created and ascribed **gender roles** rather than the product of innate biology. These gender roles are also manifest in bodily forms, appearance and behaviour. What constitutes appropriate feminine and masculine behaviour and dress has changed enormously over the years – one needs only to think of the elaborate ruffs worn by men in Europe in the Elizabethan Age or the huge male wigs worn by aristocrats in the seventeenth century.

The crucial point is that over the years there have been socially constructed differences in the appearance and behaviour of men and women. Furthermore, these differences have been constructed as sexually alluring (however *non-alluring* some of these clothing styles may appear to subsequent generations!). Thus Liz Bondi (1998a) notes that although it has proved useful in social enquiry to make a distinction between sex and gender, the two are concepts are connected since gender roles imply notions of appropriate sexual behaviour. Yet, by assuming that gender roles are socially created and that the rest is simply biological, until recently, research has tended to neglect of issues of sexuality as a cultural practice (see below).

Differences between men and women are not simply a function of the way the different sexes 'perform'. As we have seen throughout previous chapters, they are also bound up with legal structures and the allocation of resources. There is, then, a material base to the

The 'buzz' of the city: information exchange, both face-to-face and technologically mediated (the Piazza, Canary Wharf). Photo Credit: Paul Knox.

discourses surrounding what it is to be 'male' and 'female'. Furthermore, it is now widely acknowledged that this system of power relations has universally worked to the advantage of men. The broad system of social arrangements and institutional structures that enable men to dominate women is generally known as **patriarchy**. Since patriarchy is dominated by heterosexual values, this is also termed **heteropatriarchy**. Consequently, areas of cities dominated by these values are often termed **heteropatriarchal environments**. As always, we should recognize the diversity of the city at this point; the categories 'man' and 'woman' are cut across by divisions of class, age, ethnicity, religion, physical capacity and so on.

Gender roles in the sociospatial dialectic

Cities both create and reflect these gender roles. They reflect the system of patriarchy and, above all, are hetero-

patriarchal environments. For example, in the sphere of formal paid work, McDowell (1995) showed how working as a merchant banker in the City of London involves performing in a gendered and embodied way. This is a tough, aggressive, male-dominated business environment. Women are accepted up to a point, provided they can perform well at their jobs, but through numerous jibes, jokes and ironic comments they are continually reminded that they are 'the other' – interlopers in an essentially male environment. Women are therefore forced to act as 'honorary males' and adopt a masculinized form of identity. This involves working long hours, shouting down the phone if necessary to achieve ambitious performance targets and enjoying an ostentatious, consumption-intensive lifestyle (when time permits). Although avoiding the uniform dark business suit that is obligatory for male merchant bankers, women dealers need to dress in such a way as to distinguish themselves from female secretaries, while at the same time not appearing either too sexually

Box 11.1

Key thinkers in urban social geography – Peter Jackson

Peter Jackson (the geographer not the director of the *Lord of the Rings* trilogy!) has had a big influence on urban social geography in the past two decades for at least three main reasons.

First, Jackson was one of a group of geographers who, in the 1980s, re-interpreted the role of the Chicago School of urban ecology. Previously, in the 1960s, urban geographers had drawn inspiration from the quantita-tive and statistical work undertaken by the School, but Jackson (and co-author Susan Smith) drew attention to the important ethnographic tradi-tion of Park and his associates (Jackson and Smith, 1984).

Second, Jackson intervened in debates on identities, drawing atten-tion to the fact that these were often quite restricted in scope. Thus debates on gender should involve consideration of masculine identities as well as feminine ones; discussion of sexuality should involve hetero-sexuals and well as gay men; and similarly, debates on ethnic minorities should also consider the neglected identities of 'whiteness' (Jackson, 1989, 1991b).

Third, Jackson was one of a group of geographers (including the enorm-ously prolific Nigel Thrift, see Box 9.4) who argued that geographers should pay more attention to the roles of con-sumption in shaping identities and city life (Miller *et al.*, 1998).

In sum, Jackson's work represents a reassertion of the importance of cul-tural geography, but one that is dif-ferent from the traditional form. This 'new' cultural geography is, among many things, concerned to under-stand the social construction of many accepted and taken-for-grant cate-gories used in everyday social dis-course. It is concerned to understand 'cultural politics', the idea that cul-ture is not simply concerned with issues of aesthetic style or taste, but involves access to power and social resources (see Mitchell, 2004).

Key concepts associated with Peter Jackson (see Glossary)

Anti-essentialism, cultural politics, identities, identity politics, racism, sexuality, social constructionism.

Further reading

Jackson, P. (1989) *Maps of Meaning* Unwin Hyman, London

Jackson, P. (1991) The cultural politics of masculinity: towards a social geography *Transactions of the Institute of British Geographers*, **16**, 199–213

Jackson, P. and Smith, S. (1984) *Exploring Social Geography* Allen and Unwin, London

Miller, D., Jackson, P., Thrift, N., Holbrook, B. and Rowlands, M. (1998) *Shopping, Place and Identity* Routledge, London

Mitchell, D. (2004) Peter Jackson, in P. Hubbard, R. Kitchin and G. Valentine (eds) *Key Thinkers on Space and Place* Sage, London

Links with Other Chapters

Chapter 4: Qualitative Methods and Observational Fieldwork in Urban Areas

Chapter 7: Box 7.3 Robert Park

Chapter 8: Social Closure, Racism and Discrimination

Chapter 9: Place, Consumption and Cultural Politics

---

alluring or too masculine. Interestingly, this is also an environment characterized by heterosexual values and homophobia (see section on sexuality below). In a similar vein there are various types of gender culture in UK public sector organizations that serve to restrict the opportunities open to women (see Table 11.1).

Another example can be taken from certain city streets; these are sometimes dominated by gangs of young men displaying aggressive masculine attitudes. Women passing through such an environment may be subject to leering eyes, snide remarks or overt sexist comments. Such fears are especially pronounced at night – hence there have been attempts by women's organizations to reassert their right to occupy city spaces through 'Take Back the Night' campaigns. Although some men may also feel uncomfortable in many areas of the city, in general, they do not feel the same degree of threat, fear or sense of exclusion (despite the fact that according to research young men are the most likely to be caught up in random acts of violence).

The case of the street gang may be an extreme one but practically *all* urban spaces tend to be dominated by masculine heterosexual norms. Hence, the public space of cities has typically been seen as the domain of

Table 11.1 Types of gender culture in British public sector organizations

Type of culture	Characteristics
'Gentlemen's club'	➤ Woman's role seen as homemaker and mother ➤ Man's role seen as breadwinner ➤ Polite and welcoming to women who conform ➤ Women perform a caring/servicing role ➤ Ignores diversity and difference
'Barrack yard'	➤ Hierarchical organization ➤ Bullying culture ➤ No access to training and development ➤ Clear views on positions within the organization
'Locker room'	➤ Exclusive culture ➤ Relationships built on outside sporting and social activities ➤ Participation in sporting activities important to culture of organization
'Gender blind'	➤ Acknowledges no difference between men and women ➤ Ignores social and cultural diversity ➤ Separation of work from home and life experience ➤ Women as perfect mother and super manager ➤ Denies existence of reasons for disadvantage
'Smart machos'	➤ Economic efficiency at all costs ➤ Preoccupation with targets and budgets ➤ Competitive ➤ Ruthless treatment of individuals who cannot meet targets
'Paying lip service'	➤ Feminist pretenders ➤ Equal opportunities rhetoric but little reality ➤ Espouses view that *all* women make good managers ➤ Women can empathize
'Women as gatekeepers'	➤ Blocks come from women ➤ Division between home-oriented women and career women ➤ Large numbers of women in support roles ➤ Few women in senior positions ➤ Women have a sense of place ➤ Belief in patriarchal order ➤ Pressure on women in senior positions

Source: adapted from Maddock and Parkin (1994) cited in Booth *et al.* (1996), p. 176.

men while the domestic suburban sphere has been seen as the province of women. However, in recent years feminist commentators have pointed out that the private domestic sphere is also dominated by patriarchy and masculine values (Bondi, 1992). Legally enshrined as an essentially private space, state regulation of domestic spheres has been relatively limited such that gender inequalities are replicated in such environments. For example, despite much media speculation about the existence of the 'caring and sharing' 'New Man' who has a career but also takes a full share of domestic chores, there is now overwhelming social science evidence that women have a double-burden of domestic and formal labour market obligations (Pinch and Storey, 1992; Hanson and Pratt, 1995).

At a more extreme level, patriarchy is manifest in domestic violence; for example, according to the Surgeon General in the United States, the battering of women by domestic partners is the single largest cause of injury to women (Duncan, 1996). Thus in the United States roughly 6 million women are abused and 4000 women are killed by their partners or ex-partners each year. While the extent of these problems should not be under-estimated, it is also important to recognize that there has been change in recent decades. Thanks in no small part to social research, society is now much more aware of issues such as sexual harassment, rape and physical abuse, and advances in policy have been made. Thus most cities now have rape crisis helplines, women's centres, battered-women's shelters and well-women clinics. In addition, issues of access to child-care and equal pay have moved up the political agenda, even if progress in these spheres has not met up with the rhetoric of politicians. We should also recognize that cities have been centres of emancipation for women as well as imprisonment. Wilson (1991) argues that in the twentieth century cities enabled many women to come together to form feminist associations to fight patriarchal systems of oppression.

11.2 Sexuality and the city

Human sexual behaviour has been profoundly shaped by the nature of cities while at the same time, people's sexual activity has had a major influence upon the structure of those cities. A principal reason for this aspect of the sociospatial dialectic is that sexual activity is not just a primitive biological urge but is also a form of learned behaviour that is profoundly affected by cultural values. It is for this reason that we find such wide variations in sexual practices and attitudes towards sex in different cultures throughout the world. The term **sexuality** refers to ideas about sex. It therefore involves not only the character of sexual practices but also their social meanings. Implicit within the use of the term sexuality therefore is an acknowledgement of this socially constructed and culturally determined character of human sexual behaviour.

For most of the past 2000 years attitudes towards sex in Western societies have been primarily shaped by Christianity. The dominant view has been that sexual activity should only take place within marriage and primarily for the production of children. Sexual activity for vicarious pleasure outside of marriage was regarded as sinful. Other categories of sexual activity classed as deviant, abnormal, immoral, unnatural or sick under this moral code include: homosexuality, transvestitism, fetishism, sadomasochism, those engaging in cross-generational sex and those engaging in sex for money. However, in recent years an increasing number of minorities with sexual preferences that transgress this code have been struggling to achieve public recognition and legitimacy.

Compared with the vast amount of work on class, gender and ethnicity, geographers have traditionally paid little attention to issues of sexuality. In this respect geography has been no different from other social sciences. Indeed, Bell (1991) has argued that sexuality is 'the last area of discrimination'.

Prostitution and the city

Cities have often provided opportunities to transgress prevailing moral codes and one important manifestation of this has been prostitution – the granting of sexual favours in exchange for monetary reward (usually but not exclusively by women for men). This is often termed the world's 'oldest profession' but the actual term 'prostitute' came into prominence only in the eighteenth century. In older societies, sexual favours outside marriage were often granted by women who were courtesans, mistresses or slaves. The crucial point is that these women were often known to those procuring the sexual favours. With the development of large cities, prostitution changed in character in that the women and their clients are frequently unknown to each other. The reason for this change is fairly obvious: in small-scale agrarian societies people were much more likely to be familiar with each other, whereas in cities there was a much greater chance of anonymity. In addition, the economic destitution brought about by the early industrial cities meant that prostitution was the only effective means whereby some women could earn any income.

Urbanization and prostitution

Although precise figures are impossible to obtain, there is no doubt that prostitution was rife in many nineteenth-century cities. Ackroyd notes that in the Victorian period the number of female prostitutes in London was the source of endless speculation – with estimates ranging up to 90 000 (Ackroyd, 2000; see also Perkin, 1993). Most did their 'business' walking on the streets, often soliciting trade in an explicit and sometimes verbally aggressive way. Indeed, so offended was the author Charles Dickens by the behaviour of one of these women that he had her arrested for using indecent language. Prostitutes were used by men of all social classes, and shops in the Strand and Haymarket area of London advertised 'beds to let', often for limited periods of a few hours.

Henry Mayhew in his survey of social conditions in London in the 1850s divided prostitutes into six groups who frequented different parts of London (although he omitted upper-class courtesans). First, there were the 'kept mistresses' and 'prima donnas'. Kept mistresses were widespread but especially concentrated in the St Johns Wood area. Prima donnas were of lower rank and frequented the smart shopping area known as the Burlington Arcade, as well as fashionable parks, theatres and concert halls. Second, there were women who lived together in well-kept lodging-houses – they clustered in the Haymarket area. The third group of women living in low lodging-houses was concentrated down in the poor East End of London. As might be expected, the fourth group of 'sailors' women' frequented public houses in the dockland areas, such as Whitechapel and Spitalfields. A fifth group of women inhabited the park areas, while the final group of 'thieves' women' were especially concentrated in the Covent Garden area.

Despite the continual danger of physical assault from their clients and the very high risk of contracting sexually transmitted diseases (it was estimated that in London 8000 would die each year), prostitutes could earn far more on the streets than they could through work in the poorly paid industries of the time (see Chapter 2). There was, however, a double standard in operation. While prostitutes were frequently persecuted, the men who used them were not. A crucial point about prostitution, therefore, is that it reflects patriarchal gender relations (i.e. the inequalities in power relations between men and women) (Bondi, 1998b).

Sex workers in contemporary cities

Prostitution has continued to be a source of much conflict in Western cities in the twentieth century. In the United Kingdom for example, many 'red light' zones have been established in run-down inner-city environments. The residential structure of these areas

is frequently dominated by ethnic minorities who have been deeply offended by the sight of women seeking 'custom' on the streets. This has led to campaigns to expose 'curb crawlers/cruisers' (i.e. those who drive through these residential areas looking for prostitutes). Such campaigns can involve cooperation between the police and local community groups, although often the consequence is that prostitution often gets displaced elsewhere in the city.

It seems clear that the majority of women are coerced into prostitution through economic disadvantage, and often experience considerable physical and psychological harm from both their pimps and clients. However, as Duncan (1996) notes, there is a small proportion of 'sex workers' who challenge the broad moral condemnation of their trade by society and who seek their right to exercise some degree of choice over their lives in a way they think fit.

We should recognize at this point that the legal and social restrictions surrounding prostitution are still full of hypocrisy and contradiction. In the United Kingdom, for example, while prostitution is legal in principle, prohibitions on soliciting on the streets or living off 'immoral earnings' place heavy restrictions on the activity. As Duncan (1996) notes, widespread condemnations of prostitutes as being exploited by men or suffering from 'false consciousness' exclude prostitutes from the freedom to control their own bodies in safe conditions free from police harassment. It is little wonder then that the discourses surrounding prostitution also display ambiguity and contradiction; either portraying women as victims of sex-hungry predatory males or autonomous providers of social services. In either case there are firm discourses linking sexuality and gender (Bondi, 1998b). Perhaps what makes prostitution so threatening in this context is that it challenges the public/masculine, private/feminine dualisms that structure the city.

Homosexuality and the city

Homosexuality involves emotional and sexual attraction between people of the same sex. Homosexuality has existed in all cultures. For example, Ackroyd (2000) notes that there was a thriving homosexual community of 'sodomites' in medieval London centred around brothels. However, it was only in the mid-nineteenth century that the term homosexuality was devised, thereby denoting homosexuals as a distinct and separate section of the population. It has been argued that the term 'homosexuality' is so specific to this time when social scientists were bent on classifying human sexual activity into discrete categories that it is inappropriate to apply this term to analyze attitudes towards same-sex attraction in older societies (Bristow, 1997). Furthermore, the association of the term with persecution in the twentieth century has led many to prefer alternative terms such as 'gay' or '**queer**'.

Homosexuality has remained an offence in most Western societies until recently. In the United Kingdom it was only after the Sexual Offences Act of 1967 that homosexual activity was decriminalized, provided the activity took place in private between consenting adults over the age of 21. The age of consent for homosexuality in the United Kingdom was reduced to 18 in 1994 and 16 in 1998. Nevertheless, persecution continues; in 1988 two men were arrested for kissing at a bus stop in Oxford Street, London on the grounds that it incited 'public offence' (Jeffery-Poulter, 1991). Knopp's (1987) study of the gay community in Minneapolis showed that between 3500 and 5000 men were arrested in the city between 1980 and 1984 on the charge of 'indecent conduct'.

The social construction of sexuality

There is much controversy over the extent to which homosexuality is some innate biological imperative or a socially constructed phenomenon. Psychoanalysts including Freud did much to highlight the ways in which sexual behaviour was not just related to the need for biological reproduction. However, the scholar who did most to draw attention to the socially constructed nature of sexuality – including homosexuality – was Michel Foucault (1984). His ideas about sexuality must be seen in the light of his broader concerns with the ways in which social institutions such as schools and prisons produced ways of controlling people. As noted in Chapter 3, he argued that central to this control are the discourses – sets of shared understandings – that dominate these institutions.

Discourses regarding sexuality were therefore part of these ways of controlling desire. In the context of sexuality, Foucault uses notion of discourse to understand the way in which the term 'homosexuality' was devised to denote a form of social disease. Although this was intended to delineate a group who should be controlled, by giving the group a name, it raised the consciousness of a minority of the need for emancipation. Foucault's ideas have been much criticized, especially for their neglect of gender and race, but they have been extraordinarily influential in the study of sexuality.

Homosexual urban ecology

Cities have had a profound impact upon the development of homosexuality. In general, cities have provided greater anonymity and tolerance of alternative lifestyles compared with the hostility towards gays and lesbians manifest in rural communities, especially where fundamentalist views are dominant. Nevertheless, the continuing danger of prosecution and police harassment means that within cities homosexuals have tended to meet in secret bars and clubs (Chauncey, 1995). These bars serve three main functions; first they facilitate liaisons for sexual purposes; second, they enable the exchange of news and gossip; and third, they provide a place of introduction for new entrants into the gay world (Hooker, 1965; Wright, 1999). It has been found that these bars tend to cluster together in areas that display a high tolerance and permissiveness towards deviant forms of behaviour in general. However, some of these clubs have special warning lights or bells to warn customers not to stand too close to persons of the same sex when the police were in the vicinity.

Other studies have highlighted the requirements of successful 'tearooms' or 'cottages' – public toilets and areas of parkland that facilitated cruising and sexual liaisons: first, these sites had to be in areas that minimized the risk of recognition; second, they had to be situated near major transport routes to allow easy access and dispersal; third, the sites had to be sufficiently exposed to facilitate recognition of potential customers; and, finally, the toilets needed a vantage point to keep watch for the police, homophobic people or unwanted members of the public (Ponte, 1974; Troiden, 1974).

Since the 1960s there have been profound changes in the nature of gay and lesbian spaces within cities that reflect broader social and political changes in society. Arguably, there is now greater tolerance towards homosexual activity by 'straight' sections of society, although the extent of this tolerance should not be exaggerated. Studies still reveal substantial prejudice among a majority of the total population and there are still high rates of physical assault against gays and lesbians. For example, one study of Philadelphia revealed levels of reported victimization among lesbians to be twice that for women in general (cited in Valentine, 1996). Gay men also experience high levels of violent attack – even in areas of cities with a high proportion of gays (Myslik, 1996). In addition there has been a marked right-wing, anti-feminist, anti-gay, backlash in the United States in recent years.

Gay spaces

Weightman (1981) was among the first to draw attention to areas in US cities with distinctive gay lifestyles. He noted that gays were playing a leading role in the process of gentrification in some inner-city or transitional areas, often displacing the poorer residents. Without doubt the most famous of these residential districts is the Castro district of San Francisco that was mapped by Castells and Murphy (1982; Castells, 1983). The origins of this district can be traced back to the Second World War. Gays and lesbians serving in the armed forces were often released from military service in San Francisco and preferred to establish homes in the city, rather than move back and face the prejudice of their home communities.

Another important gay space is the Marigny district of New Orleans studied by Knopp (1990). As in the San Francisco case, this is located in a culturally mixed, relatively tolerant area. Knopp describes how property developers and speculators rapidly exploited the demand for property in this area by the gay population. In particular, they exploited the 'rent gap' (already discussed in Chapter 6) by artificially inflating the values of properties by bribing private appraisers. Paradoxically, this led to an influx of lower-income gays into a predominantly middle-class area. Knopp notes that many of middle-class gays in the area became concerned

over the preservation of historically important areas within the city, rather than broader issues affecting the gay community. Other urban expressions of gay sexuality have been studies in Manchester (Hindle, 1994) and Newcastle (Lewis, 1994) in the United Kingdom. In addition, Binnie (1995) presents a fascinating analysis of the gay community in Amsterdam. He notes that 3000 of the 25 000 jobs in the tourist industry in the city were dependent upon gay tourism, and in recognition of this, the city authorities actively promoted the gay area. However, this advertising campaign was eventually withdrawn because of concerns over alienating the tourist industry as a whole.

One of the most potentially significant developments in the sphere of gay spaces is the recent discovery that their presence in cities is strongly correlated with economic growth (see Florida, 2002). There is some uncertainty over the precise causal mechanisms at work here, but it would seem that communities that are tolerant of diversity and welcoming to 'outsiders' tend to foster the sort of creativity and innovation that is necessary to survive in an age increasingly dominated by high-technology and cultural industries (see also Box 11.2).

Lesbian spaces

Male homosexuality has received much more analysis than lesbianism, largely because the lesbian subcultures are smaller and less visible than male gay subcultures. Nevertheless, Winchester and White (1988) were able to chart dimensions of a lesbian ghetto in Paris (see Fig. 11.1). Egerton (1990) has also documented lesbian spaces in the form of squats, housing cooperatives and housing associations. These are attempts to create safe areas for women and sites of political resistance but they have sometimes been subject to violence from homophobes and misogynists. In general, women have fewer financial resources than men and they also face the threat of male violence. Lesbians therefore display a desire for relatively inexpensive housing as well as a concern with personal safety.

These factors, combined with the pressures of a predominantly heterosexual society, mean that lesbian residential areas are less overt than gay spaces and often have an 'underground' character. In addition, it would seem that lesbian spaces can be unstable resulting from

Box 11.2

Key thinkers in urban social geography – Richard Florida

Sadly, few geographical texts become best-sellers on a scale with pulp fiction (although one of the best-selling geography books of recent years, David Harvey's *The Condition of Postmodernity*, is rumoured to have sold over 100 000 copies). One remarkable exception is Richard Florida's *The Rise of the Creative Class* (2002). Rather like Garreau's book *Edge Cities* (see Box 2.1), Florida's work has escaped from the narrow confines of academia to reach a wider audience. Both books are helped by a highly accessible and engaging style that makes them something of an 'airport read'. Furthermore, sales cannot have been hindered by the fact that both books give their readers a somewhat flattering view of themselves. However, whereas Garreau's book is essentially based on informed journalism, Florida's work combines both social theory and detailed empirical research. Consequently, Florida's work has been highly influential among planners and urban policymakers concerned to regenerate urban areas. In this respect Florida's work echoes the world-famous business strategy 'guru' Michael Porter who has advised many governments on the creation of industrial agglomerations (what he terms 'clusters'). Thus both Florida and Porter are active on the highly lucrative 'celebrity' lecture circuit.

Florida's analysis is based on the following measures for the 50 largest metropolitan regions in the United States:

➤ *Bohemian Index:* authors, designers, musicians, artists, etc.

➤ *Melting Pot Index:* the proportion of the population that is foreign born.

➤ *Gay Index:* households in which a householder and unmarried partner were both of the same sex (in this sample male).

➤ *Culture Index:* galleries, museums, libraries, etc.

➤ *'Cool' Index:* percentage of population aged 22–29; night clubs, bars, etc.

➤ *Talent Index:* percentage of population with a bachelor's degree.

➤ *Tech-pole Index:* a measure of high-technology industry.

Florida finds that Bohemians are highly concentrated in a few select cities, including New York, Los Angeles, San Francisco, Seattle, Washington, DC, Boston, Nashville, Austin and Minneapolis-St Paul. Furthermore, while all the above measures are correlated, the Bohemian Index is highly correlated with both the Talent and Tech-pole Indices.

Following from these results, the 'Florida Thesis' argues that central to the generation of innovative high-technology industries, the engine of growth for cities, is the capacity of areas to attract talented individuals (human capital). These form a new 'creative class' of engineers, scietists, business managers, scholars and the like who live new forms of lifestyle characterized by self-actualization through both work and leisure. Such people are attracted to environments characterized by cultural diversity and artistic creativity. This amounts to more than the provision of cultural facilities such as concert halls and art galleries. Even more important is a tolerant, libertarian atmosphere in which 'alternative' so-called Bohemian lifestyles can flourish. It appears that an open attitude to diverse lifestyles breeds the capacity for innovation in a range of industries that are crucial in the new knowledge economy (see also Box 6.2 on David Ley). Some would argue that what is defined as 'cool' and alternative has now been incorporated into consumer capitalism.

A major criticism of Florida's work concerns the nature of causality; given the complex nature of regional economies it is not certain whether Bohemian enclaves are a cause or effect of economic growth. Many would also argue that his ideas give political elites carte blanche to provide services for younger skilled persons while cutting back on facilities for the disadvantaged.

Key concepts associated with Richard Florida (see Glossary)

Bohemia, creative cities, creative class, cultural industries.

Further reading

Florida, R. (2002) *The Rise of the Creative Class* Basic Books, New York

Florida, R. (2002) Bohemia and economic geography *Journal of Economic Geography*, **2**, 55–74

Florida, R. (2004) *Cities and the Creative Class* Routledge, London

Links with Other Chapters

Chapter 3: What is culture?

Chapter 5: Box 5.1 The Growth of 'Urban Entrepreneurialism'; Box 5.2 The Resurgence of Barcelona

Chapter 7: Box 7.1 The Growth of Bohemian Enclaves

Chapter 11: Box 11.3 The Resurgence of Dublin

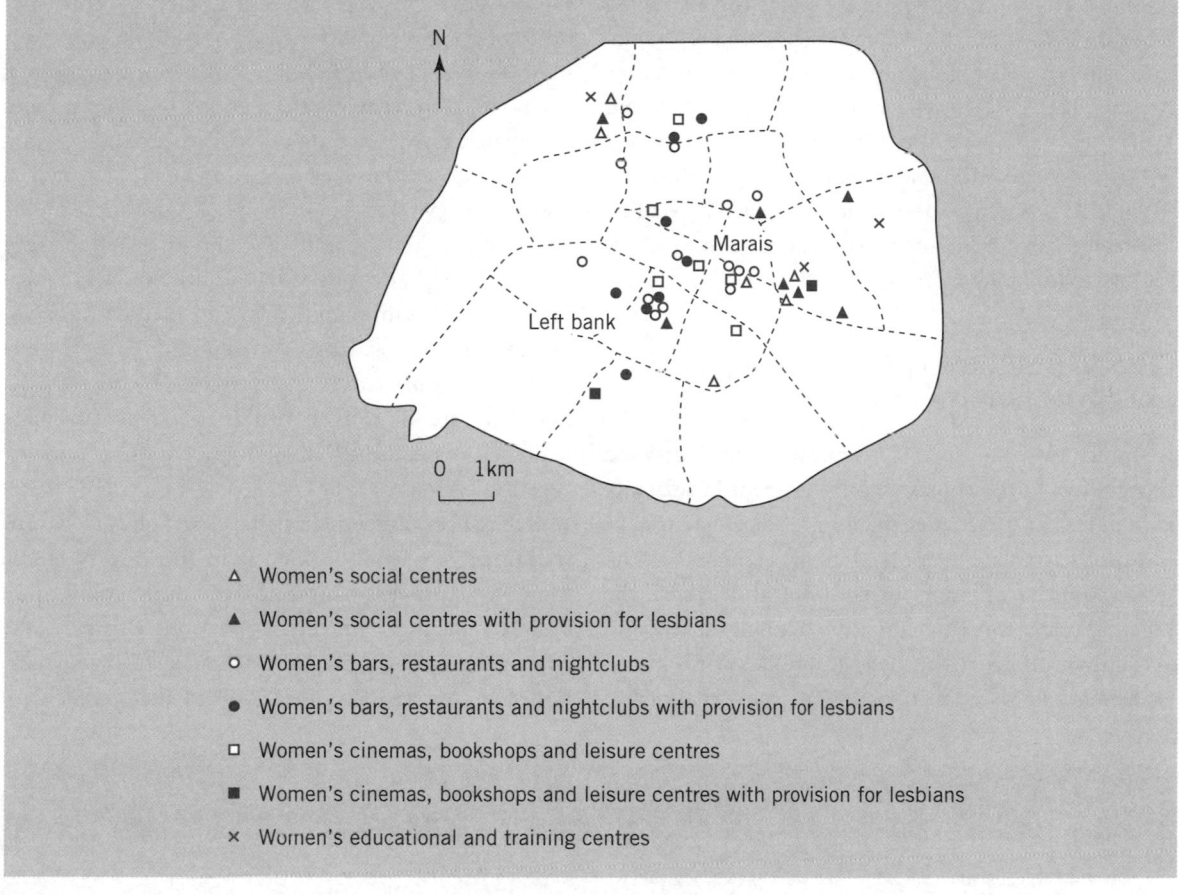

Figure 11.1 Lesbian facilities in Paris, 1984–1985.
Source: Winchester and White (1988), Fig. 1, p. 48.

underlying tensions in dense clique-ridden environments. Castells (1983) argued that gay enclaves in cities
were a spatial expression of men's desire to dominate.
He argued that women have a greater sense of mutual
solidarity and affection than men and less need for
territorial expression. However, critics have pointed
out that this argument perpetuates notions of essential differences between men and women and ignores
the processes that serve to exclude women from some
parts of cities. In recent years lesbians have expressed
resentment at the idea that their form of oppression
can be equated in a simple manner with that of gays.
Indeed, their struggle is seen by some as posing a much
greater threat to heteropatriarchy than the gay rights
movement.

Such mapping of gay spaces in cities has greatly
extended our knowledge of urban social geography but
it has been argued that they have been limited in their
contribution to an understanding of the *processes* that
lead to such geographical clusterings (Wright, 1999).
To begin with, most of these areas cannot be thought
of as exclusively gay. Furthermore, there are many
gays and lesbians who live outside such regions. Thus,
Valentine (1995) argues that this work on **gay ghettos**
tends to ignore that fact that many lesbians and gay
men conceal their identities at certain times to avoid
discrimination or persecution in certain contexts.

The crucial point is that such a focus on gay spaces
tends to conceptualize them as different when, in reality, *all* spaces in cities are constructed in a sexualized
manner. For example, research has focused upon the processes through which certain public spaces such as hotels
and restaurants become constructed as heterosexual
(Kirby and Hay, 1997). It has been argued that even the

domestic environment within the home is constructed around a heterosexual-based notion of sexual reproduction with typically a master bedroom and various rooms for children (McDowell, 1995, 1996; Johnston and Valentine, 1995; see also Chapter 9). But the most telling criticism of the spatial mapping approach is that it tends to perpetuate an essentialist view of sexuality, i.e. the view that homosexuality is entirely some natural biological imperative (Wright, 1999).

Queer politics: lipstick lesbians and gay skinheads

Inspired by the work of Foucault, an increased recognition of the constructionist element in sexuality is manifest in **queer theory** and in radical gay organizations such as Queer Nation or Lesbian Nation. **Queer politics** attempts to confront and expose the heterosexually constructed nature of public spaces and social institutions (the usually pejorative word 'queer' is deliberately used in a somewhat ironic manner to acknowledge the repressive character of constructions of 'homosexuality'). In part, this strategy was motivated by a dissatisfaction with more traditional gay political movements that tended to adopt a more essentialist view, since this tended to exclude certain people such as those who are bisexual. This radical queer strategy is manifest in mock weddings, public 'kiss-ins' and gay shopping expeditions. Another subversive strategy attempts to destabilize notions of fixed sexual identities through parody and exaggeration. Not only has that classic icon of masculinity – the cowboy – been appropriated as a dress code by many gays, but there are those who behave as 'hypermasculine males' (gay skinheads) and 'hyperfeminine females' (lipstick lesbians).

It has been argued that the geographical concentration of gay people could, as in the case of ethnic minorities, provide a base for political mobilization against repression and discrimination. However, the effectiveness of such a separatist strategy has been questioned by many. It has been argued that rather than

---

Box 11.3

Key trends in urban social geography – The resurgence of Dublin

Like Barcelona, Dublin is a European city that has, in the past two decades, transformed itself from a decaying, poor, area into a vibrant urban growth zone. Again, as in the case of Barcelona, this urban resurgence reflects the economic benefits that have accrued to the wider nation through membership of the European Union. A combination of structural funds and special dispensation to instigate a taxation regime that is highly favourable to major corporations have transformed Ireland from one of Europe's poorest nations to its richest.

Dublin, a city of 1.4 million, and over 40 per cent of the republic's population, has been at the forefront of this economic growth. A highly educated relatively young population (half are under 25) has fuelled development, such that Dublin is now a major centre for many industries including software development. Most famously, the winding lanes of Dublin's Georgian Temple Bar area have been transformed into an 'urban nightscape' (see also Box 10.2) with bars, restaurants and the like. As in the case of Barcelona, this has led to social tensions; Temple Bar has become notorious for excessive alcohol consumption by young adult British tourists on 'stag' and 'hen' nights. Dublin also exhibits the gentrification, immigration, rapidly increasing house prices and social polarization that often accompanies rapid economic growth.

Richard Florida has championed Dublin as an example of a creative city (see also Box 11.2) but recent surveys suggest that growth and immigration have been accompanied by increasing intolerance and discrimination in the Irish republic. It may of course be that those experiencing the downside of economic growth are those who are most resentful of immigration and diversity.

Key concepts associated with Dublin (see Glossary)

Civic boosterism, creative cities, deregulation, growth coalitions, social polarization, urban entrepreneurialism.

Further reading

Ellis, G. and Kim, J. (2001) City profile: Dublin *Cities*, **18**, 353–364

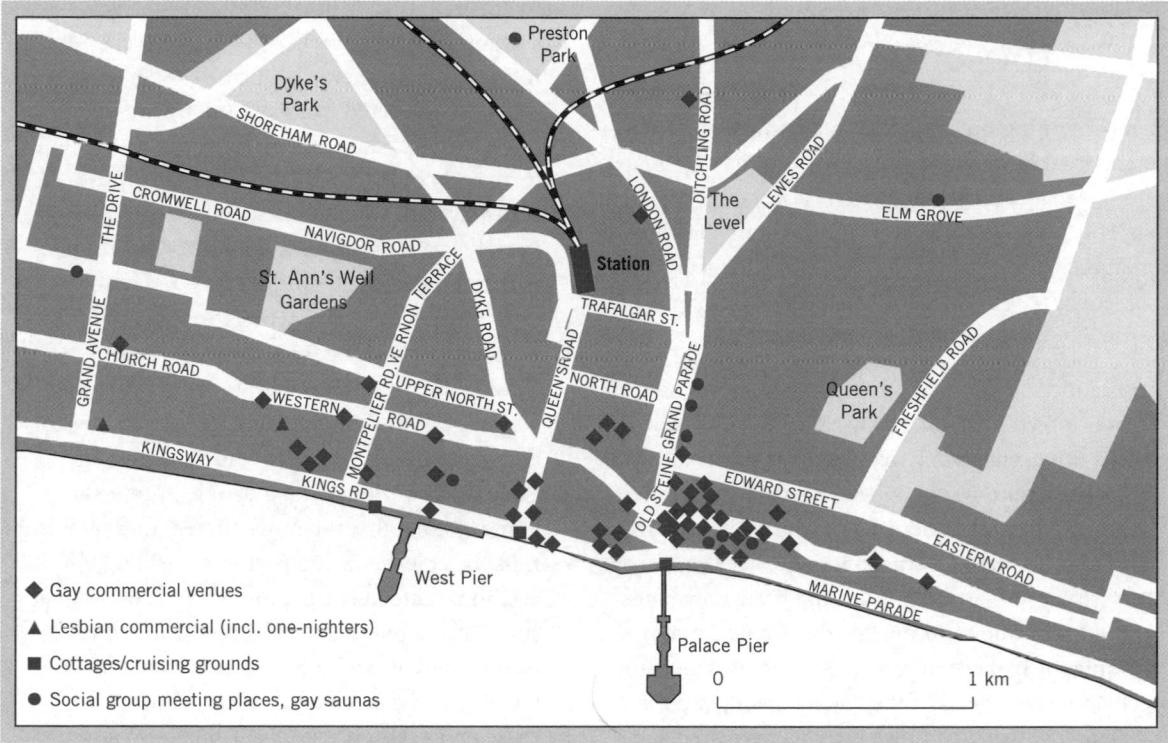

Figure 11.2 Gay commercial and social venues in Brighton in the 1990s.
Source: Wright (1999), Fig. 5.6, p. 182.

enabling homosexuality to be seen as natural as hetero-sexuality, such gay enclaves have helped to maintain the notion of gay lifestyles as separate, different, deviant and sinful. Thus, Castells noted that gay ghettos can be sites of liberation but they can also be likened to a prison. With the development of HIV and AIDS in the gay community, such areas assumed a new role as focal points for support networks and health-care services.

Yet another change in gay spaces in recent years has been the commercialization of such area as entre-preneurs have sought to exploit the high incomes of some gay households. For example, the Soho area of London has developed into a commercial area for gays in what has been described as the world's Pink Capital. Brighton also has a thriving gay commercial scene as shown by Fig. 11.2. Clubs, bars and shops catering explicitly for gays have heightened the visibility of gay lifestyles but they have tended to cater for certain types of gays – those who are youthful and

wealthy – with those who are older and poorer are somewhat excluded. Thus a new form of economic class division has been imposed upon divisions based on sexuality. Another type of exclusion has centred around the body; the exaggerated masculinity and body-building activities of 'hypermasculine' gays has tended to separate them from those individuals less endowed with the 'body beautiful'. There has also been a growing recognition of the divisions within the category 'lesbian':

In large urban centers across Canada and other Western countries, the 1980s have heralded the subdivision of activist lesbians into specialized groupings: lesbians of color, Jewish lesbians, working class lesbians, leather dykes, lesbians against sado-masochism, older lesbians, lesbian youth, disabled lesbians and so on.

(Ross 1990, cited in Chouinard and Grant, 1996, p. 179)

11.3 Disability and the city

In most writing on cities the able-bodied character of citizens is taken for granted and this, Imrie (1996) argues, amounts to an **ableist geography**. Park *et al.* (1998) note that 'Human geography has in the past found little room for studies of disability' (p. 208). The character of the problem has been movingly revealed by Vera Chouinard, a professor disabled by rheumatoid arthritis:

> Recently in a feminist geography conference session I was forced to stand because the room was filled. This was arguably my own 'fault' as I arrived late (having had to walk a long distance from another session), but after about half an hour the pain in my feet, leg and hips was so intense that I was forced to ask a young women if she and her companions could shift one chair over so that I could sit down (someone had left their seat, so there was an empty chair at the far end of the row). I apologized for asking but explained that I was ill, very tired and in a lot of pain. She turned, looked very coldly at me and simply said 'No, the seats are being used'. She may well have been right, but I suddenly no longer felt like part of a feminist geography session: I was invisible . . . and I was angry. Fighting a juvenile urge to bop her on the head with my cane, I began to see feminist geography through new eyes; eyes which recognised that the pain of being 'the other' was far deeper and more complex that I ever imagined, and that words of inclusion were simply not enough. . . . My sense of myself, as a disabled, academic woman has also been shaped by more subtle aspects of daily life. Walking on the university campus and in other public places, I am constantly conscious of frequent looks (often double and even triple-takes). I realize that it is unusual to see a relatively young woman walking slowly with a cane or using a scooter and the looks reflect curiosity, but they are a constant reminder that I am different, that I don't 'belong'. It is painful for me to acknowledge this. I guess that is why I have learned to look away: to the ground, to the side . . . anywhere that lets me avoid facing up to being the 'other'.

> It is remarkable how thoroughly ableist assumptions and practices permeate every facet of our lives, even though we often remain relatively sheltered from and insensitive to these forms of oppression. Yet disability in some form will come to each and every one of us someday, and when it does, and ableism rears its ugly head, one finds a topsy-turvy world in which none of the old rules apply and many 'new' ones don't make sense. People develop new ways of relating to you often without recognizing it. For instance, some of my students will not call me at home, despite instructions to do so, because I am 'sick'. Other students shy away from working with a disabled professor: Some assume that the most 'successful' supervisors must be able-bodied; others are unwilling to accommodate illness by, for example, occasionally substituting phone calls for face-to-face meetings or meetings at my home rather than the office. Of course this is not true of all students, but these practices are pervasive enough to hurt every day to make it just a little harder to struggle to change relations, policies, practices and attitudes.

> (Chouinard and Grant, 1996, p. 173)

We have quoted at length from this article, not only because Chouinard writes in such an articulate manner, but also because it makes additional impact in referencing the world of academe; a world in which one might have expected more enlightened attitudes. However, it is encouraging to report that in recent years geographical research has at least *begun* to address issues of disability (see Hahn, 1986; Golledge, 1993; Imrie, 1996; Butler and Bowlby, 1997; Dear *et al.*, 1997; Gleeson, 1998). A fundamental issue here is just what is meant by 'disability'. The United Nations has made the following distinctions between impairment, disability and handicap:

➤ *Impairment*: any loss or abnormality of psychological, physiological or anatomical structure or function.

➤ *Disability*: any restriction or lack (resulting from impairment) of ability to perform an activity in the manner or within the range considered normal for a human being.

➤ *Handicap*: a disadvantage for a given individual, resulting from an impairment or disability, depending upon age, sex, social and cultural factors for that individual (cited in Wendell, 1997).

However, an immediate difficulty with such a set of definitions is how to define what is 'normal', for incapacity and disability form a wide continuum of human capabilities. For example, as Longhurst (1998) shows, although not generally regarded as impaired or disabled, some pregnant women can encounter problems with the design of shopping malls. There is, however, a more fundamental limitation with the UN approach, since it attempts to make a distinction between *physical* definitions of impairment and disability and *social* definitions of handicap.

The social construction of disability

The social construction of 'disability' is bound up with the attitudes and structures of oppression in an able-bodied society, rather than the failing of a particular individual. In societies that view the role of medicine as primarily one of making sick people well, disability is often seen as something unhealthy. However, it follows from a social constructionist perspective that if sufficient facilities were provided, then disability would be regarded as something akin to short-sightedness (the wearing of glasses not generally being regarded as a disability). This is an excellent example of how social science can challenge taken-for-granted assumptions. What seems at first like 'common sense' can often be a function of a particular way we have been taught to think about the world.

The socially constructed nature of disability was strikingly revealed by a classic study of attitudes towards blindness. In the United States it was seen primarily as an experience of loss requiring counselling, in Britain as a technical issue requiring aids and equipment; and in Italy as a need to seek consolation and salvation through the Catholic Church (Oliver, 1998). Different societies therefore produce varying definitions of impairment and disability. In the United States, for example, following the politicization of disability rights, there has emerged a multimillion dollar disability industry. Disability has

Table 11.2 Ten media stereotypes of disabled people

Pitiable and pathetic: charity adverts; Children in Need; Tiny Tim, Kevin Spacey's role in *The Usual Suspects*

Object of violence: films such as *Whatever Happened to Baby Jane?*

Sinister or evil: Dr No; Dr Strangelove; Richard III; Christopher Walken's role in *100 Things to Do in Denver When You're Dead*

Atmosphere: curios in comics books or films (e.g. *The Hunchback of Notre Dame*).

Triumph over tragedy: e.g. the last item in the news.

Laughable: the butt of jokes, e.g. Mr Magoo.

Bearing a grudge: Laura in *The Glass Menagerie*.

Burden or outcast: the Morlocks in *The X-Men* or in *The Mask*.

Non-sexual or incapable of full relationships: Clifford Chatterley in *Lady Chatterley's Lover*.

Incapable of fully participating in everyday life: absence from everyday situations and not shown as integral and productive members of society.

Source: adapted from *Guardian* 13 October, 1995.

therefore become a major source of income for doctors, lawyers, rehabilitation professionals and disability activists.

Some argue that this social constructionist approach goes too far in that it ignores factors such as pain or impaired vision that are '. . . part of the bodily experience of the disabled' (French, 1993, p. 124), factors over which society has little or no control (even if they are made worse by societal oppression). But however one conceptualizes the issue, what *is* clear is that disabled people are stereotyped by a predominantly able-bodied society. This is illustrated by Table 11.2, which shows common stereotypes of disability represented in films, books, plays and the media.

Disability in urban settings

Cities frequently display numerous barriers to mobility and access for disabled people. Typical problems include high curbs, steep steps, the absence of ramps for wheelchairs, narrow doors and the absence of information in Braille. In addition, elevators/lifts for those who are impaired or disabled are often in unattractive locations (e.g. service elevators/lifts next to kitchens), badly signposted and with inaccessible buttons. Public transport systems can also pose problems for people with disabilities. Often the problem boils down to one of cost, with inadequate finances being available for adequate conversion of premises for disabled access.

As Imrie (1996) notes, the problem is also one of dominant attitudes in an 'ableist' society. All too often, architects, planners and the public at large have assumed

---

Table 11.3 A typology of approaches to access for disabled people by local authorities in the UK

'Averse'
➤ Operates with a biomedical model of disability, or disability as being derived from physical and/or mental impairment
➤ Major concern is to secure investment to support the local economy. Attitude is that insisting on access will scare away developers and much needed investment
➤ Access is seen as a minority issue that only affects a small proportion of the population
➤ Very few statements on access in local plan
➤ No budget to support access projects
➤ Never uses planning conditions to secure access
➤ No Access Officer, or where they do exist, usually performed part time by someone in Building Controls of Local Plans
➤ Local political system unaware of access issues and provides little support or encouragement
➤ Access groups either non-existent or poorly organized
➤ Typically a rural local authority and/or area with severe economic problems

'Proactive'
➤ Operates with a social model of disability or disability as a form of discrimination
➤ Access is seen as an issue to be considered because of the directives of government and the Royal Town Planning Institute
➤ Access is seen as one on many competing demands on officer time
➤ Appointed access officer, usually on a part-time basis
➤ Some small funds for access issues
➤ Brings to the attention of developers all the statutory requirements of access
➤ Seeks to negotiate with developers and persuade them to give more than is required by statute
➤ Major concern is to secure development but not at the expense of neglecting statutory duties on matters, for instance, such as access
➤ Some awareness by local council of access issues but remains peripheral and rarely discussed by local politicians
➤ Access groups usually exist but are often weak and poorly resources
➤ A mixture and range of localities

'Coercive'
➤ Conceives of disability as an equal opportunities concern
➤ Access is seen as a right for all people
➤ The local economy will benefit by providing access. people with disabilities are consumers too
➤ Insists on access provision
➤ Will seek to use all available planning instruments
➤ Will not hesitate to use planning conditions relating to access
➤ Regular meetings between planners and local access groups
➤ Full-time access officer, well networked within and between departments
➤ Active support from local politicians with key councillors
➤ Active access groups
➤ Typically a left-wing, city authority

Sources: Adapted from Imrie (1996), pp. 128 and 138.

that disability leads to immobility. Consequently, the needs of disabled people get ignored, yet it takes only a few minutes in a wheelchair to realize the scale of the problems created by most buildings or public transport systems. The barriers are therefore social and psychological as much as physical. Society's reluctance to meet the costs of ensuring accessibility for disabled persons reflects a wider set of social values towards disability. Restricted accessibility prevents disabled people from fully participating in social and economic life, such as in the world of employment. A study in Ontario, Canada, found that 80 per cent of disabled persons lived in relative poverty because of their exclusion from the job market and limited support programmes from both the public and private sectors (Chouinard and Grant, 1996).

Fortunately, in many cities the issue of disability is now being taken much more seriously. For example, in the United States the Americans with Disabilities Act of 1992 has required businesses and public institutions to provide wheelchair access. In the United Kingdom, although progress is patchy, many local authorities now have Disability Officers tasked with improving disabled access in city centres. These improvements are in part a response to the increased activities of various disability rights movements (Table 11.3). These are now many in number; for example the British Council of Organizations of Disabled People has over a hundred constituent member organizations (Campbell and Oliver, 1996). These groups have begun to campaign around issues of income, employment, civil rights and community living rather than the older issues of institutional care. As such, they reflect one of the **new social movements** focused around what is termed **identity politics** (i.e. people with a particular identity rather than traditional class-based politics).

The gay, lesbian and disability rights movements illustrate well some of the key dilemmas in an era in which we recognize that identities are multiple and unstable. In such a context it is politically expedient at times to adopt what is termed **strategic essentialism**, unification around a single dimension of identity, such as gender, sexuality, race or disability, to achieve particular objectives. While this can bring political strength, it can also lead to vulnerability and internal tensions, for claims of universal solidarity can lead to exclusion and alienation. How to reconcile the competing claims of mutual interest and difference is one of the key issues of the twenty-first century.

Box 11.4

Key films related to urban social geography – Chapter 11

The Adventures of Pricilla, Queen of the Desert (1994) A hilarious story of cross-dressing in the Australian outback, reflecting city–country relations and sexuality.

The Celluloid Closet (1995) A remarkable documentary on the changing ways in which Hollywood has treated gay issues over the years. Highly recommended.

Go Fish (1994) A lively romantic comedy centred around lesbian lifestyles in Chicago.

My Beautiful Launderette (1985) A film about sexuality, racism and neoliberalism in Thatcherite Britain in the early 1980s.

Passion Fish (1992) A warm, touching and moving consideration of issues of disability, race and sexuality set in the southern United States.

Philadelphia (1993) The first mainstream Hollywood movie to confront issues of AIDS and the gay community. Moving, thought-provoking and life-affirming.

Priest (1994) A powerful and moving film about the conflict between religion and sexuality for a gay young priest.

The Wedding Banquet (1993) A fairly light romantic comedy by the much-lauded director Ang Lee, but a film that provides amusing and provoking insights into issues of ethnicity and sexuality.

Chapter summary

11.1 People's bodily appearance is not just a function of innate biology, it is also socially constructed through various signs and systems of meaning. The spaces of cities provide a powerful environment that forces people to conform to certain standards of appearance and dress while at the same time also providing opportunities for people to transgress prevailing social norms.

11.2 Urban areas have had an important impact upon the development of sexuality. Cities provide opportunities for people to transgress dominant codes of sexual behaviour. The continuation of prostitution in cities reflects patriarchal gender relations. Cities have provided opportunities for the territorial expression of gay lifestyles. As in the case of ethnic segregation, this has provided opportunities for empowerment but may also been seen as reflecting continuing discrimination.

11.3 The social construction of 'disability' also reflects the continuing patterns of oppression in an able-bodied society. Both the Gay Rights Movement and the Disability Rights Movement may be seen as forms of new social movements.

Key concepts and terms

ableist geography
corporeality
embodied knowledge
embodiment
gay ghettos
gender
gender roles

heteropatriarchal environments
heteropatriarchy
homosexuality
identity politics
new social movements
patriarchy
performativity

queer
queer politics
queer theory
sexuality
strategic essentialism

Suggested reading

A good introduction to the relationships between geography and studies of bodies is Robin Longhurst's essay 'The body and geography' (1995: *Gender Place and Culture*, 2, 97–105) while for a broader social science overview see Chris Shilling's *The Body and the Social* (1993: Sage, London). Issues of gender are discussed in the Women and Geography Study Group's *Feminist Geographies: Exploring Diversity and Difference* (1997: Longman, London), Susan Hanson and Geraldine Pratt *Gender, Work and Space* (1995: Routledge, London), Chris Booth, Jane Darke and Susan Yeandle (eds) *Changing Places: Women's Lives in the City* (1996: Paul Chapman, London) and some of the chapters in

Doreen Massey's *Space, Place and Gender* (1994: Polity Press, Cambridge). David Bell and Gill Valentine (eds) (1995) *Mapping Desire: Geographies of Sexuality* (1995, Routledge, London) is an excellent guide to the issues of sexuality discussed in this chapter. See also Nancy Duncan (ed.) *Bodyspace: Destabilizing Geographies of Gender and Sexuality* (1996: Routledge, London), Rosa Ainley (ed.) *New Frontiers of Space Bodies and Gender* (1998: Routledge, London), Heidi Nast and Steve Pile (eds) *Places Through the Body* (1998: Routledge, London), Steve Pile *The Body and the City* (1996: Routledge, London) and Diane Richardson and Steven Seidman (eds) *Handbook of Lesbian and Gay Studies* (2002: Routledge, London). The issue of prostitution in Western cities is discussed in Philip

Hubbard *Sex and the City: Geographies of Prostitution in the Urban West* (1999: Ashgate, Aldershot). An excellent geographically oriented text on disability is Rob Imrie's *Disability and the City: International Perspectives* (1996: Paul Chapman, London). Another highly influential book in this field is J. Swain, V. Finkelstein, S. French and M. Oliver (eds) *Disabling Barriers – Enabling Environments* (1993: Sage, London) together with J.L. Davis (ed.) The *Disability Studies Reader* (1997: Routledge, London), T. Shakespeare (ed.) *The Disability Reader: Social Science Perspectives* (1998: Cassell, London) and R. Butler and H. Parr (eds) *Mind and Body Spaces: Geographies of Disability, Illness and Impairment* (1999: Routledge, London). You might also like to examine the special edition of the journal *Area* (2000, Vol. 32) edited by Steven Cummins and Christine Milligan on the theme of geographies of health and impairment together with the special issue of *Health and Place* – Chris Philo and D. Metzel 'Introduction to theme section on geographies of intellectual disability: outside the participatory mainstream?' (2005: **11**, 77–85).

Chapter 12

Residential mobility and neighbourhood change

Key questions addressed in this chapter

► Why do people move within cities?

► What patterns are generated by these moves?

► What effect do these moves have upon residential structures?

Although it is widely accepted that the shaping and reshaping of urban social areas is a product of the movement of households from one residence to another, the relationships between residential structure and patterns of residential mobility are only imperfectly understood. This is a reflection of the complexity of these relationships. While migration creates and remodels the social and demographic structure of city neighbourhoods, it is also conditioned by the existing ecology of the city: a classic example of the sociospatial dialectic. The process is undergoing constant modification as each household's decision to move (or not to move) has repercussions for the rest of the system. Chain reactions of vacancies and moves are set off as dwellings become newly available, and this movement may itself trigger further mobility as households react to changes in neighbourhood status and tone.

The basic relationship between residential mobility and urban structure is outlined in Fig. 12.1, which emphasizes the circular and cumulative effects of housing demand and urban structure on each other. Mobility is seen as a product of *housing opportunities* – the new and vacant dwellings resulting from suburban expansion, inner-city renewal and rehabilitation, etc. – and the housing *needs* and *expectations* of households, which are themselves a product of income, family size and lifestyle. Meanwhile, as Fig. 12.2 shows, residential mobility can also be interpreted within the frame of broader structural changes.

Given a sufficient amount of mobility, the residential structure of the city will be substantively altered, resulting in changes both to the 'objective' social ecology and to the associated neighbourhood images that help to attract or deter further potential movers. *Households, then, may be seen as decision-making units whose aggregate response to housing opportunities is central to ecological change.* It therefore seems logical to begin the task of disentangling the relationship between movement and

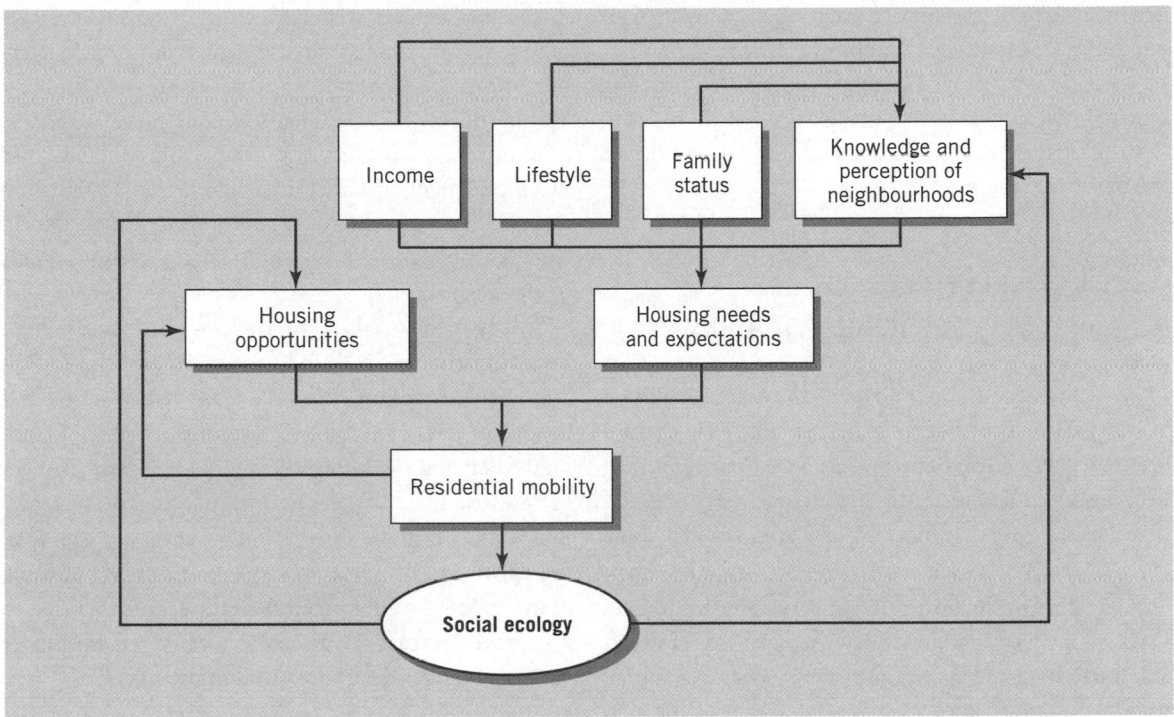

Figure 12.1 Relationships between housing demand, residential mobility and the social ecology of the city.

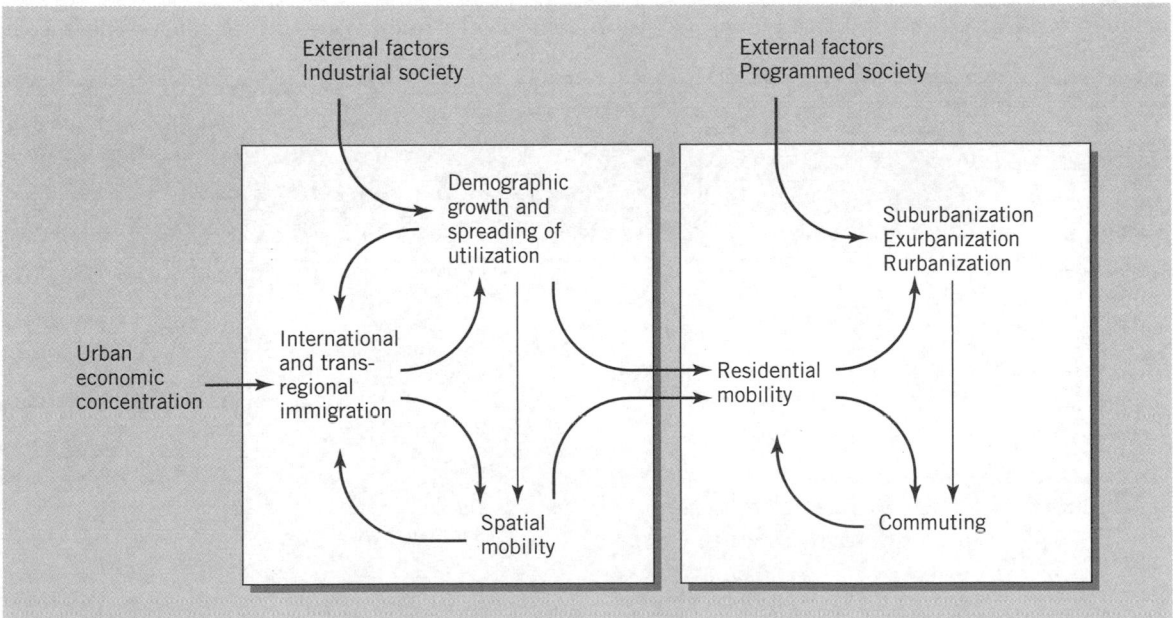

Figure 12.2 The system formed by mobility flows and the structuration of space.
Source: Bassand (1990), Fig. 1, p. 80.

urban structure by seeking to establish the fundamental parameters of household mobility. How many households do actually move in a given period? Do particular types of households have a greater propensity to move than others? And are there any spatial regularities in the pattern of migration?

12.1 Patterns of household mobility

In fact, the amount of movement by households in Western cities is considerable. In Australia, New Zealand and North America, between 15 and 20 per cent of all urban households move in any one year. Having said this, it is of course important to recognize that some cities experience much higher levels of mobility than others. Cities in the fast-growing West, South and Gulf Coast of the United States – Bellevue, Scottsdale, Riverside and Thousand Oaks, for instance – have an annual turnover of population that is double that of the likes of Scranton, Johnstown and Wilkes-Barre/Hazeltown in the slow-growing North East. In Europe, rates of mobility also vary a good deal, but in general they range between 5 and 10 per cent per year (Table 12.1).

It is also important to recognize that the magnitude of this movement stems partly from economic and social forces that extend well beyond the housing markets of individual cities. Some of the most important determinants of the overall level of residential mobility are the business cycles that are endemic to capitalist economies. During economic upswings the increase in employment opportunities and wages leads to an increase in the effective demand for new housing which, when completed, allows whole chains of households to change homes.

Changes in social organization – particularly those involving changes in family structure and the rate of household formation, dissolution and fusion – also effect the overall level of mobility by exerting a direct influence on the demand for accommodation. Long-term changes in the structure of the housing market itself are also important. In many European countries, for example, the expansion of owner-occupied and public housing at the expense of the privately rented sector has led to a general decrease in mobility because of the higher costs and longer delays involved in moving.

Movers and stayers

Residential mobility is a selective process. Households of different types are not equally mobile. Some have a propensity to move quite often; others, having once gained entry to the housing system, never move at all,

Table 12.1 Intra-urban migration rates, 1980 (or latest available year), selected European cities

	Movers per 1000 population		Movers per 1000 population
Aberdeen	73.8	Lisbon	39.9
Amsterdam	93.3	Lyons	71.9
Berlin (West)	116.3	Manchester	66.4
Brussels	57.0	Munich	72.9
Copenhagen	116.0	Newcastle	70.9
Cork	21.3	Oporto	39.9
Dijon	70.3	Paris	59.9
Dublin	20.3	Rome	70.5
Geneva	100.6	Tampere	120.7
Gothenburg	97.1	Turku	113.5
Hamburg	85.9	Sheffield	70.5
Helsinki	119.4	Stockholm	57.5
Köln	83.0	Vienna	48.8
Lausanne	109.4	Zurich	103.1
London	76.2		

Source: derived from White (1984), Table 2.

thus lending a degree of stability to the residential mosaic. This basic dichotomy between 'movers' and 'stayers' has been identified in a number of studies, and it has been found that the composition of each group tends to be related to the lifestyle and tenure characteristics of households. In particular, younger households have been found to move more frequently than older households; and private renters have been found to be more mobile than households in other tenure categories. People can also be conceptualized as 'locals' or 'cosmopolitans', depending on the type and intensity of their attachments to their immediate social environments, and this distinction has been shown to have a significant bearing on intra-urban mobility.

In addition, there appears to be an independent duration-of-residence effect, whereby the longer a household remains in a dwelling the less likely it is to move. This has been termed the principle of 'cumulative inertia', and is usually explained in terms of the emotional attachments that people develop towards their dwelling and immediate neighbourhood and their reluctance to sever increasingly strong and complex social networks in favour of the unknown quantity of the pattern of daily life elsewhere. In contrast, the actual experience of moving home probably reinforces the propensity to move. Movers 'are more oriented to future mobility than are persons who have not moved in the past and are better able to actualize a moving plan and choice' (Van Arsdol et al., 1986, p. 266).

Migration data

Spatial regularities in the migration patterns of movers are difficult to establish because of the problems involved in obtaining and analyzing migration data. Census data, although reliable, rarely include sufficient information about the origin of migrants; and few countries outside the Netherlands and Scandinavia have registers of households that can be used to plot household movements. Questionnaire surveys provide an obvious alternative, but they involve the expenditure of a large amount of time and money in order to obtain a sufficiently large sample of migrants. In North America, many researchers have resorted to data based on changes of address worked out from telephone directories; but European researchers, faced with larger numbers

of households who do not have land-line telephones, have often had to rely on town directories and electoral lists, both of which are known to be rather incomplete sources of information. Difficulties have also been experienced in analyzing migration data. In addition to the pitfalls of the ecological fallacy, these include the statistical problem of multi-collinearity and the practical problems involved in handling the large, complex data sets associated with migration studies.

Patterns of in-migration

Nevertheless, it is possible to suggest a number of important regularities in people's migration behaviour. It is useful to distinguish first between the spatial behaviour of intra-urban movers and that of in-migrants from other cities, regions and countries. Furthermore, in-migrants can be usefully divided into high- and low-status movers. The latter were particularly influential in shaping the residential structure of cities earlier this century. There are some cities, however, where low-status in-migrants continue to represent a significant component of migration patterns, as illustrated by population movements in Cincinnati, for example, where the in-migrants have been mainly poor whites from Appalachia; Australian cities such as Melbourne and Sydney, where the in-migrants are mainly foreign born; and some European cities, where the in-migrants are *gastarbeiter*. As we saw in Chapter 8, the impact of these in-migrants on urban social ecology is often finely tuned in relation to the national and regional origins, religion and ethnic status of the migrants involved.

High-status in-migrants are similar to low-status in-migrants in that the majority are drawn into the city in response to its economic opportunities. Their locational behaviour, however, is quite different. The majority constitutes part of a highly mobile group of the better-educated middle classes whose members move from one city to another in search of better jobs or career advancement. Some of these moves are voluntary and some are made in response to the administrative fiat of large companies and government departments. The vast majority, though, follow the same basic pattern, moving to a rather narrowly defined kind of neighbourhood: newly established suburban developments containing housing towards the top end of the price range.

Such areas are particularly attractive to the mobile elite because the lack of an established neighbourhood character and social network minimizes the risk of settling among neighbours who are unfriendly, too friendly, 'snobbish' or 'common': something that may otherwise happen very easily, since out-of-town households must usually search the property market and make a housing selection in a matter of days. Moreover, housing in such areas tends to conform to 'conventional' floor and window shapes and sizes, so that there is a good chance that furnishings from the previous residence will fit the new one. Nevertheless, once established in the new city, it is common for such households to make one or more follow-up or 'corrective' moves in response to their increasing awareness of the social ambience of different neighbourhoods and the quality of their schools and shops.

Intra-urban moves

This brings us to the general category of intra-urban moves that make up the bulk of all residential mobility and that therefore merit rather closer consideration. Indeed, a good deal of research effort has been devoted to the task of searching for regularities in intra-urban movement in the belief that such regularities, if they exist, might help to illuminate a key dimension of the sociospatial dialectic: the relationships between residential mobility and urban ecology.

Distance and direction

One of the most consistent findings of this research concerns the *distance moved*. In virtually every study, the majority of moves has been found to be relatively short, although the distances involved clearly depend to a certain extent on the overall size of the city concerned. This tendency for short moves notwithstanding, variability in distance moved is generally explained best by income, race and previous tenure, with higher-income, white, owner-occupier households tending to move furthest.

Directional bias has also been investigated in a number of migration studies, but with rather less consistent results. While it is widely recognized that there is a general tendency for migration to push outward

from inner-city neighbourhoods towards the suburbs, reverse flows and cross-currents always exist to complicate the issue. The most significant regularities in intra-urban movement patterns, however, relate to the relative *socioeconomic status* of origin and destination areas. The vast majority of moves – about 80 per cent in the United States – take place within census tracts of similar socioeconomic characteristics.

A parallel and related tendency is for a very high proportion of moves to take place within tenure categories. In other words, relocation within both 'community space' and 'housing space' usually involves only short distances. Where transitions do occur between tenure categories, a great deal depends on the ecology of housing supply. It follows from these observations that, while intra-urban mobility may have a significant impact on the spatial expression of social and economic cleavages, the overall degree of residential segregation tends to be maintained or even reinforced by relocation processes.

Household movement and urban ecology

Putting together these empirical regularities in an overall spatial context, we are presented with a three-fold zonal division of the city. The innermost zone is characterized by high levels of mobility, which are swollen by the arrival of low-status in-migrants. Similarly, high levels of mobility in the outermost zone are supplemented by the arrival and subsequent follow-up mobility of higher-status in-migrants. Between the two is a zone of relative stability containing households whose housing needs are evidently satisfied. Here, turnover is low simply because few housing opportunities arise, either through vacancies or through new construction. It is probably the existence of such a zone that accounts for longer-distance moves and that helps to explain the sectoral 'leap-frogging' of lower-middle class and working class households to new suburban subdivisions and dormitory towns.

The generalizations made here must be qualified in cities where there is a significant amount of public sector housing, since the entry and transfer rules for public housing are completely different from those in the rest of the housing market. In general this does not distort the overall pattern of household movement, although it is likely that different elements of the pattern will be linked to particular sectors of the housing market. In Glasgow, for example, where the privately rented and owner-occupied sectors are truncated by a massive public sector (in 2001, well over 55 per cent of the city's households lived in publicly owned dwellings), the overall pattern of residential mobility still exhibits the 'typical' components of short distance relocation within the neighbourhood of origin and of outward sectoral movement over larger distances. But a closer examination of migration flows reveals that these components are derived in composite fashion from the various flows within and between the main tenure categories.

The determinants of residential mobility

If the outward configuration of intra-urban mobility is difficult to pin down, its internal dynamics can be even more obscure. The flows of mobility that shape urban structure derive from aggregate patterns of demand for accommodation that spring in turn from the complex deliberations of individual households. An understanding of how these deliberations are structured is thus likely to provide some insight into the relocation process, and a considerable amount of attention has therefore been given by geographers to two important aspects of household behaviour:

➤ the decision to seek a new residence;

➤ the search for and selection of a new residence.

This two-stage approach is adopted here. First, attention is focused on the personal, residential and environmental circumstances that appear to precipitate the decision to move, and a conceptual model of the decision to move is outlined. Subsequently, attention is focused on how this decision is acted upon, highlighting the effects of differential access to, and use of, information.

Reasons for moving

In any consideration of migration it is important to make a distinction between *voluntary* and *involuntary* moves. As Rossi showed in his classic study of migration in Philadelphia, involuntary moves make up a significant proportion of the total. In Philadelphia,

almost a quarter of the moves were involuntary, and the majority of these were precipitated by the property demolitions and evictions (Rossi, 1980). Similar findings have been reported from studies of other cities, but remarkably little is known about the locational behaviour of affected households.

In addition to these purely involuntary moves is a further category of 'forced' moves arising from marriage, divorce, retirement, ill-health, death in the family and long-distance job changes. These frequently account for a further 15 per cent of all moves, leaving around 60 per cent as voluntary moves.

Survey data show that the decision to move home voluntarily is attributed to a number of quite different factors. It must be acknowledged, however, that the reasons given for moving in the course of household interviews are not always entirely reliable. Some people have a tendency to rationalize and justify their decisions, others may not be able to recollect past motivations; and most will inevitably articulate reasons that are simpler and more clear-cut than the complex of factors under consideration at the time of the move. Nevertheless, survey data are useful in indicating the major elements that need to be taken into consideration in explaining movement behaviour. Typically, the reasons given for moving involve a mixture of housing, environmental and personal factors.

Among the more frequently cited housing factors associated with voluntary moves are complaints about dwelling and garden space, about housing and repair costs and about style obsolescence. Environmental factors encompass complaints about the presence of noxious activities such as factories, about noisy children and about the incidence of litter, garbage and pet dogs. Personal factors are mostly associated with forced moves, but some voluntary moves are attributed to personal factors, such as a negative reaction to new neighbours. Figure 12.3 illustrates a general classification of the reasons for household relocation.

Space needs and life-course changes

Of the more frequently cited reasons for moving, it is generally agreed that the most important and widespread is related to the household's need for dwelling space.

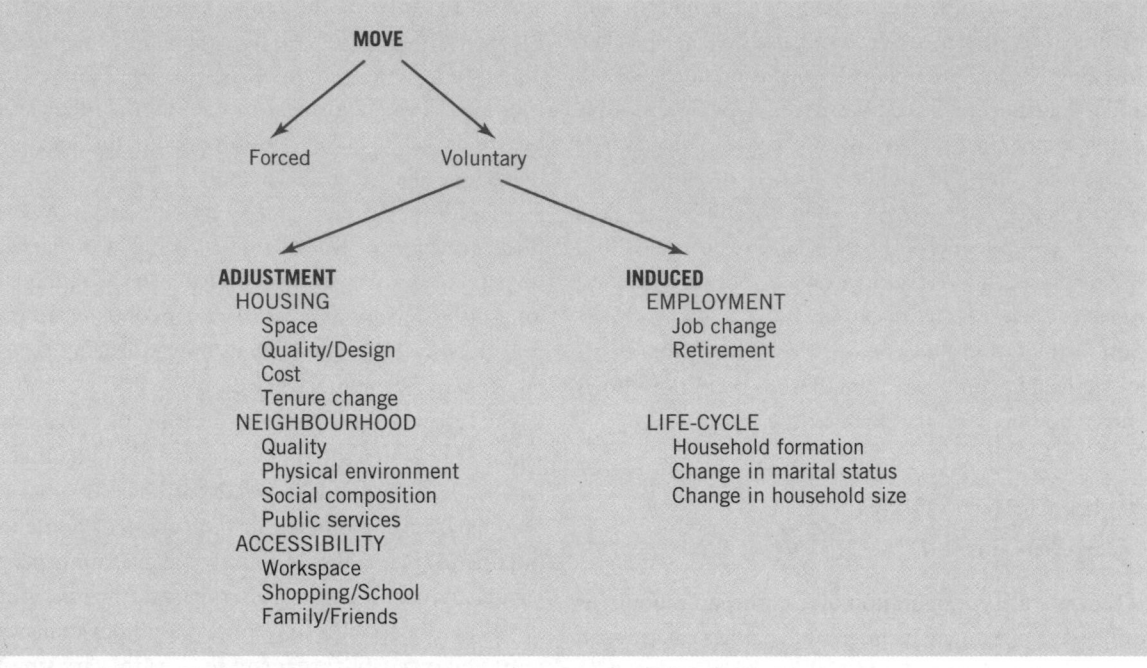

Figure 12.3 A classification of reasons for household relocation.
Source: Clark and Onaka (1983), Fig. 2, p. 50.

Box 12.1

Key debates in urban social geography – The value of the life-cycle model

In his now classic study, Rossi linked residential differentiation with changes in the *family life cycle* (or life course). As people get older, the things they require from the environment (both the dwelling and the neighbourhood) change, and, if these needs cannot be satisfied in the immediate environment, then movement is often the result. The family life-cycle model draws attention to the fact that the family is not a static entity and has therefore played an important part in social research (see below).

1. The honeymoon period	Married couples, without children
2. The nurturing period	Childbearing families eldest aged 0–2
3. The authority period	Families with preschoolchildren
4. The interpretative period	Families with schoolchildren
5. The interdependent period	Families with teenagers
6. The launching period	Families launching young adults
7. The empty-nest period	Middle-aged parents, no children home
8. The retirement period	Ageing family members, retired

Adapted from Duvall, E.M. (1971) *Family Development* Philadelphia, J.B. Lippincott.

Although undoubtedly influential, the family-life cycle model has been criticized on a number of grounds:

➤ The model only describes 'traditional' nuclear families.

continued

- The model takes no account of those who are single parents, who remarry, become divorced or widowed, or have a second family.
- By starting with marriage, the model takes no account of the overlap between families, and neither does it take into account the family from which the newly-weds have come.
- The model assumes families are isolated and self-sufficient.

There is much talk in the popular media about the decline of the 'traditional' nuclear family. It is certainly true that since 1980 family life in Western cities had been through a period of unprecedented change: there are fewer marriages, fewer children, more couples are getting divorced and more people living in single-person households. To illustrate, some of the major changes for the UK are shown below:

Changes in household composition in Britain, 1961 to 1998

Types of household	1961	1971	1981	1991	1998
Couples					
Dependent children	38	35	31	31	25
No dependent children	10	8	8	8	7
No children	26	27	26	28	28
Lone parents					
Dependent children	2	3	5	6	7
No dependent children	4	4	4	4	3
Multi-families	3	1	1	1	1
One person	11	18	22	27	28

The numbers of households with couples with dependent children fell from 38 per cent to 25 per cent between 1961 and 1998 while the number of lone parents with dependent children rose from 2 per cent to 7 per cent. The most striking change, however, has been the increase in one-person households. This change reflects a number of factors: a greater capacity for younger people to leave their parental home earlier; higher rates of divorce; a reduced preference for marriage and an increasingly aged population in which a high proportion of people are widowed. However, most commentators suggest that it is far too early to talk about the decline of the traditional family. For example, in the United Kingdom in 1997 four-fifths of dependent children had two parents and 9 in 10 parents were married. Contacts between relatives seem to have declined but families still continue to be the main source of care for elderly family members when in they are need. Many writers have pointed out that there is little evidence of families doing less. Thus the percentage of elderly people in institutions in Britain remained remarkably constant throughout the twentieth century.

Key concepts associated with the family life cycle (see Glossary)

Community care, domestic economy, domestication, family status, heteropatriarchal environment, nuclear family, patriarchy, primary relationships.

More than half of the movers in Rossi's study cited complaints about too much or too little living space as contributing to their desire to move (with 44 per cent giving it as a primary reason). Subsequent surveys have confirmed the decisive importance of living-space in the decision to move and, furthermore, have established that the crucial factor is not so much space *per se* but the relationship between the size and composition of a household and its *perceived* space requirements. Because both of these are closely related to the family life-course, it is widely believed that life-course changes provide the foundation for much of the residential relocation within cities. The importance of the family life-course as an explanatory variable is considerably reinforced by its relationships with several other frequently cited reasons for moving, such as the desire to own (rather than rent) a home and the desire for a change of environmental setting.

Changes in household structure and the fragmentation of lifestyles in contemporary cities make it difficult, however, to generalize about relationships between residential mobility and family life-course in the way that was possible in the 1960s (contrast Abu-Lughod and Foley, 1960, with Gober *et al.*, 1991). We can say, though, that a marked residential segregation tends to emerge as households at similar stages in their life-course respond in similar ways to their changing domestic and material circumstances. This, of course, fits conveniently with the results of the many descriptive studies (including factorial ecology studies) that have demonstrated a zonal pattern of family status.

The generally accepted sequence to these zones runs from a youthful inner-city zone through successive zones of older and middle-aged family types to a zone of late youth/early middle-age on the periphery. It must be acknowledged, however, that such a pattern may be the result of factors other than those associated directly with the dynamic of household life courses. Households often undergo changes in their family status at the same time as they experience changes in income and social status, so it is dangerous to explain mobility exclusively in terms of one or the other.

Quite different factors may also be at work. Developers, for example, knowing that many households prefer to live among families similar to their own age and composition as well as socioeconomic status, have reinforced family status segregation by building apartment complexes and housing estates for specific household types, with exclusionary covenants and contracts designed to keep out 'non-conforming' residents. It is thus quite common for entire condominiums to be inhabited by single people or by childless couples. The extreme form of this phenomenon is represented by Sun City, a satellite suburb of Phoenix, Arizona, where no resident under the age of 50 is allowed, and where the whole townscape is dominated by the design needs of the elderly, who whirr along the quiet streets in golf caddy-cars, travelling from one social engagement to the next. In contrast, much of the family status segregation in British cities with large amounts of public housing can be attributed to the letting policies of local authorities, since eligibility for public housing is partly a product of household size (see pp. 147–48).

The decision to move

The first major decision in the residential mobility process – whether or not to move home – can be viewed as a product of the *stress* generated by discordance between household's needs, expectations and aspirations on the one hand and its actual housing conditions and environmental setting on the other (Fig. 12.4).

People's housing *expectations* and *aspirations* are thought by behaviouralists to stem from the different frames of reference that people adopt in making sense of their lives and, in particular, in interpreting their housing situation. These frames of reference are the product of a wide range of factors that includes age, class background, religion, ethnic origin and past experience of all aspects of urban life. What they amount to is a series of lifestyles – privatized, familistic, cosmopolitan and so on – each with a distinctive set of orientations in relation to housing and residential location. Behaviouralists recognize three traditional lifestyle orientations in urban cultures: family, career and consumption.

Family-oriented people are also home-centred and tend to spend much of their spare time with their children. As a result, their housing orientations are dominated by their perceptions of their children's needs for play space, a clean, safe environment, proximity to child clinics and school, and so on.

Careerists have a lifestyle centred on career advancement. Since movement is often a necessary part of this process, careerists tend to be highly mobile; and since they are, by definition, status-conscious, their housing orientations tend to be focused on prestige neighbourhoods appropriate to their jobs, their salary and their self-image.

Consumerists are strongly oriented towards enjoying the material benefits and amenities of modern urban society, and their housing preferences are therefore dominated by a desire to live in downtown areas, close to clubs, theatres, art galleries, discotheques, restaurants and so on.

This typology can be criticized for its middle-class tenor, since it overlooks the 'lifestyle' of the large number of households whose economic position reduces their housing aspirations to the level of survival. Many working-class households view their homes as havens

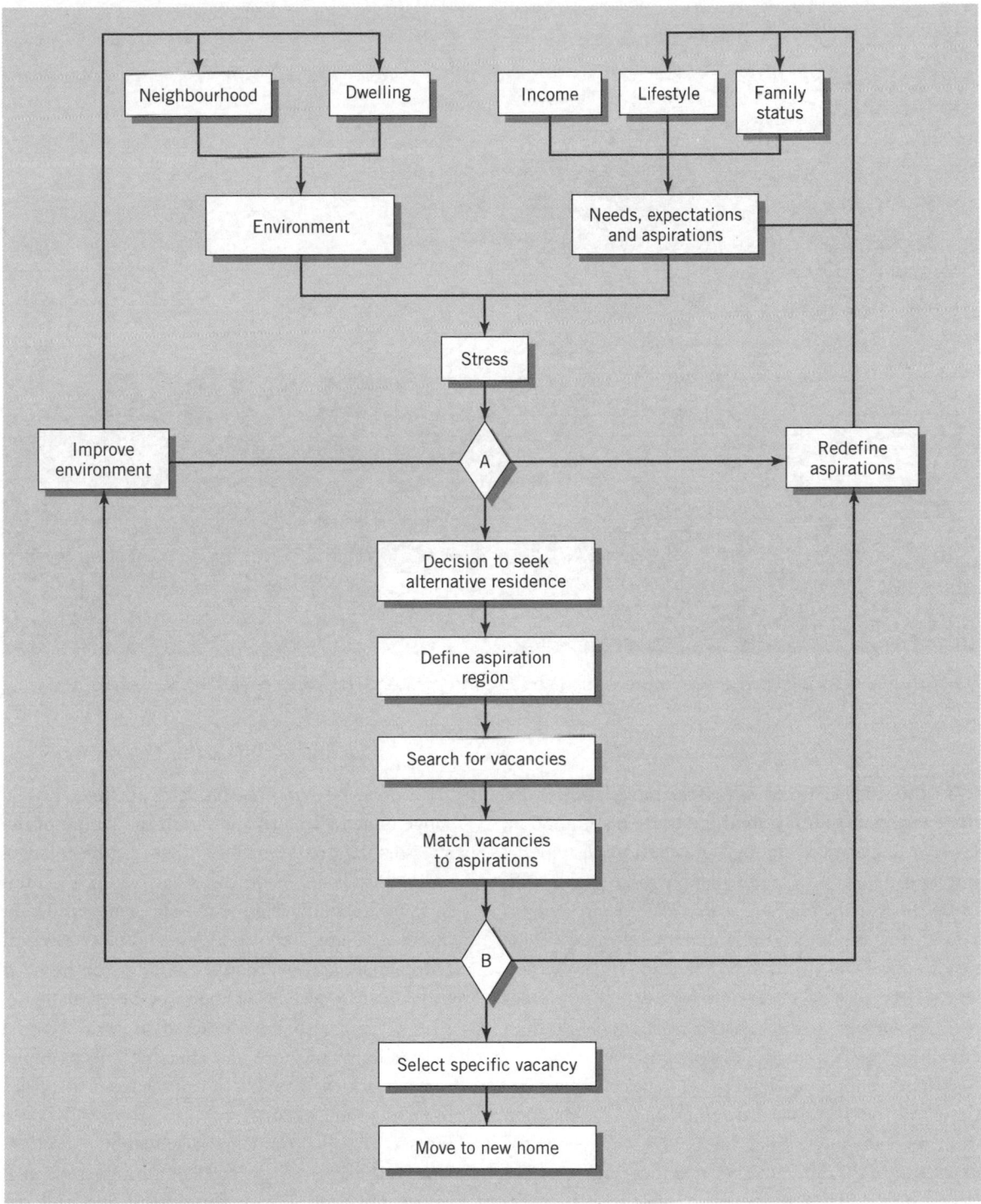

Figure 12.4 A model of residential mobility.
Source: Robson (1975), p. 33.

As high-rise falls into disrepute in the social sector, the affluent return to revivified waterfront redevelopments. Photo Credit: Andrew Vowles.

from the outside world rather than as platforms for the enactment of a favoured lifestyle. This limitation, however, is allowed for by the behaviouralist model outlined in Fig. 12.4. Quite simply, households with more modest incomes are expected to aspire only to housing that meets their minimum absolute needs. These needs are generally held to be a function of family size, so that the idea of life-course related mobility plugs in conveniently to the behaviouralist model without equating stress automatically with mobility.

Whatever the household's expectations and aspirations may be, the crucial determinant of the decision to move is the intensity of the stress (if any) generated as a result of the difference between these and its actual circumstances. The point where tolerable stress becomes intolerable strain will be different for each household but, once it is reached (at point 'A' on Fig. 12.4) the household must decide between three avenues of behaviour:

➤ *Environmental improvement.* This includes a wide range of activities, depending on the nature of the stressors involved. Small dwellings can be enlarged with an extension, cold dwellings can have central heating, double glazing and wall cavity insulation installed, dilapidated dwellings can be rewired and redecorated, and over-large dwellings can be filled by taking in lodgers. Neighbourhood or 'situational' stressors can also be countered in various ways: inaccessibility to shops and amenities, for example, can be tackled through the purchase of a car or by petitioning the local authority to provide better bus services. Environmental degradation and intrusive land-users can be tackled through residents' associations and action committees; and undesirable neighbours can be harassed or ostracized. As with other aspects of residential behaviour, these strategies vary in their appeal according to household circumstances. Owner-occupiers, for example, are much more likely to opt for neighbourhood activism than renters.

➤ *Lowering aspirations.* This is an alternative means of coming to terms with existing housing conditions. It appears to be a common strategy, since survey data show that for every household that moves there are two or three more who report that they would like to move if they could. Lowering aspirations may involve a change in lifestyle or a reformulation of plans: the decision to have children may be deferred, for example. More commonly it is simply a psychological matter of 'dissonance reduction' – learning to like what one has and to become indifferent to what one knows one cannot get. Not surprisingly, the older people get, the more proficient they become at dissonance reduction.

➤ *Residential relocation.* This, as we have seen, is the course chosen by a large minority of households. The decision to move, however, leads to a second important area of locational behaviour: the search for and selection of a new residence.

The search for a new residence

Whether the decision to move is voluntary or involuntary, all relocating households must go through the procedure of searching for suitable vacancies and then deciding upon the most appropriate new home. The chief interest of geographers in this procedure lies in the question of whether it is spatially biased and, if so, whether it is biased in different ways for different groups of households. In other words, do spatially biased search procedures contribute to the changing social ecology? Although information on the way people behave in looking for a new home is rather fragmentary, the general process is conveniently encompassed within the decision-making framework of the behavioural model (Fig. 12.4). Accordingly, it is useful to break household behaviour down into three stages:

➤ the specification of criteria for evaluating vacancies;
➤ the search for dwellings that satisfy these criteria;
➤ the final choice of a new dwelling.

Specifying the desiderata of a new home

In behaviouralist terms, the household's first step in coping with the problem of acquiring and organizing information about potential new dwelling is to define, consciously or subconsciously, its **aspiration region**. Quite simply, this is a conception of the limits of acceptability that a household is prepared to entertain as an alternative to its current accommodation. These limits may be defined in terms of the desired *site* characteristics – attributes of the dwelling itself – and/or the desired *situational* characteristics – the physical and social environment of the neighbourhood, its proximity to schools, shops, etc.

The lower limits of the aspiration region are commonly defined by the characteristics of the dwelling the household wants to leave, while the upper limits are set by the standards to which the household can reasonably aspire. In many cases these will be determined by income constraints, but there are important exceptions: some householders, for example, may not want to take on a large garden, regardless of house price; others may rule out affordable dwellings in certain areas because the neighbourhood does not conform with their desired lifestyle.

In general, the criteria used by households in specifying their aspiration region reflect their motivations in deciding to move. Living-space, tenure, dwelling amenities, environmental quality and social composition are among the more frequently cited criteria. Interior aspects of the dwelling, the social characteristics of the neighbourhood and accessibility to various facilities are more important in attracting people to a new home than in propelling them away. It also appears that movers of different types tend to differ quite a lot in the criteria they use. Households moving to single-family houses are more likely to be concerned with situational characteristics than those moving to apartments, for example. Households moving to suburban houses tend to be particularly concerned with the layout of the dwelling and its potential as an investment, whereas for those moving to downtown houses the aesthetics of dwelling style and the neighbourhood environment tend to be more important criteria.

The existence of differently conceived aspiration regions is, of course, a function of the different needs and aspirations that prompt households to move in the first place. Their significance to the relocation process lies not only in the consequent differences in households' evaluation of particular housing opportunities, but also in the fact that households set out from the very start to look for vacancies with quite different housing goals in mind.

Searching for vacancies

The general objective of the search procedure is to find the right kind of dwelling, at the right price, in the time available. It must be acknowledged that there are some households that do not have to search deliberately because their decision to move has come after accidentally discovering an attractive vacancy. These 'windfall' moves may account for as many as 25 per cent of all intra-urban moves. The majority of movers, though, must somehow organize themselves into finding a suitable home within a limited period of deciding to relocate.

Most households organize the search procedure in locational terms, focusing attention on particular neighbourhoods that are selected on the basis of their perceived *situational* characteristics and the household's evaluation of the probability of finding vacancies satisfying their *site* criteria. Moreover, faced with the problem of searching even a limited amount of space, it is natural that households will attempt to further reduce both effort and uncertainty by concentrating their search in areas that are best-known and most accessible to them.

The upshot is that households concentrate their house-hunting activities within a limited **search space** that is spatially biased by their familiarity with different districts. In behaviouralist terminology, this search space is a subset of a more general **awareness space**, which is usually regarded as a product of:

➤ people's **activity space** or action space (the sum of all the places with which people have regular contact as a result of their normal activities); and

➤ information from secondary sources such as radio, television, newspapers and even word-of-mouth.

Both elements are subject to a mental filtering and coding that produces a set of imagery that constitutes the operational part of the individual's awareness space. The subset of this space that constitutes the search space is simply the area (or areas) that a household feels to be relevant to its aspiration region, and it is spatially biased because of the inherent bias in both activity spaces and mental maps.

It follows that *different subgroups of households, with distinctive activity spaces and mental maps, will tend to exhibit an equally distinctive spatial bias in their search behaviour.* In particular, we may expect the more limited activity spaces and more localized and intensive images of the home area to restrict the search space of low-income households to a relatively small area centred on the previous home, while more mobile, higher-income households will have a search space that is more extensive but focused on the most familiar sector of the city between home and workplace.

The *information sources* used to find vacant dwellings within the search space can also exert a significant spatial bias. Since different types of households tend to rely on different sequences and combinations of sources, this results in a further process of sociospatial sorting. Overall, the most frequently used sources of information about housing vacancies are newspaper advertisements, real estate agents, friends and relatives, and personal observation of 'for sale' signs – although their relative importance and effectiveness seems to vary somewhat from one city to another.

Each of these information sources tends to be biased in a different way. Personal observation, for example, will be closely determined by people's personal activity space, while the quantity and quality of information from friends and relatives will depend a lot on social class and the structure of the searcher's social networks. Real estate agents also exert a considerable spatial bias in their role as mediators of information. This has been shown to operate in two ways: first, each business tends to specialize in limited portions of the housing market in terms of both price and area; second, while most estate agents have a fairly accurate knowledge of the city-wide housing market, they tend to over-recommend dwellings in the area in which they are most experienced in selling and listing accommodation and with which they are most familiar. As a result, households that are dependent on real-estate agents for information are using a highly structured and spatially limited information source. The critical issue in the present context, however, is *the relative importance and effectiveness of different information sources for different households.*

Accessibility to information sources is also related to another important issue affecting residential behaviour: the problem of *search barriers*. There are two important aspects of this issue: barriers that raise the costs of searching or gathering information, and barriers that explicitly limit the choice of housing units or locations available to households. Factors related to search costs include, for example, lack of transporta-

tion for searching and lack of child-care facilities while searching, as well as lack of knowledge about specific information channels. Factors that limit housing choice include financial constraints, discrimination in the housing market, and the housing quality standards of rent assistance programmes.

Time constraints

The differential use and effectiveness of different information sources for different households clearly serves to increase the degree of sociospatial sorting arising from residential mobility, while at the same time making it more complex. Another important

---

Box 12.2

Key thinkers in urban social geography – Ron Johnston

Ron Johnston deserves a special mention in any book on urban social geography for one simple reason – no one else has written more on the topic. For almost 40 years he has produced a torrent of books and articles, an outpouring that shows no signs of diminishing (contrast Johnston, 1971, with Johnston *et al.*, 2004, see also Box 8.3).

Ron Johnston's work reflects the enormous shifts in the character of urban social geography over the past four decades. For example, in his early work he was one of the new breed of quantitative geographers who linked multivariate studies of city structure to the empirical traditions of the Chicago School of human ecology via the methodology of 'factorial ecology' (Johnston, 1971). A little later he adapted behavioural notions to the study of residential mobility in cities, linking mental maps with migration patterns. Subsequently, he linked urban change with the traditions of political economy, focusing in par-

ticular upon local political systems (Johnston, 1978).

Despite all his writings on urban social geography, Johnston is perhaps best acknowledged for his extensive contributions to two other fields. First, for his work on political geography (e.g. Johnston *et al.*, 2001) and, second, and perhaps even more importantly from the perspective of geography, for his charting of the evolution of the subject in the second half of the twentieth century. This latter project has been manifest most prominently in *Geography and Geographers: Anglo-American Human Geography since 1945*, first published in 1979 and now into its sixth edition (now co-authored with James Sidaway, 2004).

Key concepts associated with Ron Johnston (see Glossary)

De jure territories, factorial ecology, jurisdictional partitioning, segregation.

Further reading

Johnston, R.J. (1971) *Urban Residential Patterns: An Introductory Review* Bell and Sons Ltd, London

Johnston, R.J. (1978) *Political, Electoral and Spatial Systems* Oxford University Press, London

Johnston, R.J. and Sidaway, J.D. (2004) *Geography and Geographers: Anglo-American Human Geography Since 1945* Arnold, London

Johnston, R.J., Pattie, C.J., Dorling, D.F.L. and Rossiter, D.J. (2001) *From Votes to Seats: The Operation of the UK Electoral System Since 1945* Manchester University Press, Manchester

Johnston, R., Poulsen, M. and Forrest, J. (2004) The comparative study of ethnic residential segregation in the USA, 1980–2000 *Tijdschrift voor Economische en Sociale Geografie*, **95**, 550–569

Links with Other Chapters

Chapter 4: Studies of Factorial Ecology

compounding factor in this sense is the constraint of *time* in the search procedure. Both search space and search procedures are likely to alter as households spend increasing amounts of time and money looking for a new home. When time starts to run out, the search strategy must change to ensure that a home will be found. Anxiety produced by a lack of success may result in a modification of the household's aspiration region, a restriction of their search space, and a shift in their use of information sources; and the pressure of time may lead people to make poor choices.

On the other hand, the longer the search goes on, the greater the household's knowledge of the housing market. Each household therefore has to balance the advantages of searching and learning against the costs – real and psychological – of doing so. Survey data in fact show a consistent tendency for the majority of households to seriously consider only a few vacancies (usually, only two or three) before selecting a new home, an observation that may appear to undermine the utility of developing elaborate models and theories of search behaviour. Nevertheless, this phenomenon can itself be explained with a behaviouralist framework: households are able to reduce the element of uncertainty in their decision-making by restricting serious consideration to only a few vacancies. Moreover, most households begin with an aspiration region that is quite narrowly defined (either because of income constraints or locational requirements), so that what appears to be an inhibited search pattern is in fact a logical extension of the decisions formulated in the preceding stage of the search procedure.

Choosing a new home

Households that find two or more vacancies within their aspiration region must eventually make a choice. Theoretically, this kind of choice is made on the basis of household *utility functions* that are used to give a subjective rating to each vacancy. In other words, vacancies are evaluated in terms of the weighted sum of the attributes used to delineate the aspiration region. These weights reflect the relative importance of the criteria used to specify the aspiration region, and so they will vary according to the preferences and predilections of the household concerned.

In practice, however, the constraints of time, coupled with the limitations of human information-processing abilities and a general lack of motivation, mean that a real choice of the kind implied in behaviouralist theory is seldom made: people are happy to take any reasonable vacancy, so long as it does not involve a great deal of inconvenience.

The behavioural model allows for those households that are unable to find vacancies within their aspiration region in the time available to them (point 'B' on Fig. 12.4) to change their strategy to one of the two options open to them at point 'A' on the diagram: environmental improvement or a redefinition of aspirations.

Finally, we must recognize that there are many households in every city whose residential location is constrained to the point where behavioural approaches are of marginal significance. The most obvious subgroups are the working poor, the elderly, the very young, the unemployed and the transient. Other subgroups whose residential choice is heavily constrained include households who have special needs (e.g. large families, single-parent families, non-married couples, former inmates of institutions and 'problem' families), households that cannot relocate because of personal handicaps, family situations or medical needs; and households that are unwilling to move because of the psychological stress of moving from familiar environments.

12.2 Residential mobility and neighbourhood change

Although the behavioural approach provides important insights into the spatial implications of mobility, the emphasis on individual decision-making tends to divert attention from the aggregate patterns of neighbourhood change that result as similar households make similar choices. In this section, therefore, some consideration is given to the macro-scale generalizations that have been advanced about processes of mobility and neighbourhood change.

One scheme that has already been introduced and discussed is the zonal patterning of socioeconomic status associated with the sequence of invasion–succession–dominance postulated by Burgess in his classic model of ecological change. The dynamic of this model, it will be recalled, was based on the pressure of low-status in-migrants arriving in inner-city areas. As this pressure increases, some families penetrate surrounding neighbourhoods, thus initiating a chain reaction whereby the residents of each successively higher-status zone are forced to move further out from the centre in order to counter the lowering of neighbourhood status.

Notwithstanding the criticisms of ecological theory per se, with its heavy reliance on biotic analogy, the concept of invasion–succession–dominance provides a useful explanatory framework for the observed sequence of neighbourhood change in cities where rapid urban growth is fuelled by large-scale in-migration of low-status families. The classic example, of course, was Chicago during the 1920s and 1930s, although many of the industrial cities in Britain had undergone a similar process of neighbourhood change during the nineteenth century. More recently, the flow of immigrants to London, Paris and larger Australian cities such as Melbourne and Sydney and the flow of *gastarbeiter* to the industrial cities of northwestern Europe has generated a sequence of change in some

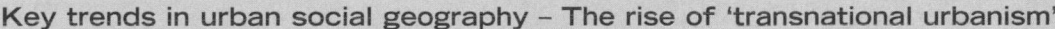

Box 12.3

Key trends in urban social geography – The rise of 'transnational urbanism'

Transnational urbanism is a term coined by Michael Peter Smith to encapsulate contemporary forms of urbanism resulting from globalization. In particular, the term relates to changing patterns of immigration into major 'world cities' such as London, New York and Los Angeles. In the past migrants were thought of as either temporary (sending back income to their families in their country of origin) or else permanent (and thus tending to sever ties with their country of birth, and often assimilating into the country of residence). However, increasingly migrants tend to be semi-permanent, maintaining links with more than one nation. Such divided loyalties are enhanced by new forms of global communication that permit media in many languages to be sent widely around the world (e.g. multilanguage cable and satellite television systems).

The controversial futurologist Ohmae summed this up in his book on *The End of the Nation State* (1995) as follows (p. 35):

There is a genuinely cross-border civilization nurtured by exposure to common technologies and sources of information, in which horizontal linkages within the same generation in different parts of the world are stronger than traditional vertical linkages between generations in particular parts of it.

Transnational migration often involves two types of workers at extreme ends of the social spectrum. On the one hand there are the highly skilled managerial, scientific and technological elites who work for major corporations, educational institutions, medical facilities and governmental bodies. At the other extreme there are the relatively low-paid workers who engage in employment that is often unpopular in the host nation (or at lower wages than the indigenous workers are prepared to tolerate) – cleaning, restaurant work, taxi driving, etc. However, given increasing labour shortages in many Western societies, many transnational migrants are taking up intermediate level occupations such as nurses, plumbers and electricians.

When transnational migrants display structural rather than behavioural assimilation, they may cluster, displaying 'ethnoscapes'.

Key concepts associated with transnational urbanism (see Glossary)

Behavioural assimilation, diaspora, ethnoscape, global cites, structural assimilation, world cities.

Further reading

Ohmae, K. (1995) *The End of the Nation State: The Rise of Regional Economies* Free Press, New York

Rofe, M. (2003) 'I want to be global': theorizing the gentrifying class as a global elite *Urban Studies* **40**, 2511–2526

Smith, M.P. (2001) *Transnational Urbanism: Locating Globalization* Blackwell, Oxford

Links with Other Chapters

Chapter 1: Box 1.3 Growing Cosmopolitanism in Western Cities: The Example of London

Chapter 2: Box 2.2 Manuel Castells; Globalization: Knowledge Economies and the Informational City

Chapter 8: Box 8.1 The Latinization of US Cities

neighbourhoods which also fits the invasion/succession model. Nevertheless, this model is of limited relevance to most modern cities, since its driving force – the inflow of low-status migrants – is of diminishing importance; the bulk of in-migrants is now accounted for by middle-income families moving from a suburb in one city to a similar suburb in another.

High-status movement, filtering and vacancy chains

An alternative view of neighbourhood change and residential mobility stems from Homer Hoyt's (1939) **sectoral model** of urban growth and socioeconomic structure. Hoyt's ideas were derived from a detailed study of rental values in 142 US cities that was undertaken in order to classify neighbourhood types according to their mortgage lending risk. This study led him to believe that the key to urban residential structure is to be found in the behaviour of high-status households. High-status households, he argued, pre-empt the most desirable land in a growing city, away from industrial activity.

With urban growth, the high-status area expands axially along natural routeways, in response to the desire among the well-off to combine accessibility with suburban living. This sectoral movement is reinforced by a tendency among 'community leaders' to favour non-industrial waterside sites and higher ground; and for the rest of the higher-income groups to seek the social cachet of living in the same neighbourhood as these *prominenti*. Over time, further sectoral development occurs as the most prominent households move outwards to new housing in order to maintain standards of exclusivity. In the wake of this continual outward movement of high-status households, the housing they vacate is occupied by middle-status households whose own housing is in turn occupied by lower-status households (a process termed **filtering**). At the end of this chain of movement, the vacancies created by the lowest-status groups are either demolished or occupied by low-status in-migrants. Subsequently, as other residential areas also expand outwards, the sectoral structure of the city will be preserved, with zonal components emerging as a secondary element because of variations in the age and condition of the housing stock.

The validity of Hoyt's sectoral model has been much debated. Empirical studies of the emerging pattern of elite residential areas and tests of the existence of sectoral gradients in socioeconomic status have provided a good deal of general support for the spatial configuration of Hoyt's model, although the relative dominance of sectoral over zonal components in urban structure is by no means a simple or universal phenomenon.

It is the *mechanism* of neighbourhood change implied in Hoyt's model that is of interest here, however. The basis of this mechanism is the chain of moves initiated by the construction of new dwellings for the wealthy, resulting in their older properties filtering down the social scale while individual households filter up the housing scale. In order for this filtering process to operate at a sufficient level to have any real impact on urban structure, there has to be more new construction than that required simply to replace the deteriorating housing of the elite.

According to Hoyt, this will be ensured by the *obsolescence* of housing as well as its physical deterioration. For the rich, there are several kinds of obsolescence that may trigger a desire for new housing. Advances in kitchen technology and heating systems and the innovation of new luxury features such as swimming pools, saunas and jacuzzis may cause 'functional obsolescence', while more general social and economic changes may cause obsolescence of a different kind: the trend away from large families combined with relative increase in the cost of domestic labour, for example, has made the mansion something of a white elephant.

Changes in design trends may also cause obsolescence – 'style obsolescence' – in the eyes of those who can afford to be sensitive to architectural fads and fashions. Finally, given a tax structure that allows mortgage repayments to be offset against taxable income, dwellings may become 'financially obsolescent' as increases in household income and/or inflation reduce the relative size of mortgage repayments (and therefore of tax relief).

Driven on to new housing by one or more of these factors, the wealthy will thus create a significant number of vacancies that the next richest group will be impelled to fill through a desire for a greater quantity and/or quality of housing. This desire can be seen not only as the manifestation of a general preference for better housing but also as a result of the influence of changing housing needs associated with the family life cycle. In addition, the social and economic pressures resulting from proximity to the poorest groups in society may prompt those immediately above them to move as soon as the opportunity presents itself, either by moving into vacancies created by the construction of new housing for other or by moving out into subdivisions specially constructed for the lower-middle classes.

Obstacles to filtering

In practice, however, the dynamics of the housing market are rather more complex than this. To begin with, **vacancy chains** may start in other ways than the construction of new housing. A substantial proportion of vacancies arise through the subdivision of dwelling units into flats and the conversion of non-residential property to residential uses. Even more occur through the death of a household, through the move of an existing household to share accommodation with another, and through emigration outside the city.

Similarly, vacancy chains may be ended in several ways other than the demolition of the worst dwellings or their occupation by poor in-migrants. Some vacancies are rendered ineffective through conversion to commercial use, while others may be cancelled out by rehabilitation or conversion schemes that involve knocking two or more dwellings into one. Vacancy chains will also end if the household that moves into a vacant dwelling is a 'new' one and so leaves no vacancy behind for others to fill. This may arise through the marriage of a couple who had both previously been living with friends or parents, through divorced people setting up separate homes, or through the splitting of an existing household with, for example, a son or daughter moving out to their own flat.

The concept of filtering has important policy implications, since it can be argued that facilitating new house-building for higher-income groups will result in an eventual improvement in the housing conditions of the poor through the natural process of filtering, without recourse to public intervention in the housing market. This argument has a long history, dating to the paternalistic logic of nineteenth-century housing reformers who used it to justify the construction of model housing for the 'industrious' and 'respectable' working classes rather than the poorest sections of society to whom their efforts were ostensibly directed. Subsequently, it became the central plank of government housing policy in many countries. Up to the 1930s, Britain relied almost entirely on the filtering process to improve the housing conditions of the working classes, while it still remains the basis of US housing strategy. The effectiveness of such policies, however, depends on the length of the vacancy chains that are set in motion by the construction of new housing. Relatively few studies have been able to furnish detailed empirical evidence, and their results are rather inconclusive.

Vacancy chains

The evidence that is available seems to suggest that an upward filtering of households does arise from the construction of few homes for the wealthy. Nevertheless, closer inspection of the results shows that the benefits to poor families (in terms of vacant housing opportunities) are not in proportion to their numbers, *suggesting that filtering is unlikely to be an important agent of neighbourhood change in poor areas.* Moreover, the fact that a large proportion of the vacancy chains end through the formation of 'new' households while only a small proportion end through demolitions also suggests that the filtering mechanism rarely penetrates the lower spectrum of the housing market to any great extent.

In summary, filtering offers a useful but nevertheless partial explanation of patterns of neighbourhood change. Among the factors that can be identified as inhibiting the hypothesized sequence of movement arising from new high-status housing are:

➤ the failure of high-income housing construction to keep pace with the overall rate of new household formation and in-migration;

➤ the structure of income distribution which, since higher-income groups constitute a relatively small class, means that the houses they vacate in preference for new homes are demanded by a much larger group, thus maintaining high prices and suppressing the process of filtering;

➤ the inertia and non-economic behaviour of some households; this includes many of the behavioural patterns discussed above, although the most striking barrier to the filtering process is the persistence of elite neighbourhoods in symbolically prestigious inner-city locations;

➤ the existence of other processes of neighbourhood change – related to invasion/succession, household life-courses, gentrification – whose dynamic is unrelated to the construction of new, high-income housing.

Chapter summary

12.1 Patterns of intra-urban residential mobility, though complex and varied, reveal a number of broad regularities. Moves of residence within cities are typically over short distances with a tendency to move outwards towards suburban areas. People move for a complex mixture of voluntary and involuntary reasons and the choice of new residence depends upon channels of information about vacancies and the housing opportunities at the time of the move.

12.2 The aggregate effects of residential mobility can have profound effects upon urban social geography. Whereas the Burgess concentric ring model suggests pressure from new migrants is the main 'push' for outmigration, Hoyt's sectoral model suggests that the 'pull' or filtering effect of properties vacated by the more affluent is the primary mechanism at work. There is some evidence for filtering but this is only a partial explanation for neighbourhood change.

Key concepts and terms

activity (or action) space
aspiration region (or space)
awareness space

ecological fallacy
filtering
search space

sectoral model
vacancy chain

Suggested reading

A thorough review of the issues covered in this chapter is provided by Martin Cadwallader in his book on *Migration and Residential Mobility* (1992: University of Wisconsin, Madison; see especially Chapters 5 and 6), which also provides a discussion of the overall theoretical frameworks within which residential mobility can be understood. Useful literature reviews can be found in M. Munro's chapter in Michael Pacione's edited volume on *Social Geography: Progress and Prospect* (1987: Croom Helm, Beckenham), and in Patricia Gober's review of the literature on urban housing demography (*Progress in Human Geography*, 1992: 16, 171–189). Coverage of behavioural models and analyses is provided by William A.V. Clark's edited volume, *Modelling Housing*

Market Search (1982: Croom Helm, Beckenham). The most recent and sophisticated analysis of household migration is William A.V. Clark and Frans M. Dielemann's *Households and Housing: Choices and Outcomes in the Housing Market* (1996: Center for Urban Policy Research, Rutgers University, Rutgers, NJ). See also W.A.V. Clark and Y. Huang 'The life course and residential mobility in British housing' (*Environment and Planning A*, 2003: 35, 323–339). For mobility patterns within the rented sector see P. Kemp and M. Keoghan 'Movement into and out of the private rental sector in England' (*Housing Studies*, 2001: 16, 21–37) and for mobility within social housing in the UK see H. Pawson and G. Bramley 'Understanding recent trends in residential mobility in council housing in England' (*Urban Studies*, 2000: 37, 1231–1259).

Urban change and conflict

Key questions addressed in this chapter

➤ Why are the main sources of conflict between differing neighbourhoods in cities?

➤ What are the main patterns of service allocation in cities?

➤ What are the main reasons growing social inequality in Western cities?

13.1 Externality effects

Much of the stress that is central to behaviouralists' approaches to urban social geography derives from households' desire to maximize the net **externalities** of urban life. Externalities, sometimes called **spillover** or **third-party** effects, are unpriced by-products of the production or consumption of goods and services of all kinds. An externality effect exists if the activity of one person, group or institution impinges on the welfare of others. The classic example is the factory that pollutes local air and water supplies in the course of its operations, bringing *negative* externalities to nearby residents. In contrast, well-kept public parks produce *positive* externalities for most nearby residents.

The behaviour of private individuals also gives rise to externality effects. These can be divided into 'public behaviour' externalities and 'status' externalities. The former include people's behaviour in relation to public comportment (e.g. quiet, sobriety and tidiness), the upkeep of property and the upbringing of children. Status externalities relate to the 'reflected glory' (or otherwise) of living in a distinctive neighbourhood.

Externality effects therefore can take a variety of forms. They are, moreover, very complex in operation. Consider, for example, the behaviour of a household in adding an imitation stone façade to the exterior of their house, adding new coach lamps as finishing touches to their work. For one neighbour this activity may generate a positive externality in the form of improved environmental quality; but for another, with different tastes in design, it may produce an equally strong negative externality effect.

The costs of proximity and the price of accessibility

For the geographer, much of the significance of externality effects stems from the fact that their intensity is usually a function of relative location. David Harvey made a useful distinction between the *price of accessibility* to desirable urban amenities and the *costs of proximity* to the unwanted aspects of urban life (Harvey, 1973). Both, however, are a product of relative location, and it is clear that the spatial organization of social groups in relation to one another and to the urban infrastructure therefore determines the net intensity of the externality effects that they enjoy.

As a general rule, of course, those with the greatest wealth, the most power and the best knowledge will be best placed to reap the benefits of positive externalities and to fend off activities that generate negative externalities. The location of public facilities such as transport routes, hospitals and sports centres is often intended to ameliorate the regressive nature of locational advantage resulting from private competition but the 'hidden mechanisms' of group conflict tend to ensure that the inhabitants of the riches and most powerful neighbourhoods enjoy a large net benefit as a result of decisions affecting the location of public goods and the organization of public services.

Competition and conflict over externalities

It is clear, then, that the pattern of externality fields can exert a powerful influence on people's welfare. Because of this, many commentators regard the social geography of the city as the outcome of competition and conflicts that are worked out in society as a whole between unequally endowed groups seeking to obtain more or less exclusive access to positive externalities and to deflect negative externality fields elsewhere: 'much of what goes on in a city . . . can be interpreted as an attempt to organize the distribution of externality effects to gain income advantages' (Harvey, 1973, p. 58). Harvey is particularly concerned to show that the form, location and focus of such conflict depends, ultimately, on long-term urban structure changes and broader class conflicts, a point that is also emphasized by Johnston:

Because changes to the urban fabric introduce new sources of positive and negative externalities, they are potential generators of local conflicts. In the face of such proposed changes, the main protestors are usually those with most to lose: property owners, who perceive possible falls in land values, and parents, who identify potential deterioration in an area's schools. In general, it is the more affluent property owners who have the most to lose, and who, because of their ability to purchase legal and technical advice and their greater knowledge of, and links to, the political systems within which such conflicts are adjudicated, are most likely to prevent changes likely to injure their interests. Such conflicts are usually played out locally, *but their existence is part of the dynamic of capitalist cities.* Alterations in land use are needed if investors are to achieve profits, and if the losers in the conflicts over changes are the less affluent, then the price paid for those changes is substantially carried by them. *Local conflicts are part of the general contest between classes within capitalist society.*

(Johnston 1984, emphases added)

We must also recognize that attitudes towards externality effects are also related to the cultural norms, religious institutions and family kinship networks displayed by different ethnic communities. For example, Takahashi (1998) shows considerable stigmatization of people with AIDS among both Latino and Vietnamese communities in California. This in turn leads to hostility among such communities towards treatment facilities for AIDS victims. In part, fear of this phenomenon was bound up with hostility to the perceived 'invasion' of 'immoral' Western practices. In addition, a lack of sympathy towards homeless people can be related to a strong ethic of self-reliance in these communities. As Takahashi notes, it is important not to simply condemn neighbourhood opposition as selfish and reactionary, but to understand the underlying cultural norms that lead to such outcomes.

In the long run, one of the principal outcomes of the resolution of locational conflicts is the creation of a set of **de facto territories** on the basis of income and ethnicity as people respond by relocating to neighbourhoods where they can share their positive externalities with one another and are able to avoid, as much possible, those who impose negative externalities. The residents of such territories also attempt to improve and preserve their quality of life through collective action: competing

through formal and informal neighbourhood groups and local political institutions to attract the utility-enhancing and keep out the utility-detracting.

One of the most common community strategies in this context is that of *voicing* claims over a particular

---

Box 13.1

Key trends in urban social geography – The emergence of clusters of asylum seekers and refugees

One of the most important – but so far little researched – developments in Western cities in recent years has been the growth of residential districts with concentrations of refugees and asylum seekers (i.e. those seeking either permanent or temporary settlement in another country after fleeing from persecution in their native land). Most Western nations have a long history of accepting asylum seekers (e.g. as in the case of Jewish people fleeing Nazi persecution in the 1930s). However, in recent years the numbers seeking political asylum in the West have risen considerably in response to increasing political upheaval and the actions of various despotic regimes around the globe (e.g. Iraq, Afghanistan, Somalia and Algeria). In response, most Western nations have taken measures to restrict the numbers of asylum seekers. In the United Kingdom for example, immigration policy has incorporated increased controls on entry, restrictions on the civil rights of asylum seekers and reduced welfare payments. As in other Western nations, these measures (together with reduced conflicts around the globe) have been associated with a rapid decline in asylum applications during 2005.

In the United Kingdom attempts have been made to disperse asylum seekers throughout the nation and this policy has sometimes led in the

reception areas to social tensions and local protest groups. Anxieties have been intensified by newspaper references to a 'flood' of asylum seekers threatening to 'swamp' neighbourhoods. One especially notorious case was the Sighthill area of Glasgow where the arrival of 3500 Kurdish asylum seekers resulted in racist attacks and the murder of a Turkish Kurd (Hubbard, 2005). However, subsequent investigations revealed this to be a mugging rather than a racist attack. Indeed, Sighthill residents were so upset by their portrayal in the media as racist bigots that they joined in a march with asylum seekers to petition the local council for better housing conditions. Attempts to foster better community relations in Sighthill revealed that many of the newcomers were qualified doctors, architects and electricians, and consequently in possession of valuable skills.

Hubbard (2005) argues that although opposition to asylum centres by local neighbourhood groups is often couched in terms of threats to national 'English' culture, these discourses often disguise a racist agenda based around imagineries of 'whiteness' (see also Chapter 8). Opposition to designated centres for asylum seekers has been especially fierce in rural areas with little tradition of multiculturalism. However, there are also anecdotal reports from London that

some ethnic minorities are moving outwards to suburban areas in response to growing numbers of asylum seekers in inner-city districts.

Key concepts associated with asylum seekers and refugees (see Glossary)

Clustering, community action, diaspora, ethnic group, exclusion, NIMBY, othering, purified communities, racism, 'turf' politics.

Further reading

Day, K. and White, P. (2002) Choice or circumstance: the UK as the location of asylum applications by Bosnian and Somali refugees *Geojournal*, **65**, 15–26

Hubbard, P. (2005) Accommodating Otherness: anti-asylum centre protest and the maintenance of white privilege *Transactions of the Institute of British Geographers*, **30**, 52–66

Robinson, V., Anderson, R. and Musterd, S. (eds) (2003) *Spreading the Burden? A review of policies to disperse asylum seekers* The Policy Press, Bristol

Links with Other Chapters

Chapter 8: Segregation and Congregation

Chapter 12: Box 12.3 The Rise of 'Transnational Urbanism'

issue, whether by organizing petitions, lobbying politicians and bureaucrats, writing to newspapers, forming local resident groups, picketing or distributing handbills and posters. This '**voice strategy**' can sometimes extend to illegal activities, such as personal violence, damage to property, sit-ins and deliberate violations of antidiscrimination laws. Alternatively, some communities are able to use formal channels of participation as a profitable strategy when they find themselves in conflict with other communities or institutions.

Another option available to communities is that of resignation to the imposition of negative externalities (also termed the '**loyalty**' **option**). This is especially common where communities feel that voicing strategies are regularly ignored or overruled and that participation is ineffective. Many people who disapprove of city plans, for example, simply resign themselves to the 'inevitable' because they feel unable, individually or collectively, to exert any real influence on policy-makers. The final strategy is that of relocation (also termed the '**exit**' **option**), which is a household rather than a community strategy, and which brings us back to the idea of a continually evolving geography of de facto territories.

13.2 Accessibility to services and amenities

In every city there are a large number and a great variety of services and amenities – parks, schools, restaurants, theatres, libraries, fire stations, shops, doctors' clinics, hospitals, day-care centres, post offices, riverside walks and so on – that are **point-specific services** (i.e. tied to specific locations) and that therefore exhibit externalities with **tapering** effects (i.e. decreasing intensity with distance from a fixed point). To these we must add certain place-specific disamenities and the noxious activities associated with some services: refuse dumps and crematoria, for example; and we must recognize that what constitutes an amenity to some (a school, or a soccer stadium, for example) may represent a disamenity to others. From another perspective, it is clear that some externalities apply only to users while others apply to whole neighbourhoods. Finally, we

must recognize that the *intensity* of the externality effects will also vary according to people's preferred distance from particular services or amenities (Fig. 13.1). We are thus faced with a very complex set of phenomena.

The externalities associated with physical proximity to services, amenities and disamenities not only prompt competition and conflict between households within different housing markets but also give rise to collective political strategies, including the formation of coalitions between different institutions and organizations and the propagation of distinctive de facto communities whose mutuality involves lifestyles that are dependent to some extent on accessibility to specific amenities.

These coalitions and communities represent the major protagonists in much of the conflict over the preservation and fortification of the relative quality of life in different urban settings. Thus **community-based politics** (also termed '**turf**' **politics**) have become a major feature of contemporary cities. It should be noted, however, that overt conflicts associated with neighbourhood activism have become more frequent with the extension of owner-occupation and as larger-scale housing and construction projects have replaced smaller-scale activities as the dominant aspect of urban development. It follows that a good deal of neighbourhood activism is directly associated with construction and development activity around the urban fringe. Those with the greatest stakes in a particular local setting (i.e. owner-occupiers and parents of school-aged children) are most likely to become involved in neighbourhood activism, and the dominant types of conflicts tend to be associated with the public regulation of privately initiated patterns of urban development, with publicly initiated construction projects (e.g. new highways, street widenings and urban renewal projects), and with the quality of public services.

Understanding patterns of service delivery and amenity location is not simply a matter of competition and conflict over which households win proximity to the most desirable services and amenities and which communities are able to 'capture' new services and amenities. The geography of many services and amenities is also a product of other factors: the 'fabric' effects of the urban environment, the internal organization and politics of particular professions and service-delivery agencies,

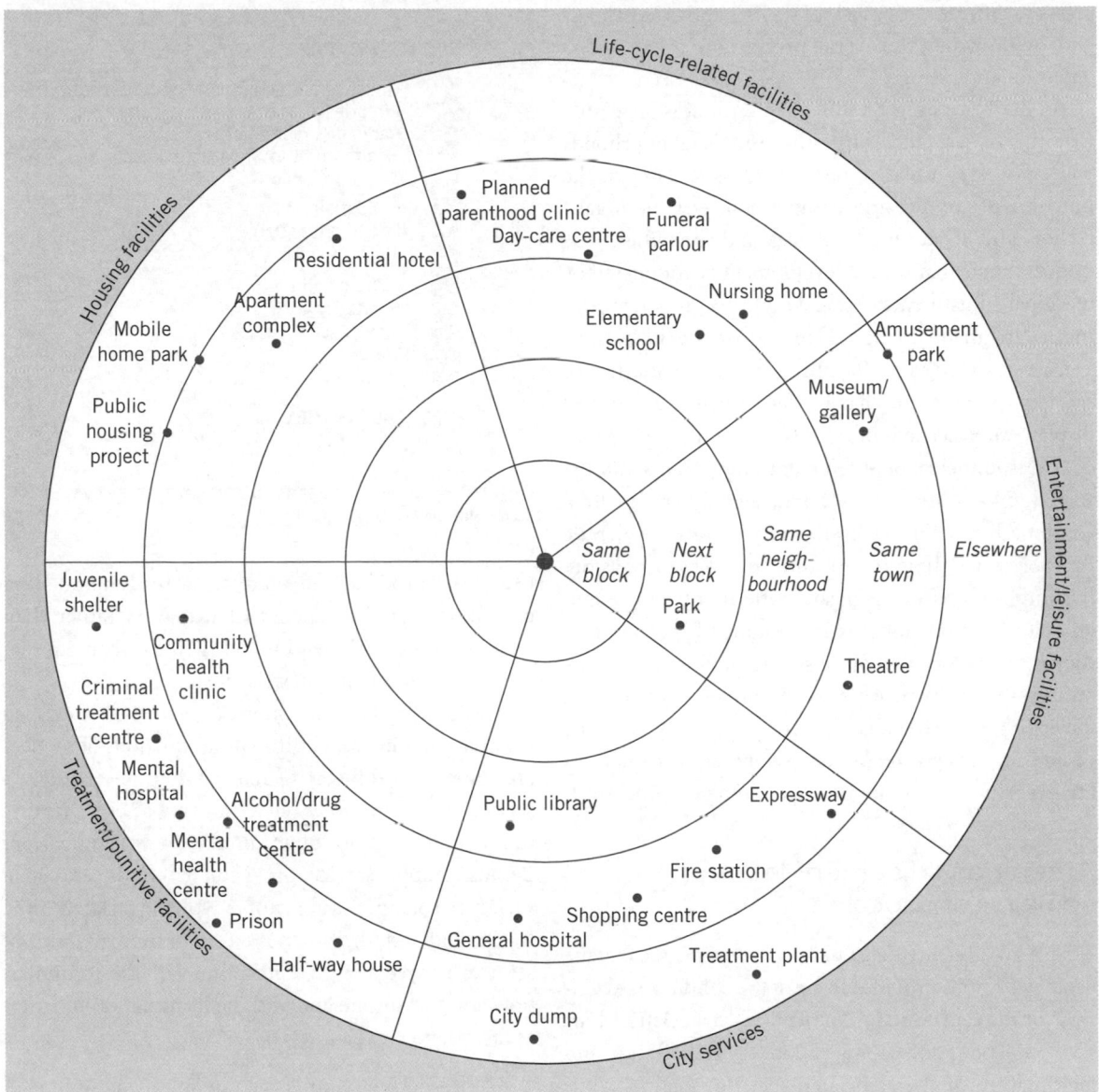

Figure 13.1 Preferred residential distance from different public facilities.
Source: Smith (1980).

and the functional linkages that exist between certain services and other activities, for example. This, of course, means that aggregate patterns of service and amenity provision rarely exhibit clear or unambiguous relationships with urban social ecology. Nevertheless, it is clear that the general tendency is for the most affluent, most powerful and most active communities to capture a disproportionate share of the positive externalities associated with urban services and amenities.

The aggregate effects of aggregate patterns

Given that much of what takes place within the urban arena involves the resolution of conflicts over spatial organization and the relative location of certain 'goods' and 'bads', the question remains as to the relationships between, on the one hand, the outcome of these conflicts (segregated communities, de facto and *de jure*, with access to different 'packages' of services and externalities) and, on the other, the overall processes of urban differentiation and change.

At the beginning of this chapter, it was suggested that the richest and most powerful neighbourhoods will enjoy a cumulative net benefit as a result of the outcome of conflict and competition over the organization and location of services and amenities. The implication is that urban development is dominated by marginal adjustments in social ecology as new developments are made to fit the mould of the *status quo*. Yet several studies of the intra-urban distribution of services and amenities have in fact concluded that they display unpatterned inequality, attributing the lack of correspondence between community status and the spatial organization of services and amenities to a combination of idiosyncratic events and bureaucratic decision rules. It is important, however, to consider the implications of aggregate patterns of service and amenity location not just in relation to the way that they intensify or ameliorate socioeconomic differentiation but also in relation to the broader sociospatial dialectic. What is at issue here is the way that the economic and class relationships inherent to capitalism are perpetuated in cities through ecological processes.

Amenities, disamenities and social reproduction

From a broad structuralist perspective we can see that the ecology of cities provides some of the conditions necessary for the **reproduction** of the necessary relationships between labour and capital and for the stabilization and **legitimation** of the associated social formation. Thus we find 'a white collar labour force being "reproduced" in a white collar neighbourhood, a blue collar labour force being reproduced in a blue collar neighbourhood, and so on' (Harvey, 1975, p. 363).

An essential factor in this reproduction is the differential access to scarce resources, especially educational resources, between neighbourhoods, since it helps preserve class and neighbourhood differences in 'market capacity' (the ability to undertake certain functions within the economic order) from one generation to another (Pacione, 1997). At the same time, the locations of 'compensatory' services and amenities not only

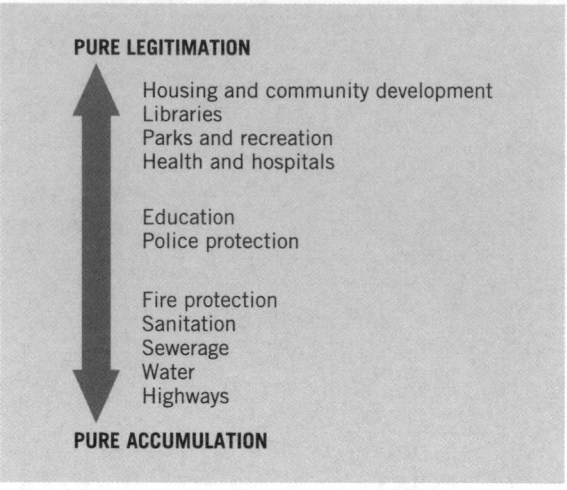

Figure 13.2 A conceptual ranking of services. *Source*: after Staeheli (1989), p. 242.

helps to reproduce and maintain a ready population of workers (at the expense of taxpayers rather than employers) but also helps to defuse the discontent that their position might otherwise foster.

The conceptual and empirical distinctions between the **accumulation** and legitimation functions of services are sometimes difficult to make, while the politics of service provision rarely relate in overt or explicit ways to functional notions of accumulation or legitimation. Because public service provision is contingent on a variety of sociopolitical factors involving different time frames and periodicities, it is useful to think in terms of services arrayed along a continuum with the accumulation function at one end and the legitimation function at the other (Fig. 13.2).

> Services located on the accumulation end of the continuum can be thought of as important for the accumulation of capital and are provided primarily in keeping with the needs of capital; roads, water and sewer systems are typical accumulation services because they allow the initial development of land and preserve its subsequent exchange value. . . . As such, their provision is greatly influenced by higher levels of government and by the needs of large, mobile capital. This influence . . . is a smoothing effect on the distribution of accumulation services.

... In contrast, legitimation services have a larger discretionary component. Parks and libraries, for instance, are generally provided for the benefit of the general, 'classless' public; their amenity value to capital is real, but indirect and possibly will accrue over a longer time span. Because such services are provided for the public, it is common for small groups and individuals to be involved in the decisionmaking process; the politics of consumption characterize this process. The variable cast of characters and concerns involved in the provision of legitimation services means that the demand for these services is likely to be uneven.

... Finally, some services blend accumulation and legitimation concerns, and are therefore situated in the middle of the continuum. These services are necessary for the accumulation of capital, and so certain aspects of their provision will be similar in most municipalities. However, services in the middle of the continuum, such as police protection and local schools, also have a large discretionary element that may lead to more variability.

(Staeheli, 1989, p. 243)

Patterns of service and amenity provision, then, are at once the product of the social formation and an element in its continuing survival.

13.3 Urban restructuring: inequality and conflict

As we saw in Chapter 1, cities throughout the developed world have recently entered a new phase – or, at least, begun a distinctive transitional phase – in response to changing economic, political, social and cultural conditions. The continual restlessness of urbanization has been accentuated by the imperatives of restructuring cities in order not only to accommodate to these changing conditions but also to exploit new technologies and new sociocultural forces.

Among the chief features of this restructuring have been the decentralization of jobs, services and residences from traditional city centres to suburban settings and 'edge cities' within expanded metropolitan frameworks; the decline of traditional inner-city employment bases in manufacturing, docks, railways, distribution and warehousing; the recentralization of high-level business services in CBDs; the gentrification of selected inner-city neighbourhoods; the localization of residual populations of marginal and disadvantaged groups and of unskilled migrants and immigrants in other inner-city neighbourhoods; the emergence of a 'new politics' of fiscal conservatism; the emergence of a new politics of race; the emergence of a 'new cultures' of material consumption and differentiated lifestyles; the feminization of poverty; and the intensification of economic and social polarization.

Meanwhile, the need to accommodate a new mix of industry and employment within the fabric of a pre-existing built environment has led to localized conflicts over development and land conversion processes. It is beyond the scope of this book to deal systematically with these issues, or to do justice in depth to any one of them. It must suffice, therefore, to illustrate just a few aspects of the sociospatial consequences of urban change and restructuring.

Decentralization and accessibility to services and amenities

The restructuring of metropolitan form in response to the ascendancy of the automobile has brought to an end the traditional notion that jobs, shops, schools, health services and community facilities will be within ready walking distance of homes. Even by 1960, over 90 per cent of the households in the most recently developed parts of metropolitan California had at least one car, and between 40 and 45 per cent had two or more. By 1970, comparable levels of car ownership had been achieved in most other metropolitan areas of the United States, while in Europe the spread of car ownership was at last beginning to accelerate rapidly. One result of this trend has been that employers, retailers and planners have tended to make their location decisions on the assumption of perfect personal mobility.

As discussed in Chapter 3, the prime example of this is the ascendancy of suburban shopping centres and shopping malls. Their magnetic power has rearranged not only the commercial geography of urban areas but also the whole social life of the suburbs. Malls have become the most popular gathering places for suburban teenagers, and adults use them to stroll and promenade, much as continental Europeans have used their city centres on Sundays and summer evenings. Americans now spend more time at shopping malls than anywhere outside their homes and workplaces.

Accessibility and social inequality

The benefits of increased personal mobility are enjoyed disproportionately by the middle-class, the middle-aged and the male population, however. In the United States, for example, although the overall ratio of motor vehicles to households increased from 1.3 to 3.0 between 1960 and 2003, only the more affluent households actually had an increase in car ownership. Less affluent households, as a group, were worse off in 2003 than in 1960. Indeed, surveys have shown that, despite the advent of the 'automobile age', around three out of ten urban residents lack direct personal access to a motor vehicle. Many of these individuals are old, poor or black, and a good many are inner-city residents.

Women are also significantly deficient in access to motor vehicles for, although their household unit may own a car, its use by other members of the household is likely to render it unavailable for much of the time. Furthermore, the urban form that the automobile has triggered – low-density development spread over a wide area – has made it very difficult to provide public transport systems that are able to meet the needs of carless suburban women, the elderly and the poor.

Women are particularly vulnerable to constraints on locational accessibility. Labour-market changes that have concentrated women in a limited range of occupations are compounded by the spatial concentration of female-dominated jobs and by the constraints of gender roles in contemporary society. Even affluent suburban housewives with access to a car are limited in their opportunities because of the limited time available between their fixed 'duties' as a homemakers: providing breakfast for the family and driving the children to school, preparing lunch, picking up the children from school, chauffeuring them to sporting or social engage-

ments, and preparing dinner. In addition, homemakers may have to be at home to accept deliveries, supervise repair workers or care for a sick child. Women without cars, of course, suffer much greater restriction on their quality of life.

Some of the most severe accessibility problems arising from urban decentralization are experienced by suburban single parents who must inevitably have lifestyles that differ from those of two-parent families. They have much less flexibility in employment and in recreation, for example, having to dovetail their activities to the time schedules of their children. Parents with school-age children may find full-time employment impossible without after-school and holiday-care facilities, while those with preschool children are dependent on preschool facilities and day nurseries. In both cases the proximity of relatives and friends is often critical. Late opening of commercial and other enterprises is also vital if a single parent is to incorporate visits to shops, banks, libraries and offices in the weekly activity pattern.

The predicament of suburban women is not simply a consequence of urban decentralization. It is intimately related to a whole nexus of the trends outlined at the beginning of this book. The intersection of economic, demographic, social and cultural trends, for example, has allocated to women a pivotal role in the consumption-oriented suburban lifestyles that dominate the logic of contemporary urbanism. In short, women are trapped economically, socially and culturally, as well as ecologically. Meanwhile, the suburban environment has evolved in ways that reinforce inequality between the sexes, contributing, among other things, to emotional strain and the erosion of 'community'.

Redevelopment and renewal

One of the longest-running aspects of urban restructuring has been the physical redevelopment and renewal of worn-out and outmoded inner-city environments. In removing the most inefficient factories and the worst slums from city centres, urban renewal has undoubtedly contributed not only to economic regeneration but also to the 'common good' in terms of environmental quality and public health (Pagano and Bowman, 1995; Fainstein, 1997; Couch and Percy, 2003; Ford, 2003). But in rehousing the residents of clearance areas and replacing the

Box 13.2

Key debates in urban social geography – The role of automobility

A relatively new concept that is having an increasing impact within human geography and urban studies is that of 'automobility'. This is a very broad-ranging notion that refers to the way in which the automobile (and more importantly all the systems that support the car) have a dominating influence upon our lives in Western societies. John Urry, a leading exponent of automobility studies, has identified six aspects within the concept:

➤ The car as a *manufactured object* (i.e. it is produced by iconic twentieth-century firms).

➤ The car as an *item of consumption* (i.e. it is a status symbol that can confer power and identity).

➤ The car as *part of an industrial complex* linked to other industries (i.e. road and housing construction, automobile accessories).

➤ The car as an *instrument of quasi-private mobility* (i.e. because of its dominance it subordinates all other modes of transport).

➤ The car as *agent of culture* (i.e. it dominates many normal social discourses).

➤ The car as an *agent of environmental and resource depletion* (i.e. it produces pollution together with noise and congestion).

This concept of automobility has links with 'large systems theory'. The large system that underpins the car involves many elements ranging from massive oil corporations and all their diverse activities, to huge manufacturing concerns, vast infrastructure in the forms of roads and bridges, together with activities such as advertising, sales, sports and leisure. The car has clearly been of great importance in affecting the spatial structure of Western cities through its influence upon suburbanization.

Some sceptics might see these elements of automobility as 'obvious' and 'common sense' but the concept draws our attention to the fact that all this infrastructure, and the way it dominates our thinking, need not necessarily be the case. Our dependence upon the automobile reflects a wide range of political, economic and technological decisions made over the years.

Key concepts related to automobility (see Glossary)

Commodity fetishism, consumption, decentralization, edge cities, Fordism, positional good.

Further reading

Beckman, J. (2001) Automobility: a social problem and theoretical concept *Environment and Planning D*, **19**, 593–607

Katz, J. (1999) *How Emotions Work* University of Chicago Press, Chicago (Chapter One, Pissed Off in L.A.)

Sheller, M. and Urry, J. (2000) The city and the car. *International Journal of Urban and Regional Research*, **24**, 737–757

Thrift, N. (2004) Driving in the city *Theory, Culture and Society*, **21**, 41–59

---

built environment, planners have managed to preside over some spectacular débâcles.

The principal charge against them in this context is the dismantling of whole communities, scattering their members across the city in order to make room for luxury housing, office developments (including, in many instances, new accommodation for the urban bureaucracy), shopping areas, conference centres and libraries. A secondary charge – urban blight – stems from the discrepancy between the ambitions of planners and what can actually be achieved within a reasonable future. During the intervening period, neighbourhoods scheduled for renewal are allowed to slide inexorably down a social and economic spiral. No landlord will repair a condemned house if he can help it; tenants who can afford it will move out; shopkeepers will close and drift away; and the city council, waiting for comprehensive redevelopment, will meanwhile defer any 'unnecessary' expenditure on maintenance. Schools, public buildings, roads and open spaces become rundown, matching the condition of the remaining population of the poor and the elderly.

This dereliction has been extensive in many cities. Combined with the actual demolition of condemned property, the result has been that large areas have been laid waste and thousands of families have been displaced. Moreover, the problem has been compounded since, as the worst parts of the cities' housing stock has been cleared, the bureaucratic offensive has gone on to condemn housing that was relatively sound, turning slum clearance from a beneficent if blunt instrument into a bureaucratic juggernaut.

Planning problems: the British experience

By removing the structure of social and emotional support provided by the neighbourhood, and by forcing people to rebuild their lives separately amid strangers elsewhere, slum clearance has often imposed a serious psychological cost upon its supposed beneficiaries. At the same time, relocatees typically face a steep increase in rents because of their forced move 'upmarket'. In Britain, most slum clearance families have been rehoused in the public sector, but this has also brought disadvantages that, for some households, outweigh the attractions of more modern accommodation at subsidized rents.

Slum clearance families must face the vicissitudes of a housing bureaucracy whose scale and split responsibilities tend to make it insensitive to their needs. Since they are 'slum dwellers', the new accommodation that is offered to them is likely to be in low-status public housing estates. Even the offer of accommodation in new maisonettes or high-rise apartments may compare unfavourably with the tried-and-tested environment of old inner-city neighbourhoods. The open spaces, pedestrian pathways and community centres regarded as major advantages by planners may seem of minor importance to their users; while some such 'amenities' serve only as focal points of vandalism, souring the whole social atmosphere.

Moreover, because most new residential planning has been guided by the objective of fostering 'community' feelings, problems of a different nature can be precipitated by the lack of privacy on new estates. Apartments and maisonettes tend to be worst in this respect, since common stairways, lifts and desk access mean that interaction with uncongenial neighbours is unavoidable. On the other hand, the planned and regulated environment of New Towns and new estates has little of the richness of opportunity associated with older neighbourhoods. Finally, it is worth noting that not everyone from clearance areas ends up being relocated

in sound accommodation, let alone satisfactory or desirable housing. This has principally been the case with ethnic minorities who, because of the combination of economic constraints and racial discrimination, have been forced to double up in other ghettos.

The chief *beneficiaries* of urban renewal are the dominant political and economic elites of the city. The former benefit from the existence of a much more lucrative tax base with which to finance public services, as well as the feelings of civic pride generated by redevelopment schemes and the symbolization of power that they represent. One particularly well-documented example of this is Newcastle upon Tyne, where a unique Victorian townscape, as well as the less appealing housing of the terraced streets off the Scotswood Road, was replaced by a city centre that earned the leader of the council the title of 'Man of the Year' from the *Architect's Journal* and led the city's politicians proudly to boast of the city as the 'Brasilia of the North'.

The dominant business elite, meanwhile, benefits in much more tangible ways. The redevelopment of British city centres has served to benefit monopoly capital by wiping out small retailers, thus giving the big stores and large supermarkets the market they require. But it is the speculative developers of property whose interests have been best served by urban renewal. Obtaining sites that have been cleared at public expense, they have been encouraged by planners to develop them for 'higher' uses – offices, hotels, conference centres and shopping precincts. Such developments have been highly lucrative, and it is therefore not surprising to find that, in many cities, developers have 'worked' the planning system in order to secure ever greater profits.

Service sector restructuring

In parallel with the restructuring and reorganization of industrial production that has transformed the economic base of cities everywhere, there have been some interdependent changes in the structure of national and local welfare systems that have resulted in significant changes to the geography of urban service provision. The combination of economic recession and the globalization of manufacturing and of financial and business services has led to a retreat from the public provision of welfare services, an increase in public–private cooperation, and an emphasis on accumulation-oriented services that enable cities to compete more effectively within an international urban system (Pinch, 1997). There has been a great deal of substitution between different forms of service provision (i.e. domestic, voluntary, commercial, subsidized commercial, community-based, city-based, state-based, etc.), which has in turn resulted not only in new patterns of service provision and relative accessibility but also, in some instances, in new sociospatial phenomena (see also Box 13.3).

Box 13.3

Key trends in urban social geography – The rise of social entrepreneurship

As we saw in Chapter 1, words do not simply describe the world – reflecting it like a mirror – they can actually create a view of the world through the various assumptions that they incorporate (both explicitly and implicitly). This is one of the reasons why politicians and policy-makers frequently use new words when introducing a new policy initiative; the old words may be tainted with associations that are regarded as outmoded and anachronistic. Thus we find that a whole new set of words has been introduced to bolster welfare reform in the past few decades such as 'workfare', 'stakeholders' and 'compacts'. In addition, rather than be called 'clients' the recipients of services have sometimes been renamed customers or consumers.

One key new concept is that of *social enterprise*. There is no easy definition of this concept but essentially *social enterprises are commercial ventures that attempt to achieve social benefits for a defined community*. The latter may be a community within a geographically defined area – such as a local neighbourhood – or it may be a community of common interests such as the long-term unemployed, disabled or single parents. In some respects social enterprises are like traditional charities, except that they tend to operate with a greater commercial element. Social enterprises are therefore hybrid organizations that work on the boundaries of the true private sector (where the aim is to

gain maximum profit) and the public sector (where there is direct control and funding by government). Collectively, social enterprises make up what is sometimes termed the *social economy* (or 'third sector'). There have long been commercial ventures within the third sector but they have taken on increased importance in recent years in meeting mainstream welfare needs as governments have attempted to restrict the scale of welfare states.

Closely related to social enterprise is the concept of *social entrepreneurship*. This term takes the notion of entrepreneurship from the sphere of private enterprise and applies this to individuals who are innovators in the sphere of social policy. These individuals are seen as capable of bringing drive, innovation and creativity to a field in need of reform, providing new services together with increased efficiency. The key feature of the social entrepreneur therefore is the ability to bring together diverse 'partners' in the welfare field to 'invest' in innovative solutions. The concept of social entrepreneurship is a good example of how the discourses of business and the market have progressively infiltrated all aspects of contemporary social life.

There can be little doubt that some people have made tremendous innovations in the sphere of social policy. Examples are Andrew Mawson, who established the Bromley-by-Bow centre in London, and Greg Macleod,

founder of the New Dawn organization in Sydney, Cape Breton, Nova Scotia, in Eastern Canada. The latter is a multifaceted non-profit organization that aims to improve the quality of life in a region that has experienced considerable economic hardship over the years. However, a reliance upon key individuals can raise a number of problems (Scott *et al.*, 2000). First there is the issue of accountability; decisively acting individuals can easily ride roughshod over other 'stakeholders' in the welfare system (e.g. co-workers, volunteer workers, welfare recipients, service funders). Second, it is not always clear that leading individuals are crucial to either the success of failure of particular organizations. Finally, particular individuals can become overloaded with unrealistic expectations about their capacity to change and reform welfare systems.

Key concepts associated with social entrepreneurship (see Glossary)

Governance, 'hollowing out', not-for-profit sector, postwelfare state.

Further reading

Amin, A., Cameron, A. and Hudson, R. (2002) *Placing the Social Economy* Routledge London

Scott, D., Alcock, P., Russell, L. and Macmillan, R. (2000) *Moving Pictures: Realities of Voluntary Action* Policy Press, Bristol

Deinstitutionalization and residualization

Perhaps the best-known example of public sector restructuring upon urban form is the way that the **deinstitutionalization** of mental health services has contributed to urban homelessness (Dear and Wolch, 1989). Deinstitutionalization involves the closure of large institutions that provide long-term care for needy groups – such as the mentally-ill, those with learning difficulties, the elderly or severely disabled – and the development of a variety of community-based forms of care. The latter include smaller purpose-built facilities or it can involve care within private households by families or friends supplemented by teams of community-based professionals such as home nurses, doctors, social workers and probation officers.

Deinstitutionalization was introduced for humane and progressive reasons in an attempt to overcome the stigma and poor conditions associated with many large institutions. However, in an era of fiscal retrenchment, the policy has often been seen as a way of saving money. As a result, large institutions have been closed very rapidly without the development of sufficient community-based facilities. The policy was also based on the assumption that there was a large reserve of volunteers (usually women) who were prepared to care for those released from large institutions – an assumption that was often mistaken. Those without family support have often ended up in rented accommodation and, since this tends to be located in inner-city areas, these display the greatest concentrations of ex-psychiatric patients.

The most extreme consequences of deinstitutionalization are to be found in California. The rapid closure of psychiatric hospitals in the state led to the creation of **service-dependent ghettoes** (also termed the 'asylum-without-walls', Wolch, 1981). Former in-patients have become restricted to poorer-quality neighbourhoods as zoning legislation has kept community-based facilities outside more affluent areas. For-profit community-based services have been developed but some of these are now relatively large with over a hundred occupants and are beginning to reproduce some of the institutional features of the older mental hospitals. Fiscal retrenchment has served to exacerbate the problems of mentally-ill persons by restricting funds for community-based forms of care. In addition, the gentrification and urban renewal of some inner city areas have led to curbs on these community-based facilities. The consequence has been increased homelessness, with people sleeping on the streets in 'cardboard cities'. Sometime the wheel has turned full circle with mentally-ill or ex-psychiatric patients ending up in hospital or prison – a process of **reinstitutionalization** (Dear and Wolch, 1987).

Deinstitutionalization is an example of one of the most common responses of governments to fiscal pressures – to reduce various types of welfare services. In some cases, such as public sector housing in the United Kingdom, cuts have meant that this is no longer available as common facility for urban populations but has been restricted to the very poorest in society – a policy known as **residualization**. However, in the case of many other welfare services it is the very poorest in society who have been most severely affected by restrictions in spending. The reason for this is that marginalized groups often lack political power and are therefore easier targets for expenditure reductions than services such as pensions, which affect a wider proportion of the population. One group that has experienced cuts in spending in both the United Kingdom and the United States in the recent years have been single-parent mothers. Indeed, it seems clear that women in general have been disproportionately affected by reductions in welfare spending. The reasons for this are twofold: first, women are the main recipients and users of many welfare services; and second, women constitute the bulk of the workers in services that have been cut.

Privatization

Privatization is another important form of public sector restructuring. This involves a complex set of processes, as revealed by Table 13.1. The most obvious form of privatization is **asset sales**, when public sector assets are sold to the private sector. For example, one of the most important manifestations of asset sales in the United Kingdom has been the sale of local authority housing, as discussed in Chapter 6. Privatization can also take the form of **contracting-out** – awarding tasks that were previously undertaken by the public sector to private sector organizations. Often these contracts are awarded

Table 13.1 Forms of privatization

Method of funding	Type of ownership	
	Private ownership	Public ownership
Market funding	Sale of assets	Commercialization/ corporatization
State funding	Contracting-out	Public provision

Source: adapted from Stubbs and Barnett (1992).

on the basis of secret bidding known as **competitive tendering**. In this case the funding is still by the public sector on non-market criteria but the provision is by the private sector. A final form of privatization is where the organization is still operated by the public sector but has to operate on commercial or market-based criteria – a process known as **commercialization**.

Contracting-out and commercialization are examples of processes designed to make the public sector services geared more to yardsticks of cost-efficiency and flexibility than need or equity. Just as private-sector services have restructured in response to changing economic circumstances, changing technologies, and new managerial strategies, so have public-sector services (Table 13.2). As Pinch (1989, 1997) points out, a very wide spectrum of change has been imprinted onto the geography of public service provision, including partial self-provisioning, **intensification**, capitalization, **rationalization**, **subcontracting**, substituting expensive employees with cheaper ones, **centralization**, **materialization**, **domestication** and spatial relocation. The patterns of winners and losers from this public sector restructuring are complex, but there is growing evidence that overall it is the most powerless, marginalized and poor groups that have suffered most from public sector restructuring.

Workfare

Another important aspect of public sector restructuring in recent years is what has become known as **workfare** – a policy which Peck claims 'has acquired the status of a political mantra' (Peck, 1998, p. 158). The policy was originally regarded as one of making unemployed people work in order to receive their benefits (i.e. work + welfare = workfare) – the so-called 'hard' or 'first generation' form of workfare (also known as 'earnfare').

However, the term workfare is now commonly used in a more general sense to indicate the so-called 'soft' or 'second generation' workfare – that income support for welfare recipients is conditional upon them participating in a range of activities designed to increase their employability. These options can include: subsidized work schemes, education, training programmes, supervised job search initiatives or community work. Workfare is also used in an even more general sense to indicate new policies designed to regulate the behaviour of welfare claimants. This includes the following type of measures (Peck, 1998):

➤ withdrawing benefits or enforcing mandatory community service upon those who do not find work or training within a specified time period;

➤ reducing benefits and allowances;

➤ intensified anti-fraud measures;

➤ denying benefits to children born to mothers who are already receiving aid;

➤ withdrawing benefits from children who have unexposed absences from school ('learnfare');

➤ denying benefits to teenage parents living independently who are not in receipt of or registered for a high-school diploma;

➤ offering employers tax breaks and subsidies to recruit welfare recipients.

Although workfare is very much a US innovation, it has spread in various forms throughout the Western world. One variant of the policy has emerged in the United Kingdom in recent years. In his first public address after being elected in May 1997, Prime Minister Tony Blair said that New Labour would be 'tough on those who are offered the chance to work but do not want to take it' (*Observer*, 1 June, 1997). He went on to

Table 13.2 Forms of service sector restructuring

Private sector	Public sector
1 Partial self-provisioning	
Self-service in retailing	Child-care in the home
Replacement of services with goods	Care of elderly in the home
Videos, microwave ovens, etc.	Personal forms of transport
	Household crime-prevention strategies, neighbourhood watch, use of anti-theft devices, vigilante patrols
2 Intensification: increases in labour productivity via managerial or organizational changes with little or no investment or major loss of capacity	
Pressure for increased turnover per employee in retailing	The drive for efficiency in the health service
	Competitive tendering over direct labour operations, housing maintenance, refuse collection
	Increased numbers of graduates per academic in universities
3 Investment and technical change: capital investment into new forms of production often with considerable job loss	
The development of the electronic office in private managerial and producer services	Computerization of health and welfare service records
	Electronic diagnostic equipment in health-care
	Distance learning systems through tele-communications video and computers
	Larger refuse disposal vehicles, more efficient compressed loaders
4 Rationalization: closure of capacity with little or no new investment or new technology	
Closure of cinemas	Closure of schools, hospitals, day-care centres for under fives, etc.
	Closure or reduction of public transport systems
5 Subcontracting of parts of the services sector to specialized companies, especially of producer services	
Growth of private managerial producer services	Privatization or contracting-out of cleaning, laundry and catering within the health service
	Contracting-out of refuse disposal, housing maintenance, public transport by local government
6 Replacement of existing labour input by part-time, female or non-white labour	
Growth of part-time female labour in retailing	Domination of women in teaching profession?
	Increased use of part-time teachers
7 Enhancement of quality through increased labour input, better skills, increased training	
In some parts of private consumer services	Retraining of public-sector personnel
	Community policing?
8 Materialization of the service function so that the service takes the form of a material product that can be bought, sold and transported	
Entertainment via videos and televisions rather than 'live' cinema or sport	Pharmaceuticals rather than counselling and therapy?
9 Spatial relocation	
Movement of offices from London into areas with cheaper rents	Relocation from larger psychiatric hospitals into decentralized community-based hostels
	Relocation of offices from London to realize site values and to reduce rents and labour costs
10 Domestication: the partial relocation of the provision of the functions within forms of household or family labour	
Closure of laundries	Care of the very young and elderly in private houses after reductions in voluntary and public service
11 Centralization: the spatial centralization of services in larger units and the closure or reduction of the number of smaller units	
Concentration of retailing into larger units	Concentration of primary and secondary hospital care into larger units, that is, the growth of large general hospitals and group general practices
Closure of corner shops	

Source: Pinch (1989), p. 910.

say, 'The best form of welfare for people of working age is work' (*Guardian*, 2 June, 1997). This approach culminated in April 1998 in the New Deal, a scheme of assisted job search and short-term preparation for work underpinned by the threat of benefit withdrawal. The policy has a strong local component being based of cooperative partnerships between central government, local authorities, employers, training agencies and the voluntary sector. The programme is focused upon those in the 18–24 age group who have been unemployed for more than six months but also has provision for single parents, the disabled and the over 25s. Welfare claimants – now termed 'jobseekers' – must demonstrate active steps towards seeking work and enhancing their employability. There is a four month period of job guidance and counselling before one of four options must be chosen: (a) six months of waged work for an employer whose wage bill is subsidized by the government (plus one day a week training), (b) full-time study in return for benefits, (c) six months in the voluntary sector in return for dole money (again including one day a week training) or (d) work for an environmental task force where benefits are topped up with a lump sum payment. Upon introducing the policy, Chancellor Gordon Brown said there would be 'no fifth option'. Welfare is therefore now less of an entitlement and more of a reciprocal obligation.

Many claims are made for workfare: that it increases the skills of the workforce; that it discourages dependency; that it reduces unemployment; that it encourages employers to create new jobs; that it reduces welfare costs; and that it is fairer in that it ensures rights are matched by obligations. Inevitably, many of these aspirations have not been fulfilled. To begin with the New Deal is based on two major, and highly flawed, assumptions. First, it is assumed that many do not want to work or at least have behavioural norms that are not conducive to employment. Hence it is assumed they may have become defeated or are insufficiently organized to conduct an effective job search. In other words it is a 'supply-side' explanation (i.e. focusing upon the characteristics of the workforce, rather than a 'demand-side' explanation based on the characteristics of the job market). This leads to the second assumption of the New Deal – that there are sufficient jobs available for the participants in the scheme. In fact, as might be expected,

the areas with the highest levels of unemployment, have the lowest numbers of job opportunities. Thus about a half of all the clients of the New Deal are in 18 out of the total of 471 local authorities participating in the scheme. Even more remarkably, one-third of the client group is located in just five main cities; London, Birmingham, Glasgow, Liverpool and Manchester. As Peck notes: 'unemployment is not five times higher in the Yorkshire coalfields because of some deficiency in the northern work ethic; it is a straightforward reflection of job availability' (Peck, 1999, p. 366).

Many other criticisms have been directed at the New Deal based on the practical difficulties of implementing the scheme. On a practical level, there should be parity among the four options but these differ considerably in terms of improving employability and job prospects. However, in reality demand for the employer option tends to exceed supply. Access to the most favoured option is therefore based on a perception of who is most fit for the labour market; those deemed less employable are more likely to be sidelined into 'sink schemes'. It has also been suggested that workfare schemes such as the New Deal will lead to an erosion of the labour force skill base as workers are forced into relatively menial low-skill jobs. There is certainly evidence of a substantial 'churning' effect as young people are recycled through many different jobs. Early evidence also suggests that many young people are dropping out of hastily constructed New Deal training schemes or else are moving from one training scheme to another without finding work. In addition, it is argued that the bargaining power of existing workers is greatly undermined by the continuous flow of new workers through the system. The 'crowding-out' of the low wage labour market creates a downward pull on wages thereby lowering standards for the poorest even further. Yet another concern is the fact that workfare schemes seem to coerce into work people who – through no fault of their own – are simply not up to the exigencies of the labour market. It is even suggested that, by coercing single mothers into work, such schemes may put the welfare of young children at risk.

The New Deal is full of the market-driven ideology that underpins a great deal of contemporary welfare reform, focusing upon reducing costs, increasing competition and relying upon local public–private partnerships. Those implementing the scheme are under great pressure to meet targets for reducing unemployment levels. Yet the language of dependency culture that surrounds work-fare schemes seem to put the 'blame' for unemployment upon the individual, thereby scapegoating some of the weakest members of society. The aims may be laudable but, overall, it appears to be a short-term work-first scheme designed to achieve rapid labour force integration and unemployment reduction at least cost. Thus, as Jessop (2002) argues, the changing role of welfare involves strengthening the structural competitiveness of the national economy by subordinating social policy to the needs of the labour market.

Social polarization

Social polarization and the spatial segregation of the poor is of course a well-worn theme in urban social geography. It is clear, however, that economic restructuring and social polarization, in tandem with social and demographic changes, have heightened economic inequality along class and racial cleavages (O'Loughlin and Friedrichs, 1996; Kodras, 1997; O'Connor *et al.*, 2001).

The 'new poor', in other words, represent a distinctive component of the new urban geography that has been produced by restructuring. Most striking among the polarized landscapes of contemporary cities are 'impacted ghettos', spatially isolated concentrations of the very poor, usually (though not always) racial minorities that have been drained of community leaders and positive role models and that are dominated numerically by young unmarried mothers and their children. Less visible, but more decisively excluded, are the 'landscapes of despair' inhabited by the homeless: micro-spaces that range from vest-pocket parks and anonymous alleyways to squalid shelters and hostels (Dear and Wolch, 1991).

These phenomena raise a wide variety of conceptual, theoretical and practical issues. Though it is beyond our scope to pursue them all, one issue that should be raised here is that of attributing causality to the deprivation inherent to social polarization. Table 13.3 outlines six main explanations of deprivation, ranging from the concept of a 'culture of poverty' – which sees urban deprivation as a pathological condition – to the concept of an 'underclass' of households that have become detached from the formal labour market.

The idea of a **culture of poverty** is seen as being both an adaptation and a reaction of the poor to their marginal position in society, representing an effort to cope with the feelings of helplessness and despair that develop from the realization of the improbability of achieving success within a capitalist system. In short, it

Table 13.3 Differing explanations of urban deprivation

Theoretical model	Explanation	Location of the problem
1. Culture of poverty	Problems arising from the internal pathology of deviant groups	Internal dynamics of deviant behaviour
2. Transmitted deprivation (cycle of deprivation)	Problems arising from individual psychological handicaps and inadequacies transmitted from one generation to the next	Relationships between individuals, families and groups
3. Institutional malfunctioning	Problems arising from failures of planning, management or administration	Relationship between the 'disadvantaged' and the bureacracy
4. Maldistribution of resources and opportunities	Problems arising from an inequitable distribution of resources	Relationship between the underprivileged and the formal political machine
5. Structural class conflict	Problems arising from the divisions necessary to maintain an economic system based on private profit	Relationship between the working class and the political and economic structure
6. Underclass	Minority groups isolated from formal labour market and from mainstream society	Spatial mismatch of jobs and labour; feminization of poverty; suburbanization of role models

results in a vicious cycle of lack of opportunity and lack of aspiration. There is, however, considerable room for debate as to whether culture is more of an effect than a cause of poverty and, indeed, whether the values, aspirations and cultural attributes of the poor in Western cities really are significantly different from those of the rest of society.

The idea of **transmitted deprivation** is really concerned with explaining why, despite long periods of full employment and the introduction of improved welfare services, problems of deprivation persist. According to this model, the answer lies in the cyclical process of transmission of social maladjustment from one generation to another. Thus, while it is acknowledged that low wages, poor housing and lack of opportunity are important factors, the emphasis is on the inadequacies of the home background and the upbringing of children.

The idea of *institutional malfunctioning* shares some common ground with the managerialist school of thought, since the behaviour of bureaucrats is given a central role in explaining the persistence of deprivation. Here, however, it is not so much the 'gatekeeping' role of bureaucrats that is emphasized as the administrative structure within which they work. Thus, it is argued, the formulation of public policy in separate departments concerned with housing, education, welfare, planning and so on is inevitably ineffective in dealing with the interlocking problems of deprivation. Moreover, such organizational structures are vulnerable to interdepartmental rivalries and power struggles that can only reduce their overall effectiveness.

The idea of a *maldistribution of opportunities* and resources can be accommodated within pluralist political theory, with deprivation being seen as the result of failures of participation and representation of certain interests in the political process.

The idea of *structural inequality* sets problems of deprivation as inevitable results of the underlying economic order and of structural changes in labour markets, etc. that are attached to the overall restructuring of the economy and of the built environment.

The idea of an **underclass** borrows from several of these perspectives, emphasizing the effects of economic restructuring and sociospatial change in isolating racial minorities not only from the economic mainstream but also from the social values and behavioural patterns of the rest of society. The existence of large numbers from minority groups with only weak connections to the formal labour force is attributed largely to the spatial mismatch between people and jobs that has intensified as many of the low-skill jobs traditionally found in inner-city areas have been relocated, to be replaced mainly by jobs requiring higher skills. The development of a distinctive context of values and attitudes is attributed largely to the feminization of poverty resulting from an increase in teenage unwed mothers (itself a product of a combination of economic and social trends), combined with the suburbanization of more affluent, better-skilled households. It has proven difficult, however, to establish the nature of the linkages between labour markets, poverty, migration, household structure, race, gender, attitudes and behaviour; while the term 'underclass' itself has been criticized because of the way it has been used as a pejorative label for the 'undeserving' poor by some commentators.

All this leaves us some way short of a clear and comprehensive explanatory framework for deprivation and social polarization. What is clear, however, is that the degree of sociospatial polarization in contemporary cities has brought a disturbing dimension of urban social geography to a new prominence. Riots, civil disorder, social unrest and social disorganization are by no means new to cities, but they appear to have developed to unprecedented levels.

Take, for example, the inner-city neighbourhoods of the Bronx, where relict and dilapidated neighbourhoods have come to represent 'burnt out' settings where social disintegration has fostered extremely high levels of poverty, substance abuse (Fig. 13.3a), violent deaths, low birthweight infants, and deaths from HIV (Fig. 13.3b) (Wallace and Fullilove, 1991). The combination of such aetiologies with continuing discrimination and a newly racialized politics (see Omni and Winant, 1993) has begun to precipitate rebellion, as manifested by rioting (Fig. 13.4). Thus we enter a new round of the sociospatial dialectic, with events such as the Los Angeles riot of April 1992, which accounted for 52 deaths and between $785 million and $1 billion in property damage (Oliver *et al.*, 1993), leading to a widespread 'hardening' of the built environment, with 'fortress' and 'bunker' architecture, the loss of public urban spaces, the 'militarization' of social

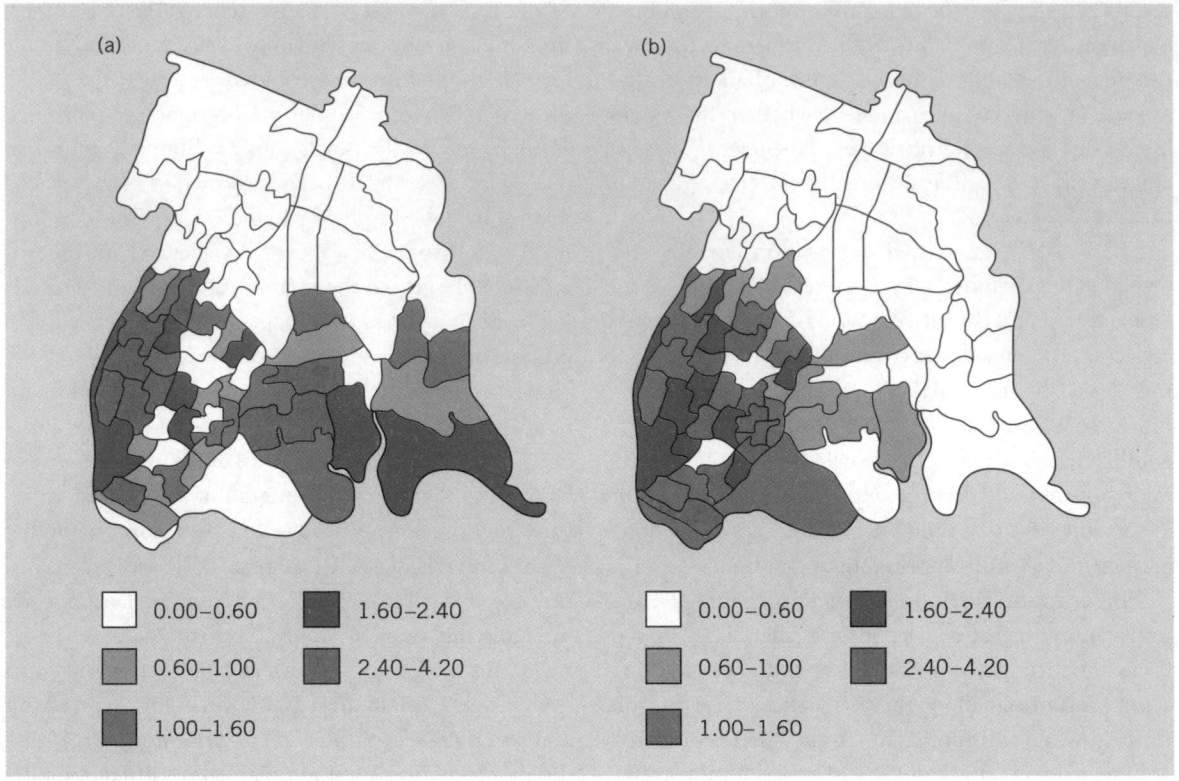

Figure 13.3 The Bronx, New York: (a) annual number of drug-related deaths, by quintiles; (b) HIV deaths per 100 000 population, cumulative to 1988.
Source: Wallace and Fullilove (1991), pp. 1707, 1708.

control, intensified surveillance, gated streets, private security forces and intensified sociospatial segregation (Davis, 1992).

The informal urban economy

One response to social polarization has been the non-recording of economic activity – creating an **'informal' economy** (also known as a 'hidden' economy). Sometimes no official record is made because the activity is illegal, such as bribery, prostitution or drug dealing, but in other cases the activity may be perfectly legal but no record is made in order to evade paying taxes or else to avoid the withdrawal of state benefits. These unrecorded exchanges often occur between people who are strangers but also between neighbours, friends and relatives who know each other. A further complication is that sometimes these exchanges may involve goods or services rather than money.

Because the activity is by definition covert, very little is known about the informal economy compared with other aspects of city life. For example, the hidden economy is often synonymous with those on the economic margins of society, yet there is a great deal of 'white-collar' informal activity that receives much less attention.

There is a common assumption that the hidden economy is on the increase in developed Western cities but this is difficult to prove conclusively (some staggering estimates put the total economy built around drug dealing as worth $300 billion at 2000 prices – approximately equivalent to the size of the petroleum economy). Another common assumption (which one might also call a prejudice) is that the hidden economy has increased because of the growth of long-term male unemployment in Western cities. Thus it is claimed that unemployed people are 'getting by' through being active in the informal economy.

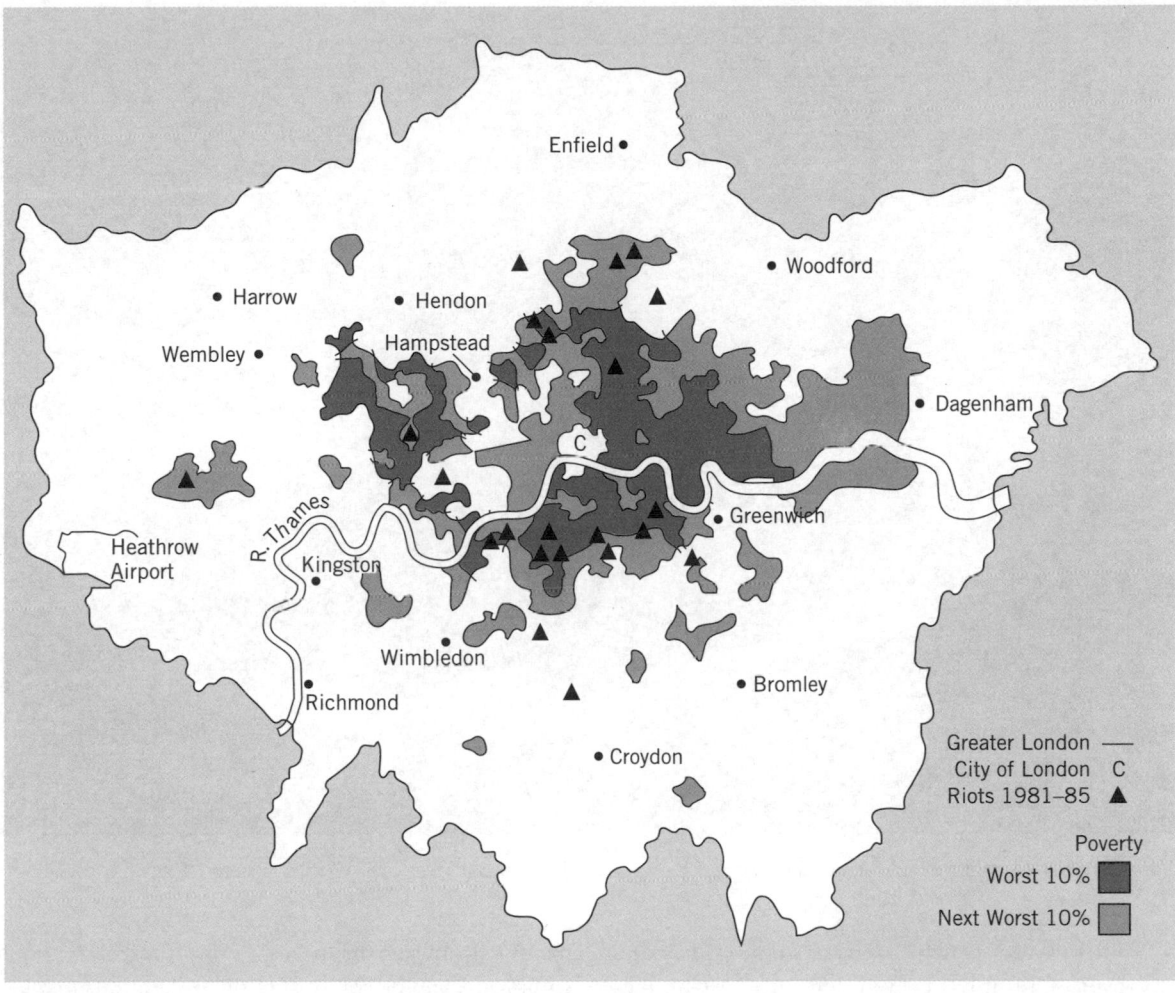

Figure 13.4 Poverty and urban riots, Greater London.
Source: Diamond (1991), p. 218.

Although this is obviously a difficult topic to research, once again, we find no clear social science evidence for this 'common-sense' assumption. In fact, to the contrary, it appears that it is the formally *employed* who are most likely to engage in informal economic activity (Pahl, 1984). The reason for this becomes 'obvious' with a little thought; to initiate and maintain economic activity, even of the unrecorded type, takes resources such as money, knowledge, equipment and personal contacts. These are things the employed tend to have in abundance but which unemployed often lack.

Pahl's work raises a related issue, that of the **domestic economy**, work that is undertaken within households.

Pahl argued that in order to understand social polarization it is important to take account of informal work in and around the household, such as cleaning, decorating and home improvements (termed '**self-provisioning**'), as well as paid work in the formal economy. Pahl claimed that in the United Kingdom the majority of households have been able to mitigate a deteriorating position in the formal labour market through informal work and self-provisioning. In addition, households with relatively modest incomes may not be so badly off compared with, say, high-earning, two-career, professional households because the former do not have to purchase services such as child-care, house maintenance, car servicing, gardening and so on.

The informal urban economy: fake Gucci handbags for sale in Venice. Photo Credit: Paul Knox.

Pahl provides convincing case-study evidence for his theories but it has proved difficult to extend these ideas more widely throughout the United Kingdom as a whole because of the extreme demands his thesis makes upon empirical data (see also Pinch, 1993). In addition, it has been argued that the increasing residualization, commercialization and privatization of public services has also encouraged domestic self-provisioning, as families are forced to become more self-reliant and provide services for themselves (Pinch, 1989). Again, this is a theory that requires further research.

Urban social sustainability

A final key issue underpinning urban social geography is **sustainability**. This term has a variety of meanings but most are broadly in accord with the definition of the influential Bruntland Report – namely, develop-

ment that meets the needs of the present without compromising the ability of future generations to meet their needs (World Commission on Environment and Development, 1987).

As we saw in Chapter 2, ever since the creation of large cities with the advent of industrial capitalism, there have been grave concerns about the environmental consequences of urbanization. However, in the early twenty-first century these fears have taken on added dimensions. These problems are most acute in the developing world with its very rapid rates of urbanization and, often, intense poverty. However, the majority of the earth's resources are being consumed by urbanites of the developed world – the focus of this book – where there are also many problems of waste, pollution, noise and traffic congestion. In response, we have witnessed the development of numerous ecological movements and green political parties. Writers such as Ulrich Beck (1992) have also had a key influence; he argued that we

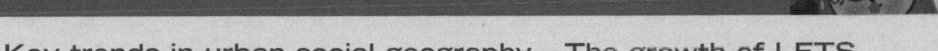

Box 13.4

Key trends in urban social geography – The growth of LETS

One form of alternative economic activity that is legal and recorded but does not involve use of the official national currency is the system of micro-credit used in some city neighbourhoods, commonly known as a *local economic trading system (LETS)*. People usually pay a small fee to join such a scheme but once within, all economic transactions are based on credit related to a local currency. This currency is in turn usually based on particular goods or units of service, so when exchanges are made no money is involved. All transactions are recorded by a local treasurer or accountant and regularly published so that members know whether they are in credit or deficit to other members. It is an inherent characteristic of these schemes that they tend to be on a small scale in localized areas of cities.

The purpose of such schemes is to help those on the margins of society who do not possess sufficient funds to participate fully in the formal economy using the national currency. However, these schemes also have an important symbolic value in that they involve cooperative forms of economic activity that can question the ethics of the conventional capitalist, market-driven, system. Micro-credit schemes were used in the slump of the inter-war period but were revived in the 1980s in Canada. Although inevitably small in scale, they have now spread throughout most developed countries.

Key concepts associated with LETS (see Glossary)

Postwelfare society, Schumpeterian welfare state, self-provisioning.

Further reading

Lee, R. (1996) Moral money, LETS and the social construction of local economic geographies in Southeast England *Environment and Planning A*, **28**, 1377–1394

Lee, R., Leyshon, A. and Williams C.C. (eds) (2003) *Alternative Economic Spaces* Sage, London

Williams, C.C. (1996) Local Exchange and Trading Systems (LETS): a new source of work and credit for the poor and unemployed *Environment and Planning A*, **28**, 1395–1415

---

have entered the age of the '**risk society**' in which technologies such a nuclear power and genetic engineering pose far greater risks than in the past.

Within this broad social climate it is possible to distinguish two main perspectives on environmental issues (O'Riordan, 1981; Luke, 1997). First there is the dominant conceptualization, the **technocentric approach** (sometimes also called the **ecological modernization** approach). This approach is based on the assumption that environmental problems should be tackled without upsetting the broad capitalist economic framework that currently guides world development. This approach therefore stresses the capacity of existing institutions to adapt to environmental issues and the capacity of modern science and technology to meet these challenges. It argues that economic growth is the key to better welfare for citizens and this should be driven primarily by market forces, regulated in the interests of the environment.

In opposition to the technocentric approach are various types of **ecocentric approach** united by a belief that ecological problems can be addressed only by changing the capitalist system and its inexorable dependency of economic growth and consumption. In addition, ecocentrists tend to see existing state institutions as working in the interests of big business and therefore incapable of reining in such forces. The solution is often seen in smaller decentralized, self-reliant political units within which there can be greater participatory democracy. However, some argue that over-arching forms of governance are needed to coordinate local actions to meet environmental needs (Pepper, 1996). David Harvey argues from a radical perspective that:

> . . . the environmental justice movement has to radicalize the ecological modernization discourse itself. And that requires confronting the underlying processes (and their associated power structure, social relations, institutional configurations, discourses and belief systems) that generate environmental and social injustices. Here, I revert to another key moment in the

argument advanced in *Social Justice and the City* (pp. 136–137): it is vital, when encountering a serious problem, not merely to try and solve the problem in itself but to confront and transform the processes that gave rise to the problem in the first place. Then, as now, the fundamental problem is that of unrelenting capital accumulation and the extraordinary asymmetries of money and political power that are embedded in that process.

(Harvey, 1996, p. 97)

As emphasized by the Habitat II Summit in Istanbul in 1996, the issues of cities and sustainability are inextricably linked. The consumption-intensive, car-dependent, decentralized, suburban city forms that have become dominant in Western cities in the second half of the twentieth century, although apparently popular with many, are clearly wasteful of resources (see also Box 13.2). The search is therefore on to find more sustainable urban forms, although, given the huge existing investment in suburbanization, there can be no immediate turn-around in city design (even if a majority desired such a shift).

Despite the popularity of gentrification among some groups in recent years, it is highly unlikely that we will witness a return to the high-density urban forms of the classic industrial city. More feasible, therefore, is what Blowers (1994) calls the 'MultipliCity' (see Figure 13.5). This is characterized by smaller relatively dense urban settlements, linked by greater use of public transport. Creating such an urban form will require a considerable shift in attitudes among many peoples in countries where car ownership is entrenched (Breheny, 1995). The important implication is that sustainability is not just about resources in isolation, it is intimately connected with social and economic issues. Nevertheless, it is possible to isolate a separate notion of **urban social sustainability** that involves notions of

Urban social sustainability: the city as a focal point for sociability, knowledge exchange and creativity (Bellinzona, Switzerland). Photo Credit: Paul Knox.

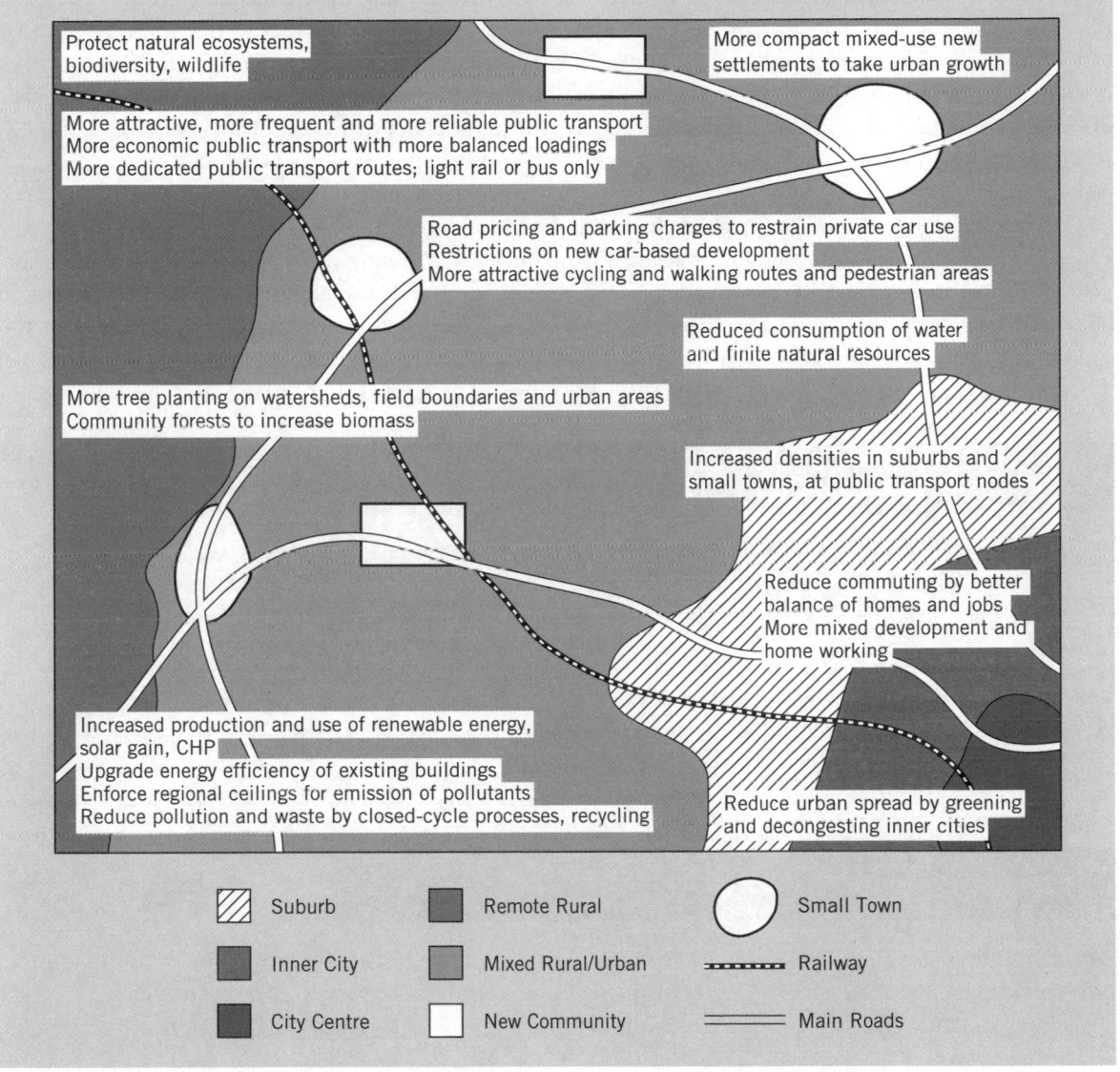

Figure 13.5 Some of the changes that will need to be made to ensure the future sustainability of city regions.
Source: Blowers (1999), Fig. 6.12, p. 282.

equity, community and urbanity. Yiftachel and Hedgcock define such a notion as follows:

> Urban social sustainability is defined here as the continuing ability of a city to function as a long-term viable setting for human interaction, communication and cultural development. It is not necessarily related to the environmental and economic sustainability of a city, although the links often exist between the three areas. A socially sustainable city is marked by vitality, solidarity and a common sense of place among its residents. Such a city is characterized by a lack of overt or violent intergroup conflict, conspicuous spatial segregation, or chronic political instability. In short, urban social sustainability is about the *long term survival* of a viable urban social unit.
>
> (Yiftachel and Hedgcock, 1993, p. 140, authors' emphasis)

As we have seen throughout this book, a number of forces – including the globalization of capitalism, extraordinarily rapid technological change and post-modern consumer culture – have brought about an era of acute economic instability and social insecurity. Jobs, communities, families and identities all seem to be increasingly under threat.

Richard Sennett (1994) has argued that this widespread sense of instability means attachment to *place* – whether it be the nation, region, city or neighbourhood – has increased. Thus, attachment to these spaces provides something that is perceived to be more stable than the insecurities associated with employment and the marketplace. Featherstone and Lash (1999, p. 2) note that in this context cities may take on a new role for '. . . they offer the potential of an open public space built around the values of diversity, urbanity and experience.' Thus, Sennett's notion of a cosmopolitan public space holds out the prospect of people developing new forms of sociability based on tolerance. As we have seen in previous chapters, there are currently many forces leading to exclusion, inequality and *in*tolerance in cities. Nevertheless, it is our fervent hope that the issues discussed in this book will help foster attitudes, actions and policies that encourage urban social sustainability in the future.

Chapter summary

13.1 The various costs and benefits associated with access to services greatly affect the quality of life of urban residents. The struggle over access to scarce resources leads to coalitions of interest, often based around neighbourhoods. The aggregate effect of service allocations is one of 'unpatterned inequality'.

13.2 Decentralization, urban renewal and the restructuring of the public sector have greatly affected access to services in urban areas. Social polarization is an endemic feature of many Western cities and is a complex phenomenon that has consequently given rise to a variety of explanations. A key challenge for the future is how to achieve urban social sustainability.

Key concepts and terms

accumulation	externalities	self-provisioning
asset sales	informal economy	service-dependent ghetto
'asylum-without-walls'	intensification	social polarization
centralization	legitimation	spillovers
commercialization	local economic trading systems	subcontracting
community-based politics	(LETS)	sustainability
competitive tendering	'loyalty' option	tapering
contracting-out	materialization	technocentric approach
culture of poverty	outreach services	third-party effects
de facto territories	place- (or point) specific services	transmitted deprivation
deinstitutionalization	privatization	'turf' politics
domestication	rationalization	underclass
domestic economy	reinstitutionalization	urban social sustainability
ecocentric approach	reproduction	'voice' option
ecological modernization	residualization	workfare
'exit' option	'risk society'	

Suggested reading

For recent changes in welfare states see Christopher Pierson and Frances Castles's edited volume *The Welfare State Reader* (2000: Polity Press, Cambridge) Geographical aspects of welfare restructuring are covered in Steven Pinch's *Worlds of Welfare: Understanding the Changing Geographies of Social Welfare* Provision (1997: Routledge, London). Also of relevance in this context is Roger Burrows and Brian Loader (eds) *Towards a Post-Fordist Welfare State?* (1994: Routledge, London) and the special issue of *Environment and Planning A*, 2003: Volume 35, number 5, on geographies of care and welfare edited by Lynn A. Staeheli and Michael Brown. On urban restructuring, inequality, and conflict, see Robin Law and Jennifer Wolch, 'Social reproduction in the city: restructuring in time and space' in *The Restless Urban Landscape* edited by Paul Knox (1993: Prentice Hall, Englewood Cliffs). This extended essay provides an excellent overview of the interdependence of economic restructuring, the restructuring of households, of communities, and of the welfare state, relating them all to changing urban activity patterns. Useful empirical studies of various aspects of restructuring can be found in John Mollenkopf *Power, Culture, and Place* (1988: Russell Sage Foundation, New York (see especially Chapters 7–12), Susan Fainstein, Ian Gordon and Michael Harloe, *Divided Cities: New York and London in the Contemporary World* (1992: Blackwell, Oxford) and John Mollenkopf and Manuel Castells' *Dual City. Restructuring New York* (1991: Russell Sage Foundation, New York). For an overview of urban problems in the UK see Michael Pacione's edited volume *Britain's Cities: Geographies of Division in Urban Britain* (1997: Routledge, London). The classic text on deinstitutionalization is Michael Dear and Jennifer Wolch's *Landscapes of Despair* (1987: Princeton University Press, Princeton). The literature on deprivation is vast so only a few examples can be given in the space available: for issues of poverty and the efforts being made to redress them in a European context see Ali Madanipour, Goran Cars and Judith Allen (eds) *Social Exclusion in European Cities: Processes, Experiences and Responses* (1998: Jessica Kingsley, London); for poverty issues in a British context see Chris Philo (ed.) *Off the Map: The Social Geography of Poverty in the UK* (1995: Child Poverty Action Group, London); David Byrne's thought-provoking *Social Exclusion* (1999: Open University Press, Buckingham) and John Mohan 'Geographies of welfare and social exclusion' (*Progress in Human Geography*, 2000: **24**, 291–300). There is a rapidly growing literature on the issue of urban sustainability: for an introduction see the chapters in Mike Breheny (ed.) *Sustainable Development and Urban Form* (1992: Pion, London). See also M. Kenny and J. Meadowcroft *Planning Sustainability* (1999: Spon, London), A. Layard *et al. Planning for a Sustainable Future* (2001: Spon, London) and Stephen M. Wheeler *Planning for Sustainability* (2004: Routledge, London). Entertaining and inspiring as ever is Mike Davis' work on discourses of ecological disaster in Los Angeles *Ecology of Fear: Los Angeles and the Imagination of Disaster* (1998: Metropolitan Books). For a radical manifesto outlining a progressive democratic city see Ash Amin *et al.* (2001) *Cities for the Many Not the Few* (2001: Policy Press, Bristol). Another vision of the future city is the report UK Urban Task Force led by the cult architect Sir Richard Rogers *Towards an Urban Renaissance* (1999: Department of the Environment, Transport and the Regions, London).

Chapter 14

Whither urban social geography?: Recent developments

Key questions addressed in this chapter

➤ What is meant by the 'Los Angeles School' of urban geography?

➤ What are the advantages and limitations of the LA School's work?

➤ What are the relationships between cinema and the city?

As we have seen throughout this book, urban social geography is today a highly diverse set of studies. This diversity was memorably captured over a decade ago by one of the leading figures of the 'Quantitative Revolution' in geography, Peter Haggett, who likened urban geography itself to a city:

> a city with districts of different ages and vitalities. There are some long-established districts dating back to a century ago and sometimes in need of

repair; and there are areas which were once formidable but are no longer so, while others are being rehabilitated. Other districts have expanded recently and rapidly; some are well built, other rather gimcrack.

(Haggett, 1994, p. 223)

In this final chapter we illustrate some of this diversity by considering two newly constructed 'districts' of the subdiscipline. First we look at the controversy surrounding the so-called 'Los Angeles' or 'California' School of urban geography. This debate may be envisaged as a reconstruction of an older 'district', namely, whether we can conceive of a **paradigmatic city**, i.e. one that can be used to generalize about future developments elsewhere. The second 'district' is related to the first, but may be regarded as essentially a new construction. Michael Dear, a key member of the 'LA' School has written that 'the urban grows increasingly to resemble televisual and cinematic fantasy' (Dear, 2000, p. 166). The second focus of this chapter is therefore appropriately upon recent work on the links between cinema and the city.

14.1 Los Angeles and the 'California School'

The diverse and extensive writings of the Chicago School of urban sociology in the early part of the twentieth century meant that Chicago came to be regarded as the classic example of an industrial city. In a similar fashion, in recent years an extensive body of work by a group of scholars based in California has meant that Los Angeles has come to be seen as the archetypal 'postmodern' city (e.g. Scott, 1988; Soja, 1989, 1996; Dear and Flusty, 1998). Indeed, these scholars – largely based in the Graduate School of Architecture and Urban Planning at University College, Los Angeles – have self-consciously portrayed Los Angeles as a harbinger of future urban forms – encouraging others to portray them as the '**Los Angeles School**' (or '**California School**'). As in the case of the earlier Chicago School, there is a great deal of diversity in the writings of this school, but they do share a number of underlying themes.

Underpinning much of the work of the California School have been attempts to link the sprawling suburbs of Los Angeles with regulationist-inspired notions of a new regime of accumulation (see Chapter 2). A new regime of flexible accumulation is argued to be manifest in California in high-technology agglomerations (Scott, 1986), dynamic, fluid, creative industries such as those producing movies (Christopherson and Storper, 1986) and industrial clusters based around illegal or low-paid workers. However, critics have argued that these notions of industrial restructuring are too broad and economic in focus to provide a satisfactory explanation of the myriad small-scale processes involved in neighbourhood formation (Savage and Warde, 1993).

Another main theme to emerge from this work is the newness of Los Angeles. Although the origins of the current city can be traced back to the original settlement by Spanish missionaries in 1781, Los Angeles displays little of the industrial legacy of the classic industrial city. The new urban forms created on the west coast of the United States were highlighted as early as 1945 by Harris and Ullman. They drew attention to reduced significance of a central business district and the presence of many competing decentralized centres in their **multiple nuclei model**. This dispersion was of course made possible by the widespread adoption of the automobile for transportation. As Reiff says of Los Angeles:

> There are few experiences more disconcerting than walking along a wide LA street without the reassuring jangle of keys in your pocket. These streets are so unshaded, their sidewalks appearing wider because they are so empty.
>
> (Reiff, 1993, p. 119)

This dispersion and lack of recognizable pattern has been a key theme developed by the Los Angeles School. What Soja (1989) has described variously as the **postmodern global metropolis, cosmopolis** and **postmetropolis** (1997) is seen as a physically and socially fragmented entity. Contrary to the popular stereotype, Los Angeles is not a city without a centre. Indeed, there is a recent strong element of recentralization in the form of the command centres linked into the new global economy but the city also consists of numerous subcentres and **edge cities** (Garreau, 1992). These are not the exclusively affluent suburbs of an earlier era but show enormous variations in character, some being industrial or commercial and others being relatively poor and/or with distinctive ethnic minorities. Soja (1992) develops this theme into the concept of **exopolis** – a city that has been turned inside-out. In such an environment it is difficult for individuals to have a sense of belonging to a coherent single entity (see also Box 14.1). The fragmentation and diversity of postmodern culture is therefore manifest in the physical structure of the landscape. The term **galactic metropolis** has also been coined to describe such cities (Lewis, 1983; Knox, 1993). The reason for this label is that the commercial centres in such cities look more like stars spread about a wider galaxy rather than a single recognizable centre. The most extreme portrayal of the city as a decentralized random expression of elements is Dear and Flusty's (1998) highly imaginative notion of **keno capitalism**. In this model the city structure varies between elements of edge cities, consumption, spectacle, gated communities and global command centres in a random manner (rather like the selection of cards in the game of keno) (see also Box 14.2).

Box 14.1

Key thinkers in urban social geography – Edward Soja

Latham has argued that Edward Soja is perhaps unique among contemporary urban social geographers in his capacity to both inspire and irritate (Latham, 2004)! The inspiration comes from the breadth and depth of his scholarship, the scale and sophistication of his erudition. The irritation comes from what some see as pretentious overgeneralization that signifies relatively little. What is in no doubt is that Soja has raised the profile of urban social geography, especially to researchers in other disciplines closely related to human geography.

Like David Harvey (Box 1.1), Soja has been greatly inspired by the work of French scholar Henri Lefebvre (see Box 9.3). Indeed, in Soja's case the influence would seem to have been even greater for, like Lefebvre, Soja has been concerned to put the production of space – what Soja terms 'spatiality' – at the centre of social theory. In *Post-modern Geographies* (1989) Soja argues that spatiality is a fundamental organizing principle of postmodern culture that serves to disguise the underlying power relations of a globalized capitalism.

In his second book *Thirdspace* (1996) Soja outlines six 'discourses'

that encompass contemporary developments in Los Angeles:

➤ *The Post-Fordist Industrial Metropolis*: the city of new industrial spaces and sweatshops.

➤ *The Exopolis*: the city turned 'inside out' through fragmentation and decentralization.

➤ *The Cosmopolis*: the city shaped by globalization.

➤ *The Fractal City*: the city divided by multiple ethnicities and identities.

➤ *The Carceral Archipelago*: the city of privatized space and surveillance.

➤ *The SimCity*: the city of the digital knowledge economy.

In response to those who argue that this is 'top down' theorizing that ignores the lively plurality of postmodern culture, Soja replaced his early sociospatial dialectic (that underpins this book, see Soja, 1980) with a 'trialectic'. As with Lefebvre, this concept integrates three conceptions of space: 'firstspace' (the material world); 'second space' (imagined representations of spatiality); and 'thirdspace', a conceptualization that integrates the previous two concepts. However, some claim this approach is

vague and unspecified (see Barnett, 1997: Merrifield, 1999).

Key concepts associated with Edward Soja (see Glossary)

Cosmopolis, exopolis, hyperreality, hyperspace, Los Angeles School, postmodern global metropolis, third space.

Further reading

Barnett, C. (1997) Review of Thirdspace: journeys to Los Angeles and other real and imagined places *Transactions of the Institute of British Geographers*, **22**, 529–530

Latham, A. (2004) Ed Soja, in P. Hubbard, R. Kitchin and G. Valentine (eds) *Key Thinkers on Space and Place* Sage, London

Merrifield, A. (1999) The extraordinary voyages of Edward Soja: inside the 'trialetics of spatiality' *Annals of the Association of American Geographers*, **89**, 345–348

Soja, E.W. (2000) *Postmetropolis: Critical Studies of Cities and Regions* Blackwell, Oxford

Another important theme running through the work of the California School has been the development of protective measures by the more affluent sections of society in an environment in which violence and crime is routine (Davis, 1990; Christopherson, 1994). This may be seen as a response to the increasing social polarization mentioned previously. One manifestation of these defensive measures is the growth of so-called **bunker architecture** (also termed 'citadel', 'fortified' and 'paranoid' architecture): urban developments with gates, barriers and walls, security guards,

infra-red sensors, motion detectors, rapid response links with police departments and surveillance equipment such as CCTV. These form what Davis terms a **'scanscape'**, all designed to exclude those regarded as undesirable. These systems are often established in residential areas but exclusionary measures may also be undertaken in shopping malls and city centres. Thus, in some central parts of Los Angeles park benches are curved in such a way as to inhibit people from sleeping on them over night. It is little wonder then that Davis (1990) talks of the 'militarization' of

Critique of the LA School

Although the work of the Los Angeles School has undoubtedly been highly influential, the key question raised by their approach is: just how valid is Los Angeles as a general model of future urban developments? Although members of the California School hedge their comments with numerous caveats, acknowledging the unique history of Los Angeles, it is clear that they see the city as representative of future urban forms. There are, however, many other cities throughout the Western world with stronger residual elements from previous periods as well as different political and administrative regimes. For example, McNeill (1999) notes that European cities have different urban morphology and stronger traditions of Left-leaning social democratic governments from those of their North American counterparts. As Hall (1998) notes, it is not so much the urban forms as the *processes* involved in their creation that are the most important element in the work of the Los Angeles School and one should not assume that Los Angeles is the forerunner of future urban forms elsewhere. In addition, one should not exaggerate the newness of the city. For example, in the ring of barrios and ghettos that surround the newly fortified downtown of Los Angeles, Davis (1992) denotes a ring with similar attributes to the classic zone of transition of an earlier period in Chicago. These inner areas act as classic reception areas for immigrants to Los Angeles and also manifest the classic teenage gangs of an earlier era (although the latter are now much more extensive, stretching out into suburban areas).

A second major criticism, not just of the California School but of all theories that emphasize the influence of postmodernism upon urban forms, is the silence over the issue of *race* (Jacobs, 1996). This is a criticism that emerges from postcolonial theory. Although postmodern perspectives claim to give expression to the diversity of identities in cities, it is argued that they remain a central vision in which immigrant groups are slotted in to perform a role in the new global economy. Yet postcolonial theory suggests that there is no one central vision but also many views from outside the centre. As Jane M. Jacobs notes:

> Within social polarization arguments the complex politics of race is translated into a variant form of

Panopticism in the city: CCTV cameras. Photo Credit: Andrew Vowles.

city life. However, it is not just the marginalized who are kept under surveillance in the contemporary city. New technologies centred around credit and loyalty cards, computers and pay-for-service facilities, mean that corporations and governments can access vast amounts of information about people's travel and consumption habits.

class differentiation produced by the now more thoroughly globalised and deceptively aestheticised unevenness of capitalism. A fractured, positional and often angry politics of difference is (mis)recognised as a static, structural outcome of advantage and disadvantage. Through this manoeuvre, the politics of race is cast off from the history of the constitution of difference and racialised subjects are denied the kind of agency captured by theorisations of a politics of identity. It is not simply that there is not enough race in these accounts of the postmodern city, it is that the cultural politics of racialisation is deactivated.

(Jacobs, 1996, p. 32)

A third major critique of the LA School argues that their dystopian vision of a fragmented metropolis fractured by the processes of globalization simply does not fit the facts. Indeed, Gordon and Richardson (1999) argue that the School has a cavalier attitude towards factual evidence in general. For example, they argue that Soja's assertion that planned peripheral communities such as in San Bernadino lack jobs and that consequently the inhabitants of these areas need to commute long distances does not seem to equate with commuting times of under 45 minutes. In addition, the often-quoted assertion that LA is striken with social polarization is countered by the evidence of substantial social mobility between social groups in the city. Thus Gordon and Richardson argue that many Latino immigrants in LA have moved into home-ownership and form a new middle class. In addition, Jennifer Wolch's association of increasing homelessness with the forces of globalization is also criticized for ignoring many other contributory factors including deinstitutionalization, the destruction of poor-quality housing and the drugs problem. The assertion that LA has become the global command centre in the world economy is also seen as mistaken since all the major banks have left Los Angeles. Indeed, it is argued that the revival of the LA CBD can also be exaggerated, for the downtown area has very high rates of vacancies.

This vein of criticism is taken up by Curry and Kenney (1999). They argue that Los Angeles' emblematic status as a new global metropolis has been severely eroded by a series of setbacks. They claim that much of the city's manufacturing activity in the second half of the twentieth century was dependent upon defence expenditure on aerospace during the Cold War. When this ended growth faltered and the manufacturing activities were remarkably rigid (and decidedly non-post-Fordist) in adapting to new market opportunities.

Los Angeles: a paradigmatic city?

There is insufficient space here to discuss the numerous other criticisms of the School's interpretation of Los Angeles made by Curry and Kenney or their rebuttal (see Scott, 1999; Storper, 1999, for details). However, a key wider question raised by the work of the LA School is how useful is the notion of a **paradigmatic city** (i.e. a city that displays more clearly than others the key urban patterns and processes of a particular period, see also Nijman, 2000). Curry and Kenney argue that the concept of a paradigmatic city seems to run counter to the LA School and other postmodernists' emphasis upon diversity, fragmentation and lack of structure. Furthermore, paradoxically, Dear and Flusty's notion of keno capitalism may be seen as perpetuating the detached modernist gaze upon cities of which they have been so critical (Beauregard, 1999). Thus it is argued that there is little of the cultural richness of the city in their extensive use of new terms to describe various social groups in the postmodern city (e.g. the cybergeoisie and protosurps!) (Lake, 1999). Implicit in much of the work of the Los Angeles School is a critique of the work of the Chicago School which has dominated urban geography throughout much of the twentieth century. However, it has been argued that this critique (like many others before) concentrates upon the spatial aspects of the Chicago School's approach, and in particular the Burgess concentric zone model (see Chapter 7), thereby ignoring the rich vein of case study material of local Chicago subcultures obtained through ethnographic research methods (qualitative methods such as participant observation and unstructured interviews) (Jackson, 1999).

Amin and Graham (1997) have argued that the tendency to take a limited range of urban contexts as paradigmatic of general change is not confined to adherents of the California School but also applies to those who proffer concepts of world cities (e.g. London, New York and Tokyo) or 'creative cities' (e.g. Barcelona). Echoing a growing chorus of other urban analysts, they argue that cities are increasingly intersections of multiple webs of economic and social, life, many of which do not interconnect. This they call the **multiplex city** 'a juxtaposition of contradictions and diversities, the theatre of life itself' (Amin and Graham, 1997, p. 418). Boyer (1995) makes a similar distinction between the 'figured city' – the isolated planned city for affluent groups – and the 'disfigured city' – the neglected, unmanaged, spaces of the city inhabited by poorer groups. Increasingly, then, we cannot generalize about *the* city.

Box 14.2

Key thinkers in urban social geography – Michael Dear

Michael Dear has made a significant contribution to the study of urban social geography in two distinctive and interrelated ways. First, he has undertaken a series of path-breaking surveys of the social and geographical consequences of deinstitutionalization – the closure of long-stay psychiatric institutions – and the associated phenomenon of homelessness. This work is best demonstrated in his early book *Not On Our Street: Community Attitudes Towards Mental Health Care* (1982) (with Martin Taylor) and two later books (both with Jennifer Wolch), *Landscapes of Despair: From Deinstitutionalization to Homelessness* (1987) and *Malign Neglect: Homelessness in an American City* (1993). This work is in many respects a model of how to undertake policy relevant, theoretically informed, social research. It draws upon a wide range of perspectives and differing types of social survey to show the consequences of cutbacks in welfare spending. Furthermore, Dear has been highly active in drawing policy-makers' attention to these issues.

Much of the above work is set in southern California where deinstitutionalization has been implemented in its most extreme form. This context provides the setting for Dear's second main contribution, as a key member of the so-called 'Los Angeles School' of urban geography. As with the work of fellow school member, Ed Soja, this aspect of Dear's work has aroused considerable controversy (see also Box 14.1). A number of complex issues underpin these debates.

First there is the issue of just how representative is Los Angeles of future urban forms? Here the work of Dear and his colleagues seems to be contradictory. At times they play down the generality of their claims but at other times they seem unable to resist the temptation to proclaim Los Angeles as a paradigm for new global urban forms.

Second, there is the issue of just how relevant is postmodernism for an understanding of these changes. Dear thrives on using new textual forms and this has led to a wide range of new terms – some of which have proved irritating to his contemporaries.

Third, and related to the above two points, it is argued that for all the talk of 'postmodern urbanism' the LA School represents a Marxian overarching perspective in which various geographically and socially fragmented peoples are at the mercy of large corporations. In so doing it is argued that such work ignores issues of ethnicity and cultural difference.

Whether the LA School will assume the status of the Chicago School of urban ecology is therefore far from certain. Nevertheless, it is well worth reading Dear's work and the controversies it has generated.

Key concepts associated with Michael Dear (see Glossary)

'Asylum-without-walls', deinstitutionalization, keno capitalism, Los Angeles School, postmodernism, reinstitutionalization.

Further reading

Dear, M. (2000) *The Postmodern Urban Condition* Blackwell, Oxford

Dear, M. (ed.) (2001) *From Chicago to LA: Making Sense of Urban Theory* Sage, Thousand Oaks, CA

Dear, M. (2003) The Los Angeles School of Urbanism: an intellectual history *Urban Geography*, **24**, 493–509

Dear, M. and Flusty, S. (1998) Postmodern urbanism. *Annals of the Association of American Geographers*, **88**, 50–72

McNeill, D. and Tewdr-Jones, M. (2004) Michael Dear, in P. Hubbard, R. Kitchin and G. Valentine (eds) *Key Thinkers on Space and Place* Sage, London

Links with Other Chapters

Chapter 3: Postmodernism in the City

Chapter 13: Service Sector Restructuring

14.2 Cinema and the city

A relatively new, but rapidly growing, field of research within urban social geography concerns the inter-relationships between cinema and the city. This is an important sphere of enquiry for two main reasons. First, films are one of the most important forms of cultural expression in contemporary Western societies. Second, as we have seen previously in this book, cities play a crucial role in social life today. The result is a complex set of linkages between cities and cinema. These mutual interactions are reflected in the much quoted observation by the French philosopher Baudrillard that, 'The American city seems to have stepped right out of the movies' (Baudrillard, 1989, p. 56). On the one hand, cities have greatly influenced the development of film. The earliest movies reflected urbanism at the beginning of the twentieth century. These films gave their audiences new perspectives on the cities in which they lived, with overhead views, tracking shots and close-ups. Indeed, some have drawn parallels between the cinema-goer and Benjamin's notion of the urban *flaneur*, the casual stroller observing the diversity of the city (see also Box 7.2). Thus, both the movie watcher and the city walker are often close physically, but distant socially. On the other hand, while city life continues to dominate films, there is growing evidence that films and the film industry are having an increasingly important impact upon the evolution of new urban forms. Each of these interrelationships will be considered in turn in the chapter.

Films as texts

There has been much complex debate within film theory over whether motion pictures have any special status as forms of cultural representation. For example, it is sometimes noted that, in comparison with many art forms, such as paintings, sculpture or opera, which are obviously artificial representations of the world, films can look as if what is happening on the screen is authentic and 'real'. This is sometimes referred to as the *haptical* quality of film (from haptic meaning 'to touch'). Indeed, what happens in motion pictures is in many cases all too real, especially when it has documentary overtones. For example, John Houston's

film *The Misfits* contains harrowing scenes portraying the capture of wild horses. The terror and the panic on the part of the horses is real and shocking (as indeed, it also transpires, was the horrified reaction of star Marilyn Monroe). As the writer J.M. Coetzee describes:

> Despite all the cleverness that has been exercised in film theory since the 1950s to bring film into line as just another system of signs, there is something irreducibly different about the photographic image, namely that it bears in or with itself an element of the real. That is why the horse capturing sequences of *The Misfits* are so disturbing: on the one side, out of the field of vision of the camera lens, an apparatus of horse wranglers and directors and writers and sound technicians united in trying to fit the horses into places that have been prescribed for them in a fictional construct called *The Misfits*; on the other side, in front of the lens, a handful of wild horses that make no distinction between actors and stuntmen and technicians, that don't know about and don't want to know about a screenplay by the famed Arthur Miller in which they are or are not, depending on one's point of view, the misfits, who have never heard of the closing of the western frontier but are at this moment experiencing it in the flesh in the most traumatising way. The horses are real, the stuntmen are real, the actors are real; they are all at this moment, involved in a terrible fight in which the men want to subjugate the horses to their purpose and the horses want to get away; every now and then the blonde woman screams and shouts; it all really happened; and here it is, to be relived for the ten-thousandth time before our eyes. Who would dare to say that it is just a story?
>
> (Coetzee, 2003, p. 67)

Nevertheless, in recent years this view of film has been much criticized and is referred to as the 'naturalistic fallacy'. What has become clear is that films are highly selective in what they represent. Films are therefore important cultural texts, full of socially constructed narratives and discourses – what in the context of motion pictures are often called **tropes**. A trope is thus a regular form of pattern that emerges in storytelling (such as the victory of the individual over 'the system' in Hollywood movies). Study of these tropes reveals much about films, the film industry and the society they represent (and especially the urban contexts in which they are shot).

The influence of the city on film

Cities have had an enormous influence upon film. For example, some of the most popular and influential representations of city life in motion pictures are in a form known as *film noir*. This type of film, characteristic of the 1940s and 1950s, may be seen as a manifestation of an art form called expressionism, in which painters had sought to convey human emotions (as in Van Gogh's *Starry Night* or Munch's *The Scream*). German artists in particular, including Ernst Kirchner, Erich Heckel and Karl Schmidt-Rotluff, sought to portray a generation emotionally scarred after the First World War. This they did by representing cities and their inhabitants as dark and disturbing. After the Second World War, expressionism was carried to Hollywood by those escaping Nazi fascism and its aftermath. Émigré film directors in Hollywood continued to represent cities as nightmarish centres of corruption, menace, vice and greed, although much of this reflected the Cold War paranoia of the era (Krutnik, 1997). Cities were represented as insecure places where things are seldom what they appear. City inhabitants were often portrayed as disturbed, dishonest and untrustworthy. Narratives were disrupted by people experiencing dreams, nightmares and hallucinations. Indeed, the mood of these films was often enhanced by dark interior scenes with basements and stairwells or exterior night-scenes in backalleys.

The private detective – as popularized in the novels of Dashiell Hammett and Raymond Chandler – is often a central character in such films. As Raymond Chandler famously wrote of the detective genre, 'Their characters live in world gone wrong. . . . The streets were dark with something other than the night' (Chandler, 1973a, p. 7). However, in Chandler's hands the detective is a modern-day knight in shining armour for, 'Down these mean streets a man must go who is himself not mean' (Chandler, 1973b, p. 198). The private detective is an especially useful device in *film noir* because, during the course of performing his duties (and the private eye is usually male), he is able to move quickly between the many different strata of the city, the rich and poor, honest and corrupt. As we have seen previously in this book, these strata are associated with different residential areas of the city, and the detective in *film noir* is always on the move between these areas. However, paradoxically, with the extensive use of dark interior studio scenes in these films, we get a little sense of the city's geography. Key films of this type are *The Maltese Falcon* (John Houston, 1941) based on Hammett's novel of the same name and Chandler's *The Big Sleep* (Howard Hawks, 1946).

The influence of film on the city

As is widely known, Los Angeles contains the suburb of Hollywood, the centre that dominates the finance, production and distribution of the majority of the world's films. As such, no city has figured more in motion pictures. Despite the popularity of *film noir* in the past, the dominant image of Los Angeles in movies in other films has been rather different. The mood is predominantly hedonistic and characterized by sunshine, surfing, beaches, palm trees, conspicuous consumption and widespread personal mobility based on the automobile and freeways. Inevitably many films have attempted to puncture this hegemonic vision with insights into social inequality and the urban underclass, violence, traffic congestion, drug-taking and alienation.

According to Walton (2001), Los Angeles illustrates the way in which film can influence local politics and urban development. He argues that popular understanding of the evolution of the city, together with the activities of green activists and environmental legislation in California, have all been influenced by *Chinatown* (1974). This film presented a view of the development of Los Angeles which, although arguably capturing certain underlying truths about the evolution of US cities, was in all main respects a work of fiction. To understand the influence of this film in more detail we first need to understand the reality of urban development in Los Angeles.

Los Angeles has developed into a sprawling metropolis of over 20 million inhabitants in what is essentially an arid landscape. As with all cites, central to this development has been an adequate supply of water. It was therefore necessary for the city to appropriate water from over 230 miles away to the east in the Owen's Valley of the Sierra Nevada mountains and transport this back to the city. This was accomplished via an elaborate system of aqueducts (as depicted in the LA-based films such as *Terminator 3*). This extraction of groundwater began in 1905 but the resulting drought and lowering of the water table brought about protests from local farmers, culminating in 1924 in the occupation and bombing of a local aqueduct (Walton, 2001). Although the protests were unsuccessful, the plight of the Owen's Valley farmers became something of a *cause célèbre* among social activists in the first half of the twentieth century. Nevertheless, the City of Los Angeles gradually purchased farms within the Valley to satisfy the seemingly insatiable need of the city for water.

The story then moves forward to 1974 and the release of *Chinatown* directed by the brilliant and controversial director Roman Polanski. This 'retro' detective movie is set in a mythical Los Angeles of 1937, and is complete with art deco backdrops and classic cars of the period. In the tradition of *film noir* the central figure is an ex-cop turned private eye (Jack Nicholson) who gradually unravels a plot in an environment riven with corruption. City developers are bent on secretly acquiring, by criminal methods, land for urban development at reduced cost (in this case adjacent San Fernando Valley rather than distant Owen's Valley). Officials in the Los Angeles Department of Water and Power (LADWP) are bribed into secretly dumping water from city reservoirs at night in a time of drought with the aim of intensifying water shortages and thereby winning public support for dam construction, aqueducts and urban expansion. Local farmers who are deprived of water are made bankrupt and their land is subsequently acquired at artificially low cost by a dummy syndicate serving to disguise the interests of the developers. During the course of his investigations the Nicholson character also unearths a subplot of incest which Walton (2001) argues serves as a sexual metaphor for the literal rape of the land. The label 'Chinatown' serves as a trope for

Nicholson's old police beat and signifies the hidden dark corrupt of the city that is beyond police control or civil governance. The upshot of the narrative is that the public are relatively powerless and duped about the essentially exploitative and immoral character of urban development.

Quite apart from setting the centre of the conflict in the San Fernando Valley rather than Owen's Valley, the main fictional component of *Chinatown* is the corruption of the LADWP. However, understandably, many who have watched the film have tended to believe that the story bears some approximation to the truth. Also understandable, therefore, is the sensitivity of the LADWP to the maligning of their reputation, especially as the film seems to have prompted green activists in California to petition for limits on groundwater extraction. Such has been the controversy that the LADWP refused to allow a film company to use the Owen's Valley aqueduct to make a semi-documentary drama entitled *Ghost Dancing* about the bombing of the water supply (Walton, 2001). The extent to which *Chinatown* captures the spirit of urban development in US cities is a matter for debate, but what the film makes clear is the power of movies to capture the popular imagination and even influence local politics.

Another illustration of the power of cinema to influence the city is the Hollywood Redevelopment Project (Stenger, 2001). This is a plan to redevelop over 1000 acres, based around Hollywood Boulevard, into a centre for tourism based around the film industry. The aim is to recapture the halcyon 'Golden Age' of Hollywood of the 1930s and 1940s through a new Entertainment Museum, the renovation of existing movie theatres, and the construction of new multiplexes, hotels and retailing outlets. Thus Hollywood will once again become a premier centre for movie*going* as well as movie*making* (Stenger, 2001).

Film as business

When considering contemporary manifestations of cities in movies, it is important to remember that film-making is above all a business. Film production usually involves great expense, even without the huge fees commanded by international superstars. Mainstream commercial movies require large numbers of skilled personnel, expensive props, special effects or location shoots. Furthermore, even well-accepted formulas can easily flop at the box office. This means that film production is an inherently risky business, although if the right target is hit the commercial rewards can be enormous. Consequently, film producers often adopt various strategies to minimize the risk of commercial failure (the mixture depending on the target audience): audience testing of pre-production versions of films; making sure there is a happy ending; giving people a view of themselves of which they approve; avoiding any radical questioning of dominant societal viewpoints; and the use of humour to diffuse difficult situations.

'Curtisland'

The influence of commercial imperatives can clearly be seen in the representations of British city-life in films scripted by Richard Curtis. These are the most commercially successful films ever produced in the UK and include *Four Weddings and a Funeral*, *Notting Hill* and *Love Actually* (see Table 14.1). In addition Curtis assisted with the script of the highly successful *Bridget Jones' Diary*.

City life in London as portrayed by Richard Curtis (in what might be termed 'Curtisland') has a number of distinctive characteristics. To begin with, his films tend to offer a rather sanitized 'picture postcard' view of the city. This is more than simply reinforcing the identity of London with some shots of central tourist images, a device used in many films located in cities. In addition, residential parts of London are typically represented as an 'urban village'. In the opening sequence of the film *Notting Hill*, for example, the key character William Thacker (played by Hugh Grant) walks to work through narrow traffic-free streets full of small shops and market traders. There are none of the homogeneous retail stores that characterize the average British high street (Mazierska and Rascaroli, 2003). Neither is there any hint of street crime, dirt, pollution or problems with public transport. Indeed, Notting Hill is portrayed as full of parks, squares and greenery. Although disabled and gay characters appear in the film, there is no reference to the ethnic diversity of Notting Hill or, indeed, the one feature that has made the area world famous (at least before the success of the film) – the annual Caribbean-style street carnival. The predominantly upper-middle class central characters in 'Curtisland' are represented in a way that makes them easily identifiable as English to a wider international audience, yet at the same time in such a way that is acceptable or at least not offensive to the home audience. Hugh Grant is typically amicable and self-effacing but has difficulty in expressing his emotions (in sharp contrast to some of the US characters).

City branding

One final manifestation of the power of films is the intense competition that exists among cities to become the locations for films that require setting in a particular place. Many cities therefore have highly proactive film departments devoted to attracting movies. Various inducements can be provided: finance, tax breaks, the provision of discounted municipal services, help with finding suitable locations and arranging for streets to

Table 14.1 The highest grossing British-based films of recent years

Film	World gross up to Dec 2003 (£ million)	Approx. cost of production (£ million)
Notting Hill	215.6	24.9
Bridget Jones' Diary	166.3	15.4
The Full Monty	152.6	2.1
Trainspotting	84	5
Bend It Like Beckham	45.2	3.1
Billy Elliot	42.1	3
Four Weddings and a Funeral	31.3	3.5
Lock Stock and Two Smoking Barrels	4.1	1

Source: www.the-numbers.com/movies

be closed during shooting. Some cities have attempted to become a particular 'brand' displaying certain attributes; for example, Philadelphia markets itself as a paradigmatic late nineteenth-century US city. Thus it played the role of late nineteenth-century New York in the film *The Age of Innocence*. Philadelphia has also been successful in attracting other films in recent years; *Philadelphia* (of course!), *Sixth Sense* and *Twelve Monkeys*.

Some have disputed the economic benefits granted to cities by the encouragement of motion picture shoots. Critics point out that since many of the skilled staff are brought in from outside the employment gains for the city concerned are often quite small and of short

duration during shooting. In the case of *The Sixth Sense* filmed in Philadelphia, for example, only about 30 local people were employed for three months, about the same as for a large restaurant (Swann, 2001). To put things in perspective it should be noted that 35 000 jobs were lost with the closure of the nearby naval dockyard. Another problem is that once the novelty of figuring in a film wears off, local residents can begin to resent the inconvenience of frequent street closures and restrictions on movements in public spaces.

Nevertheless, the desire of cities to take an active role in film-making shows no sign of diminishing. One possible explanation for this phenomenon is a growing

Box 14.3

Key debates in urban social geography – The impact on cities of 9/11

An extensive, disturbing, but crucially important, debate currently raging within urban studies is the impact of the terrorist destruction of the New York World Trade Centre twin towers on 11 September, 2001. This debate has widened to encompass the significance of other recent terrorist bombings throughout the world (e.g. in Delhi, Jakarta, Kabul, London, Karachi, Bali, Madrid, Moscow, Mumbai (Bombay) and Nairobi). The literature has also extended to consider the significance of other recent acts of urban violence and conflict throughout the world such as manifest in the antiglobalization protests that have dogged recent international summit meetings of the 'International Monetary Fund (IMF), World Trade Organization (WTO) and the leading G8 economies.

The debates are complex but a few themes are apparent:

➤ Many have interpreted recent acts of terror and urban conflict as fundamentalist resistance to forces of globalization and modernization that have been intensified by the pluralistic mixing of cultures

through processes of transnational urbanism (see Box 12.3).

➤ Many see the response of the United States and its allies to renewed terrorism (e.g. the invasion of Afghanistan and Iraq, increased surveillance of cities, armed compounds for elites and threatened groups) as a reassertion of global forces of capitalism. This may be seen as a radical extension of the 'militarization' of city life as described by Mike Davis in the context of Los Angeles (see Box 5.3).

➤ Some have argued that, for various political and commercial reasons, fears have been intensified by politicians and by the media. They asset that binary distinctions between 'them' and 'us' have been reasserted into diverse pluralistic societies, thus threatening civil liberties, social cohesion and democratic life. However, the greatest challenge to civil liberties lies with the actions of terrorists themselves and the scale of the threat from future terrorist attacks should not be underestimated.

Key concepts related to the impact on cities of 9/11 (see Glossary)

Binaries, othering, panopticon, 'scanscape', surveillance, transnational urbanism.

Further reading

Graham, S. (2002) Special collection: reflections on cities, September 11 and the 'war on terrorism' – one year on *International Journal of Urban and Regional Research*, **26**, 589–590

Graham, S. (ed.) (2004) *Cities, War and Terrorism: Towards an Urban Geopolitics* Blackwell, Oxford

Savitch, H. and Ardashev, G. (2001) Does terror have an urban future? *Urban Studies*, **38**, 2515–2533

Links with Other Chapters

Chapter 3: What is Culture?

Chapter 13: Urban Social Sustainability

realization of the importance of skilled, talented people in fostering economic growth in spheres such as high-technology sectors and the creative industries (of which films are an integral part). Films that can raise the international profile of a city can play a key role in attracting such talent (see also Box 11.1).

14.3 Conclusion: whither urban social geography?

It is appropriate to conclude this book by taking a step back and looking at the state of urban social geography in the early twenty-first century. This book has demonstrated that, to say the least, urban social geography is a highly diverse enterprise, both in terms of its subject matter and the diversity of methodological approaches used. Another metaphor to replace Haggett's 'city of districts' might be useful in this context – that of a river! Urban social geography can be thought of as a river growing in size and strength having been fed by many tributaries over the years. The original source was the work of the Chicago School of human ecology. Later this stream of thought was fed by regional science and urban sociology. More recently a major tributary has been cultural studies. Some of these influences, such as quantitative geography, have lessened in importance; others, such as the study of urban ethnography, have become resurgent. As the subject matter has become more diverse, rather like an older river, it has subdivided and become reconstituted in new forms.

However, it is at this point that perhaps the hydrological metaphor breaks down. Whereas older rivers tend to slow down as they develop, urban social geography has shown increased dynamism in recent years. To be sure, in the late 1960s and early 1970s, urban social geography was also at the forefront of methodological developments in human geography, reflecting first quantitative approaches and later political economy. However, in the late 1970s and early 1980s, under the influence of the massive restructuring of Western economies at this time, the study of economic processes and industrial geography became dominant and at the cutting edge of methodological developments. In the late 1980s and early 1990s when Western societies grew again in affluence, cultural studies grew in importance, and there was initially a shift of emphasis in human geography towards analysis of diverse issues such as consumption, identities and postcolonialism. However, in the last decade cities have once again taken central stage in key debates in human geography. We suggest that the main reason for this development is that cities throw into sharp focus many contemporary issues, questions and dilemmas that bring together the two main sources in contemporary human geography: the so-called 'new economic geography' and 'cultural studies'.

➤ Although the issue is much disputed, many argue that cities have become key centres for innovation and economic growth as they try to foster a 'creative class' (Florida, 2002).

➤ Cities are crucial in the creation of new forms of identities in Western societies as diverse groups interact in a relatively close setting.

➤ Cities raise crucial questions about the social sustainability of changes in Western society. These questions have been raised at various times throughout this book and we hope that it will encourage you to study them in greater depth.

Chapter summary

14.1 The 'Los Angeles School' has portrayed the city as a harbinger of urban developments elsewhere, although this claim (and whether Los Angeles can be a envisaged as a general model of urban development) is much disputed.

14.2 Cities have had a big influence upon the evolution of motion pictures and have also been affected by motion pictures through the ways in which cities are represented.

Key concepts and terms

bunker architecture
'California School'
cosmopolis
edge cities
exopolis
film noir

galactic metropolis
keno capitalism
Los Angeles School
multiple nuclei model
multiplex city
paradigmatic city

postmetropolis
postmodern global metropolis
'scanscape'
trope

Suggested reading

The work of the 'California School' of Urban Geo-graphy is to be found in Allen J. Scott and Ed W. Soja (eds) *The City: Los Angeles and Urban Theory at the End of the Twentieth Century* (1996: University of California Press, Berkeley). See also Michael Dear's *From Chicago to LA; Making Sense of Urban Theory* (2001: Sage, Thousand Oaks) and his article 'The Los Angeles School of Urbanism: An Intellectual History' (*Urban Geography*, 2003: **24**, 493–509). A key text on the relationships between cinema and the city is David B. Clarke (ed.) *The Cinematic City* (1997: Routledge, London). See also Stuart Aitken and L. Zonn (eds) *Place, Power, Situation and Spectacle: a Geography of Film* (1994: Rowman and Littlefield, Lanham). Also worth examining is Mark Shiel and Tony Fitzmaurice (eds) *Cinema and the City* (2001: Blackwell, Oxford) together with their other edited volume *Screening the City* (2003: Verso, London). Another excellent book is Ewa Mazierska and Laura Rascaroli (eds) *From Moscow to Madrid: Postmodern Cities, European Cinema* (2003: I.B. Tauris, London). Stephen Barker's little volume *Projected Cities* (2002: Reaktion Books, London) explores the relationships between film and the urban landscape with particular reference to Europe and Japan. Michael Dear considers the relations between film and architecture in a postmodern context in Chapter 9 of *The Postmodern Urban Condition* (2000: Blackwell, Oxford).

Glossary

There can be little doubt that contemporary social theory contains a great deal of jargon! One might easily conclude that this terminology is pretentious, unnecessarily duplicating existing terms that are perfectly adequate, or else elevating simple ideas with grandiose labels. However, this conclusion would be a mistake, for underpinning the vast majority of new terms are serious attempts to come to grips with the complexity of contemporary social change. The very best social science makes us rethink our existing 'common-sense' and taken-for-granted notions about the world, but in order to do this new words and forms of language are often needed. The main problem is, of course, keeping pace with the sheer number and diversity of these concepts.

To assist you in this task the following glossary contains definitions of all the main concepts used in this book (together with a few others that are commonly used in geography and urban studies). Wherever possible we have attempted to convey the essence of the concepts in an accessible manner. One word of warning, however, this has often involved considerable simplification – concepts such as *ideology* and the *underclass* really deserve books to themselves. Furthermore, these concepts should not be regarded as definitive, or set in stone. Concepts are never static but are continually evolving as people dispute their meaning. There is a danger in this context that precise, seemingly watertight definitions can be constraining or misleading. Nevertheless, one has to start somewhere and we hope this glossary, together with the rest of the book, will encourage you to follow up some of these ideas in greater detail. We have incorporated extensive cross referencing to related concepts to help you in this task (all words in italics are defined below).

Ableist geography Geographical studies that assume people are able-bodied, thereby ignoring the problems faced by those with disabilities. Such studies contribute to the continuing oppression of the disabled.

Accumulation A term associated within *Marxian theory* to refer to the processes by which capital is acquired. The term alludes to a system in which the ownership of wealth and property is highly concentrated and not just to a system based on profit-making.

Action (or activity) space A term used in behavioural studies of residential mobility to indicate the sum of all the areas in a city with which people have direct contact. See *awareness space, aspiration region.*

Aestheticization (of everyday life) Originally used to denote situations where issues of class conflict were obscured by appeals to high art. Now used in a broader sense to indicate the increasing importance of signs or appearances in everyday life. Especially applied to processes of consumption and material objects (including buildings) which are seen as indicating of the social position of the user. See also *exchange value, use value, positional good, symbolic capital.*

Agency The capacity of people to make choices and take actions to affect their destinies. Often played down in *structuralism* and deterministic theories. Also termed *human agency.* Contrast with *economic determinism.* See *reflexivity.*

Ageographia Sorkin's term to indicate that the postmodern city may be likened to a theme park centred around Disney-like simulations. See *Disneyfication, hyperreality, postmodernism, simulacra.*

Alienation A term used generally to indicate the ways in which people's capacities are dominated by others. Used in Marxist theory to indicate the loss of control that workers have over their labour and the things they make in a capitalist *mode of production.*

Alterity   A term used in *postcolonial theory* to indicate a *culture* that is radically different from and totally outside that to which it is opposed. Disputed by those who argue that all cultures evolve in relation to one another. See *hybridity*.

Ambivalence   A term used in *postcolonial* theory to describe the mixture of attraction and repulsion that characterizes the relationship between the colonizer and the colonized. It is argued that all colonial relationships are ambivalent because the colonizers do not want to be copied exactly. Imitation can lead to *mimicry*. Ambivalence tends to decentre power since it can lead to *hybridity* on the part of those in power.

Andocentrism   An approach that privileges men and downplays women. See *feminism, sexism*.

Anomie   A situation in which people are less affected by conventions and established social norms. Often associated with the isolation of urban life. See *Gesellschaft*.

Anti-essentialism   Rejection of the idea that there is some underlying essence to phenomena such as truth or natural identity. The opposite of *essentialism*.

Appropriation   The taking-over of elements of imperial culture by postcolonial societies. See *ambivalence, hybridity, mimicry*. Also used in a broader sense to denote the borrowing and mixing together of different cultural elements. Especially applied to popular music and its numerous subgenres. Also known as *syncretism*. See *bricolage*.

Area-based urban policy   Policies targeted at particular urban areas (e.g. enterprise zones).

Areal differentiation   Another term for *residential* (or *sociospatial*) *differentiation*. May also be used in a general sense to refer to areas with commercial or industrial activity rather than just the social fabric of cities.

Aspiration region (or space)   A term used in behavioural studies of residential mobility to indicate the areas of a city to which a potential mover aspires – the product of both *activity space* and *awareness space*.

Asset sales   The sale of publicly owned organizations (such as utilities) and assets (such as public housing) to the private sector. See *privatization*.

Assimilation   The process whereby a *minority group* is incorporated into the wider society (or *charter group*). Can be *behavioural assimilation* or *structural assimilation*. May explain degrees of *segregation*.

'Asylum-without-walls'   Another term for the *service-dependent ghetto*.

Authenticity   The idea that there is a pure, basic *culture*. Disputed by the notion of *ambivalence*. Used in postmodern theory to distinguish 'reality' from copies of the real known as *simulacra*. See also *hybridity*.

Authority constraint   A term used within Hagerstrand's *time–geography* to indicate the influence of laws and customs upon daily lives. See *capability constraint, coupling constraint*.

Automobility   A concept that looks at the way in which the automobile and all the systems that support it have a dominating influence upon the lives of people in Western societies.

Awareness space   All the areas of a city of which a person or household has knowledge resulting from both direct contact (*activity space*) and indirect sources information (e.g. newspapers, estate agents). See *aspiration region*.

Balkanization   A metaphor for the administrative subdivision of US cities into numerous local governments. Also known as *metropolitan fragmentation*.

Banlieue   Poor-quality suburban areas of French cities occupied by immigrants. Also termed *bidonvilles*.

Behavioural approach   An approach that examines people's activities and decision-making processes within their perceived worlds. See *behaviouralism*.

Behavioural assimilation   The process whereby a *minority group* adopts the *culture* of the wider society (or *charter group*). Contrast with *structural assimilation*.

Behaviouralism   An approach in psychology that recognizes that human responses to stimuli are mediated by social factors. Contrast with *behaviourism*.

Behaviourism   An early approach in psychology that examined the responses of people to particular stimuli. Tended to ignore mediation by social factors. See *behaviouralism*.

'Betweenness' (of place)   The argument that the character of regions is dependent upon the subjective interpretations of people living within these areas, as well as the perceptions of those living outside. See *place, social constructionism*.

Bidonvilles   Poor-quality suburban areas of French cities occupied by immigrants. More recently termed *banlieue*.

Binaries   Two-fold categorizations that succeed in dividing people and concepts (e.g. male/female, healthy/sick, sane/mad, heterosexual/homosexual, true/false, reality/fiction, authentic/fake). Can lead to *exclusion* or *objectification*.

Biological analogy   The application of ideas from the plant and animal world to the study of urban residential patterns. See *Chicago School, human ecology, social Darwinism*.

Biotic forces   A term used by the *Chicago School* to indicate the competitive economic forces within cities that lead to *residential differentiation* and *segregation*. See *biological analogy, social Darwinism*.

'Blockbusting'   The practice undertaken by some estate agents (realtors) of introducing black purchasers into predominantly white areas in the hope that the latter will sell up and move out at deflated prices, thereby enabling the agents to resell the properties to new black families at higher prices.

BOBOs (BOhemian–BOurgeois)   Brooks' term for what he regards as a new class of people inhabiting inner cities who combine Bohemian lifestyles of hedonism with bourgeois traits of hard work. See *Bohemia, creative class*.

Bohemia (Bohemian enclaves)   Districts of cities inhabited by large proportions of people with 'alternative', libertarian lifestyles that contrast with conventional 'bourgeoisie' lifestyles (i.e. enjoyment rather than work and household arrangements other than the nuclear family). See also *creative class*.

Boomburb   A rapidly growing suburban area outside traditional urban centres. See *edge cities, technoburbs*.

Borderlands   Geographical and metaphorical spaces on the margins of dominant cultures where new hybrid forms of *identity* can emerge. See *hybridity, liminal space, heterotopia, third space*.

Bricolage   A French word meaning an assembly of miscellaneous objects (hence the English term bric-a-brac) used by the structuralist Levi-Strauss to indicate the apparently bewildering mixture of elements to be found in primitive cultures. Also used in cultural studies to describe the complex mixtures of elements found in new cultural forms such as popular music and its various genres. See *appropriation, creolization, enculturation, hybridity, nomadization, transculturation*.

Bunker architecture   Buildings designed to exclude certain sections of society (usually those thought to be undesirable or threatening by the more affluent). See *gated communities, 'scanscape'*.

California School   The group of scholars who have interpreted the contemporary urban forms of Los Angeles as emblematic of city structures in a postmodern or post-Fordist society. Also termed the *Los Angeles School*. May also refer to the explanations for industrial agglomeration derived from transactions cost analysis and *regulation theory*. See *new industrial spaces*.

Capability constraint   A term used in Hagerstrand's *time–geography* to indicate physical and biological constraints on daily activity. See *authority constraint, coupling constraint*.

Carceral city   A term coined by Michel Foucault to describe a city in which power is decentred and in which people are 'imprisoned' or controlled by various types of discourse (medical, sociological, psychological, etc.) as manifest in various institutions (the family, schools, prisons, hospitals, etc.). (From the Latin word *carcer* meaning 'prison' – hence the English term incarcerate.) See *interpellation, micropowers*.

Cartesian approach   The argument developed by Enlightenment philosopher Rene Descartes that the observer can be separated from the observed.

Casualization   The increasing use of various non-core workers such as part-time, temporary and agency workers. Also termed *numerical flexibility*.

Centralization   The spatial regrouping of activities into larger units. May refer to reductions in numbers of service units of the welfare state or movements back into central cities. Contrast with *decentralization*.

Charter group   The majority group with the dominant *culture* of a society.

Chicago School   The group of sociologists working in Chicago in the early part of the twentieth century. Noted for their studies of urban subcultures and the application of ideas from the plant and animal world to the study of residential patterns (known as *human ecology*). May also refer to a group of economists based in Chicago in the late twentieth century advocating monetarist economic policies.

Circuit of production   The process of capitalist exploitation (also known as *accumulation*) in which capital or money (M) is invested in commodities (C) and labour power (LP)

and the means of production (MP) to produce more commodities (C') which are then sold to acquire more money
(M') in the form of profits. See *time–space compression*.

Citizenship   The relationship between individuals and the
community and/or the state.

Civic boosterism   Attempts by local governments to
develop their local economies by attracting inward
investment and through partnerships with private sector
sources of capital. Also termed *civic entrepreneurialism*. See
*coalition building, growth coalitions, governance, urban
entrepreneurialism*.

Civic entrepreneurialism   See *civic boosterism*.

Civil society   All the elements of society outside government including private-sector businesses, the family and the
voluntary sector.

Class   Material differences between groups of people. See
economic status.

Classical Marxism   The ideas formulated by Karl Marx in
the nineteenth century. Contrast with *neo-Marxism*.

Clustering   The tendency for people with similar
attributes such as *class* or *ethnicity* to live close to each other
in cities. May also be termed *segregation* and *residential
differentiation*. In extreme cases may comprise a *ghetto*. Also
applied to geographical agglomerations of firms. See *new
industrial spaces*.

Coalition building   Formal and informal interest groups
in cities combining to achieve political objectives. Linked to
regime theory. See *growth coalitions*.

Cognitive distance   A measure of the perceived (rather
than just physical) distance that people feel from features
in an urban area taking account of mental maps and the
symbolic features of the environment.

Collective consumption   Usually refers to goods and
services provided by the public sector. Less often refers
to services that, literally, have been consumed by a group
of people in a collective manner (such as a lecture). The
term originated in a neo-Marxist (or *Marxian*) theory
formulated by Manuel Castells, which argues that there
are certain services that are crucial for the maintenance
of capitalism but that are too expensive for provision by
individual capitalist enterprises and therefore require provision through non-market means via the public sector.
The theory also attempts to define cities as essentially places

for the consumption of public services – a notion that has
been much criticized. See *neo-Marxism, public goods*.

Colonial discourse   The social practices and attitudes
associated with *colonialism*. See *discourse, imperialism,
postcolonial theory*.

Colonialism   The rule of one territory by another country
through the creation of new settlements. The product of
imperialism. See *postcolonial theory*.

Colony   A territory ruled by another. Also used in an
urban context to indicate a minority residential cluster that
is a temporary phenomenon before the group is integrated
into the wider society. Contrast with *enclave, ghetto*.

Commercialization   The tendency for publicly owned
organizations to behave like private sector companies (such
as through the imposition of user charges). Also termed
proprietarization. See also *corporatization*.

Commodification   The use of private markets rather than
public sector allocation mechanisms to allocate goods and
services. Also termed *recommodification* and *marketization*.

Commodification (of culture)   The ways in which local
cultural forms are being supplanted by mass-produced
cultural forms. See *McDonaldization*.

Commodity fetishism   The obsession of people with the
acquisition of consumer goods. The term recognizes that
these material objects not only have *use value* but also have
a symbolic value that reinforce social status or lifestyle.
A key element of *neo-Fordism*. See *aestheticization*.

Community   A much-used term with little specific meaning
but usually refers to a social group characterized by dense
networks of social interaction reflecting a common set of
cultural values. Often, but not necessarily, geographically concentrated. See *'ethnic village', Gemeinschaft, neighbourhood*.

Community action   Political movements based in a
local area usually defending a residential district against
the intrusion of unwanted activities (sometimes termed
community-defined politics, or *'turf' politics*). See *externality*.

Community care   Care for the needy in local communities
either in small decentralized facilities or in private households – both supported by teams of community-based professionals. Associated with *deinstitutionalization*. A policy
much criticized for inadequate funding and resources –
hence the term care 'in' the community but not 'by' the
community.

'Community lost'   The argument that urbanization has destroyed *community* life. Contrast with *'community saved'*, *'community transformed'*.

Community politics   See *community action.*

'Community saved'   The argument that communities still exist in urban areas. See *'ethnic village'*. Contrast with *'community lost'*, *Gesellschaft.*

'Community transformed'   The argument that new forms of *community* life have been created in suburban areas.

Competitive tendering   A process through which contracts are awarded on the basis of competitive (usually secret) bidding by a variety of agencies according to specified criteria of such as cost, quality and flexibility. See *contracting-in.*

Compositional theory   A theory that examines the impacts of ethnicity, kinship, neighbourhood and occupation on behaviour in residential areas of cities. Similar to *subcultural theory*. Contrast with *behaviourism.*

Concentric zone model   Burgess's idealized model of city structure based on Chicago in the 1920s in which social status increases in a series of concentric zones leading out from the city centre. See *Chicago School, human ecology.* Contrast with *sectoral model.*

Congregation   The residential clustering of an ethnic minority through choice (rather than involuntary *segregation* brought about by structural constraints and discrimination).

Constitutive otherness   The process whereby identities and cultural meanings evolve in relation and opposition to other identities and cultures. See *alterity, ambivalence, appropriation, binaries, culture, hybridity, mimicry, objectification, scripting, othering.*

Consumption   The purchase and utilization of goods and services. See *commodity fetishism.*

Contextual theory   A broad trend in social analysis characterized by a desire to understand the settings or contexts within which human behaviour takes place. These approaches seek to understand how people are influenced by, but at the same time create, these contexts. See *situatedness, structuration theory.*

Contingent workers   Workers who display *numerical flexibility*. Contrast with *functional flexibility.*

Contracting-in   A situation in which a contract is won by a subdivision of the parent organization putting the work up for tender. This is often said to be kept in-house. See also *market testing.*

Contracting-out   A situation in which one organization contracts with another external organization for the provision of a good or service. Often associated with *competitive tendering* but this need not be the case. May also be termed *subcontracting, distancing* or outsourcing. See also *market testing.*

Contractualization   The use of contracts to govern the relationships between organizations and subdivisions within organizations. Increasingly used to allocate public services to private sector companies, voluntary organizations or internal departments within the public sector. See *contracting-in, contracting-out, internal markets.*

Corporatization   An extreme form of *commercialization* in which publicly owned organizations behave in an identical manner to private sector companies.

Corporatism   Forms of social organization in which certain interest groups, usually certain sectors of business and organized labour, have privileged access to government. Characterized by collaboration to achieve economic objectives. See *neo-corporatism, welfare corporatism.*

Corporeality   A term that recognizes that body images are not just the result of biological differences but are socially constructed through various signs and systems of meaning. Contrast with *essentialism.*

Cosmopolis   Ed Soja's term for Los Angeles as an expression of postmodern urban form. See *exopolis, galactic metropolis, heteropolis, Los Angeles School, postmodern global metropolis.* Also Sandercock's term for a process of city planning that attempts to create a utopian city that both accommodates and encourages diversity and difference.

Counter-culture   A *subculture* that is opposed to the dominant values in a society. See *counter-site, heterotopia.*

Counter-site   A space that is outside the mainstream of society and reflects a *counter-culture*. See *heterotopia.*

Coupling constraint   A term used in Hagerstrand's *time–geography* to indicate the constraint on human activity resulting from the need to interact on a face-to-face basis with other people. See *authority constraint, capability constraint.*

Creative cities   Cities characterized by innovation in both manufacturing and services resulting from collective learning through the interactions of diverse peoples in overlapping social networks. See *new industrial spaces*.

Creative class   Richard Florida's term for diverse artistic and professional groups who are playing a key role in promoting economic growth in urban areas. See *BOBOs, Bohemia, creative cities, new industrial spaces*.

Creolization   Originally used to denote the racial inter-mixing and cultural exchange of indigenous peoples with colonizers but also used to denote cultural mixing. See *hybridity*.

Cross-tabulations   Data showing the interrelationships between two or more variables (e.g. the relationship between housing tenure and average household income). See *microsimulation*.

Crowding theory   The idea that high-density living in urban areas leads to strains and tensions that can lead to aggression, withdrawal and high rates of mental and physical illness. An approach which tends to ignore the mediating effects of *culture* upon human behaviour. See *behaviourism, determinist theory*.

Cultural capital   Ways of life and patterns of consumption that make people distinct and appear superior or dominant. See *positional good*. Also used to indicate skills and knowledge (as distinct from economic capital). Also termed human capital.

Cultural imperialism   A term used by Iris Young to indicate the way in which society asserts that certain types of behaviour are 'natural' by marking out certain types of non-conforming behaviour as 'other', 'deviant' and 'non-natural'. See *othering*.

Cultural industries   Industries in 'creative' spheres such as performing arts, design, advertising, entertainment, media and publishing. The term is also used in a theory that argues that cultural elements such as popular films, music and books have become mass produced in the same way as consumer goods such as cars.

Cultural intermediaries   Bourdieu's term for key staff working in *cultural industries* such as popular music who interpret and promote in a selective manner certain styles of cultural 'product' and associated lifestyles. A form of *social gatekeeper*.

Culturalization of the economy   The idea that economic life has been transformed or has been penetrated by cultural forces. Manifest in the increased importance of non-material goods (films, popular music, magazines, etc.) and the increasing importance of image or sign values in material objects (e.g. designer clothing, spectacles). Contrast with *economization of culture*. See also *aestheticization (of everyday life), commodification (of culture), cultural industries, cultural mode of production, semiotic redundancy*.

Cultural materialism   A theoretical perspective that recognizes that the spheres of culture and the economy are closely interrelated with no one sphere dominating the other. Attempts to avoid *economic determinism*. See *post-Marxism*.

Cultural mode of production   The thesis that issues of *culture* have become dominant in contemporary economies through factors such as: the rise of cultural industries; the *aestheticization* of material objects; and the use of notions of culture in modern management practices.

Cultural myopia   The tendency to assume that the arrangements within a nation or *culture* are the only set of possible arrangements or that these are a superior approach to social organization.

Cultural politics   A term that indicates that issues of *culture* are not just concerned with aesthetics, taste and style, but also involve issues of power and material rewards bound up with competing 'ways of life'. See *identity politics*.

Cultural practices   A term that draws attention to the fact that culture is not simply about ideas but is also a product of material practices (i.e. it involves both thought and action). See *discursive practices, material practices*.

Cultural production   The notion that culture is an integral part of everyday social practices. Involves the idea that culture is not simply a reflection of an underlying material base but is a major element shaping society. See *cultural mode of production, economic determinism*.

Cultural studies   A complex set of developments in social analysis which pay attention to the complexity of cultural values and meanings. See *culture* and *'cultural turn'*.

Cultural transmission   The idea that values and norms are transmitted from one generation to the next in local subcultures. See *culture of poverty, neighbourhood effect, subculture, transmitted deprivation*.

'Cultural turn'   The tendency for many social sciences to pay greater attention to issues of *culture*. Also termed the *linguistic turn* because of the attention given to language and the ways in which ideas are represented. See *post-structuralism*, *deconstruction*.

Culture   This may be broadly interpreted as 'ways of life'. It consists of the values that people hold, the rules and norms they obey and the material objects they use. Also commonly regarded as systems of shared meanings (see *discourse*).

Culture of poverty   The argument that poverty results from a distinctive *culture*. Closely related to the notion of *transmitted deprivation*.

Culture of property   The way in which the housing market in a nation or region is socially constructed by social institutions and social behaviour related to factors such as *class* and *ethnicity*.

Cyberspace   A term devised by William Gibson in his novel *Neuromancer* and now used in a very broad sense to indicate developments in the sphere of advanced telecommunications. Similar to *telematics*.

Dasein   A term used within *structuration theory* to indicate the time span of people's lives. Contrast with *durée* and *longue durée*.

Decentralization   The movement of first people and later employment and services out of inner-city areas into suburban districts and then into more distant commuter hinterlands beyond city limits. May also refer to the fragmentation and geographical dispersal of organizational structures within manufacturing, services and the public sector. May be associated with *devolution* but the two policies are distinct. See *deconcentration*, *delegation*, *edge cities*, *tapering*.

Decision rules   The criteria used by bureaucrats (usually but not necessarily in the public sector) to allocate resources in cities. Used to simplify decisions that have to be made frequently, they may not be made explicit. Also termed *eligibility rules*. See *managerialism*, *social gatekeepers*, '*street-level*' *bureaucrats*.

Deconcentration   Another name for *decentralization*. See *delegation*, *devolution*.

Deconstruction   A form of analysis that examines the various *discourses* represented by various forms of representation (known as *texts*). These meanings are regarded as continually changing through the interactions of the reader/viewer and the text in question.

De facto territories   Areas that may be defined by reference to factors such as common interests or lifestyles (rather than just in legal terms). Contrast with *de jure territories*.

Defederalization   The devolution of responsibilities for welfare policies from federal government to states in the United States. Associated with capped budgets and a series of policies known as *workfare*. See *decentralization*, '*hollowing out*', *postwelfare state*.

Defensible space   The argument that recent housing developments lack spaces that people can identify with, survey or exert control over.

Dehospitalization   A term preferred by some to *deinstitutionalization* in recognition of the fact that community-based care can involve small institutional settings. See *community care*.

Deindustrialization   The decline in manufacturing activity both in terms of jobs and contribution to national output. See *postindustrial city*.

Deinstitutionalization   The closure of institutions providing long-term care for needy groups and their replacement by various alternative forms of care including purpose-built or converted smaller facilities and care within private households by families supported by teams of community-based professionals such as nurses, doctors and social workers. See *community care*, *rationalization*, *reinstitutionalization*, *self-provisioning*, *domestication*.

de jure territories   Geographical areas defined according to the law (i.e. with legal powers as in political and administrative regions). Contrast with *functional urban areas*. See *jurisdictional partitioning*.

Delegation   A form of *decentralization* in which certain functions and managerial responsibilities are delegated to neighbourhood offices but where local autonomy tends to be severely restricted by central responsibility for expenditure and targets. Contrast with *deconcentration*, *devolution*.

Delocalization   The destruction of local economic and social relationships by forces of *globalization*. May result in new and distinct hybrid forms rather than all places becoming similar. See *glocalization*, *global–local nexus*, *McDonaldization*.

Demunicipalization   Attempts by central governments to reduce the powers and responsibilities of local governments. Applied especially to the sale of local authority housing in the UK in combination with restrictions on new public sector housing construction. See *governance, ghettoization, residualization*.

Deregulation   Policies designed to increase competition by breaking up state monopolies and introducing a number of private agencies to provide goods and services. May also be applied to the deregulation of labour markets through policies to erode workers rights and to increase labour flexibility. See *commodification, marketization*.

Design determinism   Studies of the impact of the physical environment and architectural design upon human behaviour. See *behaviourism, crowding theory, defensible space*.

Deskilling   Strategies to reduce the skill levels and knowledge required in particular occupations.

Determinist theory   An approach that draws upon behaviourist notions to argue that city living affects behaviour. See *Gesellschaft, 'psychological overload'*.

Deterritorialization   The destabilized nature of identity and meaning within postmodern society; in particular decline of forms of identity associated with particular spaces such as neighbourhoods, regions and nations. Also used a general term for urban restructuring. Contrast with *reterritorialization*. See *deconstruction, McDonaldization*.

Deviant subculture   See *deviant subgroup*.

Deviant subgroup   A group within society that has values and norms substantially different from the majority population. May be expressed in *residential differentiation*. Also termed *deviant subculture*. See *culture*.

Devolution   The subdivision of welfare organizations into separate units, each with its own budget. Usually associated with devolution of responsibilities and with enhanced performance monitoring of the units. See also *decentralization*.

Dialectic   A form of reasoning or analysis involving the use and possible reconciliation of opposites. See *sociospatial dialectic*.

Diaspora (diasporic group)   The movement, either voluntarily or forced, of people from their homeland to a new territory.

Difference   A term that recognizes the ways in which differences between categories are socially constructed in relation to one another. See *binaries*. Contrast with *essentialism*.

Disciplinary regimes   Processes through which social control is exercised: socialization, the construction of dominant discourses and surveillance. See *disciplinary society*.

Disciplinary society   A society in which control is exercised through socialization processes as manifest in schools hospitals and factories. See *interpellation, micropowers*.

Discourse   Sets of meanings that are indicated by various texts that form a way of understanding the world. See *deconstruction*.

Discursive practices   The words, signs, symbols and ideas that are used to represent *material practices*.

Disfigured city   The city that is unplanned and inhabited by deprived groups. Contrast with *figured city*.

Disneyfication   The conscious creation of the 'theme park' city characterized by a superficial veneer of culture and often a sanitized view of history that ignores social conflict. See *imagineering, simulacra*.

Distance-decay effect   The tendency for those who live furthest away from the sources of goods and services to consume them less often. This is usually attributed to the increased travel costs or the increased time involved in visiting the source of supply. Also known as *tapering*.

Distanciation   The tendency for interactions and communications between people to be stretched across time and space through the use of books, newspapers, telephones, faxes and the Internet. Also termed space–time distanciation.

Distancing   Another term for *contracting-out* – a situation when one organization contracts with another external organization for the provision of a good or service.

Domestication   The use of family and household labour. Has been forced upon some households (and usually women within them) through the run-down of state provision. See *community care*.

Domestic economy   Work done within households (either informally by the family of other members of the household or formally through directly purchased services). See *domestication, self-provisioning*.

Dominance   A term used by the *Chicago School* to indicate the process whereby certain land uses and types of people come to dominate particular parts of cities. Also used in a general sense to indicate power relations. See *human ecology*.

Double hermeneutic   The need for researchers to be aware of their own values as well as those of the people they are studying. See *hermeneutics, situatedness*.

Dual cities   Large metropolitan centres characterized by disparities in wealth and status and/or a trend towards increasing social inequality. See *global cities, social polarization*.

Durée   A term used within *structuration theory* to indicate the time span of daily routines. See *daesein, lifeworld, longue durée*.

Ecocentric approach   Various types of ecological movement united by a belief that environmental problems can be addressed only by a fundamental change in the capitalist system involving greater decentralized participatory democracy. Contrast with *technocentric approach*.

Ecological approach   A term used to denote spatial or geographical analysis of cities. May also refer to *human ecology*, the application of ideas concerning the distribution of plants and animals to the study of urban social geography. See *Chicago School*.

Ecological fallacy   The potential mistakes that can arise when making inferences about individuals from data based on aggregate information (such as for residential areas within cities). See *individualistic fallacy*.

Ecological modernization   See *technocentric approach*.

Economic determinism   Theories that attempt to relate social changes directly to underlying economic changes in society and that play down the ability of people to make decisions to affect their destinies. Contrast with *voluntarism*.

Economic status   The name frequently given to one of the main dimensions of urban residential structure as shown by *factorial ecology* – variations in the extent of wealth. See *ethnic status, family status, social rank*.

Economies of scale   Factors that cause the average cost of a commodity to fall as the scale of output increases. There are two main types, see *external economies of scale, internal economies of scale*. A crucial part of *Fordism*.

Economies of scope   Factors that make it cheaper to produce a range of commodities rather than to produce each of the individual items on their own. See *external economies of scope* and *internal economies of scope*. A crucial part of *neo-Fordism*.

Economization of culture   The idea that cultural life is increasingly dominated by commercialism (the cash nexus) and the logic of capital *accumulation*. Contrast with *culturalization of the economy*.

Edge cities   A term coined by the journalist Joel Garreau to describe recent urban developments outside large metropolitan areas characterized by decentralized and functionally independent nodes of offices and shopping malls. See *decentralization, exopolis*.

Eligibility rules   The criteria used by *social gatekeepers* to determine who has access to scare resources in cities. These may be explicit or tacit. Usually applied to public officials such as housing managers but may also be applied to the private sector (e.g. estate agents and bankers). Also called *decision rules*.

Elsewhereness   The tendency for shops and other spaces within cities to copy images from other places in other times. See *placelessness, simulcra*.

Embeddedness   The notion that economic behaviour is not determined by universal values that are invariant (as in *neoclassical economics*) but is intimately related to cultural values that may be highly specific in time and space. Also termed social embeddedness. See *culture, situatedness*.

Embodied knowledge   Ideas and concepts that attempt to avoid the mind/body division of the *Cartesian approach* and recognize that knowledge emerges from people in particular contexts. Also termed *local knowledge*. See *embeddedness, situatedness*.

Embodiment   The process through which the body is socially constructed through wider systems of meaning. See *corporeality, embodied knowledge*.

Embourgeisement thesis   The argument that working-class people moving into suburban areas adopt middle-class lifestyles based around consumption and the *nuclear family*. See *commodity fetishism*.

Empowerment zones   An urban regeneration policy in the US characterized by collaboration between public bodies, private enterprises and community groups. See *enterprise zones*.

Enabling state   A key element of the new mode of governance and urban entrepreneurialism in which the direct role of the state is reduced and replaced by greater partnership between government and business interests.

See *coalition building, contracting-out, 'hollowing-out', regime theory.*

Enclave   The name for a residential cluster of an ethnic minority that is a long-term phenomenon, although generally not as segregated as a *ghetto*. Contrast with *colony*.

Enculturation   A broad term for the assimilation and integration of new cultural forms. Derived from anthropology but also used in cultural studies. See *appropriation*.

Enlightenment project (movement)   The broad trend in Western intellectual thought, beginning in the Renaissance, that attempted to analyze and control society through principles of scientific analysis and rational thought. See *Cartesian approach, modernism, social engineering*.

Enterprise zones   Zones in which special incentives such as tax exemptions or reduced planning regulations are used to encourage economic development.

Entrepreneurial cities   Cities characterized by active policies to ensure economic development. Part of a new era of *urban entrepreneurialism*. See *governance, growth coalitions*.

Environmental conditioning   The argument that people's behaviour is strongly influenced by their social environment. Often applied to explain the lack of social and intellectual skills of those brought up in environments lacking in sensory stimulation. See *behaviouralism, cultural transmission*. Contrast with *behaviourism*.

Essentialism   The notion that there are basic, unvarying, elements that determine, or strongly affect, the behaviour of people and social systems (e.g. the idea that there are inherent differences in the behaviour of men and women, or basic immutable laws of economics that govern capitalist societies). Contrast with *anti-essentialism*. See also *social constructionism*.

Ethnic group   A minority group whose members share a distinctive *culture*. This is conceptually distinct from the notion of a *racial group* but in practice the two are intimately linked. See *ethnicity*.

Ethnic status   The name frequently given to one of the main dimensions of urban residential structure as shown by *factorial ecology* – variations in the extent of *ethnicity*. See *family status, social rank*.

Ethnicity   The culture and lifestyle of an *ethnic group*, often manifest in a distinctive residential areas of a city. Contrast with *racial group*. See *ghetto*.

'Ethnic village'   A minority groups that exhibits *residential differentiation* within a city and a distinctive *culture* characterized by dense social networks.

Ethnoburb   A suburban area characterized by the clustering of a particuar ethnic group. See *boomburb, edge cities, technoburb*.

Ethnocentrism   The assumption that one culture is superior to others. Usually applied to Western assumptions of technological and moral superiority. Called Eurocentrism when European culture is seen as superior.

Ethnography   The study of *culture*, especially the values and norms of minority ethnic groups. Often linked to qualitative research methods such as participant observation and semi or unstructured questionnaires.

Ethnoscape   Appadarai's term for the diverse landscape of mobile groups to be found in many contemporary cities: tourists, immigrants, refugees, exiles, guest workers and the like.

Exchange value   The amount that a commodity such as housing can command on the market. Related to, but conceptually distinct from, *use value*.

Exclusion   Social processes whereby certain kinds of people are prevented from gaining access to various types of resources (including public services). These may be non-material resources such as prestige. See *eligibility rules, social closure*.

Exclusionary closure   Another name for processes whereby powerful groups exclude other groups from wealth, status and power. May be called *social closure*.

Exclusionary zoning   Planning policies that restrict certain types of activity and people from moving into a local government area. See *purified communities*.

'Exit' option   A strategy of outmigration from an area in the wake of a problem. Contrast with *'loyalty'* and *'voice'* options.

Exopolis   Ed Soja's term to describe the idea (or *discourse*) of the city as an 'inside-out' metropolis characterized by *edge cities*. See *postmodern global metropolis*.

Expressive interaction   *Secondary relationships* involving some intrinsic satisfaction (such as joining a sports club). Contrast with *instrumental interaction*.

Extensive regime of accumulation   A phase of capitalist development during which profits were enhanced primarily through increasing the amount of output and expanding the scale of the market, rather than through increasing the productivity of workers. A key concept in certain forms of *regulation theory*. Contrast with the *intensive regime of accumulation* and the *flexible regime of accumulation*.

External economies of scale   Factors that reduce the costs of production when the industry to which the firm belongs is large (e.g. the development of specialist suppliers, services and skilled workers). These factors apply irrespective of the size of the individual firm. Contrast with *internal economies of scale*.

External economies of scope   Economies of scope that arise when the industry to which the firm belongs is large (i.e. there are a large number of producers). Contrast with *internal economies of scope*.

Externality   An unpriced effect resulting from activities in cities. May be a benefit received by those who have not directly paid for it, or a cost (or disbenefit) incurred by those who have not been compensated. Also termed a *spillover* and *third-party effect*. May lead to *free-riders*.

Externalization (of production)   The tendency for firms to subcontract out work to other organizations (also termed *vertical disintegration*). Usually interpreted as a response to increasing market variability and technological change as well as a desire to reduce costs. Also related to declining internal *economies of scale*.

'Fabric effect'   A situation in which the physical structure of the housing market has an impact on the distribution of a social group in a city. Usually applied to the impact of cheaper accommodation on the location of ethnic minorities.

Factor analysis   A multivariate quantitative technique used to summarize the main patterns in a complex set of data. Technically very similar to *principal components analysis*. See *factorial ecology*.

Factorial ecology   The application of factor analysis and principal components analysis to the study of residential patterns in cities. See *ecological approach, human ecology*.

Family status   The name frequently given to one of the main dimensions of urban residential structure as shown by *factorial ecology* – variations in the extent of nuclear family lifestyles. See *ethnic status* and *social rank*.

Feminism   A broad social movement advocating equal rights for men and women. Also various forms of academic analysis that attempt to expose the diverse processes that lead women to be oppressed. See *gender, patriarchy, sexism*.

Feminist geography   Geographical analysis that is committed to achieving equal rights for men and women by exposing existing and past inequalities between the sexes.

Feminization   The increasing numbers and influence of women in certain spheres of life. Usually applied to the workplace.

Feminization (of poverty)   Social processes that lead to women becoming a relatively deprived underclass (e.g. the growth of female-headed single parent families dependent upon welfare benefits or trapped in low-paid service sector employment).

Festival retailing   Shopping complexes characterized by 'spectacular' elements. See *commodity fetishism, spectacle*.

Fetishizing   Exaggerating the importance of a particular theory, principle, concept or factor in social analysis (such as overemphasizing the role of 'space' in isolation of social processes). Used originally to indicate ways of obscuring class conflict. See *commodity fetishism*.

Figured city   The city that is planned and organized for the affluent. Contrast with *disfigured city*. See also *revanchist city*.

Film noir   Films predominantly from the Cold War era of the late 1940s and 1950s that portray cities as disturbing, corrupt, nightmarish environments.

Filtering   The thesis argued by Homer Hoyt that the primary motor behind residential mobility is the construction of new dwellings for the wealthy, thereby leading to out-migration of the more affluent from older properties and their occupation by persons of lower social class. See *sectoral model*. Contrast with *invasion* model of residential mobility.

Fiscal imbalance   Disparities between the needs of urban areas and the available resources to meet these needs. Commonly associated with central or inner city local governments in US cities. See *suburban exploitation thesis*.

Fiscal mercantilism   Attempts by local governments to increase local revenues by attracting lucrative taxable land uses. Similar to *civic boosterism* and *civic entrepreneurialism*.

Fiscal stress   See *fiscal imbalance*.

Flanerie   The act of being a *flaneur*.

Flaneur   A leisurely stroller (typically male) observing the bustle of city life. Also applied to those who browse through the Internet. See *gaze*.

Flaneuse   A female traveller/observer (usually associated with the Imperial era).

Flexibilization   A set of policies designed to increase the capacity of firms to adjust their outputs to variations in market demand. May be applied to forms of industrial organization and to labour practices as well as to both private and public sector bodies. See also *functional flexibility* and *numerical flexibility*.

Flexible accumulation   The idea that the intensive *regime of accumulation* has been replaced by a new regime in which the prime emphasis is upon flexibility of production. See also *regulation theory*, *neo-Fordism*, *flexibilization* and *flexible specialization*.

Flexible specialization   The idea that mass production using unskilled workers is being replaced by batch production of specialized products in small companies using skilled workers. Has similarities with concept of *neo-Fordism* in *regulation theory* but is highly voluntarist in approach and is less concerned with matching industrial change to wider economic forces. See *voluntarism*.

Forces of production.   The technological basis of a particular *mode of production*. See also *social relations of production*.

Fordism   A system of industrial organization established by Henry Ford in Detroit at the beginning of the twentieth century for the mass production of automobiles. In *regulation theory* the concept refers to a *regime of accumulation* which was dominant after the Second World War based on *Keynesianism*, mass production and the *welfare state*.

'Fortress cities'   Cities characterized by social inequality, crime, violence and protective strategies in local neighbourhoods designed to exclude groups regarded as dangerous. See *gated communities*, *social polarization*, *'scanscape'*, *surveillance*.

Free-riders   Those who obtain benefits in cities that they have not directly paid for. See *externality*, *suburban exploitation thesis*.

Functional flexibility   The capacity of firms (and public sector organizations) to deploy the skills of their employees to match the changing tasks required by variations in workload.

Functionalism   A type of reasoning incorporated, either explicitly or implicitly, into a great deal of social theory that is characterized by a number of limitations including: attributing 'needs' to social systems; assuming that social systems are functionally ordered and cohesive; assuming teleology in social systems (i.e. that events can only be explained by movement towards some pre-ordained end); assuming effects as causes; and assuming empirically unverified or unverifiable statements as tautological statements (i.e. true by definition). May also be used to refer to a form of managerial philosophy that advocates the subdivision of organizations around particular tasks and responsibilities.

Functionalist sociology   An approach to social theory, of which the sociologist Talcott Parsons was the principal exponent, that attempts to explain social phenomena in terms of their function in maintaining society. See *functionalism*, *system*.

Functional urban areas   Cities or urban areas defined as geographical agglomerations of people predominantly engaged in non-agricultural occupations who are integrated by overlapping journey-to-work patterns. May not correspond with *de jure territories*.

Galactic metropolis   Another term for the postmodern city in which urban areas are spread around like stars, rather than forming a single, easily identifiable, centre. See *postmodernism*, *postmodern global metropolis*.

Gastarbeiter   'Guest workers' in continental European cities but usually in Germany, and predominantly from Turkey. Assumed to be temporary workers and therefore often without full citizenship rights, many have lived in the country for many years.

Gated communities   Residential areas of cities with protective measures such as barriers, fences, gates and private security guards designed to exclude social groups deemed undesirable and dangerous. See *fortress cities*, *purified communities*, *panopticon*, *'scanscape'*, *spaces of exclusion*.

Gay ghetto   A residential area of a city characterized by a high concentration of gay people. See *ghetto*, *ghettoization*.

Gaze   The surveillance, scrutiny and analysis of peoples and places by observers (traditionally men). Often linked to the idea that these observers can provide a privileged, objective, value-free description of the world. Known as the imperial gaze when linked with *colonialism*. See also *Cartesian approach, mimetic approach*. Disputed by *social constructionism*.

Gemeinschaft Tight-knit social relationships based around family and kin which Tönnies argued were manifest in traditional agrarian environments. Contrast with *Gesellschaft*.

Gender Social, psychological and cultural differences between men and women (rather than biological differences of sex). See *feminism, heteropatriarchy, patriarchy, sexism*.

Gender roles 'Masculine' and 'feminine' ways of performing that are derived from *gender*. See *performativity*.

Genius loci The idea that there is a unique 'spirit' of a place, sometimes captured in novels, poetry and painting.

Gentrification The renovation and renewal of run-down inner-city environments through an influx of more affluent persons such as middle-class professionals. Has led to the displacement of poorer citizens. Associated with the development of gay areas in some cities.

Geographical imagination The need for geographers to understand the diversity of cultural values of those they study in different places (and to recognize the influence of their own values upon the frameworks they use to represent these people). See *contextual theory, situatedness*.

Geographical information systems See *microsimulation*.

Gerrymandering The manipulation of the boundaries of electoral subdivisions to gain political advantage.

Gesellschaft Loose-knit social relationships between people which Tönnies argued were manifest in urbanized environments. Contrast with *Gemeinschaft*.

Ghetto The geographical concentration of social groups. Tends to imply a high degree of involuntary segregation. Usually applied to ethnic minorities but may also refer to older people, gays and lesbians, single parents or those who are mentally-ill. *See colony, enclave, service-dependent ghetto*.

Ghettoization Social trends and public policies that lead to geographical concentrations of social groupings, including deprived groups, elderly people, single parents, mentally ill people or ethnic minorities, often in public sector or social housing estates. The term usually implies involuntary clustering. See *residualization, demunicipalization*.

Global cities Saskia Sassen's preferred term for *world cities* that exercise a leading role in the control and organ-ization of the world's economy, trade and finance. They are characterized by conspicuous consumption and *social polarization*.

Globalization The tendency for economies and national political systems to become integrated at a global scale. Also the tendency for the emergence of a global *culture* (i.e. universal trends that it is argued are sweeping through all nations). See *global cities*.

Global–local nexus The relationships (and tensions) between forces of *globalization* and the distinctive features of local areas (e.g. the desire of transnationals to manufacture at a global level yet be sensitive to the needs of particular local markets).

Glocalization A term coined by Eric Swyngedouw to indicate the ways in which developments in particular places are the outcome of both local and global forces. See *globalization, global–local nexus*.

'God trick' A term used by the social critic Donna Haraway to draw attention to the assumption of value-free neutrality incorporated into many scientific studies of society (i.e. the assumed capacity to see 'everything from nowhere'). Disputed in *social constructionism*. See *situatedness, poststructuralism*.

Governance All the methods by which societies are governed. The term is used to indicate the shift away from direct government control of the economy and society via hierarchical bureaucracies towards indirect control via diverse non-governmental organizations. Associated with the demise of local forms of government. May also be termed *urban governance*. See *'hollowing out', quango, quasi-state*.

Grand metaphors Theories based around metaphors that purport to provide comprehensive explanations for the world around us (e.g. Fordist/post-Fordist cities). See *totalizing narrative*.

Grands ensembles Large-scale, high-density and typically high-rise developments of social housing in suburban areas of French cities.

Grounded theory An approach that lets the data 'speak for itself' without making prior assumptions about the relationships to be found.

Growth coalitions Partnerships of private and public sector interests that implement strategies to enhance the economic development of cities and regions, largely through attracting inwards investment, mostly from the private sector but also from public funds. Also termed *civic boosterism* and *civic entrepreneurialism*. See *regime theory*. Coalitions may also be anti-growth. See *exclusionary zoning*.

Habitus   The termed coined by the social theorist Pierre Bourdieu to indicate the *culture* associated with people's *lifeworld* that involves both material and discursive elements.

Hegemonic discourse   The prevailing *ideology*, or domin-ant set of ideas in society. See *discourse, hegemony*.

Hegemony   Domination through consent, largely induced by *hegemonic discourse* that shape people's attitudes. See *interpellation*. May be reflected in the *iconography* of landscapes and buildings.

Heritage landscapes   Older elements of city structure that have been preserved through renovation or conversion to new uses.

Hermeneutics   Theories that examine the complexity of people's views, ideas and subjective interpretations of the world around them.

Heteropatriarchal environment   An area in which the values of *patriarchy* and heterosexuality are dominant (i.e. most parts of cities).

Heteropatriarchy   A term that recognizes that the system of *patriarchy* is dominated by heterosexual values.

Heteropolis   See *postmetropolis*.

Heterotopia   A term used by Michael Foucault to denote spaces comprising many diverse cultures outside, and in opposition to, the mainstream of society. Sometimes called a *counter-site*. Is analogous to a ship, part of, yet detached from, a wider culture. May also be used in a general sense to refer the culture of *postmodernism*. See also *alterity, borderlands, liminal space, spaces of resistance, third space*.

Historical materialism   The philosophy that underpins *classical Marxism* which argues that there is a material base – the means of production – which is the foundation of all social action.

'Hollowing-out'   The transfer of powers from the nation state to political units at other levels such as the supra-national or subnational level. May also refer to the transfer of powers at the local government level to private sector

organizations rather than other political jurisdictions. Also used to refer to the *contracting-out* of activities by private corporations. See *governance*.

Homeland   The geographical space to which a national or ethnic groups feels that it naturally belongs. Often associated with diasporic groups who long for return to their place of origin. May also be used to denote the family home as a place of safety and retreat.

Homosexuality   Mutual emotional and physical attraction between people of the same sex. The term is resisted by many gays and lesbians because it stems from the period when same sex attraction was seen as a social disease. See *queer*.

Housing associations   The not-for-profit voluntary sector of housing provision in the UK.

Housing submarkets   Distinctive types of housing in localized areas of cities which, through various institutional mechanisms, tend to be inhabited by people of a particular type (e.g. in terms of *class*, age or *ethnicity*). See *culture of housing, 'fabric effect', managerialism*.

Human agency   Another term for *agency*. See *voluntarism*.

Human ecology   The application of ideas from the plant and the animal world to the study of residential patterns in cities. An approach of the *Chicago School*.

Humanism   The idea that people share a common humanity (i.e. similar characteristics which can explain human behaviour). Disputed by *discourse* theory.

Hybridity   A term used in *postcolonial theory* to indicate the new forms that are created by the merging of cultures. Linked in the past with imperialist notions of racial super-iority (which were considered to be undermined by racial interbreeding) but now alludes to the fact that identities are not stable but full of *ambivalence*. Criticized for assuming that cultures can mix in an unproblematic manner through a process of *assimilation*. See *liminal space, third space*. Also termed *synergy, transculturation*.

Hyperreality   Sets of signs within forms of representation such as advertising which have internal meanings with each other, rather than with some underlying reality. May also be thought of as copies that become more important than, or take on separate meanings from, the originals they represent. See *simulacra*.

Hyperspace   An environment dominated by *hyperreality* (such as Disneyland).

Icon An image, landscape, building or other material artefact that symbolizes cultural meanings. See *iconography*.

Iconography The study of signs known as icons. Similar to *semiology* but is especially concerned with landscapes. May reflect dominant power relations and the *hegemony* in society.

Ideal type A notion derived from ideal type analysis that attempts to simplify and exaggerate key elements of reality for the sake of conceptual and analytic clarity.

Identity(ies) The elements that make up the view that people take of themselves (e.g. *class*, race, age, *place*). In *cultural studies* identity is seen as the unstable product of *discourse* – hence use of the plural term identities. Contrast with *essentialism*. See also *interpellation, subjectivity*.

Identity politics Political action based around particular identities. Often used to refer to political action other than *class* conflict (e.g. gay rights or disability action groups). May be related to *place*. See *community action*.

Ideological superstructure Sets of institutions such as schools and the family that reinforce ideas that serve the interests of the wealthy and powerful. These are distinguished from the underlying economic base. Also termed *state apparatuses*. See *relative autonomy*.

Ideology Ideas that support the interests of the wealthy and powerful. May also be used in a general sense to refer to any belief system.

Imaginative geographies The way in which we project our own attitudes and beliefs in representations of people and places. See *geographical imagination*.

Imagined communities A term coined by Benedict Andersen to describe the discourses used to construct senses of national identity.

Imagineering The conscious creation of places with characteristics similar to other places (as in Disneyland). Often seen as the creation of a superficial veneer or facade of culture. See *Disneyfication, elsewhereness, McDonaldization, placelessness, simulacra*.

Imperialism The actions and attitudes of a country that dominates distant territories. Often associated with dominant metropolitan centres. Leads to *colonialism*. See also *colonial discourse, postcolonial theory*.

Imperialist discourse The ideas that underpin and legitimize colonial rule. See *colonialism, imperialism, postcolonial society/state*.

Impersonal competition An idea emerging from the *Chicago School* of *human ecology* referring to the economic processes that distribute people into residential areas of differing wealth and status.

Index of dissimilarity A quantitative measure of the extent to which a minority group is residentially segregated within a city. See *segregation*.

Individualistic fallacy The potential mistakes that can arise when attempting to make inferences about groups of people (such as in residential neighbourhoods) based on information for individuals. Contrast with *ecological fallacy*.

Industrial cities Cities of the type that emerged in the nineteenth century dominated by manufacturing activity (sometimes termed 'smokestack industries'). Contrast with *postindustrial cities*.

Informal economy Economic activity that is unrecorded (also known as the 'hidden' economy).

Informational city Manuel Castells' term for the city structures associated with the *information economy*. See *cyberspace, spaces of flows*.

Information economy The growing importance of knowledge (both scientific, technical and fashion-related) in contemporary economies. See *aestheticization*.

Instanciation The idea that the social structures do not exist 'out there' independently of people but are continually created by people through their everyday interactions. See *structuration theory*.

Instrumental interaction Secondary relationships designed to achieve a particular objective, such as joining a business organization. Contrast with *expressive interaction*.

Instrumentalism The theory that both the central and local state serve the interests of capitalist ruling classes, who are represented by the upper-class social background of key politicians, law-makers, bureaucrats and officials. Contrast with *pluralism*.

Intensification Increases in labour productivity through managerial and organizational changes.

Intensive regime of accumulation A period of history during which profits were enhanced through increasing the efficiency with which inputs to the production system were used. Also termed *Fordism*. See *regulation theory, regime of accumulation*.

Intentionality The idea that physical objects (including buildings) have no intrinsic meanings in themselves but take on meaning in relation to their intended use.

Internal economies of scale Factors that lower the cost of production for a firm, irrespective of the size of the industry to which the firm belongs. These factors usually involve high levels of output that lead to the possibility of specialist machines that can increase rates of productivity and which thereby help to recoup the costs of installing such machinery. Contrast with *external economies of scale*. See also *Fordism*.

Internal economies of scope Factors that lower the costs of production when the number of products made within the firm increases. When internal economies of scope begin to decline they can lead to *vertical disintegration* as firms take advantage of *external economies of scope*. See *new industrial spaces*.

Internal markets Attempts to introduce market mechanisms within public sector organizations by dividing them up into separate units for the purchase and supply of services and by establishing various contracts and trading agreements between these agencies.

Interpellation The discourses that shape the view that people take of themselves (e.g. as in regard to concepts of *citizenship*). Used in connection with Marxian notions of *hegemonic discourse*. See *state apparatuses, subjectivity*.

Intersubjectivity The shared sets of meanings that people have about themselves (and where they live) resulting from their everyday experience. See *lifeworld*.

Intertextuality The continually changing meanings that result from the interactions between the reader/observer and the *text*. Part of a form of analysis known as *deconstruction*. Contrast with *mimetic approach*.

Invasion A concept derived from the study of plants and animals used by the *Chicago School* of human ecologists to refer to the process whereby a new social group may begin to 'invade' a residential district. Contrast with *filtering*. See also *succession*.

Inverse-care law The idea that welfare services such as health-care are poorest in the most needy areas. Evidence is contradictory so this is a tendency rather than a law. See *race preference hypothesis, territorial justice, underclass*.

Investment and technical change Capital investment in new forms of machinery and equipment. Often associated with employment loss.

Joint supply The idea that some goods and services have characteristics such that if they can be supplied to one person, they can be supplied to all other persons at no extra cost. See *theory of public goods*.

Jurisdictional partitioning The subdivision of nation states into political and administrative units with responsibility for the allocation of goods and services. See *Balkanization, de jure territories*.

Keno capitalism A model of city structure derived from Los Angeles that consists of a random set of elements (hence the analogy with random cards drawn in the game of keno). The antithesis of the centralized *industrial city*. See *postmodern global metropolis, exopolis*.

Keynesianism A set of policies that underpinned the *welfare state* in the 1950s and 1960s. The objective was to manage economies by countering the lack of demand in recessions through government spending – hence the term 'demand management'. This approach was undermined by inflation and high unemployment in the 1970s. A key element of *Fordism*. See *post-Keynesianism*.

Keynesian welfare state (KWS) A *welfare state* underpinned by Keynesian demand-management policies. Also characterized by universal benefits, citizens' rights and increasing standards of provision through the *social wage*. See also *Keynesianism* and *welfare statism*.

Labour theory of value Karl Marx's explanation for creation of value in capitalist societies. The idea that the value of products should not reflect their *exchange value* in markets but their *use value* – the amount of socially necessary labour that goes into their production. See *surplus value*.

Laissez-faire The *ideology* that underpinned many capitalist societies in the nineteenth century which argued that the state should not intervene in the operation of private markets. See *New Right*.

Landscapes of power Sharon Zukin's term for the spaces of cities newly restructured in the interests of powerful economic interests. See *gentrification, reterritorialization, revanchist city*.

Late capitalism The idea that capitalism has reached a phase that is fundamentally different from previous eras characterized by *globalization*, mass consumption of diverse products and a *culture* of *postmodernism* involving moral relativism. Sometimes equated with *flexible accumulation*, or *neo-Fordism*.

Late modernism Anthony Giddens' interpretation of the cultural and political practices associated with *postmodernism*. Rather than constituting a rupture with the *modernism* of the past, Giddens sees the contemporary period as a late stage of modernism characterized by a high degree of *reflexivity* among both intellectuals and citizens. Also characterized by militarism and *surveillance*.

Legitimating agent An institution that makes the capitalist system acceptable through promulgating certain ideas and/or by acting in a particular fashion (e.g. through the provision of social housing or ideas of *citizenship* in education). See *hegemony, hegemonic discourse, ideology, local state.*

Legitimation Social processes that make capitalism seem acceptable. See *legitimating agent.*

Liberalism A set of ideas that underpin the Western democracies. Characterized by a belief in the value of the individual whose rights should not be subordinated to those of society as a whole; tolerance for opposing views; and a belief in equality of opportunity rather than equality of outcomes. See also *neoliberal policies, libertarianism* and *New Right.*

Libertarianism A form of *New Right* theory which argues that, apart from preserving property rights, the state should leave individuals to do whatever they wish.

Lifeworld The routine patterns of everyday life. The concept is closely linked with *phenomenology* and focuses upon the cultural meanings that people ascribe to the spaces that they inhabit. See *habitus, time-geography.*

Liminal space An in-between space or territory in which cultures mix and interact to create new hybrid forms. See *ambivalence, borderland, heterotopia, hybridity, paradoxical space, third space.*

Lingustic turn Another term for the 'cultural turn' in the social sciences denoting increased attention paid to language and issues of *representation.*

Local economic trading systems (LETS) Groups of people in a local area involved in economic activity using a system of credit based around the exchange of goods and services instead of the national currency.

Locales Distinctive settings or contexts in which interactions between people take place. See *structuration theory, contextual theory* and *recursiveness.*

Locality studies A type of study undertaken predominantly by geographers in Britain in the 1980s that attempted to examine how global forces interacted with the characteristics of local areas.

Local knowledge Another name for *embodied knowledge.* See *situatedness.*

Local state Another term for local government. Also associated with a *Marxian theory* which interprets local governments as serving to maintain the capitalist system and the class interests behind it. See *functionalism.*

Logocentrism The belief in a world composed of a central inner meaning and logic.

'Long boom' (of Fordism) The period after the Second World War between 1945 and the mid-1970s when, according to *regulation theory*, there was in the Western economies a relatively harmonious matching of production and consumption. See *Fordism, regime of accumulation.*

Longue durée A term used in *structuration theory* to indicate the time span over which social institutions such as the family and legal system evolve. See *daesein, durée.*

Los Angeles School Another term for the *California School.* See also *postmodern global metropolis.*

'Loyalty' option A strategy of resignation and inactivity in the face of a problem. Contrast with 'exit' and 'voice' options.

Malapportionment Electoral subdivisions of unequal size. See *gerrymandering.*

Managerialism A type of analysis that focuses upon the influence of managers upon access to scarce resources and local services. Also known as *urban managerialism.* These managers are also known as *social gatekeepers* and 'street-level' bureaucrats. See *eligibility rules.*

Manipulated city hypothesis The argument that coalitions of private interests can operate through legal and institutional frameworks in cities to achieve favourable resource allocations. See *coalition building, growth coalitions, parapolitical structure, regime theory.*

Margins   Areas on the fringes of a dominant region. May also be used metaphorically to indicate cultures on the fringes of dominant cultures where new hybrid identities are being formed. See *borderlands, hybridity*.

Marketization   Transferring the allocation of goods and services from non-market to market principles. See *internal markets, commodification*.

Market testing   A process whereby various external organizations are invited to bid for contracts by an organization wishing to test the efficiency of its own internal division in supplying the good or service in question. See *contracting-in, contracting-out*.

Marxian theory   See *neo-Marxism*.

Masculinism   An approach that privileges and represents as normal the activities of men.

Materialization   Restructuring a service into a physical form that can be bought, sold and transported.

Material practices   Flows of money, goods and people across space to facilitate *accumulation* and *social reproduction*. May also be termed material spatial practices. Contrast with *discursive practices*.

Mediation   A term used in cultural studies to indicate the processes through which cultural forms and styles are appropriated and reinterpreted by certain people and before being passed on to others. See *cultural industries, cultural intermediaries*.

McDonaldization   A term coined by G. Ritzer to indicate the ways in which processes of mass consumption are eroding cultural differences throughout the world. See *globalization*.

Megacities   Manuel Castells' term for large cities in which some people are connected up to global information flows while others are disconnected and 'information poor'.

Mental map   The mental images that people form of areas. See *cognitive distance*.

Merit goods   Goods and services which are regarded as so desirable they cannot be left to private markets and are allocated by the public sector. The reason for this is that the benefits to the community exceed those to the individual, so that latter will tend to consume too little for the common good.

Mestizo   A term originally used to indicate racial mixing of European and indigenous peoples in Latin America but now used in a general sense to indicate both geographical and metaphorical spaces on the margins of dominant cultures where new hybrid forms of identity can emerge (from the Spanish term *mestizaje*). See *borderlands, creolization, heterotopia, hybridity, liminal space, third space*.

Metanarrative   A theory or conceptual framework that purports to be a superior way of looking at the world providing superior or privileged insights. Also known as a *totalizing narrative*. See also *postmodernism, deconstruction*.

Metropolitan fragmentation   The administrative subdivision of US cities into numerous local governments. Also known as *Balkanization* and *jurisdictional partitioning*.

Micropowers   Everyday interactions through which social control becomes exercised. See *disciplinary regimes*.

Microsimulation   Numerical techniques that use detailed *cross-tabulations* for observations such as individuals and households obtained at large spatial scales to simulate complex information at smaller spatial scales. Also used to integrate at a small spatial scale data from different sources and various spatial scales.

Mimetic approach   The idea that writing and other forms of representation are mirrors which reflect the world around us. Contrast with *social constructionism*.

Mimicry   A term used in *postcolonial theory* to indicate the copying of the culture of the dominant group by a colonized people. May lead to an undermining of authority through the development of *hybridity* and mockery. See *ambivalence, liminal space, third space*.

Minority group   A subgroup of society that is characterized by factors such as race, religion, nationality or *culture*.

Mixed economy of welfare   A system in which welfare needs are met by a diverse set of agencies including the voluntary and private sectors rather than exclusively by the state. Also termed *welfare pluralism*.

Mode of production   The way in which productive activity in society is organized (e.g. socialist or capitalist). It comprises the *forces of production* and the *social relations of production*. It also involves methods of *social reproduction*, the *social division of labour* and the *technical division of labour*.

Mode of regulation   An idea central to *regulation theory* which asserts that conflicts within a capitalist society are mediated by various types of norms, rules and regulations that are manifest in various types of legislation and institutions. See also *regime of accumulation*.

Modernism   A mode of thinking characterized by a belief in universal progress through scientific analysis together with the notion that social problems can be solved by the application of rational thought. See *Enlightenment project*, *social engineering*.

Modernity   The period in which modernism was the dominant mode of thinking beginning in the late eighteenth century (the Age of Enlightenment) and lasting until the late twentieth century.

Monumental architecture   Architectural forms that symbolize power and authority. See *iconography*, *monuments*.

Monuments   Elements of the landscape that have symbolic meaning, usually for national and ethnic groups (e.g. war memorials).

Morphogenesis   Processes that create and reshape the physical fabric of urban form.

Morphological regions   Areas characterized by distinctive land uses, buildings and landscapes. See *morphogenesis*.

Multiculturalism   Public policies that support the right of ethnic groups to maintain their distinctive cultures rather than assimilate into the dominant culture of the society. Also involves policies to promote equal opportunities for participation in society irrespective of ethnic background. Contrast with *assimilation*. See also *structural assimilation*.

Multinationals   Companies engaged in production and marketing in more than one country. Sometimes regarded as synonymous with *transnationals* although the latter has a somewhat different meaning.

Multiple deprivation   A situation where people are deprived in respect of a number of attributes such as income, housing, health-care and education. See *territorial social indicators*.

Multiple nuclei model   Harris and Ullman's model of urban city structure characterized by decentralization into numerous central points. See also *edge cities, exopolis, keno capitalism*. Contrast with *concentric zone model* and *sectoral model*.

Multiplex city   A metaphor based on the theatre or the cinema to indicate cities characterized by numerous webs of social and economic interaction, some of which meet in creative ways and some of which remain isolated or disconnected.

Municipal socialism   A form of local government that emerged in Victorian cities between 1850 and 1910 concerned to extend the scope of public services.

Natural areas   An idea formulated by the *Chicago School* of human ecologists which asserts that certain areas of cities have a natural tendency to reflect a particular type of land use or social grouping. See *dominance*.

Naturalization   The process whereby socially constructed differences centred around factors such as class, gender, race, age and nationality are regarded as natural and inevitable. See *essentialism*.

Neighbourhood   Territories containing people of broadly similar demographic, economic and social characteristics but without necessarily displaying elements of close community interaction. See *community*.

Neighbourhood effect   The hypothesis that residential environments both influence and reflect local subcultures. See *cultural transmission*.

Neoclassical economics   Attempts to update the ideas of the classical economists of the late eighteenth and early nineteenth centuries. Characterized by belief in the value of market mechanisms. The approach tends to focus upon microlevel individual market problems rather than wider economic issues. It looks for universal, unchanging principles of human economic behaviour and tends to ignore the social context of economic activity. Contrast with *embeddedness* and *situatedness*.

Neocolonialism   The domination of previously colonized countries by former colonial powers in the period after *colonialism*. See *postcolonial theory*.

Neocorporatism   Corporatist forms of social organization designed to increase the competitiveness of the economy. See *Schumpeterian workfare state*. Contrast with *neoliberalism* and *neostatism*.

Neo-Fordism   A new *regime of accumulation* based around flexibility which it is assumed has, or is about to, replace the Fordist *regime of accumulation* based on mass production. Similar to *flexible accumulation*. Contrast with *Fordism*. Also used more generally to refer to lower-order concepts such as labour practices and forms of industrial organization.

Neoliberal policies (also neoliberalism)   Strategies to make economies competitive by various types of *New Right* policy including *privatization* and *deregulation*. Contrast with *neocorporatism* and *neostatism*. (May sometimes be referred to neoclassical liberalism.)

Neo-Marxism   Attempts to upgrade classical Marxist theories in the light of development in social theory and society in the twentieth century. Also termed *Marxian* and *post-Marxist* theories.

Neopluralism   Attempts to update *pluralism* in the wake of extensive criticism.

Neostatism   Direct state intervention to achieve international competitiveness. Contrast with *neocorporatism* and *neoliberalism*.

Network society   The idea that societies (and the cities within them) consist of an ever-complex flow of diverse factors including knowledge, ideas, images, money, commodities, services and people. See *spaces of flows*.

New economic geography   A diverse set of approaches in geography unified by their desire to escape the abstract reasoning of *neoclassical economics* and take account of *culture*, *embeddedness* and local contingency upon economic forms and processes.

New industrial spaces.   The geographical concentration of firms involved in dense networks of subcontracting and collaboration. Often related to innovative firms in sectors such as electronics and biotechnology. Also termed 'industrial districts' (although the latter term is often applied to small districts within cities). May be linked with *flexible specialization* and *neo-Fordism*.

New Right   A set of ideas that share a common belief in the superiority of market mechanisms as the most efficient means of ensuring the production and distribution of goods and services.

New social movements   See *social movements, identity politics*.

'New Wave' management theory   A set of ideas that stress the advantages of demolishing elaborate managerial hierarchies and their replacement by 'leaner/flatter' managerial structures. Often associated with *devolution*.

NIMBY ('Not In My Backyard')   An acronym for community action groups hostile to urban development in their neighbourhood. See *exclusionary zoning, externality, 'turf' politics*.

Nomadization   The destabilization of identities. This may result from geographical movement between cultures but the term is often used metaphorically. See *authenticity*.

Non-excludability   The idea that some goods and services have characteristics such that it is impossible to withhold them from those who do not wish to pay for them. See *theory of public goods, non-rejectability*.

Non-rejectability   The idea that some goods and services have characteristics such that once they are supplied to one person, they must be consumed by all, even those who do not wish to do so. See *theory of public goods, non-excludability*.

Normative theory   A theory which deals with what ought to be. Contrast with *positive theory*.

Not-for-profit sector   A term often used in the US to denote the charitable or *voluntary sector*.

Nuclear family   A family consisting of a married couple and dependent children. Often celebrated as an ideal family form. Characteristic of many suburban areas but diminishing in importance in Western societies. See *family status*.

Numerical flexibility   The ability of firms (and public sector organizations) to adjust their labour inputs over time to meet variations in output. May be in the form of temporary, part-time or casualized forms of working.

Objectification   A form of scientific analysis inherent in *modernism* which purports to subject people to objective scrutiny but typically leads to them being regarded as different and inferior. Often associated with the use of binary categories and *exclusion*. See *binaries, gaze, othering*.

Orientalism   A term coined by Edward Said to describe the ways in which European thought constructed a view of the Orient. See *discourse, othering*.

Othering   A term used in *postcolonial theory* to indicate the discourses that surround colonized people. Also a mode of thinking that leads to people being regarded as different and inferior. A key element in the work of Foucault on those excluded from power, including prisoners, gays and the mentally-ill. See *objectification, postcolonial theory*.

Outreach services   Services that travel to the consumer (such as fire or ambulance services). Contrast with *point-specific services*.

Overdetermination A term used in *Marxian theory* which denotes that social structures and behaviour have more than one determining factor and cannot therefore be reduced to economic factors alone. Contrast with *economic determinism*. See also *relative autonomy*.

Panopticon A metaphor derived from Jeremy Bentham's nineteenth-century plan for a model prison in which a central tower would enable all inmates to be kept under continual *surveillance*. Used to describe the processes whereby people are scrutinized and controlled in contemporary society. See *disciplinary society, gaze*.

Paradigmatic city A city that illustrates more clearly than others the key features of urban patterns and processes in a particular period (e.g. Manchester in the nineteenth century, Chicago in the early twentieth century and (possibly?) Los Angeles in the late twentieth and early twenty-first centuries). See *Chicago School, Los Angeles School*.

Paradoxical space Another term for *third space*.

Parapolitical structure Informal groups that mediate between individuals and the state in the operation of politics (e.g. business organizations, trade unions, community action groups, voluntary organizations). See *coalition building, governance*.

Pariah city A city that is stigmatized in the wake of extreme social problems and financial difficulties. See *fiscal imbalance*.

Patriarchy Social arrangements such as in the form of institutional practices and prevailing social attitudes that enable men to dominate women.

Performance The ways in which identities are socially constructed through particular ways of acting and not the result of some biological essence. See *subjectivity, subjectivities, subject positions*.

Performativity The process through which identities are constructed. See *performance*. Also the practice of monitoring the performance of workers. Can involve worker productivity and efficiency in terms of output but also the extent to which workers perform certain roles as in service jobs. Used as a defining element of *postmodernism* through new forms of *governance*. Also a key element of '*new wave*' *management theory*.

Phenomenology A set of perspectives that focus upon people's subjective interpretations of the world, rather than some external objective reality. Contrast with *mimetic approach*. See *lifeworld*.

Place A term used by geographers to indicate that the characteristics of territories or spaces are socially constructed (but also have a material base). See *social constructionism*.

Placelessness The tendency for spaces in contemporary cities to be modelled on other places but in ways that produce a uniform, anonymous pastiche. See *elsewhereness, simulacra*.

Place marketing See *place promotion*.

Place promotion Policies to encourage economic development through advertising, lobbying and other incentives such as tax exemptions. Also used to encourage tourism. See *civic boosterism*.

Pluralism A theory that argues that the diverse interest groups in US cities have equal access to the democratic system and there is no systematic bias in favour of one particular group (e.g. business or labour interests). Contrast with *instrumentalism*.

Point- (or place-) specific services Services (either public or private sector) that have to be located at a particular point, such as a school or libraries. Contrast with *outreach services*.

Polarization See *social polarization*.

Polity A type of government (i.e. democratic, fascist, etc.). See *civil society*.

Polyvalency The capacity of workers to undertake multiple tasks. Another name for *functional flexibility*.

Positional good A good that displays the status of the consumer. See *cultural capital*.

Positionality The values adopted by an individual. Linked to the argument that writings are not an objective mirror of reality but reflect the cultural context in which they are produced. Contrast with the *mimetic approach*. See *contextual theory* and *situatedness*.

Positive theory A theory that is concerned with what actually exists (rather than what ought to be). Contrast with *normative theory*.

Postcolonial society/state A nation that has gained independence following a period of *colonialism*. May be associated with *appropriation, ambivalence* and *hybridity*.

Postcolonial theory An approach that examines the discourses running through Western representations of non-Western societies, both in the colonial period and in contemporary texts. A perspective that attempts to subvert the notion, embedded in these writings, that Western thought is superior. Attempts to expose *ethnocentrism*. See also *colonialism, imperialism, othering*.

Postindustrial cities Cities dominated by service activity. Often the outcome of *deindustrialization*. May exhibit *postmodern* forms of consumption and *culture*, and the *postwelfare society*.

Post-Keynesianism A term that encompasses both economic theory that has emerged to replace Keynesian theory as well as broader social and economic conditions following the demise of *Keynesian welfare state*. See *Keynesianism, postwelfare state/society*.

Post-Marxism Another name for *neo-Marxist* theory. Places greater emphasis upon cultural issues than *classical Marxism*. Attempts to avoid *economic determinism*. May also be used a catch-all phrase for various postmodern and poststructuralist perspectives that attempt to avoid being *metanarratives*.

Postmetropolis Another term for *postmodern global metropolis*. See *galactic metropolis, heteropolis*.

Postmodern global metropolis Ed Soja's term to describe the structure of Los Angeles which is seen as an archetype of new urban forms. See *galactic metropolis, California School*.

Postmodernism A term with many meanings: rejection of the idea that there is one superior way of understanding the world (see *metanarrative* and *totalizing narrative*); a type of analysis known as *deconstruction*; a style characterized by eclecticism, irony and pastiche (as in architecture but also in writing and advertising); a period of history and a cultural trend which is the logical accompaniment to the era of *neo-Fordism* or *flexible accumulation*.

Poststructuralism A type of analysis. Unlike *structuralism*, which assumes a close relationship between the *signifier* and the *signified*, poststructuralism assumes that these are disconnected and in a continual state of flux. See *deconstruction, text, intertextuality*.

Postwelfare state/society A term used to indicate the broad range of changes to the welfare state in contemporary societies including *residualization* and *privatization*. Also used to indicate broader cultural shifts such as the move towards greater *privatism* and an ethic of self-sufficiency. May be linked to notions of *neo-Fordism*. See also *Schumpeterian workfare state*.

Preindustrial city A city without an industrial base, usually in earlier historical periods before the Industrial Revolution.

Primary relationships Social ties between family members and friends. Contrast with *secondary relationships*.

Principal components analysis A quantitative technique used by geographers for summarizing large data sets that is technically very similar to factor analysis and used within *factorial ecology*.

Privatism An ideology underpinning most Western capitalist societies, based around belief in superiority of private ownership of wealth and the allocation of goods and services by market mechanisms. See *liberalism*.

Privatization A diverse set of policies designed to introduce private ownership and/or private market allocation mechanisms to goods and services previously allocated and owned by the public sector. See *asset sales, commercialization, commodification* and *marketization*.

Professionalization thesis The argument that the social structure of Western societies is displaying an increase in professional and managerial groups. Contrast with *proletarianization thesis*. Se also *creative cities*.

Pro-growth coalition. Another name for *growth coalitions*. See also *civic entrepreneurialism*.

Projective identification A tendency to define one's own culture in terms of the imagined failings of other cultures. See *binaries, othering*.

Proletarianization thesis The argument that the social structure of Western societies is becoming biased towards lower social classes. Contrast with *professionalization thesis*. See also *deskilling, social polarization*.

Property-led development The regeneration of urban areas by private speculators investing in office properties. See *urban development corporations*.

Proprietarization The tendency for voluntary or non-profit agencies to adopt the strategies of private sector organizations. See *commercialization*.

'Psychic overload' The proposition that the diversity, density and anonymity of social relationships in cities lead to anxiety and nervous disorders. Similar to *'psychological overload'*.

'Psychological overload' The notion that in urban environments people are bombarded with stimuli that may lead to aloofness, impersonality and deviant behaviour. See *behaviourism, gesellschaft*.

Public goods Goods and services which have characteristics which make it impossible for them to be allocated by private markets. May also be used in a general sense to indicate goods and services provided by the public sector. See *theory of public goods*.

Public space A space that is owned by the state or local government and in theory is accessible to all citizens but which in reality may be policed to exclude some sections of society.

Public sphere Fora in which people can discuss issues on the basis of equality (at least in theory if not in practice). See *civil society*. May literally be a space in the city (such as Speakers' Corner in London).

Purified communities A term coined by Richard Sennett to indicate the ways in which some groups attempt to segregate themselves from other groups whom they consider to be different and inferior. See *authenticity*.

Purified space Another term for *purified communities*.

Quality-of-life indices Social measures of people's lives (as a supplement to or in place of economic indices). May also be termed social indicators, and measures of *social well-being*. See *territorial social indicators*.

Quango An acronym for quasi-autonomous non-governmental organization. See *governance, 'hollowing-out'*.

Quantitative geography Studies that attempted to analyze the world in a scientific, value-free manner developing universal laws of human behaviour based on mathematical models and statistics. Contrast with *situatedness*. See also '*god trick*'.

Quasi-market A market in which goods and services are purchased for consumers by intermediaries (as when health-care is purchased by hospital administrators or physicians). See *internal markets*.

Quasi-state New institutions that undertake roles previously performed by central and local government but that are now outside traditional channels of democratic control. See *governance, quango, shadow state*.

Queer An abusive term for homosexuals that has been adopted by advocates of *queer theory*. See also *queer politics*.

Queer politics Political practices such as that advocated by the gay activist group Queer Nation including 'kiss ins' and 'mock weddings' that attempt to subvert dominant naturalized notions of sexuality.

Queer theory A theory, much inspired by the work of Michel Foucault, that attempts to expose the fluid and socially constructed character of sexual identities. The appropriation of the abusive word *queer* is meant to draw attention, in an ironic way, to the repressive character of social discourses surrounding sexuality.

Race-preference hypothesis The argument that the some ethnic groups (typically African-Americans or Latinos) receive the worst levels of both public- and private-sector service provision. See *underclass hypothesis*.

Racial group A group of people who are assumed to be biologically distinct because of some characteristic of physical appearance, usually skin colour or facial appearance. Since these differences are of no greater significance than other physical attributes such as hair colour, a racial group is one in which certain physical attributes are selected as being ethnically significant. See *racism, ethnicity, ethnic group*.

Racism A set of ideas and social practices that ascribe negative characteristics to a particular racial group who are mistakenly assumed to be biologically distinct. See *ethnic group*.

Recapitalization A broad term for various *New Right* policies designed to put the interests of business at the top of the political agenda. See *Schumpeterian workfare state*.

Recentralization A trend for the reconcentration of facilities in urban centres following a period of decentralization. See *gentrification*.

Recommodification The reallocation of goods and services from non-market to market mechanisms. Similar to *marketization, commodification*.

Recursiveness A key element of *structuration theory* that recognizes that social systems are made up of the numerous everyday interaction of people. Also termed recurrent social practices.

Rationalization   The closure of industrial capacity typically leading to employment loss. May also refer to the closure of facilities within the welfare state. See *deindustrialization*.

'Redlining'   The practice by building societies and mortgage companies of withholding loans for properties in areas of cities that are perceived to be bad risks.

Reflexive accumulation   Lash and Urry's term to describe the emerging economy characterized by services, knowledge-intensive industries and the aestheticization of products. Argued to be a superior concept to both *flexible specialization* and *flexible accumulation* which are production dominated and ignore the growth of the *cultural industries*.

Reflexivity   The capacity of people to have knowledge of the situations that face them and to make choices based on this knowledge. See *human agency*.

Regime of accumulation   An abstract concept central to *regulation theory* that claims that from time to time within capitalist societies there emerge stable sets of social, economic and institutional arrangements that serve to link production and consumption. See also *mode of regulation, Fordism, neo-Fordism*.

Regime theory   An approach that examines how various coalitions of interests come together to achieve outcomes in cities (often the promotion of local economic development by *pro-growth coalitions* of business interests). Argues that power does not flow automatically but has to be actively acquired.

Regulation theory   A set of Marxist-inspired ideas that attempt to relate changes in labour practices and forms of industrial and social organization to wider economic developments and the changing relations between nation states. See *regime of accumulation, mode of regulation, Fordism* and *neo-Fordism*.

Reification   Treating people as objects (but may also involve regarding objects as having *agency*).

Reinstitutionalization   The process whereby ex-patients of welfare institutions that have been closed, such as psychiatric hospitals, end up in other types of institution, especially prisons. See *deinstitutionalization*.

Relative autonomy   The idea embodied in certain structuralist approaches that the ideological superstructure is not rigidly determined by the economic base of society. See also *economic determinism, functionalism, overdetermination, state apparatuses*.

Relativism   The notion that truth and knowledge are relative to particular times and places. See *embodied knowledge, situatedness*.

Rent gap   The disparity between the rents currently charged for run-down inner-city areas and their potential market rents following renovation. If large can lead to urban development and *gentrification*. See also *revanchist city*.

Representation   All the ways in which societies portray themselves and the world around them.

Representations of space   Lefebvre's term for the discourses used to represent areas. See *material practices* and *spaces of representation*. Also termed representational space.

Reproduction   A metaphor derived from biology used within *Marxian theory* to refer to all the elements needed to ensure maintenance of the capitalist system. Also termed *social reproduction*. See *accumulation*.

Reserve army of labour   The idea derived from *classical Marxism* that within capitalist economies there are groups of marginalized low-income workers who are given employment in times of high demand and laid off in times of recession.

Residential differentiation   The tendency for people with distinctive characteristics and cultures to reside close to each other in cities thereby forming distinctive neighbourhoods. Also termed *sociospatial differentiation*. See *clustering, community, neighbourhood, segregation, ghetto*.

Residualization   Reductions in welfare spending so that services are limited to deprived minorities. See *ghettoization*.

Residual welfare state   A welfare system that comes into operation only as a last resort when other means of meeting welfare needs, through families, voluntary bodies and private sector agencies, fail. See *postwelfare society*.

Reterritorialization   Castells' term for the restructuring of urban space by powerful economic interests. See *gentrification, revanchist city*. Contrast with *deterritorialization*.

Revanchist city   Neil Smith's term for a city in which the powerful take their 'revenge' (from the French work *revanche*) by reasserting their authority through processes such as *gentrification, privatization* and *deregulation*.

'Risk society'   Ulrich Beck's notion that the risks in contemporary society are much greater than in the previous societies.

Rotation   A technical procedure used within factor analysis to obtain the clearest patterns within the data. Also used within *principal components analysis*. See *factorial ecology*.

Scale   A term used by geographers to indicate that geographical processes operate at different levels (e.g. global, national, regional, neighbourhood, household, etc.). It is increasingly recognized that these processes are interconnected in complex ways. See *global–local nexus*.

'Scanscape'   A term used by Mike Davis to describe the electronic surveillance strategies utilised in Los Angeles to exclude groups regarded as undesirable from certain parts of the city. See *Panopticon*.

Schumpeterian workfare state (SWS)   An emerging form of *welfare state* in which the needs of individuals are subordinated to enhancing the international competitiveness of the economy. Unlike the *Keynesian welfare state*, the SWS tends to be based on discretion, minimalism and means testing.

Scripting   The process whereby spaces (and social groups) are produced or constructed through various forms of discourse or representation. See *text*.

Search space   The region within which a potential migrant searches for a new location. See *aspiration region, awareness space, activity space*.

Secondary relationships   Relationships with people other than family and friends designed to achieve a particular purpose. See *expressive interaction, instrumental interaction*.

Sectoral model   The model of urban residential structure advocated by Homer Hoyt which suggests that class differences in residential areas are arranged in sectors. Contrast with *concentric zone model*. See *filtering*.

Segregation   The tendency for minority groups to be unevenly distributed in cities (i.e. to display *residential differentiation*). Very rarely are groups completely separate in residential terms – hence studies measure the degree of segregation. See *ghetto, index of dissimilarity, assimilation*.

Self-provisioning   A situation where individuals make their own arrangement to meet their welfare needs, rather than relying upon the state. The alternatives could be self-help, the voluntary sector or private sector agencies. See also *domestication*.

Semiology   The study of signs and their meanings. Also termed semiotics. See *signifiers, signified, text*.

Semiotic redundancy   The tendency for changes in style and fashion to make existing products undesirable even though they may currently function adequately. See *aestheticization, semiology*.

Service-dependent ghetto   Concentrations of ex-psychiatric patients and other dependent groups in inner city areas close to community-based services. Also known as the *'asylum-without-walls'*. See *community care*.

Sexism   Sets of ideas, attitudes and behaviour that ascribe one of the sexes with inferior characteristics. See *gender, feminism*.

Sexuality   Ideas and concepts about sex. Implicit in this term is recognition that human sexual activity is primarily a learned form of behaviour shaped by cultural values.

Shadow state   The tendency for the *voluntary sector* to take over services that were previously allocated by the state. The shadow state is diverse and outside traditional channels of democratic control.

Signification   The process whereby places, peoples and things are given meaning in writing and other forms of representation. See *spaces of representation*.

Signified   The cultural meaning that is indicated by the *signifier*. See also *text*.

Signifier   That which points to some wider cultural meaning. This may be a word, sign or material object. See *signified*.

Signifying practices   Social activities that are full cultural meaning. See *representation, signification, signifier, signified, signifying systems*.

Signifying systems   The idea that culture involves signs, symbols and material activities that provide a way of interpreting and understanding the world.

Simulacra   Images or copies of the 'real' world that are difficult to distinguish from the original reality they purport to represent. May be thought of as copies without originals that take on a 'life of their own'. A key element in postmodern culture. See *postmodernism, hyperreality*.

Situatedness   An approach that recognizes that all writings and other forms of representation emerge from people with particular values and in cultures that are distinct in time and space. An approach that denies that there are invariant patterns of human behaviour across time and space, as assumed in much *neo-classical economics*. Also referred to as situated knowledge.

Social Area Analysis   The work undertaken in the 1950s primarily by Shevky and Bell which attempted to relate measures of social change to the geographical structure of cities. Influenced by the *Chicago School* of *human ecology* and in turn influenced *factorial ecology*. May also be used as a general sense to indicate geographical analysis of city structure.

Social closure   Another name for processes whereby powerful groups exclude others groups from wealth, status and power. May be called *exclusionary closure*. See *purified communities*.

Social constructionism   An approach that asserts that most of the differences between people are not the result of their inherent characteristics but the way in which they are treated by others in society. Can be applied to differences related to *ethnicity* and *gender* together with the characteristics of places and technologies. See *place*, *racism* and *sexism*. Contrast with *essentialism*.

Social Darwinism   The application of ideas about natural competition in the plant and the animal world to the study of urban social geography. See *Chicago School, human ecology*.

Social distance   Differences between people based on factors such as *class*, status and power leading to separation in social life. May be the result of mutual desire or predominantly the wishes of the powerful. Often expressed in terms of physical distance and *residential differentiation*.

Social division of labour   The social characteristics of the people who undertake different types of work (e.g. age, *ethnicity, gender*). See also *technical division of labour*.

Social economy   The part of the economy made up of *social enterprises* that is not part of the true private or public sectors. See *social entrepreneurs*.

Social engineering   The belief that society can be improved by rational comprehensive planning based on scientific principles (as in comprehensive slum clearance and urban redevelopment schemes).

Social enterprises   Organizations that use commercial activities to achieve social benefits for defined communities and interest groups. Part of the *social economy* or *third sector*.

Social entrepreneurs   Innovators in social policy; key drivers of social enterprises.

Social gatekeepers   Professionals, managers and bureaucrats (in both the private and public sectors) who determine access to scarce resources and facilities (e.g. housing, mortgage finance, welfare benefits). See *decision rules, managerialism, 'street-level' bureaucrats*.

Social movements   Pressure groups and organizations with varying degrees of public support petitioning for change, often outside of conventional political channels. Sometimes termed *new social movements* and *urban social movements*. These formed an important part of the theory of *collective consumption*.

Social polarization   Growing inequalities between groups in society. May refer to increases in the poorest, the most wealthy, or both (i.e. a disappearing middle class forming a social structure shaped like an hour-glass).

Social rank   The name frequently given to one of the main dimensions of urban residential structure revealed by *factorial ecology* studies, *class*-based variations in the material wealth of inhabitants. See *family status, ethnic status, multiple deprivation*. May produce results similar to *territorial social indicators*.

Social relations of production.   The various legal, institutional and social arrangements in society that permit the capitalist *mode of production* to function. See also *forces of production*.

Social reproduction   All the various elements that are necessary to reproduce the workforce and the consumers needed to keep a capitalist society functioning (e.g. the family, schools, health services, *welfare state*, etc.). A key part of *Marxian theories* which stress the role of the welfare state in overcoming the problems of capitalism. Much criticized in the past for *functionalism*.

Social wage   The public services and activities undertaken by the state (such as the regulation of labour markets) to maintain the welfare of citizens. See *welfare statism*.

Social well-being   See *quality-of-life indices*.

Sociobiology   Explanations of human behaviour based on genetic factors relating to biology. Disputed by those who adopt *social constructionism*.

Sociospatial dialectic   Ed Soja's term for the mutually interacting process whereby people shape the structure of cities and at the same time are affected by the structure of those cities.

Sociospatial differentiation Another name for *residential differentiation*.

Space A term often used in a general sense to indicate geography, location or distance, but also used specifically by human geographers to acknowledge the socially constructed nature of environments. Also termed *place*. See *'betweenness' of place, purified communities, social constructionism, spatiality*.

Spaces of exclusion Areas in which certain groups of people are excluded by other more powerful groups. Often based on stereotyped notions of other groups. See *gated communities, othering, purified communities*.

Spaces of flows Manuel Castells' term to describe the spatial structures associated with the *information economy*. See *cyberspace, distanciation, time–space compression*.

Spaces of representation A term used by Lefebvre to indicate the various ways in which new spatial practices can be planned or imagined in cities. See *material practices, representations of space*.

Spaces of resistance Areas of cities that challenge dominant, majority, ways of life through fostering 'alternative' lifestyles. See *counter-site, heterotopia, liminal space, paradoxical space, third space*.

Spatial autocorrelation Interdependence, resulting from spatial contiguity, among so-called 'independent' variables used in multivariate techniques such as *factor analysis* and multiple regression leading to unstable and unreliable results.

'Spatial fix' The periodic (usually temporary) resolution to problems of integrating production and consumption in capitalist economies as manifest in urban structures.

Spatiality Also known as sociospatiality, a term used by geographers to acknowledge the socially constructed and material nature of space (as with the term *place*). See *space, social constructionism*.

Spatialized subjectivities A term that recognizes the explicit role of space in the formation of *subjectivities* and *identities*.

Spatial science Another name for *quantitative geography*.

Spectacle The idea that social life is increasingly dominated by images. See *commodity fetishism*. Also may refer to tendency to promote cities through grand events and spectacular landscapes. See *Disneyfication, festival retailing*.

Spillover Another name for an *externality*.

Splintering urbanism Increasing inequalities in cities resulting from differential access to new communications technologies (such as the Internet). See *cyberspace, social polarization*.

Standpoint theory The controversial argument that women can provide a deeper understanding of the world through their involvement in child-rearing and *social reproduction*. Also used in a general sense to indicate theories that recognize the *situatedness* of theory and the need to champion the oppressed.

State apparatuses A term used within structuralist theories to refer to key elements of the ideological superstructure such as the church, family and education system. See *ideological superstructure, micropowers, structuralism*.

Strategic essentialism The temporary adoption of essentialist attitudes by deconstructionists to achieve political objectives. See *essentialism, deconstruction, discourse, social constructionism*.

'Street-level' bureaucrats Managers who have direct contact with the public such as housing inspectors and police officers. See *decision rules, managerialism*.

Structural assimilation The process whereby a *minority group* is incorporated into the *class* and occupational structure of the wider society (or *charter group*). Contrast with *behavioural assimilation*.

Structuralism A theoretical approach derived originally from the study of languages which involves delving below the surface appearance of human activity to examine the underlying structures that affect human behaviour. See *poststructuralism*.

Structuration theory A theory expounded by Anthony Giddens that attempts to bridge the divide between voluntarist and determinist theories. See *voluntarism, economic determinism*.

Structure Used in a general sense to indicate a broad over-arching framework. Also a key part of *structuration theory* which refers to the rules, norms and resources that individuals draw upon to carry out their lives. See *system, recursiveness*.

Structured coherence A term coined by David Harvey to indicate the ways in which urban regions assume distinctive characteristics which are the products of local systems of production, local labour markets and the associated modes of consumption and life style. A *Marxian* explanation that argues struggles over the labour process are the key (but not the only) process at work in cities.

Subaltern classes   A term originally devised by Gramsci to denote subordinate groups in society who are subject to the hegemony of ruling classes (from the army term 'subaltern' meaning ranks below the officer class). Used in *postcolonial theory* to highlight people subject to *colonialism*.

Subcontracting   A situation in which one organization contracts with another for the provision of a good or service. Also termed *contracting-out*.

Subcultural theory   An approach that examines the influence of factors such as *class*, *ethnicity* and *family status* upon behaviour in cities arguing that new subcultures are spawned by urban living.

Subculture   A group with values and norms different from the majority *culture* in society. Often expressed in *residential differentiation* in cities. May be termed a *deviant subgroup*.

Subject formation   The process whereby *identities* are socially constructed. See *social constructionism*. Contrast with *essentialism*.

Subjectivities   A term similar to *subjectivity* but which explicitly recognizes the context-dependent, and therefore continually changing, nature of the concept. See *spatialized subjectivities*.

Subjectivity   The continually changing views that people take of themselves and the world around them. In *cultural studies* these views are seen as the product of *ideology* and *discourse* and not some stable factor resulting from innate characteristics (as argued in *essentialism*). Similar to identities but is a more dynamic concept resulting from the interactions of the self, experience and discourses in different contexts. See *subjectivities*, *spatialized subjectivities*.

Subject positions   Ways of acting and thinking that are implicit with various discourses about people classified in some way (e.g. on the basis of *class*, age or *gender*). These interact with *subjectivities* to form *identity*.

Subsidiarity   The idea that national-level decision-making should be devolved to the most appropriate level (usually downwards to local communities). See *devolution* and *decentralization*.

Suburban exploitation thesis   The argument, mainly applied to the US, that residents in relatively wealthy suburban local governments are consuming services in poorer inner cities which they are not fully paying for (such as roads and policing). Related to *fiscal imbalance*, *free-rider*, *metropolitan fragmentation*.

Succession   A term derived from the study of plants and animals used by the *Chicago School* of *human ecology* to refer to the process whereby a new social group begins to dominate a residential district after initial *invasion*. See also *dominance*, *natural areas*, *social Darwinism*.

Superorganic (culture)   The controversial view of Carl Sauer and the Berkeley School of cultural geography, derived in part from evolutionary theory, that the culture of a region should be regarded as a single over-arching entity struggling with other cultures.

Superstructure   A term derived from *classical Marxism* to indicate all the elements of society outside of the system of production including the state and legal system. Similar to the notion of *civil society*. See *state apparatuses*.

Surplus value   A key element of Karl Marx's *labour theory of value* – the difference between the wages paid to workers and the prices the goods they produce can command through market exchange. See *exchange value*, *use value*.

Surveillance   The scrutiny and control of subordinate peoples. See *gaze*, *interpellation*, *Panopticon*, '*scanscape*'.

Sustainability   A much-contested idea with many different interpretations but generally alludes to economic development in a manner that can be sustained in the long run for future generations. See *ecocentric approach*, *technocentric approach*, *urban social sustainability*.

Symbolic capital   Goods and services that reflect the social position, taste and distinction of the owner. May also be reflected in imposing buildings also known as *monumental architecture*. See *aestheticization*, *cultural capital*, *positional good*, *symbolic distancing*.

Symbolic distancing   The tendency for people to display their social position through various forms of ostentatious consumption (including residential location and housing type). See *symbolic capital*.

Syncretism   See *appropriation*.

Synekism   Ed Soja's term for processes leading to the creation of regional networks of settlements which excel at innovation, both social and economic. See *creative cities*.

Synergy   Another name for *hybridity*.

System A term used in many different ways according to the approach in question but generally used to refer to the interdependent parts of a larger entity. In *structuration theory* the system is the outcome of all the actions undertaken by people. See also *structure, recursiveness, reflexivity*.

Tapering Another name for *distance-decay effect*.

Taylorism A set of ideas developed by US engineer Frederick Taylor to manage the labour processes that were adopted by Henry Ford in the early twentieth century to mass produce automobiles in Detroit. Also termed the 'principles of scientific management'. These involved simplification of tasks, managerial control of workers and the utilization of 'time and motion' studies to determine the most efficient ways of working. See *Fordism*.

Technical division of labour The types of work that need to be undertaken within an industrial system. Contrast with *social division of labour*.

Technoburb A relatively independent suburban area or edge city characterized by high-technology industries. See *edge cities, new industrial spaces*.

Technocentric approach An approach to *sustainability* that argues that environmental problems can be met without fundamentally disturbing the capitalist system. Also called *ecological modernization*. Stresses the capacity of existing institutions to adapt and meet environmental problems and the ability of science and technology to meet these challenges. Contrast with *ecocentric approach*.

Technological determinism The notion that technology exists as some independent external force that impinges 'upon' society. Disputed by social constructionist who argue that technologies are an integral part of society (i.e. the product of economic, political and cultural processes).

Telematics Services that link computer and digital media equipment to new forms of satellite and fibre-optic telecommunications channels. See *cyberspace*.

Territoriality A term with various interpretations, including the idea that humans have an innate desire to occupy a specific territory to satisfy needs of safety, security and privacy and to enable the expression of personal *identity*. Sometimes called the 'territorial imperative'. A form of explanation based on *sociobiology* that is disputed by those who take a *social constructionist* approach. Also a concept on postmodern thought that involves any institution that represses people's desires (such as the family). May also be a strategy to achieve political power by mobilizing the support and resources in geographical areas such as urban neighbourhoods, cities or regions.

Territorial justice The allocation of resources across a set of areas in direct proportion to the needs of the areas. See *territorial social indicators*.

Territorial social indicators Measures of social disadvantage (or need) that relate to particular types of geographical region such as residential areas within cities or local government areas. May be used to evaluate degrees of *territorial justice*.

Text A key concept in *cultural studies* which refers to any form of representation that conveys social meanings – not just the written word but also paintings, landscapes and buildings. See *discourse* and *deconstruction*.

Theory of public goods A theory that states that some goods and services have characteristics that make it impossible for them to be allocated by private markets. See *joint supply, non-rejectability, non-excludability*.

Third sector The part of the economy that lies between the private and public sectors comprised of voluntary organizations and social enterprises. See *social economy*.

Third-party effect Another name for an *externality*.

Third space The mixture of meanings that emerges when two cultures interact, as under *colonialism*. See *ambivalence, hybridity, liminal space, paradoxical space*.

Time–geography The work originated by Torsten Hagerstrand that examines the joint influences of time and space upon people's daily lives. See *authority constraint, capability constraint, coupling constraint*.

Time–space compression David Harvey's term to indicate the ways in which various processes including technological change have speeded up processes of capital accumulation.

Time–space convergence The idea that new transport systems are leading to much greater mobility and a 'shrinking world'. Contrast with *distanciation*.

'Tipping point' A situation when a new *minority group* migrating into a residential area becomes such a significant presence that they provoke a sudden and rapid exist of the remainder of the original population. See *'blockbusting', invasion*.

Totalizing discourse See *totalizing narrative*.

Totalizing narrative   A theory that purports to be a privileged way of interpreting the world, providing superior insights. May also be termed a *metanarrative*.

Transculturation   Another term for the reciprocal inter-action of dominant and subordinate cultures as depicted by *hybridity*.

Transitional cities   Cities in economies that have moved from centrally planned command economies under com-munist regimes to market-led capitalist economies.

Transmitted deprivation   The idea that poverty results from poor skills and low aspiration levels that result from poor parenting.

Transnationals   Companies whose production, distribu-tion and marketing operate in more than one country. May also refer to companies whose operations are integrated at a global level. See *globalization, global–local nexus*.

Transnational urbanism   New urban forms created by migrants who maintain diverse connections with more than one nation. Contrast with previous distinctions between permanent migrants (who might undergo *assimilation*) or temporary migrants who might remain excluded from the host society. Has been linked with the growth of *transnationals* and flexible forms of capitalism that require a mobile workforce. May be elite groups or the relatively disadvantaged. See also *diaspora*.

Trope   A regular pattern or convention in story-telling (such as the victory of the individual over 'the system' in Hollywood movies).

'Turf' politics   Another name for *community action*.

Underclass   The poorest and most disadvantaged in society. Often linked with the *culture of poverty* explanation. Also used to denote the growing numbers of the poor and the changing character of poverty. See *social polarization*.

Underclass hypothesis   The argument that the poorest groups in society receive the worst levels of both public and private sector service provision. See *race-preference hypothesis*.

Urban development corporations (UDCs)   Quasi-public sector bodies in the UK that encourage private sector investment in run-down urban areas through the provision of infrastructure such as reclaimed land and transport networks. See *property-led development*.

Urban entrepreneurialism   A new period of *governance* in cities characterized by competition between cities to encourage economic development. May also be termed *civic entrepreneurialism* and be linked with the *'hollowing-out'* of the central state. See *civic boosterism, growth coalitions, regime theory*.

Urban governance   All the methods and institutions by which cities are governed. The term is commonly used to indicate the shift away from direct government control of cities via hierarchical bureaucracies towards indirect control via diverse non-governmental organizations. Associated with the demise of local forms of government. May also be termed *governance*. See *quango, quasi-state*.

Urban growth coalitions   See *growth coalitions*.

Urban managerialism   Urban-based versions of the managerialist thesis. See *managerialism*.

Urban morphology   The physical structure of the urban environment. See *morphogenesis*.

Urban nightscapes   Districts of cities catering for the nocturnal leisure-based activities of young adults (e.g. clubs, pubs, bars and cinemas). Linked with *civic boosterism* and increasing *surveillance*.

Urban social areas   Residential districts within cities in which people with similar characteristics tend to live near one another.

Urban social movements   See *social movements*.

Urban social sustainability   Social life within cities that is relatively free of inequality and conflict and that can be sustained in the long run. A component of *sustainability*.

Use value   The utility of a commodity (such as housing) to the consumer. Related to but distinct from *exchange value*.

Vacancy chains   The chains of movement resulting from properties becoming available through factors such as new building, the subdivision of properties, and the death or outmigration of existing occupants. See *filtering*.

Vertical disintegration   A situation in which companies and organizations subcontract work out to other (usually small) organizations. Contrast with *vertical integration*. See also *contracting-out*.

Vertical integration   A structure in which functions are integrated into a large organization in a complex inter-dependent hierarchy. Contrast with *vertical disintegration*.

'Voice' option   A strategy of overt campaigning by a community action group. Contrast with *'exit'* and *'loyalty'* options. See *community action, 'turf' politics*.

Voluntarism   The use of the voluntary sector to meet welfare needs. See also *shadow state*. This term may also refer to a type of social analysis that envisages people as capable of making decisions to evolve in an almost infinite range of possible directions. This approach therefore plays down the constraints upon people. See also *human agency*. Contrast with *economic determinism*.

Voluntary organizations   Interest groups and pressure groups in cities (e.g. work-based clubs, religious organizations, community groups, welfare organizations). Only a small proportion are likely to be overtly politically active at any given time.

Voluntary sector   May refer to voluntary organizations in general but more usually to the diverse set of non-profit-making agencies attempting to meet welfare needs such as charities, charitable trusts and pressure groups.

'Weightless world'   Diana Coyle's term to indicate the increasing importance of knowledge and non-material products in modern economies. See *culturalization of the economy*.

Welfare corporatism   A society characterized by corporatist forms of collaboration in which certain groups can gain privileged access to government to derive benefits of various types (e.g. contracts, tax concession). Usually applies to big business or organized labour rather than the most deprived. See *corporatism*.

Welfare pluralism   A system in which welfare needs are met by a diverse set of agencies including those from the voluntary and private sectors rather than relying upon universal provision by state agencies. Also known as the *mixed economy of welfare*. See *contracting-out, privatization*.

Welfare state   A set of institutions and social arrangements designed to assist people when they are in need through factors such as illness, unemployment and dependency through youth or old age.

Welfare statism   The notion that the state should have responsibility to ensure an adequate standard of living for its citizens through policies designed to achieve full employment, relatively high minimum wages, safe working conditions and income transfers from relatively affluent majorities to deprived minorities.

Wirthian theory   The highly influential ideas of Louis Wirth, which suggest that social life in cities (i.e. 'urbanism') is characterized by increased rates of crime, illness and social disorganization which are largely a product of the increasing size and heterogeneity of urban life. See *'psychic overload'*.

Workfare   A system in which those who are unemployed have to undertake work in order to receive benefits. Also associated with a number of other policies designed to regulate the behaviour of welfare recipients.

World cities   See *global cities*.

Worlding   The discourses used to represent colonized territories (see *colonial discourse*). May be used to describe the ways in which any place is represented.

Zeitgeist   The spirit of the age (i.e. the prevailing *ideology*, or *hegemonic discourse*).

Zone in transition   The name given by the *Chicago School* of *human ecology* for the concentric ring between the city centre and working-class residential areas. Characterized by a mixture of industry and poor-quality rented accommodation, often inhabited by immigrants and various forms of 'social deviant'. Also termed transition zone.

Zoning   See *exclusionary zoning*.

References

Abu-Lughod, J. (1969) Testing the theory of Social Area Analysis: the ecology of Cairo, Egypt *American Sociological Review*, **34**, 198–212

Abu-Lughod, J. and Foley, M.M. (1960) Consumer strategies, in N. Foote *et al.* (eds) *Housing Choices and Constraints* McGraw-Hill, New York

Ackroyd, P. (2000) *London: The Biography* Chatto and Windus, London

Aglietta, M. (1979) *A Theory of Capitalist Regulation* New Left Books, London

Aitken, S. (1990) Local evaluations of neighbourhood change *Annals, Association of American Geographers*, **80**, 247–267

Allen, J. (1998) Europe of the neighbourhoods: class, citizenship and welfare regimes, in A. Madanipour, G. Cars and J. Allen (eds) *Social Exclusion in European Cities: Processes, Experiences and Responses* Jessica Kingsley, London

Amin, A. and Graham, S. (1997) The ordinary city *Transactions of the Institute of British Geographers*, **22**, 411–429

Amin, A. and Thrift, N. (1992) Neo-Marshallian nodes in global networks *International Journal of Urban and Regional Research*, **116**, 571–587

Amin, A.T. and Thrift, N. (eds) (2002) *Cities: Reimagining the Urban* Polity Press, Cambridge

Anderson, B. (1983) *Imagined Communities: Reflections on the Origins and Spread of Nationalism* Verso, London

Anderson, K.J. (1988) Cultural hegemony and the race definition process in Chinatown, Vancouver: 1890–1980 *Environment and Planning D: Society and Space*, **6**, 127–149

Appadurai, A. (1996) *Modernity at Large: Cultural Dimensions of Globalization* University of Minnesota Press, Minneapolis, MN

Archer, K. (1997) The limits to the Imagineered City: sociospatial polarization in Orlando *Economic Geography*, **73**, 322–336

Atkinson, J. (1985) The changing corporation, in D. Clutterbuck (ed.) *New Patterns of Work*, Gower, Aldershot

Badcock, B. (1989) Smith's rent gap hypothesis: an Australian view *Annals of the Association of American Geographers*, **79**, 125–145

Badcock, B. (1995) Building upon the foundations of gentrification: inner city housing development in Australia in the 1990s *Urban Geography*, **16**, 70–90

Baerwald, T. (1981) The site selection process of suburban residential builders *Urban Geography*, **2**, 351

Banfield, E.C. and Wilson, J.Q. (1963) *City Politics* Harvard University Press, Cambridge, MA

Barnett, C. (1997) Review of Thirdspace: journeys to Los Angeles and other real and imagined places *Transactions of the Institute of British Geographers*, **22**, 529–530

Bassand, M. (1990) *Urbanization: Apportionment of Space and Culture* Graduate School and University Centre, CUNY, New York

Baudrillard, J. (1988) *Selected Writings* Polity Press, Cambridge

Baudrillard, J. (1989) *America* Verso, London

Beauregard, R.A. (1999) Breakdancing on Santa Monica Boulevard *Antipode*, **31**, 396–399

Beck, C. (1992) *Risk Society: Towards a New Modernity* Sage, London

Bell, C.R. and Newby, H. (1976) Community, communion, class and community action, in D.T. Herbert and R.J. Johnston (eds) *Social Areas in Cities* Vol. 2: *Spatial Perspectives on Problems and Policies* Wiley, Chichester

Bell, D. (1973) *The Coming of Post-industrial Society* Heinemann, London

Bell, D. (1991) Insignificant other: lesbian and gay geographies *Area*, **23**, 323–329

Bell, D. and Valentine, G. (eds) (1997) *Consuming Geographies: We Are What We Eat* Routledge, London

Berman, M. (1995) 'Justice/just us': rap and social injustice in America, in A. Merrifield and E. Swyngedouw (eds) *The Urbanization of Injustice* Lawrence and Wishart, London

Berry, B.J.L. and Kasarda, J.D. (1977) *Contemporary Urban Sociology* Macmillan, New York

Bhabha, H.K. (1994) *The Location of Culture* Routledge, London

Binnie, J. (1995) Trading places: consumption, sexuality and the production of queer space, in D. Bell and G. Valentine (eds) *Mapping Desire: Geographies of Sexuality* Routledge, London

Blom, P. (2004) *Encyclopedie: The Triumph of Reason in an Unreasonable Age* Fourth Estate, London

Blomley, N. (1989) Interpretive practices, the state and the locale, in J. Wolch and M. Dear (eds) *The Power of Geography* Unwin Hyman, London

Blomley, N. (1993) Editorial: Making space for law *Urban Geography*, **14**, 6

Blomley, N. and Clark, G. (1990) Law, theory and geography *Urban Geography*, **11**, 433–446

Blowers, A. (1994) *Planning for a Sustainable Environment* Earthscan, London

Blowers, A. (1999) The unsustainable city?, in S. Pile, C. Brook and C. Mooney (eds) *Unruly Cities?: Order, Disorder* Open University Press, Routledge, London

Blum, V. and Nast, H. (1996) Where's the difference? The heterosexualization of alterity in Henri Lefebvre and Jacques Lacan *Environment and Planning D: Society and Space*, **14**, 559–580

Bogardus, E. (1962) Social distance in the city, in E. Burgess (ed.) *The Urban Community* University of Chicago Press, Chicago, IL

Bondi, L. (1991) Gender divisions and gentrification: a critique *Transactions of the Institute of British Geographers*, **16**, 290–298

Bondi, L. (1992) Gender, symbols and urban landscapes *Progress in Human Geography*, **16**, 157–170

Bondi, L. (1998a) Gender, class and urban space: public and private space in contemporary urban landscapes *Urban Geography*, **19**, 160–185.

Bondi, L. (1998b) Sexing the city, in R. Fincher and J.M. Jacobs (eds) *Cities of Difference* The Guilford Press, New York

Booth, C. (1903) *Life and Labour of the People of London* Macmillan, London

Booth, C., Darke, J. and Yeandle, S. (eds) (1996) *Changing Places: Women's Lives in the City* Paul Chapman, London

Borchert, J. (1991) Futures of American cities, in J.F. Hart (ed.) *Our Changing Cities* Johns Hopkins Press, Baltimore, MD, pp. 218–250

Bottoms, A.E. and Wiles, P. (1992) Housing markets and residential crime areas, in D.J. Evans, N.R. Fyfe and D.T. Herbert (eds) *Crime, Policing and the State* Routledge, London

Bourdieu, P. (1984) *Distinction: A Social Critique of the Judgement of Taste* Routledge and Kegan Paul, London

Bourne, L.S. (1981) *The Geography of Housing* Edward Arnold, London

Boyer, C. (1992) Cities for sale: merchandising history at South Street Seaport, in M. Sorkin (ed.) *Variations on a Theme Park* Noonday, New York

Boyer, C. (1995) The great frame up: fantastic appearances in contemporary spatial politics, in H. Liggett and D. Perry (eds) *Spatial Practices* Sage, London

Boyle, M. (1995) Still on the agenda? Neil Smith and the reconciliation of capital and consumer approaches to the explanation of gentrification *Scottish Geographical Magazine*, **111**, 120–123

Bramley, G. and Lancaster, S. (1998) Modelling local and small-area income distributions in Scotland *Environment and Planning C: Government and Policy*, **16**, 681–706.

Breheny, M. (1995) The compact city and transport energy consumption *Transactions, Institute of British Geographers*, **20**, 81–101

Brenner, N.T.N. (ed.) (2002) *Spaces of Neoliberalism. Urban Restructuring in North America and Western Europe* Oxford, Blackwell

Bristow, E. (1997) *Vice and Vigilance* Gill and Macmillan, Dublin

Bryant, C.G.A. and Jary, D. (eds) (1991) *Giddens's Theory of Structuration: A Critical Appreciation* Routledge, London

Burgess, E.W. (1924) The growth of the city: an introduction to a research project *Publications, American Sociological Society*, **18**, 85–97

Burgess, E.W. (ed.) (1926) *The Urban Community* University of Chicago Press, Chicago, IL

Burgess, E.W. (1964) Natural area, in J. Gould and W.L. Kolb (eds) *Dictionary of the Social Sciences* Free Press, New York

Burgess, J. and Jackson, P. (1992) Streetwork – an encounter with place *Journal of Geography in Higher Education*, **16**, 151–157

Butler, R. and Bowlby, S. (1997) Bodies and spaces: an exploration of disabled people's experience of public space *Environment and Planning D: Society and Space*, **15**, 411–433

Butler, T. (1997) *Gentrification and the Middle Classes* Ashgate, Aldershot

Butler, T. and Robson, G. (2001) Social capital, gentrification and neighbourhood change in London: a comparison of three South London neighbourhoods *Urban Studies*, **38**, 2145–2162

Caldwell, S.B., Clarke, G.P. and Keister, L.A. (1998) Modelling regional changes in US household income and wealth: a research agenda *Environment and Planning C: Government and Policy*, **16**, 707–722.

Calhoun, J.B. (1962) Population density and social pathology *Scientific American*, **206**, 139–148

Campbell, J. and Oliver, M. (1996) *Disability Politics: Understanding Our Past, Changing Our Future* Routledge, London

Carlstein, T., Parkes, D. and Thrift, N. (1978) *Human Activity and Time Geography* Edward Arnold, London

Castells, M. (1983) *The City and the Grassroots* Edward Arnold, London

Castells, M. (1989) *The Informational City* Blackwell, Oxford

Castells, M. (1997) *The Power of Identity* Vol. 2 of *The Information Age* Blackwell, Oxford

Castells, M. (1998) *End of Millennium* Vol. 3 of *The Information Age* Blackwell, Oxford

Castells, M. (1996) *The Rise of the Network Society* Vol. 1 Blackwell, Oxford

Castells, M. and Murphy, K. (1982) Cultural identity and urban structure: the spatial formation of San Francisco's gay community, in N. Fainstein and S. Fainstein (eds) *Urban Policy Under Capitalism* Sage, Beverly Hills, CA

Central Housing Advisory Committee (1969) *Council Housing: Purposes, Procedures and Priorities* HMSO, London

Chandler, B. *et al.* (1993) Liverpool, in A. Montanari, G. Curdes and L. Forsyth (eds) *Urban Landscape Dynamics* Avebury, Aldershot

Chandler, R. (1973a) Introduction, in *Pearls are a Nuisance* Penguin, Harmondsworth, pp. 7–10

Chandler, R. (1973b) The simple art of murder, in *Pearls are a Nuisance*, Penguin, Harmondsworth, pp. 181–199

Chatterton, P. and Hollands, R. (2003) *Urban Nightscapes: Youth Cultures, Pleasure Spaces and Power* Routledge, London

Chauncey, G. (1995) *Gay New York; The Making of a Gay Male World, 1890–1940* Flamingo, Glasgow

Chouinard, V. and Grant, A. (1996) On being not even anywhere near 'The Project': ways of putting ourselves in the picture, in N. Duncan (ed.) *Bodyspace* Routledge, London

Christensen, J. (1982) The politics of redevelopment: Covent Garden, in D.T. Herbert and R.J. Johnston (eds) *Geography and the Urban Environment* Vol. 4 John Wiley, Chichester

Christopherson, S. (1994) The fortress city: privatized spaces, consumer citizenship, in A. Amin (ed.) *Post-Fordism: A Reader* Blackwell, Oxford

Christopherson, S. and Storper, M. (1986) The city as studio: the world as back lot: the impact of vertical disintegration on the location of the motion picture industry *Environment and Planning D: Society and Space*, **4**, 305–320

Clark, G. (1990) The geography of law, in R. Peet and N. Thrift (eds) *New Models in Geography* Unwin Hyman, London

Clark, G. (1992) 'Real' regulation: the administrative state *Environment and Planning A*, **24**, 615–627

Clark, W.A.V. (1991) Problems of integrating an urban society in J. Hart (ed.) *Our Changing Cities* Johns Hopkins Press, Baltimore, MD

Clark, W.A.V. and Onaka, J.L. (1983) Life-cycle and housing adjustments as explanations of residential mobility *Urban Studies*, **20**, 47–58

Clarke, D.B. (2003). *The Consumer Society and the Postmodern City* Routledge, London

Clarke, D.B. and Bradford, M.G. (1998) Public and private consumption and the city *Urban Studies*, **35**, 865–888

Coetzee, J.M. (2003) The misfits, in J. Shepard (ed.) *Writers at the Movies* Marion Boyars, London

Cohen, P. (1997) Out of the melting-pot into the fire next time: imagining the East End as city, body, text, in S. Westwood and J. Williams (eds) *Imagining Cities: Scripts, Signs, Memory* Routledge, London

Collinson, P. (1963) *The Cutteslowe Walls* Faber and Faber, London

Couch, C.F. and Percy, S. (eds) (2003) *Urban Regeneration in Europe* Blackwell, Oxford

Coyle, D. (1997) *The Weightless World: Strategies for Managing the Digital Economy* Capstone, Oxford

Crang, M. (1998) *Cultural Geography* Routledge, London

Crang, M. (2002) Between places: producing hubs, flows and networks *Environment and Planning A*, **34**, 569–574

Crawford, M. (1992) The world in a shopping mall, in M. Sorkin (ed.) *Variations on a Theme Park* Noonday Press, New York

Curdes, G. (1993) Spatial organization of towns at the level of the smallest urban unit: plots and buildings, in A. Montanari, G. Curdes and L. Forsyth (eds) *Urban Landscape Dynamics: A Multi-level Innovation Process* Avebury, Aldershot, pp. 281–294

Curry, J. and Kenney, M. (1999) The paradigmatic city: post industrial illusion and the Los Angeles School *Antipode*, **31**, 1–28

Dahl, R. (1961) *Who Governs?* Yale University Press, New Haven, CT

Dahmann, D. (1985) Assessments of neighbourhood quality in metropolitan America *Urban Affairs Quarterly*, **20**, 511–536

D'Arcy, E. and Keogh, G. (1997) Towards a property market paradigm of urban change *Environment & Planning A*, **29**(4), 685–706

Darden, J. (1980) Lending practices and policies affecting the American Metropolitan System, in S.D. Brunn and J.O. Wheeler (eds) *The American Metropolitan System: Present and Future* Winston, New York

Davies, W.K.D. (1984) *Factorial Ecology* Gower, Aldershot

Davis, M. (1990) *City of Quartz* Verso, London

Davis, M. (1992) Fortress Los Angeles: the militarization of urban space, in M. Sorkin (ed.) *Variations on a Theme Park* Noonday Press, New York

Davis, M. (1998) *Ecology of Fear, Los Angeles and the Imagination of Disaster* Metropolitan Books, New York

Davis, M. (2000) *Magical Urbanism: Latinos Reinvent the US City* Verso, London

Dear, M.J. (2000) *The Postmodern Urban Condition* Blackwell, Oxford

Dear, M. and Flusty, S. (1998) Postmodern urbanism *Annals of the Association of American Geographers*, **88**, 50–72

Dear, M. and Taylor, S.M. (1982) *Not On Our Street: Community Attitudes Towards Mental Health Care* Pion, London

Dear, M. and Wolch, J. (1987) *Landscapes of Despair* Princeton University Press, Princeton, NJ

Dear, M. and Wolch, J. (1989) How territory shapes social life, in J. Wolch and M. Dear (eds) *The Power of Geography: How Territory Shapes Social Life* Unwin Hyman, Boston, MA

Dear, M. and Wolch, J. (1991) *Landscapes of Despair* Princeton University Press, Princeton, NJ

Dear, M., Wilton, Lord, R., Gaber, S. and Takahashi, L. (1997) Seeing people differently: the sociospatial construction of disability *Environment and Planning D: Society and Space*, **15**, 455–480

Deas, I., Robson, B., Wong, C. and Bradford, M. (2003) Measuring neighbourhood deprivation: a critique of the index of multiple deprivation *Environment and Planning C*, **21**, 883–903.

de Certeau, M. (1985) *The Practice of Everyday Life* University of California Press, Berkeley, CA

Delaney, D. (1993) Geographies of judgement: the doctrine of changed conditions and the geopolitics of race *Annals, Association of American Geographers*, **83**, 48–65

Deutsche, R. (1991) Boy's town *Environment and Planning D: Society and Space*, **9**, 5–30

Diamond, D. (1991) Managing urban change: the case of the British inner city, in R. Bennett and R. Estall (eds) *Global Change and Challenge* Routledge, London

Domosh, M. (1996) *Invented Cities: The Creation of Landscape in Nineteenth Century New York and Boston* Yale University Press, New Haven, CT

Douglas, J. (1977) Existential sociology, in J. Douglas and J.M. Johnson (eds) *Existential Sociology* Cambridge University Press, Cambridge

Dowling, R. (1993) Femininity, place and commodities: a retail case study *Antipode*, **25**, 295–318

Dowling, R. (1998) Suburban stories' gendered lives: thinking through difference, in R. Fincher and J.M. Jacobs (eds) *Cities of Difference* Guilford, New York

Downing, P.M. and Gladstone, L. (1989) *Segregation and Discrimination in Housing: A Review of Selected Studies and Legislation* Library of Congress, Congressional Research Report 89-317, Washington, DC

Drake, S. and Cayton, H.R. (1962) *Black Metropolis* Harper and Row, New York

Drier, P. and Atlas, J. (1995) U.S. housing problems, politics, and policies *Housing Studies*, **10**(2), 245–270

Duncan, J.S. and Duncan, N. (2004) *Landscapes of Privilege: The Politics of the Aesthetic in the American Suburb* Routledge, New York

Duncan, N. (1996) Renegotiating gender and sexuality in public and private spaces, in N. Duncan (ed.) *Bodyspace* Routledge, London

Duncan, O.D. and Duncan, B. (1955) Occupational stratification and residential distribution *American Journal of Sociology*, **50**, 493–503

Dunford, M. (1990) Theories of regulation *Environment and Planning D: Society and Space*, **8**, 297–321

Duvall, E.M. (1971) *Family Development* Philadelphia, J.B. Lippincott

Edel, M., Sclar, E. and Luria, D. (1984) *Shaky Palaces: Homeownership and Social Mobility in Boston's Suburbanization* Columbia University Press, New York

Edwards, J. (1995) Social policy and the city: a review of recent policy developments *Urban Studies*, **32**, 695–712

Egerton, J. (1990) Out but not down: lesbians' experience of housing *Feminist Review*, **35**, 75–88

Engels, F. (1844) *The Condition of the Working Class in England* Panther, London (1969 reprint)

Entrikin, N. (1991) *The Betweeness of Place* Johns Hopkins University Press, Baltimore, MD

Fainstein, S. (1997) The egalitarian city: the restructuring of Amsterdam *International Planning Studies*, **2**(3), 295–314

Falzon, C. (1998) *Foucault and Social Dialogue* Routledge, London

Faris, R.E.L. and Dunham, H.W. (1939) *Mental Disorders in Urban Areas* University of Chicago Press, Chicago, IL

Feagin, J.R. (1997) *The New Urban Paradigm: Critical Perspectives on the City* Rowman and Littlefield, Lanham, MD

Featherstone, M. and Lash, S. (eds) (1999) *Spaces of Culture: City, Nation, World* Sage, London

Feins, J.D. and Bratt, R.G. (1983) Barred in Boston: racial discrimination in housing *Journal of American Planning Association*, **49**, 344–355

Filion, P. (1996) Metropolitan planning objectives and implementation constraints: planning in a post-Fordist and postmodern age *Environment & Planning A*, **28**(9), 1637–1660

Firey, W. (1945) Sentiment and symbolism as ecological variables *American Sociological Review*, **10**, 140–148

Fischer, C.S. (1976) *The Urban Experience* Harcourt Brace Jovanovich, New York

Fischer, C.S. (1981) The public and private worlds of city life *American Sociological Review*, **46**, 306–316

Fitzgerald, J. (1991) Class as community: the dynamics of social change *Environment and Planning D: Society and Space*, **9**, 117–128

Florida, R. (1995) Towards the learning region *Futures*, **27**, 527–536

Florida, R. (2002) *The Rise of the Creative Class* Basic Books, New York

Foggin, P. and Polese, M. (1977) *The Social Geography of Montreal in 1971* Research Paper No. 88, Centre for Urban and Community Studies, University of Toronto

Ford, L. (1994) *Cities and Buildings: Scryscrapers, Skid Rows and Suburbs* Johns Hopkins University Press, Baltimore, MD

Ford, L.R. (2003) *America's New Downtowns: Revitalization or Reinvention?* Johns Hopkins University Press, Baltimore, MD

Foucault, M. (1984) *The History of Sexuality*: Vol. 1 *An Introduction* Penguin, Harmondsworth

French, S. (1993) Disability, impairment or something in between? in J. Swain, V.K. Finkelstein, S. French and M. Oliver (eds) (1993) *Disabling Barriers – Enabling Environments* Sage, London

Friedland, R. (1981) Central city fiscal stress: the public costs of private growth *International Journal of Urban and Regional Research*, **5**, 356–375

Friedmann, J. and Woolff, K. (1982) World city formation: an agenda for research and action *International Journal of Urban and Regional Research*, **6**, 309–344

Friedrichs, J. and Alpheis, H. (1991) Housing segregation of immigrants in West Germany, in E.D. Huttman (ed.) *Urban Housing Segregation of Minorities in Western Europe and the United States* Duke University Press, London

Galster, G.C. and Hester, G.W. (1981) Residential satisfaction; compositional and contextual correlates *Environment and Behaviour* **13**, 735–758

Gans, H. (1962) *The Urban Villagers* Free Press, New York

Gans, H. (1967) *The Levittowners* Allen Lane, London

Garcia, P. (1985) Immigration issues in urban ecology, in L. Maldonado and J. Moore (eds) *Urban Ethnicity in the United States* Sage, Beverly Hills, CA

Garreau, J. (1992) *Edge City: Life on the New Frontier* Doubleday, New York

Giddens, A. (1979) *Central Problems in Social Theory* Macmillan, London

Giddens, A. (1981, 1985) *A Contemporary Critique of Historical Materialism* Polity Press, Cambridge

Giddens, A. (1984) *The Constitution of Society: Outline of the Theory of Structuration* Polity Press, Cambridge

Giddens, A. (1989) *Sociology* Polity Press, Cambridge

Giddens, A. (1991) Structuration theory: past, present and future, in G.A. Bryant and D. Jary (eds) *Giddens' Theory of Structuration: A Critical Appreciation* Routledge, London

Gleeson, B. (1998) Justice and the Disabling City, in R. Fincher and J.M. Jacobs (eds), *Cities of Difference* The Guildford Press, New York

Gober, P. McHugh, K.E. and Reid, N. (1991) Phoenix in flux: household instability, residential mobility and neighbourhood change *Annals, Association American Geographers*, **81**, 80–88

Golledge, R. (1993) Geography and the disabled: a survey with special reference to vision impaired and blind populations *Transactions, Institute of British Geographers*, **18**, 63–85

Goodman, D. and Chant, C.E. (eds) (1999) *European Cities and Technology: Industrial to Post-Industrial City* Routledge, London

Goodwin, M. and Painter, J. (1996) Local governance, the crises of Fordism and the changing geographies of regulation *Transactions of the Institute of British Geographers*, **21**, 635–648

Goodwin, M.S., Duncan, S. and Halford, S. (1993) Regulation theory, the local state and the transition of urban politics *Environment and Planning D: Society and Space*, **11**, 67–88

Gordon, P. and Richardson, H.W. (1999) Review essay: Los Angeles, City of Angels? No city of angles *Urban Studies*, **36**, 575–591

Goss, J. (1993) The 'Magic of the Mall': an analysis of form, function and meaning in the contemporary retail environment *Annals, Association of American Geographers*, **83**, 18–47

Graham, S. and Marvin, S. (2001) *Splintering Urbanism* Routledge, London

Gramsci, A. (1973) *Selections From the Prison Notebooks* (translated by O. Hoare and G. Nowell-Smith) Lawrence and Wishart, London

Green, A.E. (1998) The geography of earnings and incomes in the 1990s *Environment and Planning C: Government and Policy*, **29**, 129–147.

Gregory, D. (1989) Areal differentiation and post-modern human geography, in D. Gregory and R. Walford (eds) *Horizons in Human Geography* Barnes and Noble, Totowa, NJ

Grosz, E. (1992) Bodies-cities, in B. Colomina (ed.) *Sexuality and Space* Princeton Architectural Press, Princeton, NJ

Guy, S. and Henneberry, J. (eds) (2002) *Development and Developers. Perspectives on Property* Oxford, Blackwell

Habermas, J. (1989) *The Structural Transformation of the Public Sphere* MIT Press, Cambridge, MA

Haggett, P. (1994) Geography, in R. Johnston, D. Gregory and D. Smith (eds) *The Dictionary of Human Geography* Blackwell, Oxford

Hahn, H. (1986) Disability and the urban environment: a perspective on Los Angeles *Environment and Planning D: Society and Space*, **4**, 273–288

Hall, S. and Jefferson, T. (eds) (1976) *Resistance through Rituals: Youth Subcultures in Postwar Britain* Hutchinson, London

Hall, T. (1998) *Urban Geography* Routledge, London

Hall, T. and Hubbard, P. (1996) The entrepreneurial city: new urban politics, new urban geographies *Progress in Human Geography*, **20**, 153–174

Hamnett, C. (1991) The blind men and the elephant: the explanation of gentrification *Transactions of the Institute of British Geographers*, **17**, 173–189

Hamnett, C. (1992) 'Gentrifiers or Learnings? A response to Neil Smith', *Transactions of the Institute of British Geographers*, **17**, 116–119

Hamnett, C. (1998) *The New Urban Frontier* (book review essay) *Transactions of the Institute of British Geographers*, **23**, 412–416

Hamnett, C. (1999) *Winners and Losers: Home Ownership in Modern Britain* UCL Press, London

Hamnett, C. (2003) *Unequal City: London in the Global Arena* Routledge, London

Hanson, S. and Pratt, G. (1995) *Gender Work and Space* Routledge, London

Haraway, D. (1991) *Simians, Cyborgs and Women: The Reinvention of Nature* Routledge, London

Harvey, D. (1969) *Explanation in Geography* Arnold, London

Harvey, D. (1972) *Society, the City and the Space Economy of Urbanism* Resources Paper No. 18, Commission on College Geography, Association of American Geographers, Washington, DC

Harvey, D. (1973) *Social Justice and the City* Arnold, London

Harvey, D. (1975) Class structure in a capitalist society and the theory of residential differentiation, in R. Peel, M. Chisholm and P. Haggett (eds) *Processes in Physical and Human Geography* Heinemann, London

Harvey, D. (1978) Labour, capital and the class struggle around the built environment, in K.R. Cox (ed.) *Urbanization and Conflict in Market Societies* Methuen, London

Harvey, D. (1982) *The Limits to Capital* Basil Blackwell, Oxford

Harvey, D. (1989a) From managerialism to entrepreneurialism: the transformation of urban governance in late capitalism *Geografiska Annaler*, 71B, 3–17

Harvey, D. (1989b) *The Condition of Postmodernity* Blackwell, Oxford

Harvey, D. (1992) Social justice, postmodernism and the city *International Journal of Urban and Regional Research*, 16, 588–601

Harvey, D. (1993). From space to place and back again; reflections on the condition of postmodernity, in J. Bird *et al.* (eds) *Mapping the Futures: Local Cultures, Global Change* Routledge, London

Harvey, D. (1996) *Justice, Nature and the Geography of Difference* Blackwell, Oxford

Harvey, D. (2003) *The Capital of Modernity. Paris in the Second Empire.* Routledge, New York

Hayden, D. (2003) *Building Suburbia: Green Fields and Urban Growth, 1820–2000* Pantheon, New York

Heidegger, M. (1971) *Poetry, Language, Thought* Harper and Row, New York

Herbert, D.T. (1977) Crime delinquency and the urban environment *Progress in Human Geography*, 1, 208–239

Hesse, B. (1997) White governmentality: urbanism, nationalism, racism, in S. Westwood and J. Williams (eds) *Imagining Cities: Scripts, Signs, Memory* Routledge, London

Hiebert, D.L. and Ley, D. (2003) Assimilation, cultural pluralism, and social exclusion among ethnocultural groups in Vancouver, *Urban Geography*, 24(1), 16–44

Hindle, P. (1994) Gay communities and gay space in the city, in S. Whittle (ed.) *The Margins of the City: Gay Men's Urban Lives* Arena, Aldershot

Holston, J. (2001) *Urban Citizenship and Globalization. Global City-regions. Trends, Theory, Policy* Oxford University Press, Oxford, pp. 325–348

Hooker, E. (1965) Male homosexuals and their 'worlds', in J. Mermor (ed.) *Sexual Inversion* Basic Books, New York

Hoyt, H. (1939) *The Structure and Growth of Residential Neighbourhoods in American Cities* Federal Housing Administration, Washington, DC

Howarth, C., Kenway, P., Palmer, G. and Street, C. (1998) *Monitoring Poverty and Social Exclusion: Labour's Inheritance* New Policy Institute and Joseph Rowntree Foundation, London

Hubbard, P. (2004) Manuel Castells, in P. Hubbard, K. Kitchin and G. Valentine (eds) *Key Thinkers on Space and Place* Sage, London

Hubbard, P. (2005) Accommodating Otherness: anti-asylum centre protest and the maintenance of white priviledge *Transactions of the Institute of British Geographers*, 30, 52–66

Hudson, R. and Williams, A. (1999) *Divided Europe* Sage, London

Hughes, A. and Pinch, S. (1999) Unpublished mimeo, Department of Geography, University of Southampton

Hunter, F. (1953) *Community Power Structure: A Study of Decision Makers* University of North Carolina Press, Chapel Hill, NC

Huttman, E.D. (1991) Subsidized housing segregation in Western Europe: stigma and segregation, in E. Huttman, W. Blauw and J. Saltman (eds) *Urban Housing Segregation of Minorities in Western Europe and the United States* Duke University Press, Durham, NC

Imrie, R. (1996) *Disability and the City: International Perspectives* Paul Chapman, London

Imrie, R. and Thomas, H. (1999) *British Urban Policy and the Urban Development Corporations* (2nd edn) Paul Chapman, London

Ingersoll, R. (1992) The disappearing suburb *Design Book Review*, 26

Jablonsky, T. (1993) *Pride in the Jungle* Johns Hopkins Press, Baltimore, MD

Jackson, P. (1989) *Maps of Meaning* Unwin Hyman, London

Jackson, P. (1991a) Mapping meanings: a cultural critique of locality studies *Environment and Planning A*, 23, 219

Jackson, P. (1991b) The cultural politics of masculinity: towards a social geography *Transactions of the Institute of British Geographers*, 16, 199–213

Jackson, P. (1999) Postmodern urbanism and the ethnographic void *Antipode*, 31, 400–402

Jackson, P. and Smith, S. (1984) *Exploring Social Geography* Allen and Unwin, London

Jackson, P. and Thrift, N. (1995) Geographies of consumption, in D. Miller (ed.) *Acknowledging Consumption* Routledge, London

Jacobs, J. (1961) *The Life and Death of Great American Cities* Vintage, New York

Jacobs, J.M. (1996) *Edge of Empire: Postcolonialism and the City* Routledge, London

James, F.J., McCummings, B.I. and Tynan, E.A. (1984) *Minorities in the Sunbelt* NJ Centre for Urban Policy Research, Rutgers University, New Brunswick, NJ

James, S. (1990) Is there a 'place' for children in geography? *Area*, 22, 278–283

Jameson, F. (1984) *Post-modernism or the Cultural Logic of Late Capitalism* Duke University Press, Durham, NC

Janson, C.-G. (1971) A preliminary report on Swedish urban spatial structure *Economic Geography*, 47, 249–257

Jeffery-Poulter, S. (1991) *Peer, Queers and Commons: The Struggle for Gay Law Reform from 1950 to the Present* Routledge, London

Jessop, B. (1997) The entrepreneurial city: re-imagining localities, redesigning economic governance or restructuring capital, in N. Jewson and S. MacGregor (eds) *Transforming Cities: Contested Governance and New Spatial Divisions* Routledge, London

Jessop, B. (2002) Liberalism, neoliberalism, and urban governance: a state-theoretic perspective *Antipode*, 34, 452–472

Johnston, L. and Valentine, G. (1995) Wherever I lay my girlfriend, that's my home: the performance and surveillance of lesbian identities in domestic environments, in D. Bell and G. Valentine (eds) *Mapping Desire: Geographies of Sexuality* Routledge, London

Johnston, R.J. (1971) *Urban Residential Patterns: An Introductory Review* Bell and Sons Ltd, London

Johnston, R.J. (1978) *Political, Electoral and Spatial Systems* Oxford University Press, London

Johnston, R.J. and Sidaway, J.D. (2004) *Geography and Geographers: Anglo-American Human Geography Since 1945* Arnold, London

Johnston, R.J., Pattie, C.J., Dorling, D.F.L. and Rossiter, D.J. (2001) *From Votes to Seats: The Operation of the UK Electoral System Since 1945* Manchester University Press, Manchester

Johnston, R.J. (1984) *City and Society* (2nd edn) Hutchinson, London

Johnston, R., Poulsen, M. and Forrest, J. (2002) From modern to post-modern? Contemporary ethnic segregation in four US metropolitan areas *Cities*, 19, 161–172

Johnston, R., Poulsen, M. and Forrest, J. (2003) And did the walls come tumbling down? Ethnic segregation in four US metropolitan areas, 1980–2000 *Urban Geography*, 24, 560–581

Johnston, R., Poulsen, M. and Forrest, J. (2004) The comparative study of ethnic residential segregation in the USA, 1980–2000 *Tijdschrift voor Economische en Sociale Geografie*, 95, 550–569

Jonas, E.G. and Wilson, D. (eds) (1999) *The Urban Growth Machine* State University of New York Press, Albany, NY

Jones, C. and Porter, R. (1998) *Reassessing Foucault: Power, Medicine and the Body* Routledge, London

Judge, D., Stoker, G. and Wolman, H. (eds) (1995) *Theories of Urban Politics* Sage, London

Kearns, A. (1995) Active citizenship and local governance: political and geographical dimensions *Political Geography*, 14, 155–175

Keith, M. and Cross, M. (1992) Racism and the postmodern city, in M. Cross and M. Keith (eds) *Racism, the City and the State* Routledge, London

Killworth, P.D. *et al.* (1990) Estimating the size of personal networks *Social Networks*, 12, 289–312

King, A.D. (ed.) (1997) *Culture, Globalization and the World System: Contemporary conditions for the representation of identity* University of Minnesota Press, Minneapolis, MN

Kirby, S. and Hay, I. (1997) (Hetero)sexing space: gay men and 'straight' space in Adelaide, South Australia *Professional Geographer*, 49, 295–305

Knopp, L. (1987) Social theory, social movements and public policy: recent accomplishments of the gay and lesbian movements in Minneapolis, Minnesota *International Journal of Urban and Regional Research*, 11, 243–261

Knopp, L. (1990) Exploiting the rent gap: the theoretical significance of using illegal appraisal schemes to encourage gentrification in New Orleans *Urban Geography*, 11, 48–64

Knopp, L. (1992) Sexuality and the spatial dynamics of capitalism *Environment and Planning D: Society and Space*, 10, 651–669

Knox, P.L. (1984) Styles, symbolism and settings: the built environment and the imperatives of urbanised capitalism *Architecture et comportement*, 2, 107–122

Knox, P.L. (1986) Disadvantaged households and service provision in the inner city, in G. Heinritz and E. Lichtenberger (eds) *The Take-off of Suburbia and the Crisis of the Central City* Steiner, Stuttgart

Knox, P.L. (ed.) (1993) *The Restless Urban Landscape* Prentice Hall, Englewood Cliffs, NJ

Knox, P.L. (1994) *Urbanization* Prentice Hall, Englewood Cliffs, NJ

Knox, P.L., Agnew, J. and McCarthy, L. (2003) *The Geography of the World Economy* (4th edition) Edward Arnold, London

Kodras, J. (1997) The changing map of American poverty in an era of economic restructuring and political realignment *Economic Geography*, 72, 67–93

Krutnik, F. (1997) Something more than the night; tales of the noir city, in D.B. Clarke (ed.) *The Cinematic City* Routledge, London

Kurpick, S. and Weck, S. (1998) Policies against social exclusion at the neighbourhood level in Germany: the case of Northrhine-Westphalia, in A. Madanipour, G. Cars and J. Allen (eds) *Social Exclusion in European Cities: Processes, Experiences and Responses* Jessica Kingsley, London

Lake, R.W. (1999) Postmodern urbanism *Antipode*, 31, 393–395

Lash, S. and Urry, J. (1994) *Economies of Signs and Space* Sage, London

Latham, A. (2004) Ed Soja, in P. Hubbard, R. Kitchin and G. Valentine (eds) *Key Thinkers on Space and Place* Sage, London

Lauria, M. (1996) *Reconstructing Urban Regime Theory: Regulating Urban Politics in a Global Economy* Sage, Beverly Hills, CA

Lees, L. (1994) Rethinking gentrification: beyond the positions of economics or culture *Progress in Human Geography*, 18, 137–150

Lees, L. (1997) Ageographia, heterotopia, and Vancouver's new public library *Environment and Planning D: Society and Space*, 15, 321–337

Lees, L. (2000) A re-appraisal of gentrification: towards a geography of gentrification *Progress in Human Geography*, 24, 389–408

Lefebvre, H. (1991) *The Production of Space* Blackwell, Oxford

Leslie, D.R. and Reimer, S. (2003) Gender, modern design, and home consumption *Environment & Planning D: Society and Space*, 21(3), 293–317

Lewis, M. (1994) A sociological pubcrawl around gay Newcastle, in S. Whittle (ed.) *The Margins of the City: Gay Men's Urban Lives* Arena, Aldershot

Lewis, P. (1983) The galactic metropolis, in R.H. Platt and G. Mackinder (eds) *Beyond the Urban Fringe* University of Minnesota Press, Minneapolis, MN

Ley, D. (1974) *The Black Inner City as Frontier Post* Association of American Geographers, Washington, DC

Ley, D. (1983) *A Social Geography of the City* Harper and Row, New York

Ley, D. (1996) *The New Middle Class and the Remaking of the Central City* Oxford University Press, Oxford

Leyshon, A. and Thrift, N. (1997) *Money Space: Geographies of Monetary Transformation* Routledge, London

Liebow, E. (1967) *Tally's Corner* Little Brown, Boston, MA

Lineberry, R.L. (1977) *Equality and Urban Policy* Sage, Beverly Hills, CA

Lipietz, A. (1986) New tendencies in the international division of labor: regimes of accumulation and modes of regulation in A. Scott and M. Storper (eds) *Production, Work, Territory: The Geographical Anatomy of Industrial Capitalism* Routledge, London

Lister, R. (1998) From equality to social inclusion: New Labour and the welfare state *Critical Social Policy*, **18**, 215–226

Locke, K. (2001) *Grounded Theory in Management Research* Sage, London

Lofland, L. (1973) *A World of Strangers* Basic Books, New York

Lomnitz, L. and Diaz, R. (1992) Cultural grammar and bureaucratic rationalization in Latin American cities, in R. Morse and J. Hardoy (eds) *Rethinking the Latin American City* Woodrow Wilson Press Center, Washington, DC

Longhurst, R. (1998) (Re)presenting shopping centres and bodies: questions of pregnancy, in R. Ainley (ed.) *New Frontiers of Space, Bodies and Gender* Routledge, London

Low, S. (2003) *Behind the Gates. Life, Security and the Pursuit of Happiness in Fortress America* New York, Routledge

Luke, T. (1994) Placing power/siting space: the local and the global in the New World Order *Environment and Planning D: Society and Space*, **12**, 613–628

Luke, T. (1997) *Ecocritique: Contesting the Politics of Nature, Economics and Culture* University of Minnesota Press, Minneapolis, MN

Lynch, K. (1960) *The Image of the City* MIT Press, Cambridge, MA.

Lynch, K. (1984) Reconsidering 'The Image of the City', in R.M. Hoister and L. Rodwin (eds) *Cities of the Mind* Plenum, New York

Lynd, R.S. and Lynd, H.M. (1956) *Middletown* Harcourt Brace Jovanovich, New York

Macey, D. (1993) *The Lives of Michel Foucault* Vintage, London

Mackenzie, S. (1988) Building women, building cities: toward gender sensitive theory in the environmental disciplines, in C. Andrew and B. Milroy (eds) *Life Spaces* University of British Columbia Press, Vancouver

MacLaran, A. (1993) *Dublin* Belhaven, London

MacLaran, A. (ed.) (2003) *Making Space: Property Development and Urban Planning* Edward Arnold, London

Madanipour, A., Cars, G. and Allen J. (eds) (1998) *Social Exclusion in European Cities: Processes, Experiences and Responses* Jessica Kingsley, London

Maddock, S. and Parkin, D. (1994) Gender cultures: how they affect men and women at work, in M.V. Davidson and R.V. Burke (eds) *Women in Management: Current Research Issues* Paul Chapman Publishing, London

Madsen, P.P. and Plunz, R. (eds) (2001) *The Urban Lifeworld. Formation, Perception, Representation* Spon Press, London

Marcuse, P. (1997) The ghetto of exclusion and the fortified enclave: new patterns in the United States *American Behavioural Scientist*, **41**, 311–326

Marston, S. (1990) Who are the people?: gender, citizenship, and the making of the American nation *Environment and Planning D: Society and Space*, **8**, 449–458

Maslow, A.H. (1970) *Motivation and Personality* Harper and Row, New York

Massey, D. (1992) Politics and space/time *New Left Review*, **196**, 65–84

Massey, D. and Catalano, A. (1978) *Capital and Land* Arnold, London

Mayhew, H. (1862) *The Criminal Prisons of London* Griffin, Bohn and Co., London

Mazierska, E. and Rascaroli, I. (eds) (2003) *From Moscow to Madrid: Postmodern Cities, European Cinema* I.B. Tauris, London

McAuslan, P. (2003) *Bringing the Law back in. Essays in Land, Law, and Development* Ashgate, Burlington, VT

McCullen, S.C. and Smith, W.J. (2004) *The Geography of Poverty in the Atlanta Region*, Andrew Young School of Policy Studies, Fiscal Research Centre, Atlanta

McDowell, L. (1995) Body work: heterosexual gender performances in city workplaces, in D. Bell and G. Valentine (eds) *Mapping Desire: Geographies of Sexuality* Routledge, London

McDowell, L. (1996) Spatializing feminism: geographic perspectives, in N. Duncan (ed.) *Bodyspace: Destabilizing Geographies and Gender and Sexuality* Routledge, London

McNeill, D. (1999) Globalization and the European city *Cities*, **16**, 143–147

McNeill, D. (2004a) Barcelona as imagined community; Pasqual Maragall's spaces of engagement *Transactions of the Institute of British Geographers*, **26**, 340–352

McNeill, D. (2004b) Mike Davis, in P. Hubbard, R. Kitchin and G. Valentine (eds) *Key Thinkers on Space and Place* Sage, London

Merrifield, A. (1993) Space and place: a Lefebvrian reconciliation *Transactions of the Institute of British Geographers*, **18**, 516–531

Merrifield, A. (1999) The extraordinary adventures of Edward Soja: inside the 'trialetics of spatiality' *Annals of the Association of American Geographers*, **89**, 345–348

Merrifield, A. (2002) *MetroMarxism: A Marxist Tale of the City* Routledge, New York

Miller, A. (1987) *The Drama of Being a Child* Virago Press, London

Miller, D., Jackson, P., Thrift, N., Holbrook, B. and Rowlands, M. (1998) *Shopping, Place and Identity* Routledge, London

Mitchell, D. (2004) Peter Jackson, in P. Hubbard, R. Kitchin and G. Valentine (eds) *Key Thinkers on Space and Place* Sage, London

Moos, A. (1989) The grassroots in action: gays and seniors capture West Hollywood, California, in J. Wolch and M. Dear (eds) *The Power of Geography* Unwin Hyman, London

Moos, A. and Dear, M. (1986) Structuration theory in urban analysis: theoretical exegesis *Environment and Planning D: Society and Space*, **18**, 231–252

Morris, S. (2002) Testing the boundaries *Guardian*, 10 August, p. 8

Moudon, A.V. (1992) The evolution of twentieth-century residential forms: an American case study, in J.W.R. Whitehand and P.J. Larkham (eds) *Urban Landscapes: International Perspectives* Routledge, London, pp. 170–206

Muller, P.O. (1981) *Contemporary Suburban America* Prentice Hall, Englewood Cliffs, NJ

Mumford, L. (1940) *The Culture of Cities* Secker and Warburg, London

Murdie, R.A. (1969) *Factorial Ecology of Metropolitan Toronto, 1951–1961* Research paper No. 116, Department of Geography, University of Chicago

Murphy, L. (1995) Mortgage finance and housing provisions in Ireland 1970–1990 *Urban Studies*, **32**, 135–184

Musterd, S. (1991) Neighbourhood change in Amsterdam *Tijdschrift voor Economische en Sociale Geographie*, **82**, 30–39

Myslik, W.D. (1996) Renegotiating the social/sexual identities of places, in N. Duncan (ed.) *Bodyspace* Routledge, London

Negus, K. (1992) *Producing Pop: Culture and Conflict in the Popular Music Industry* Edward Arnold, London

Negus, K. (1999) *Music Genres and Corporate Cultures* Routledge, London

Newman, O. (1972) *Defensible Space* Macmillan, New York

Newson, J. and Newson, E. (1965) *Patterns of Infant Care in an Urban Community* Penguin, Harmondsworth

Newton, K. (1978) Conflict avoidance and conflict suppression in K. Cox (ed.) *Urbanization and Conflict in Market Societies* Methuen, London

Nijman, J. (1999) Cultural globalization and the identity of place: the reconstruction of Amsterdam *Ecumene*, **6**, 146–164

Nijman, J. (2000) The paradigmatic city *Annals of the Association of American Geographers*, **90**, 135–145

O'Connor, A., Tilly, C. and Bobo, L.D. (eds) (2001) *Urban Inequality. Evidence from Four Cities* Russell Sage Foundation, New York

Ohmae, K. (1995) *The End of the National State: The Rise of Regional Economies* Free Press, New York

Oliver, M. (1998) Theories of disability in health practice and research *British Medical Journal*, **7170**, 1446–1449

Oliver, M.L., Johnson, J.H. and Farrell, W.C. Jr (1993) Anatomy of a rebellion: a political-economic analysis, in R. Gooding-Williams (ed.) *Reading Rodney King: Reading Urban Uprising* Routledge, New York

O'Loughlin, J. (1976) The identification and evaluation of racial gerrymandering *Annals, Association of American Geographers*, **72**, 165–184

O'Loughlin, J. and Freiedrichs, J. (eds) (1996) *Social Polarization in Post-industrial Metropolises* Walter de Gruyter, New York

Omni, M. and Winant, H. (1993) The Los Angeles 'Race Riot' and contemporary US politics, in R. Gooding-Williams (ed.) *Reading Rodney King: Reading Urban Uprising* Routledge, New York

Oncu, A. and Weyland, P. (1997) *Space, Culture and Power: New Identities in Globalizing Cities* Zed Books, Atlantic Heights, NJ

Orfield, M. (2002) *American Metro Politics. The New Suburban Reality* The Brookings Institutions Press, Washington, DC

O'Riordian, T. (1981) *Environmentalism* Pion, London

Orleans, P. (1973) Differential cognition of urban residents: effects of scale on social mapping, in R. Downs and D. Stea (eds) *Image and Environment* Aldine, Chicago, IL

Pacione, M. (1982) Evaluating the quality of the residential environment in a deprived council estate *Geoforum*, **13**, 45–55

Pacione, M. (1997) The geography of educational disadvantage in Glasgow *Applied Geography*, **17**(3), 169–192

Pacione, M. (2004) Environments of disadvantage: geographies of persistent poverty in Glasgow *Scottish Geographical Magazine*, **120**, 117–132

Pagano, M.A. and Bowman, A.O. (1995) *Cityscapes and Capital: The Politics of Urban Redevelopment* Johns Hopkins University Press, Baltimore, MD

Pahl, R. (1969) Urban social theory and research *Environment and Planning A*, 143–153

Pahl, R.E. (1984) *Divisions of Labour* Blackwell, Oxford

Pain, R. (1991) Space, violence and social control: integrating geographical and feminist analyses of women's fear of crime *Progress in Human Geography*, **15**, 415–431

Pain, R.H. (1997) Social geographies of woman's fear of crime *Area*, **22**, 231–244

Park, C. (1994) *Sacred Worlds* Routledge, London

Park, D.C., Radford, J.P. and Vickers, M.H. (1998) Disability studies in human geography *Progress in Human Geography*, **22**, 208–233

Park, R.E. (1916) The city: suggestions for the investigation of human behaviour in an urban environment *American Journal of Sociology*, **20**, 577–612

Park, R.E., Burgess, E.W. and McKenzie, R.D. (1925) *The City* University of Chicago Press, Chicago, IL

Parker, S. (2004) *Urban Theory and the Urban Experience: Encountering the City* Routledge, London

Parkin, F. (1979) *Marxism and Class Theory: A Bourgeois Critique* Tavistock, London

Patterson, A. and Theobold, K.S. (1995) Sustainable development, Agenda 21 and the new local governance in Britain *Regional Studies*, **29**, 773–778

Pawson, E. and Banks, G. (1993) Rape and fear in a New Zealand city *Area*, **25**, 55–63

Peach, C. (1998) South Asian and Caribbean ethnic minority housing choice in Britain *Urban Studies*, **35**, 1657–1680

Peach, G.C.K. (1984) The force of West Indian identity in Britain, in C. Clarke, D. Ley and G.C.K. Peach (eds) *Geography and Ethnic Pluralism* Allen and Unwin, London

Peck, J. (1998) Postwelfare Massachusetts, *Economic Geography* Extra Issue for the 1998 Meeting of the Association of American Geographers, 25–29 March 1998, pp. 62–82

Peck, J. (1999) New Labourers?: making a new deal for the 'workless class' *Environment and Planning C: Government and Policy*, **17**, 345–372

Peck, J. and Jones, M. (1995) Training and enterprise councils: Schumpeterian workfare state or what? *Environment and Planning A*, **27**, 1361–1396

Pepper, D. (1996) *Modern Environmentalism: An Introduction* Routledge, London

Perkin, J. (1993) *Victorian Women* John Murray, London

Perl, E., Lynch, K. and Hormer, J. (1994) Perspectives on mortgage lending and redlining *Journal of the American Planning Association*, **60**, 344–354

Phelps, N.A. and Parsons, N. (2003) Edge urban geographies: notes from the margins of Europe's capital cities *Urban Studies*, **40**, 1725–1749

Phillips, D. (1998) Black minority ethnic concentration, segregation and dispersal in Britain *Urban Studies*, **35**, 1681–1702

Phillips, D. and Karn, V. (1991) Racial segregation in Britain: patterns, processes and policy approaches, in E.D. Huttman (ed.) *Urban Housing Segregation of Minorities in Western Europe and the United States*, Duke University Press, Durham, NC

Philo, C. (1991) Delimiting human geography: new social and cultural perspectives, in C. Philo (ed.) *New Words, New Worlds: Reconceptualising Social and Cultural Geography* Social and Cultural Geography Study Group of the Institute of British Geographers, Lampeter

Pile, S. (1993) Human agency and human geography revisited: a critique of 'new models' of the self *Transactions, Institute of British Geographers*, **18**, 122–139

Pinch, S. (1979) Territorial justice and the city: a case study of social services for the elderly in Greater London, in D. Herbert and D. Smith (eds) *Social Problems and the City: Geographical Perspectives* Oxford University Press, Oxford

Pinch, S. (1985) *Cities and Services: The Geography of Collective Consumption* Routledge and Kegan Paul, London

Pinch, S. (1989) The restructuring thesis and the study of public services *Environment and Planning A*, 21, 905–926

Pinch, S. (1993) Social polarization: a comparison of evidence from Britain and the United States *Environment and Planning A*, 25, 779–795

Pinch, S. (1997) *Worlds of Welfare: Understanding the Changing Geographies of Social Welfare Provision* Routledge, London

Pinch, S. and Storey, A. (1992) Who does what where?: a household survey of the division of domestic labour in Southampton *Area*, 24, 5–12

Ponte, R. (1974) Life in a parking lot; an ethnography of a homosexual drive-in, in E. Goode and R.R. Troiden (eds) *Sexual Deviance and Sexual Deviants* William Morrow, New York

Poulsen, M.F. and Johnston, R.J. (2000) The ghetto model and ethnic concentration in Australian cities *Urban Geography*, 21(1), 26–44

Raban, J. (1975) *Soft City* Fontana, New York

Rabinowitz, H. (1996) The developer's vernacular: the owner's influence on building design *Journal of Architectural and Planning Research*, 13(1), 34–42

Radford, J.P. (1979) Testing the model of the preindustrial city: the case of ante-bellum Charleston, South Carolina *Transactions of the Institute of British Geographers*, 4, 392–410

Ratcliff, P. (1997) Race, housing and the city, in N. Jewson and S. MacGregor (eds) *Transforming Cities: Contested Governance and New Spatial Divisions* Routledge, London

Ravetz, A. (1980) *Remaking Cities* Croom Helm, London

Rees, P. and Phillips, D. (1996) Geographical spread: the national picture, in P. Ratcliffe (ed.) *Ethnicity in 1991 Census* Volume 3 *Social Geography and Ethnicity in Britain* HMSO, London

Reibel, M. (2000) Geographic variation in mortgage discrimination: evidence from Los Angeles *Urban Geography*, 21(1), 45–60

Reiff, D. (1993) *Los Angeles: Capital of the Third World* Phoenix, London

Relph, E. (1976) *Place and Placelessness* Pion, London

Rich, R. (1979) Neglected issues in the study of urban service distributions *Urban Studies*, 16, 143–156

Ritzdorf, M. (1989) The political economy of urban service distribution, in R. Rich (ed.) *The Politics of Urban Public Services* Lexington Books, Lexington, MA

Robins, K. (1991) Tradition and translation: national culture in its global context, in J. Corner and S. Harvey (eds) *Enterprise and Heritage* Routledge, London

Robson, B.T. (1975) *Urban Social Areas* Clarendon Press, Oxford

Room, G. (1995) Poverty and social exclusion: the new European agenda for policy and research, in G. Room (ed.) *Beyond the Threshold* Policy Press, Bristol

Rose, H.M. (1970) The development of an urban sub-system: the case of the negro ghetto *Annals, Association of American Geographers*, 60, 1–17

Rossi, P. (1980) *Why Families Move* (second edition) Free Press, Glencoe, IL

Rubin, B. (1979) Aesthetic ideology and urban design *Annals, Association of American Geographers*, 69, 360

Rutheiser, C. (1996). *Imagineering Atlanta* Verso, New York.

Sack, R. (1992) *Place, Modernity and the Consumer's World* Johns Hopkins University Press, Baltimore, MD

Said, E. (1978) *Orientalism: Western Conceptions of the Orient* Routledge, London

Sandweiss, E. (1992) *Design Book Review*, 26

Sarre, P., Phillips, D. and Skellington, R. (1989) *Ethnic Minority Housing: Explanations and Policies* Avebury, Aldershot

Sassen, S. (1991) *The Global City: New York, London, Tokyo* Princeton University Press, Princeton, NJ

Sassen, S. (1997) *Losing Control? Sovereignty in an Age of Globalization* Wiley, Chichester

Sassen, S. (1999a) Digital networks and power, in M. Featherstone and S. Lash (eds) *Spaces of Culture: City, Nation, State* Sage, London

Sassen, S. (1999b) *Guests and Aliens* The New York Press, New York

Sassen, S. (2001) *The Global City* (2nd edn) Princeton University Press, Princeton, NJ

Savage, M. and Warde, A. (1993) *Urban Sociology, Capitalism and Modernity* Macmillan, Basingstoke

Sayer, A. (1989) On the dialogue between humanism and historical materialism in geography, in A. Kobayashi and S. Mackenzie (eds) *Remaking Human Geography* Unwin Hyman, London

Schein, R.H. (1997) The place of landscape: a conceptual framework for interpreting an American scene *Annals of the Association of American Geographers*, **87**, 660–680

Schmid, C.F. (1960) Urban crime areas *Sociology Review*, **25**, 527–542, 655–678

Schmid, C.F. and Schmid, S.E. (1972) *Crime in the State of Washington* Law and Justice Planning Office, Washington State Planning and Community Affairs Agency, Olympia, WA

Scott, A. (1986) Industrial organization and location: division of labor, the firm and spatial process *Economic Geography*, **62**, 215–231

Scott, A. (1988) *Metropolis: From the Division of Labour to Urban Form* University of California Press, Berkeley, CA

Scott, A.J. (1999) Los Angeles and the LA School: a response to Curry and Kenney *Antipode*, **31**, 37–44

Scott, D., Alcock, P., Russell, L. and Macmillan, R. (2000) *Moving Pictures: Realities of Voluntary Action* Policy Press, Bristol

Sennett, R. (1971) *The Uses of Disorder: Personal Identity and City Life* Allen Lane, London

Sennett, R. (1990) *The Conscience of the Eye: The Design and Social Life of Cities* Knopf, New York

Sennett, R. (1994) *Flesh and Stone: The Body and the City in Western Civilization* Faber and Faber, London

Shaw, C.R. *et al.* (1929) *Delinquency Areas* University of Chicago Press, Chicago, IL

Shaw, C.R. and McKay, H.D. (1942) *Juvenile Delinquency and Urban Areas* University of Chicago Press, Chicago, IL

Shields, R. (1989) Social spatialization and the built environment: the West Edmonton Mall *Environment and Planning D: Society and Space*, **7**, 147–164

Short, J.R., Benton, L.M., Luce, W.N. and Walton, J. (1993) Reconstructing the image of an industrial city *Annals of the Association of American Geographers*, **83**, 207–224

Short, J.R., Kim, Y., Kuss, M. and Wells, H. (1996) The dirty little secret of world city research *International Journal of Urban and Regional Research*, **20**, 697–717

Shuker, R. (1998) *Key Concepts in Popular Music* Routledge, London

Sibley, D. (1990) Invisible women? The contribution of the Chicago School of Social Service Administration to urban analysis *Environment and Planning A*, **22**, 733–745

Sibley, D. (1995) *Geographies of Exclusion: Society and Difference in the West* Routledge, London

Simmel, G. (1969) The metropolis and mental life, in R. Sennett (ed.) *Classic Essays on the Culture of Cities* Appleton Century Crofts, New York

Simonsen, K. (1991) Towards an understanding of the contextuality of mode of life *Environment and Planning D: Society and Space*, **9**, 417–431

Sjoberg, A. (1960) *The Pre-industrial City: Past Present and Future* The Free Press, New York

Skelton, T. and Valentine, G. (eds) (1998) *Cool Places: Geographies of Youth Cultures* Routledge, London

Slater, T. (2002) Looking at the 'North America City' through the lens of gentrification discourse *Urban Geography*, **23**, 131–153

Smart, B. (2000) A political economy of the new times? Critical reflections on the network society and the ethos of informational capitalism *European Journal of Social Theory*, **3**, 51–65

Smith, C.A. and Smith, C.J. (1978) Locating natural neighbours in the urban community *Area*, **10**, 102–110

Smith, C.J. (1980) Neighbourhood effects on mental health, in D. Herbert and R.J. Johnston (eds) *Geography and the Urban Environment* Wiley, Chichester

Smith, C.J. (1984) Economic determinism and the provision of human services, in A. Kirby, P. Knox and S. Pinch (eds) *Public Service Provision and Urban Development* Croom Helm, London

Smith, D.M. (1994) *Geography and Social Justice* Blackwell, Oxford

Smith, D.M. (1973) *The Geography of Social Well-being in the United States* McGraw-Hill, New York

Smith, D.M. (1977) *Human Geography: A Welfare Approach* Edward Arnold, London

Smith, N. (1984) *Uneven Development: Nature Capital and the Production of Space* Blackwell, Oxford

Smith, N. (1992) Blind man's buff or Hamnett's methodological individualism in search of gentrification *Transactions of the Institute of British Geographers*, **17**, 110–115

Smith, N. (1996) *The New Urban Frontier: Gentrification and the Revanchist City* Routledge, London

Smith, P.F. (1977) *The Syntax of Cities* Hutchinson, London

Smith, S.J. (1986) *Crime Space and Society* Cambridge University Press, Cambridge

Smith, S.J. (1988) Political interpretations of 'racial segregation' in Britain *Environment and Planning D: Society and Space*, **6**, 423–444

Smith, S.J. (1989a) Social relations, neighbourhood structure and fear of crime in Britain, in D. Evans and D.T. Herbert (eds) *The Geography of Crime* Routledge, London

Smith, S.J. (1989b) Society, space and citizenship: a human geography for the 'new times'? *Transactions, Institute of British Geographers*, **14**, 144–156

Soja, E. (1980) The socio-spatial dialectic *Annals of the Association of American Geographers*, **70**, 207–225

Soja, E. (1989) *Postmodern Geographies: The Reassertion of Space in Critical Social Theory* Verso, London

Soja, E. (1992) Inside exopolis: scenes from Orange County, in M. Sorkin (ed.) *Variations on a Theme Park: The New American City and the End of Public Space* Noonday Press, New York

Soja, E. (1996) *Thirdspace: Journeys to Los Angeles and Other Real and Imagined Spaces* Blackwell, Oxford

Soja, E. (1997) Six discourses on the postmetropolis, in S. Westwood and J. Williams (eds) *Imagining Cities: Scripts, Signs and Memory* Routledge, London

Sorkin, M. (ed.) (1992) *Variations on a Theme Park* Noonday Press, New York

Spain, D. (1992) *Gendered Spaces* University of North Carolina Press, Chapel Hill, NC

Staeheli, L. (1989) Accumulation, legitimation and the provision of public services in the American metropolis *Urban Geography*, **10**, 229–250

Stein, M. (1960) *The Eclipse of Community* Harper and Row, New York

Stenger, J. (2001) Return to Oz: the Hollywood redevelopment project or film history as urban renewal, in M. Shield and T. Fitzmaurice (eds) *Cinema and the City* Blackwell, Oxford

Storper, M. (1999) The Poverty of Paleo-Leftism: a response to Curry and Kenny, *Antipode* 31, 37–44

Strauss, A. and Corbin, J. (1998) *Basics of Qualitative Research, Techniques and Theories for Developing Grounded Theory* Sage, London

Stubbs, J.G. and Barnett, J.R. (1992) The geographically uneven development of privatisation: towards a theoretical approach *Environment and Planning A*, **24**, 1117–1135

Suttles, G. (1968) *The Social Order of the Slum: Ethnicity and Territory in the Inner City* University of Chicago Press, Chicago, IL

Swann, P. (2001) From workshop to backlot: the Greater Philadelphia Film Office in M. Shiel and T. Fitzmaurice (eds) *Cinema and the City* Blackwell, Oxford

Takahashi, L.M. (1998) Community responses to human service delivery in US cities, in R. Fincher and J.M. Jacobs (eds) *Cities of Difference* Guilford, New York

Taylor, P.J. (2004) *World City Network: A Global Urban Analysis* Routledge, London

Taylor, P.J. and Hadfield, H. (1982) Housing and the state: a case study and structuralist interpretation, in K. Cox and R.J. Johnston (eds) *Conflict, Politics, and the Urban Scene* Longman, London

Taylor, T. (1997) *Global Pop* Routledge, London

Teaford, J.C. (1997) *Post-Suburbia: Government and Politics in the Edge Cities* Johns Hopkins University Press, Baltimore, MD

Thrift, N. (1977) Time and theory in human geography, part one *Progress in Human Geography*, **1**, 65–101

Thrift, N. (1985) Flies and germs: a geography of knowledge, in D. Gregory and J. Urry (eds) *Social Relations and Spatial Structures* Macmillan, Basingstoke

Thrift, N. (1998) The rise of soft capitalism, in A. Herod, G. O'Tuathail and S.M. Roberts (eds) *An Unruly World? Globalization, Governance and Geography* Routledge, London

Tickell, A. and Peck, J. (1996) The return of the Manchester men: men's words and men's deeds in the remaking of the local state *Transactions, Institute of British Geographers*, **21**, 595–616

Tiebout, C.M. (1956) A pure theory of local expenditures *Journal of Political Economy*, **64**, 416–424

Toffler, A. (1970) *Future Shock* The Bodley Head, London

Tönnies, F. (1963) *Community and Society* 4th edition (translated by C.P. Loomis) Harper, New York (first published 1887)

Troiden, R.R. (1974) Homosexual encounters in a highway bus stop, in E. Goode and R.R. Troiden (eds) *Sexual Deviance and Sexual Deviants* William Morrow, New York

Tuan, Y.-F. (1974) *Topophilia* Prentice Hall, Englewood Cliffs, NJ

Tuan, Y.-F. (1976) Geopiety: a theme in man's attachment to nature and place, in D. Lowenthal and M.J. Bowden (eds) *Geographies of the Mind* Oxford University Press, London

US National Committee on Urban Problems (1968) *Building the American City* US Government Printing Office, Washington, DC

Valentine, G. (1992) Images of danger: women's sources of information about the spatial distribution of male violence *Area*, **24**, 22–29

Valentine, G. (1995) Out and about: geographies of lesbian landscapes *International Journal of Urban and Regional Research*, **19**, 96–111

Valentine, G. (1996) (Re)negotiating the 'Heterosexual Street': lesbian productions of space, in N. Duncan (ed.) *BodySpace* Routledge, London

Van Arsdol, M.D. Jr *et al.* (1986) Retrospective and subsequent metropolitan residential mobility *Demography*, **5**, 249–257

Vance, J.E. (1971) Land assignment in pre-capitalist, capitalist and post-capitalist societies *Economic Geography*, **47**, 101–120

Wallace, R. (1989) 'Homelessness', contagious destruction of housing, and municipal service cuts in New York City: 1. Demographics of a housing deficiency *Environment and Planning A*, **21**, 1585–1603

Wallace, R. and Fullilove, M.T. (1991) Structural factors affecting the location and timing of urban underclass growth *Environment and Planning A*, **23**, 234–264

Walton, J. (2001) Film mystery as urban history, in M. Shiel and T. Fitzmaurice (eds) *Cinema and the City* Blackwell, Oxford

Ward, C. (1978) *The Child in the City* Architectural Press, London

Ward, D. and Zunz, O. (1992) Introduction, in D. Ward and O. Zunz (eds) *Landscapes of Modernity* Russell Sage Foundation, New York

Warner, W.L. and Lunt, P.S. (1941) *The Social Life of a Modern Community* Yale University Press, New Haven

Warren, S. (1996) Popular cultural practices in the 'postmodern city' *Urban Geography*, **17**, 545–567

Warrington, M.J. (1995) Welfare pluralism or the shadow state? The provision of social housing in the 1990s *Environment and Planning A*, **27**, 1341–1360

Waterman, S. and Kosmin, B.A. (1988) Residential patterns and processes: a study of Jews in three London boroughs *Transactions, Institute of British Geographers*, **13**, 79–95

Watson, J.B. (1913) Psychology as the behaviouralist views it *Psychological Review*, **20**, 158–177

Webber, M.M. (1964) The urban place and nonplace urban realm, in M.M. Weber *et al.* (eds) *Explorations into Urban Structure* University of Pennsylvania Press, Philadelphia, PA

Weber, M. (1947) *The Theory of Social and Economic Organization* Free Press, Glencoe, IL

Weightman, B. (1981) Commentary: towards a geography of the gay community *Journal of Cultural Geography*, **1**, 106–112

Weir, S. (1976) Red line districts *Roof*, **1**, 109–114

Weisman, L.K. (1992) *Discrimination by Design* University of Illinois Press, Urbana, IL

Wellman, B. (1987) *The Community Question Re-evaluated* Research Paper 166, Centre for Urban and Community Studies, University of Toronto

Wendell, S. (1997) Towards a feminist theory of disability, in L.J. Davis (ed.) *The Disability Studies Reader* Routledge, London

Werlen, B. (1993) *Society, Action, and Space: An Alternative Human Geography* Routledge, London

White, P. (1984) *The West European City: A Social Geography* Longman, London

White, P. and Winchester, H.P.M. (1991) The poor in the inner city: stability and change in two Parisian neighbourhoods *Urban Geography*, **12**, 35–54

Whitehand, J.W.R. (1992) *The Making of the Urban Landscape* Blackwell, Oxford

Whitehand, J.W.R. and Carr, C.M.H. (2001) *The Twentieth Century Suburb. A Morphological Approach* Routledge, London

Whyte, W. (1956) *The Organization Man* Doubleday, New York

Williams, R. (1973) *The City and the Country* Chatto and Windus, London

Wilson, D. (1992) Space and social reproduction in local organizations: a social constructionist perspective *Environment and Planning D: Society and Space*, **10**, 215–230

Wilson, E. (1991) *The Sphinx in the City: Urban Life, The Control of Disorder and Women* University of California Press, Berkeley, CA

Winchester, H.P.M. and White, P. (1988) The location of marginalised groups in the inner city, *Environment and Planning D: Society and Space*, 6, 37–54

Wirth, L. (1928) *The Ghetto* University of Chicago Press, Chicago, IL

Wirth, L. (1969) Urbanism as a way of life, in R. Sennett (ed.) *Classic Essays on the Culture of Cities* Appleton-Century-Crofts, New York 0(first published 1938)

Wohlenberg, E. (1982) The geography of civility revisited: New York blackout looting *Economic Geography*, 58, 29–44

Wolch, J. and Dear, M. (1993) *Malign Neglect: Homelessness in an American City* Jesse-Bars, San Francisco

Wolch, J.R. (1981) The location of service dependent households in urban areas *Economic Geography*, 57, 52–67

Wolch, J.R. (1989) The shadow state: transformations in the voluntary sector, in J. Wolch and M. Dear (eds) *The Power of Geography* Unwin Hyman, Boston, MA

Wolch, J.R. (1990) *The Shadow State: Government and Voluntary Sector in Transition* The Foundation Centre, New York

World Commission on Environment and Development (1987) *Our Common Future* (The Brundtland Report) Oxford University Press, Oxford

Wright, D.N.M. (1999) *Sexuality, Communality and Urban Space: An Exploration of Negotiated Senses of Communities Amongst Gay Men in Brighton* Unpublished Ph.D. Thesis, Department of Geography, University of Southampton.

Wrigley, N. and Lowe, M. (eds) (1996) *Retailing, Consumption and Capital: Towards the New Retail Geography* Longman, London

Yiftachel, O. and Hedgcock, D. (1993) Urban social sustainability: the planning of an Australian city *Cities*, **May**, 139–157

Young, I. (1990) *Justice and the Politics of Difference* Princeton University Press, Princeton, NJ

Young, M. and Wilmott, P. (1957) *Family and Kinship in East London* Routledge and Kegan Paul, London

Zorbaugh, H.W. (1929) *The Gold Coast and the Slum* University of Chicago Press, Chicago, IL

Zukin, S. (1991) *Landscapes of Power. From Detroit to Disney World* University of California Press, Berkeley, CA

Zukin, S. (1995) *The Cultures of Cities* Blackwell, Oxford

Index